Food Science, Nutrition and Safety

Sukhneet Suri | Anita Malhotra

Associate Editor–Acquisitions: Neeraj Bhalla
Associate Editor–Production: Ruchi Sachdev

Published by Pearson India Education Services Pvt. Ltd, CIN: U72200TN2005PTC057128.

ISBN: 978-81-317-7110-5

First Impression, 2014
Sixth Impression, 2021
Seventh Impression, 2022

Head Office:15th Floor, Tower–B, World Trade Tower, Plot No. 1, Block–C, Sector–16, Noida 201301, Uttar Pradesh, India.
Registered Office: 7th Floor, SDB2, ODC 7, 8 & 9, Survey No.01
ELCOT IT/ ITES - SEZ, Sholinganallur, Chennai – 600119 Tamil Nadu, India.
Website: in.pearson.com, Email: companysecretary.india@pearson.com

Compositor: Mukesh Technologies
Printer: India Binding House Noida

To our parents.

Contents

Part II

NUTRITION

Foreword

Food industry, growing at a fast pace in India, heavily relies upon the knowledge related to science, nutrition and safety of food. It is imperative for all food business operators to understand various aspects of food science and nutrition so that customers receive wholesome and safe food products. An undergraduate student, aspiring to become a professional in food service industry, requires access to knowledge which is easy to comprehend, clarifies concepts and lays foundation for a sound prospective career.

The book *Food Science, Nutrition and Safety* presents diverse information for learners in an integrated yet systematic manner. The text is comprehensive, clear and student friendly. The content emphasizes on the fundamentals yet encompasses the latest developments in the field of food science, nutrition and food safety. It is also sprinkled with lucid examples from the hospitality industry.

The publication of this book is apt and timely. It meets the syllabus of several universities offering undergraduate courses in Foods and Nutrition, especially B.Sc., Hospitality Management of National Council of Hotel Management and Catering Technology. It will also be a useful textbook for students of Home Science studying Food and Nutrition.

I wish the authors success for the first edition of this book.

Dr S. Malhan
Emeritus and Founder Director
Institute of Home Economics

Preface

t gives us immense pleasure to present *Food Science, Nutrition and Safety* for readers. This book is interesting nd easy-to-comprehend. It covers syllabus requirements of related courses in B.Sc., such as Hospitality and Iotel Administration, Hotel Management (Degree and Diploma Courses), Bachelor and Associate Bachelors n DC I and DC II of Home Science as well as in other undergraduate programmes in the discipline of Food cience, Nutrition, Food Technology and Food Safety.

This book is divided into three sections: *Food Science, Nutrition* and *Food Safety and Quality*.

ood Science (Part I): The first section of this book is on food science. Chapter 1 highlights the scope of ood science in all forms of micro and macro enterprises associated with food business with special emphasis n the hospitality industry. Chapters 2, 3 and 4 elaborate the role of carbohydrates, fats and proteins in ood processing and preparation. Chapters 5 and 6 explain various techniques of food processing and sensory valuation, respectively. Applications of colloidal chemistry are discussed in Chapters 7 and 8 while Chapter 9 escribes select flavours. Last chapter in this section illustrates different types of browning and their application n food processing.

utrition (Part II): This section focuses on the concepts of under, over and optimum nutrition, and the asis of preparing nutritious meals. Chapter 11 gives an introduction to the science of nutrition and nutrints. Chapter 12 discusses about energy (calories) with relevance to the components and factors influencing nergy requirements as well as energy balance. Chapters 13, 14 and 15 dwell upon the functions, food sources, deficiency and excess of carbohydrates, fats, proteins, vitamins, minerals and water. Chapter 16 elaborates ne significance of balanced diet, food groups and dietary guidelines while the next chapter (Chapter 17) iscusses the applied aspects of planning nutritionally balanced meals. Useful tips related to conserving and nhancing nutritive value of food have been covered in Chapter 18. Newer trends in nutrition based products ave been briefly discussed in the last chapter of this section.

ood Safety and Quality (Part III): This part focuses on food safety. Chapter 20 introduces the relevance f food safety in the current scenario, Chapter 21 explores the relationship between microorganisms and od, Chapter 22 describes the causes of food spoilage and methods of food preservation and Chapter 23 ustrates food fermentation. Different types of food-borne illnesses are explained in Chapter 24 while hapters 25 and 26 provide information about food additives, contaminants and adulterants. The role of od laws and regulations, risk management as well as sanitation and hygiene in ensuring food safety is amply stified in the Chapters 27, 28 and 29. The last chapter highlights some recent concerns in food safety.

edagogical Features

his book focuses on teaching by arousing learner's curiosity. To engage the interest of students and provide odated knowledge, diverse pedagogical features have been included.

1. **Chapter objectives:** Each chapter begins with a list of objectives stating the issues about which the student is expected to develop knowledge after going through the chapter.

Chapter Objectives

After reading this chapter, the reader should be able to

- know the types and causes of food spoilage.
- identify different sources of microbial contamination.

2. **Opening Vignette:** In Section I, a brief introduction about the significance of each chapter has been given with the aim of capturing student's attention.

Have you wondered why rice grains swell upon cooking, a teaspoon of custard powder (cornstarch) thickens milk sufficiently to form a custard, semolina porridge tends to thicken when allowed to stand for some time, jams develop a smooth texture when pectin is added to them, psyllium seed husks (isabgol) remain dispersed in water while stirring but swell and settle down when allowed to stand and breadcrumbs do not dissolve in water but instead they give a crisp texture to cutlets? Carbohydrate is a common food constituent which is responsible for such changes while cooking/processing.

3. **Situations:** In Section II, the introduction is provided through 4 to 5 real life situations relating to nutrition. Students have also been asked to consider interesting situations in the introductory chapters of Sections I and III (Chapters 1 and 20). The objective is to arouse the interest of learners.

Consider the following situations:

Situation 1: Preschoolers in a slum area are not able to play late evening because most of them cannot see clearly after dark. Responding to this observation, a non-government organization started feeding them food cooked in red palm oil.

Situation 2: A group of adolescents were having pizzas, burgers and pastas very frequently. Of late some are feeling anorexic and lethargic with mild gastro-intestinal disturbances.

4. **Think Tank:** The chapters in Section III provide a think tank with introduction. Think tanks are sprinkled in several chapters. The aim is to stir learner's mind, induce curiosity and thereby improve the learning process.

THINK TANK

Lapses in Food Hygiene

Consider the following examples in the context of food safety:

- A trolley boy, whistling while carrying meat out of a cold store room, forgets to close the door.

5. **Information in Box:** To break the monotony of the text, information is presented in boxes in various chapters of the three sections.

Box 2.1 What Is a Modified Starch?

It is a starch that has been chemically modified to make it functional under the conditions of processing and storage such as high temperatures, high shear, low pH, freezing and thawing. Modified starches are E coded according to International Numbering System (INS), i.e., 1400–1415. The E numbers 1401, 1402, 1403 and 1405 are commonly used as food ingredients.

6. Good to Know and Did You Know? These boxes contain extra knowledge to update the student in the related field.

GOOD TO KNOW

Water Activity (a_w)

'Water Activity' is the quotient of water vapour pressure of the food divided by the vapour pressure of pure water at the same temperature. The a_w of pure water is 1.00, whereas that of most bacteria, yeast and mould is 0.91, 0.88 and 0.80. An a_w below 0.85 inhibits the growth and multiplication of most microbes. Based on their water activity, foods can be divided into following three categories:

DID YOU KNOW?

Bancroft's Rule

Bancroft stated that the phase in which an emulsifier is most soluble forms the continuous phase of an emulsion.

7. Useful Tip: The book envisages encouraging the learner to apply the acquired knowledge for cooking and processing. Some useful tips are given to initiate the process.

ACTIVITY

Useful Tip

If a soup tastes too salty, you can make it palatable by adding some egg white which on heating will adsorb salt on to its surface. Allow the egg white with adsorbed salt to settle at the bottom of the container from where you can remove it subsequently.

8. Activity: Suggestions for useful activity are provided to make students learn by hands-on experience.

USEFUL ACTIVITY

Take seven eggs. Separate egg white of one egg in a steel bowl and whip with a hand beater till a stiff peak stage is formed. Transfer the foam lightly into a funnel held on a measuring cylinder. Note the drainage volume (liquid) in the measuring cylinder after 20 minutes. Repeat the exercise by adding or using:

1. 10ml water
2. 2 teaspoon sugar
3. 1/2 teaspoon vinegar
4. 1 teaspoon oil/yolk
5. mechanical beater
6. a different shaped bowl

9. Tables: Many tables have been provided in the text to give information in an easy-to-learn format. These will be very handy for students when they prepare for examination.

Table 17.1 Evaluation of certain commonly available snacks and meals

Menu/Meal	Energy Giving		Body Building		Protective and Regulatory	
	Cereals	Fats/Sugars	Milk/Meat	Pulses/legumes	Fruits	Vegetables
Idli-sambhar	🙂	😐	☹	🙂	☹	😐
Aloo tikki	🙂	🙂	☹	☹	☹	☹
Vegetable burger	🙂	😐	☹	☹	☹	😐
Chicken burger	🙂	😐	🙂	☹	☹	😐

10. Illustrations and Figures: Many figures are provided in the text of each chapter with different objectives.

A. Smart Art Figures given in the text either support the discussion or give stand-alone information in a simple format.

Water (continuous phase)

Oil (discontinuous phase)

Oil-in-water emulsion (o/w)

Figure 7.3 Oil-in-water and water-in-oil emulsions

Folic Acid	• Megaloblastic macrocytic anemia • Neural tube defect (spina bifida).
Cyanocobalamin	• Megaloblastic microcytic anemia (pernicious anemia).

Figure 14.12 Common deficiency diseases of B-group vitamins

B. Simple Representation of Chemical Reactions has been given so that a student with limited knowledge of science is also able to comprehend the chemical changes in foods.

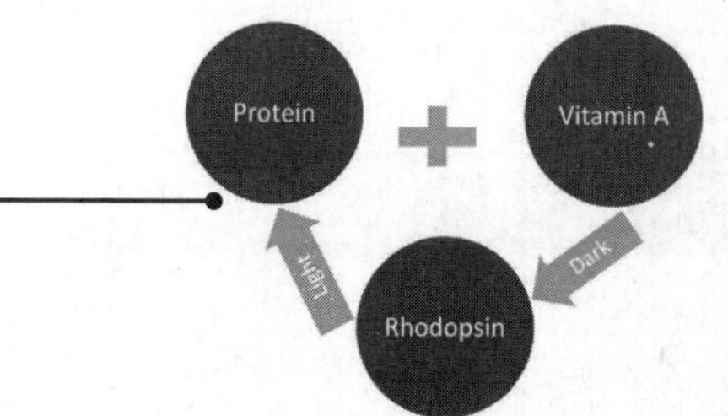

Figure 14.5 The visual cycle (the dark and light adaptation)

C. Pictures appropriate to the context have been inserted to provide zest to the text.

Figure 10.5 Desirable browning in food preparation

D. Structural Diagrams are provided wherever necessary to enhance the learning experience of a student.

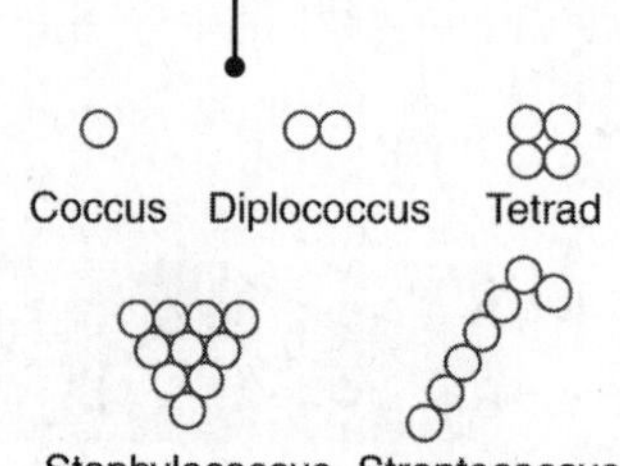

(a) Cocci

Figure 21.2 Arrangements of cocci and bacilli

11. **Know Your Progress:** To test the knowledge of the learner after going through the chapter, varied questions have been asked in the form of true/false and very short/short/long answers.
12. **Beyond the Text:** These are assignment-type activities which can be taken up by students individually or in groups and the experience shared with rest of the class. The aim is to expand the learning experience of students.

    Beyond the Text

    - Different foodstuffs have different calorie content. You can check that yourself. Compare the calorie content (energy density) of packaged soft (carbonated) drink with fruit drink and fruit juice.
13. **Glindex:** This is a unique feature of the book which explains the important terms used in the chapters along with their indexing.
14. **References:** Exhaustive list of references are given at the back for further reading.
15. **Websites:** Few interesting and informative websites are provided for each chapter for students to explore information online.

    Chapter 11

    American Dietetic Association: www.eatright.org
    American Medical Association: www.ama-assn.org
    Indian Council of Medical Research http://www.icmr.nic.in/

Instructor Resources

Interactive and innovative teaching enhances the learning experience. Therefore, instructor resources in the form of power point presentations have been developed for each chapter. These power point presentations are not just limited to the content of this book but provide further information on latest trends and current practices. Printable version of the power point presentations is also available at www.pearsoned.co.in/sukhneetsuri.

As you know, food science, nutrition and food safety are vast areas with many new developments. We have tried to include most of the relevant content within the scope and page limit of this book.

Student Resources

For the inquisitive readers who wish to know more about a particular topic and may even want to lay hands on the practical aspects of the theory, web resources have been designed for select topics in the field of food science, nutrition and food safety.

This symbol indicates access to additional resources like tables, charts and detailed information about the discussed topics. Students may access this material after registering themselves at http://wps.pearsoned.co.in/suri_fsns_1/ and access codes are available inside back cover.

We sincerely hope that this book will be a useful resource for the budding professionals and will make learning an enjoyable experience.

Acknowledgements

Writing a book is not the solitary effort of its authors. It requires the ceaseless support of many dedicated and competent people. We would like to acknowledge the experts involved in creating the ocean of knowledge in the field of food science, nutrition and safety as well as its allied areas. The long drawn process of writing this book was supported by the Pearson team, especially Neeraj Bhalla and Ruchi Sachdev. We thank Neera Bhalla for his continuous support right from the conception of this project; through the writing and review process, cover designing and the development of e-resources for students and teachers. We also acknowledge the professional expertise and untiring efforts of Ruchi Sachdev in copyediting the manuscript and steering it through the production process.

We would like to appreciate Dr Renu Sahni, Principal, Vivekananda College, and Dr Karuna Kaushik former Principal, Lakshmibai College, for their unsaid support and continuous motivation in this endeavour. Our thanks are also due for the experts who peer reviewed the chapters and our colleagues who provided endless support. The reference work required frequent visits to many libraries, we acknowledge the suppor extended by library staff of University of Delhi, National Medical Library, Indian Agriculture Research Institute, Vivekananda College, Lakshmibai College, Nutrition Foundation of India and Indian Council of Medical Research.

We appreciate the love, understanding and unstinted support of our family without which our journey of creating this book would not have been complete. The blessings of almighty supported this humble piece of work. We welcome suggestions for the improvement of the book in terms of content and/or presentation. Readers can reach us at sukhneets@yahoo.co.in or anitamalhotra28@gmail.com.

Sukhneet Sur

Anita Malhotra

About the Authors

Sukhneet Suri is an Assistant Professor in the Department of Food Technology and Nutrition at Vivekananda College, University of Delhi. She is an alumnus of the Institute of Home Economics, University of Delhi (Ph.D., Food and Nutrition). During her 17 years of undergraduate and postgraduate teaching and research, she taught at the University of Delhi, Institute of Hotel Management Catering and Technology (Pusa), Delhi Institute of Hotel Management and Indira Gandhi National Open University to name a few. Dr Suri was member of the Joint Review Mission for evaluation of Mid-day Meal Scheme by Ministry of Human Resource Development. She is presently involved in an innovative research project for under-graduate students funded by University of Delhi. Dr Suri has reviewed a book *Food Chemistry*. She has also authored course content for M.Sc. Food Technology and Preservation, Yashwantrao Chavan Maharashtra Open University, and M.Sc. Dietetics and Food Service Management, Indira Gandhi National Open University. There are research/review papers/articles, chapters in a book, *Cholesterol Facts,* and scientific presentations at national and international levels (South Africa, New Zealand and London) to her credit. She has been a resource person/consultant with Indian Railways, Ministry of Personnel Public Grievances and Pensions (Government of India) and several organizations such as National Council for Hotel Management and Catering Technology, Center for Professional Development in Higher Education, School of Open learning, Board of Technical Education, National Institute of Public Co-operation and Child Development, and Vocational Training Center.

Anita Malhotra is an Associate Professor in the Department of Food Technology and Nutrition at Lakshmibai College, University of Delhi, with more than 28 years of teaching experience. She is an alumnus of Lady Irwin College and holds a doctorate degree from the University of Delhi (Ph.D., Food and Nutrition). She has participated in about many conferences and also presented papers at national as well as international level (USA, UK, Japan, China, Malaysia and Thailand). She has many publications to her credit including course material for masters in Food Processing and Preservation and original research paper in indexed international journal, *Asia Pacific Journal of Clinical Nutrition*. She has been a nutrition panelist on TV and has organized several seminars and workshops, some funded by University Grants Commission and Indian Institute of Public Administration on issues relating to nutrition, health, media and consumer awareness. She is a member of Committee of Courses, Department of Home Science at University of Delhi and also a member of State Nutrition Council, Government of NCT of Delhi. In the year 2011, Dr Malhotra was awarded USAID's MEASURE Evaluation fellowship to participate in Asia Regional Workshop and scholarship from the Howard Foundation to participate in SCOPE Summer School at Cambridge.

Part I

Food Science

Chapter 1

Food Science: An Introduction

Chapter Objectives

After reading this chapter, the reader should be able to:

- have a basic understanding of food science and its association with other related sciences.
- know about the interdisciplinary aspects and the objectives of food science.
- appreciate the scope of food science and the current trends in this field.
- recognize the significance of food science in various activities of hospitality industry.
- associate the science of food with the changes occurring within it, as witnessed in food production and food service industries.
- know the careers and positions in institutional food management, which require professionals having knowledge about food science.

Consider the following situations:

Situation 1: Tomatoes were blanched by putting an 'x' mark with a knife on their base, immersing them in boiling water for the required time, then removing them from hot water, dipping them instantly in cold water, and removing their skin. After this process, the blanched tomatoes had a smooth surface and their natural structure was maintained. The same process was repeated for potatoes. Upon removing peels, the product did not have the desired texture.

Situation 2: During the preparation of tea, the water was brought to boiling temperature and the tea leaves were allowed to steep. However, during the preparation of coffee, the heat was turned off just before the beverage reached its boiling point. Both the beverages had optimum flavour.

Situation 3: An omelette was prepared by beating the egg white and egg yolk separately. In another method, the omelette was prepared by beating the egg white and egg yolk together. The former omelette had a fluffy texture, while the latter was flat.

Situation 4: A salad bar operator sprinkled fresh lemon juice on cut fruits. When served after an hour, the fruit salad had an attractive visual appeal. On another day, the operator left the cut fruits in a bowl in the pantry without sprinkling the lemon juice. The surface of fruits turned brown in colour, and it could not be served to customers even after 30 minutes.

Professional cooking cannot be a hit-and-trial exercise as there may not be a second chance to satisfy the customer. Understanding the behaviour of food in different environments is the fundamental thing for successful cooking, serving and storing. It also helps to handle challenging situations well. This is possible only when a food industry professional has thorough knowledge about the composition, the properties and the response of food to its environment.

1.1 Definitions

Food: It can be described as a material or a product obtained from the nature, which contains nutritive and non-nutritive components that help to maintain or improve the health of an individual, if consumed in optimal amounts. Food, as obtained from nature, is of either plant origin or animal origin. Food can be processed into endless varieties of products.

Food Science: It is an interdisciplinary science which deals with the physical, biochemical, biological, microbiological, sensory, nutritional, toxicological, organoleptic and engineering properties of foods and beverages. It helps us to relate these properties with the changes occurring in the food systems during handling, pre-preparation, processing/cooking, packaging, storage and distribution as well as with the shelf life, cost, quality, safety, nutritive value, wholesomeness and convenience. Thus, it is associated with all the changes in the food from the farm to the fork. Box 1.1 describes the technical definition given by the Institute of Food Technologists.

Box 1.1 Definition of Food Science

'Food science is the discipline in which the engineering, biological and physical sciences are used to study the nature of foods, the causes of deterioration, the principles underlying food processing and the improvement of foods for the consumption by public'.

Source: Dennis R. Heldman. *IFT and the Food Science Profession*, Food Technology, 2006.

Food science has several sub-disciplines which are as follows:

Food Physics: It refers to the physical aspects of food which influence the shape (cubes or roll), texture (creamy or dry), size and weight or volume of the food.

Food Chemistry: It deals with the molecular composition of food systems and their association with the structure, properties and changes associated with handling, processing, storage and distribution of food.

Food Microbiology: It deals with the positive and negative interactions of micro-organisms (bacteria, yeast and fungi) and their spores with food, equipment, processing plant and food handler. Thus, it is an important aspect of food processing, work area hygiene and sanitation as well as food safety.

Food Biotechnology: It involves the utilization of biologically derived molecules, structures, cells or organisms (microbes, enzymes and DNA) during food processing or for the development of newer food products/ingredients. Genetic engineering (DNA transcription/recombination and tissue culture) is an important aspect of food biotechnology.

Food Engineering: It involves the development of utensils, appliances and machines which are required for cooking or processing of food. It is also associated with the shape modulation or the rheology of food.

Nutrition: It is very intimately associated with food science. It deals with the structure, functions and properties of nutrients, i.e., carbohydrates, proteins, fats, vitamins, minerals, water and fibre, and also deals with how they interact with each other and their environment during handling, cooking, storage and service/distribution of food.

Food Technology: It is the application of above sciences to the production (agriculture or livestock), processing/cooking, packaging, distribution, storage and consumption of safe, nutritious and wholesome food.

Molecular Gastronomy: It is a sub-discipline of food science which studies the physical and chemical transformations of food (while cooking) along with social, artistic and technical components of culinary art. It helps to explore the new tools, ingredients and methods of cooking.

The study of food science thus helps to

1. gain knowledge regarding the functions and properties of food.
2. understand the composition and chemistry of food viz.-à-viz. their association with the changes observed during processing, handling, storage and distribution.
3. select food (or ingredient) which is most suitable for a particular use.
4. learn in detail the various methods of cooking (secondary processing and tertiary processing–hospitality industry) and its impact on food.
5. understand the various principles of food preservation.
6. utilize and manage judiciously the resources (time, workforce and money) required for food processing.
7. ensure the consumption of safe food by the customers.

1.2 Inter-relationship with Other Fields

The magnitude of food industry is very vast and therefore a food scientist cannot rely solely on the knowledge of food science for the preparation and sale of food. Food scientists often have to depend upon the experts from other fields of study to be successful in their profession (Figure 1.1). Some of the inter-related fields are discussed in the following sections.

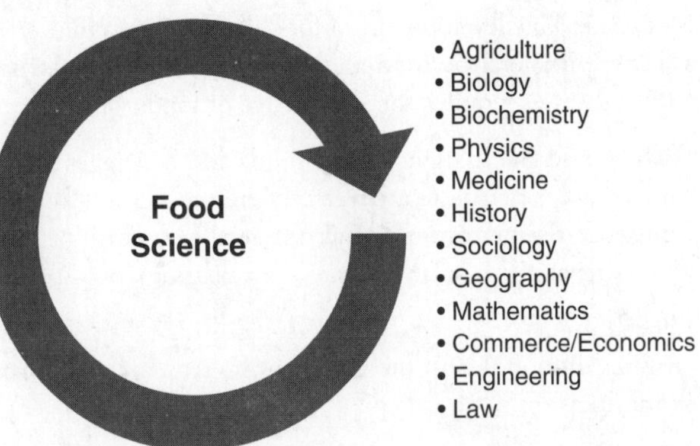

Figure 1.1 Association of food science with other fields

Agriculture Science: Several processed foods rely upon the product of a specific variety/breed or species of a plant or an animal due to their inherent characteristics

Box 1.2 Food Science-related Terms

Food Safety: It is a scientific discipline describing handling, preparation and storage of food which prevent all types of food-related health hazards.

Food Analysis: It is the careful study of various food constituents (protein, carbohydrates, fats, vitamins, minerals, water and enzymes) and those closely related to them.

Food Preservation: It is the process of treating or handling food with the objective of stopping or retarding its spoilage (loss of quantity, quality and nutritive value) and thus prolonging the shelf life of the product. Jams, marmalades, fruit butter, cordials and pickles are some of the products prepared by using the methods of preserving fruits and vegetables.

Food Fortification: It is the practice of deliberately increasing the content of an essential macro or micronutrient, particularly vitamins and minerals (including trace elements) in a food, irrespective of whether the nutrients were originally present in the food before processing or not, so as to improve the nutritional quality of the food supply and to provide a public health benefit with minimal risk to health. The addition of iodine to salt is an example of food fortification.

Food Rheology: It is the study of rheological (texture related) properties of food, such as fluidity, crispiness and tenderness.

Food Anthropology: It is the effect of religion, culture, traditions, customs, myths and misbeliefs on food consumption.

such as the flavour of Basmati rice. Juicy tomatoes would be appropriate for preparing soup, but those with good amount of pulp would be more suitable for making tomato chutney/ketchup or tomato-based gravy. Post-harvest technologies, hybridizations and genetic engineering not only influence the shelf life but also influence the properties of agricultural produce.

Biology: The composition of leaves is different from that of stalk, fruit or seed, and hence their properties/use would also vary. Similarly, the properties and composition of rice bran are different from that of rice. Because of this reason, the two products are used for entirely different purpose(s).

Biochemistry: The interaction of food with enzymes, micro-organisms and other cells (whole or part) plays an important role in the production, service and storage of food. The development of prebiotics and probiotics is the result of an understanding of basic biochemistry by a food scientist.

Physics: The tools and techniques required for measuring the physical properties of food, the development of a product (biscuit or lollipop) or the configuration of a packaging container with specific dimensions require the basic knowledge of physics.

Medicine: The development of food products for children or individuals with special needs or diseases such as lactose intolerance, food allergy (wheat or peanuts), diabetes, hypertension, obesity and other lifestyle-related diseases requires the knowledge of foods which are associated with health and disease.

History and Sociology: Several foods and beverages have been found to have some historical association with a place, a person or an event. Therefore, certain foods are associated with festival, occasions and events. Cultural variations can influence the recipe of a dish or its cooking method or the utensils/technology used for its preparation. Pulao prepared in Kashmir would be much different from that prepared in Kerala.

Geography: It influences the seasonal availability, the nutritive value, the composition and the sensory qualities of food. Thus, the choice of ingredients varies from one region to another and from one country to another.

Mathematics: Standardization of recipes and analysis of food for quality control require calculations associated with time, temperature and amount of ingredients (or reagents) used. Computing the nutritive value of a dish or a menu would also require mathematics.

Commerce and Economics: Time, money and workforce are three major resources required for food production and service. Knowledge about book keeping and accounting is a must for all food service professionals. Understanding the basic concepts such as break-even point and SWOT analysis can help to find out whether a dish or a menu is providing profit or loss to an institution.

Engineering: Engineering is closely associated with the preparation of a dish, the modulation of the structure of a natural food or a food product and the development of utensils, appliances and packaging material.

Law: Several laws and regulations govern the production, purchase and sale of foods. The laws applicable at the national level may be different from the laws applicable at the international level.

1.3 Scope of Food Science

Food science is the backbone of several billion dollar food industry. It has helped the food industry to expand and provide products to meet the most difficult requirements and the weirdest imaginations related to foods by the customers. The scope of food science is very vast and difficult to summarize in this chapter.

However, to give an idea about the magnitude of food science, a few salient achievements are enumerated below (Figure 1.2).

1. **Combating Undernutrition:** Low-cost nutritious foods such as Incaprina (USA), Miltone, Chikki, Nutro-Crispo Spicy, Nutri Crisp Sweet and Surichi Meether (India) have been developed to reduce the prevalence of protein energy malnutrition (kwashiorkor and marasmus) in the vulnerable populations (children, pregnant women and nursing mothers).
2. **Space Foods:** Novel foods have been prepared for astronauts. Space foods have a long shelf life, are usually dehydrated, are easy to reconstitute and have a small volume and mass. They are generally packaged in thermostabilized flexible packaging, which should preferably be biodegradable, reusable and edible. The foods consumed in International Space Stations have a shelf life of about 180 days, which is longer than the foods which are consumed in space shuttles. NASA has developed about 200 such products and is trying to enhance their shelf life to 2–5 years for its forthcoming space milestone in 2020.
3. **Military Feeding:** Nutrient-dense, lightweight, compact and easily reconstitutable food products with prolonged shelf life at elevated temperatures have been developed for soldiers and military personnel. They are packaged in edible or biodegradable flexible films.
4. **Fabricated Foods:** Texturized vegetable proteins and texturized soya proteins (soya chunks or granules) are the most commonly available fabricated foods. They have the texture and taste which are

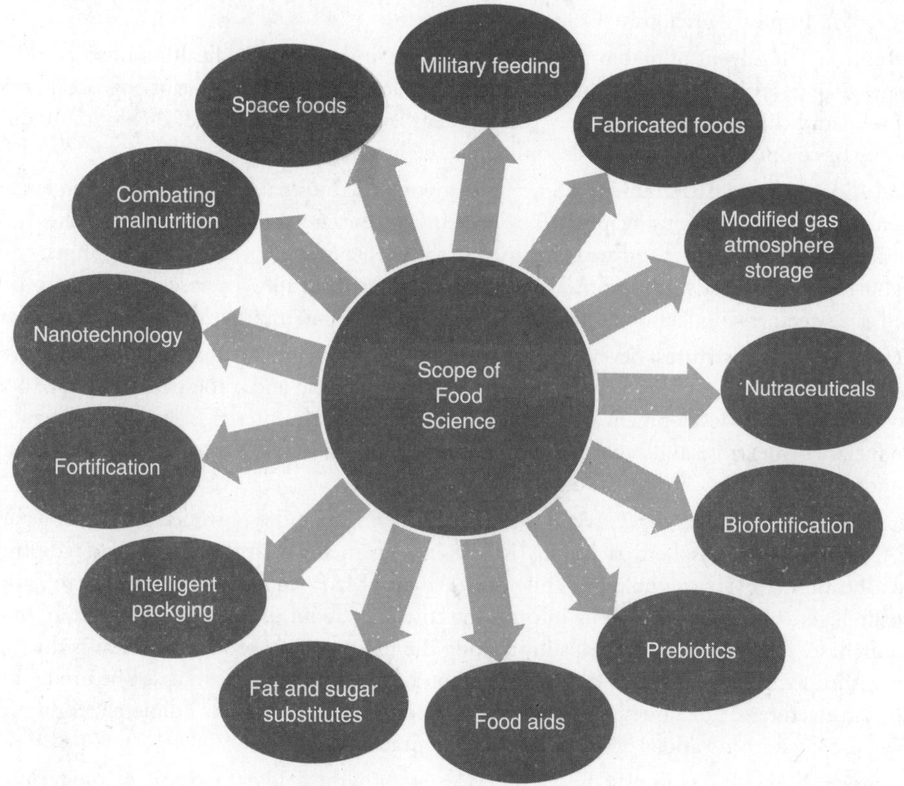

Figure 1.2 Scope of food science

similar to meat; therefore, they are often flavoured accordingly. *Food analogues* for meat, milk and eggs have been developed. For example, natural egg yolk has been replaced with the yolk prepared from vegetable oils, milk solids, emulsifiers, nutrients and additives.

5. **Modified Gas Atmosphere:** Modifying the composition of cold storage air, wherein fruits and vegetables are stored, helps to enhance the shelf life and regulate the ripening process. Quick freezing of fruit or vegetable pulps has improved markedly their non-seasonal availability; for example, the use of mango pulp in mango-based drinks.
6. **Nutraceuticals:** Most nutraceutical products have been developed with the help of food science. For example, the n-3 fatty acids such as eicosapentaenoic acid and docosahexaenoic acid have been extracted from liver of large fish such as whale and are being used for the prevention treatment and management of several degenerative diseases such as elevated cholesterol, heart diseases, certain cancers and arthritis. Dietary fibre has been isolated from several foods such as, guar-gum from broad beans and pectin from fruits and are being used for the treatment of constipation, obesity, heart diseases, diabetes and ulcers.
7. **Biofortification:** Biofortification of crops has also witnessed success with the help of food science. For example, rice with high iron content has been developed by International Rice Research Institute, Philippines and Manila. International Maize and Wheat Improvement Center, Mexico has developed maize with high-quality protein. Similarly, sweet potato and cassava with increased amount of beta-carotene have been produced by International Potato Center, Peru and International Center for Tropical Agriculture, Colombia.
8. **Prebiotics:** The advent of prebiotics has created a stir in the food and health industries. While prebiotics such as lassi, curd and yoghurt are traditional foods in India, the use of oligosaccharides such as inulin and chitin in dairy and other products has lately been introduced. Prebiotics modulate the activity of colonic microflora and also influence the health positively.
9. **Food Aids or Special Ingredients:** Specific components from natural foods are being extracted or isolated and used in various recipes. Whey protein concentrates and whey protein isolates have been extracted from whey water and are being used in the preparation of breads, cakes, muffins, croissants, yoghurt, cheese spreads, ice creams, meringues and health drinks. Lactose extracted from milk is used as a sweetness moderator along with sucrose and is used to improve the texture/bulk of biscuits.
10. **Fat and Sugar Substitutes:** Several carbohydrate and protein based fat substitutes and fat replacers have been developed, such as olestra, caprenin, salatrim, benefat and sorbestrin. They have been used successfully in the development of several bread spreads, table spreads, salad dressings and bakery products. Polydextrose and other polyols have been used to develop sugar substitutes because they cannot be digested by the human digestive tract.
11. **Recent Trends in Packaging:** Advancements in packaging include controlled atmosphere packaging (CAP), modified atmosphere packaging (MAP), intelligent and active packaging and radiofrequency identification (RFID) technology. While the CAP and MAP regulate or prevent microbial growth, the intelligent and active packaging informs the customer about the storage conditions of food prior to purchase. RFID technology helps in tracking the movement of food once it leaves the manufacturing unit for distribution and sale. Thus, if a food expires on retail shelf, it can be brought back by the manufacturer or discarded. This ensures safe consumption of food. Edible packaging films are being prepared from various proteins and carbohydrates.
12. **Nanotechnology:** It is currently the latest development in the field of food science and technology. It is being used for isolating (whole or part), modulating or manufacturing flavours, antioxidants and

nutraceuticals. Such products are known as nanostructures. Nano technology has been employed for designing membrane-based separation systems such as those required during freeze dehydration.

The list is endless and more and more innovations are being added to it on day-to-day basis. Food processing and the hospitality industry are constantly evolving to keep pace with these developments in the field of food science.

1.4 Food Science and the Hospitality Industry

The most significant use of the knowledge of food science is related to cooking of food, though it is equally important in other activities such as receiving of raw material (food) storage, pre-preparation, serving (or dishing) and even dish washing. This would be evident from the following examples:

1. **Receiving Sorting and Grading:** While receiving a food as ingredient for various dishes, it is ensured that it meets purchase specifications. For example, sorting and grading of ripened and unripened tomatoes or oranges may be required. It is equally important to check the grain size of refined flour to be used for bread making, which would be different from that used for making biscuits. The person in-charge of receiving sorting and grading raw material would perform their role more effectively only if they are aware as to which characteristic of an ingredient would influence the quality of the final product.
2. **Storage:** While the first-in-first-out (FIFO) strategy is useful for most ingredients, the same may need to be modified depending upon the state of perishability of a food product. Milk should be stored in closed translucent containers and that to be used for preparing paneer should preferably not be frozen, oil should be stored in a cool and dark place which is away from the direct sunlight, sugar should be stored in such containers that they can be emptied from the bottom and sugar should not be poured in large quantities from a height into the transport trolley (can cause blast); such and other facts are known only through the study of food science.
3. **Pre-preparation:** Proper pre-preparation is a must for preparing a dish with good sensory characteristics. Whipping of egg white foam to stiff peak stage is a prerequisite for achieving the right texture of soufflé. The use of acids or enzymes in a marinade helps to soften vegetable or meat during marination. The addition of sodium bicarbonate creates a mashed texture. Food science helps to give such useful tips for better results.
4. **Preparation:** The choice of ingredients and utensils to be used and the process of cooking are based upon the principles of food science. Several questions arise in the mind of a chef while cooking; for example: what would happen if a fat replacer is added to a salad dressing?, how can curdling of tomato soup be prevented? or why does the flavour of fat change after repeated frying? Good quality dishes can be prepared repeatedly if the chef is clear about the interaction of food with its environment.
5. **Serving/Dishing:** Hot foods are often placed on the service table, over electric or spirit '*rechauds*'. Protein-rich dishes such as paneer develop a rubbery texture when kept warm at the service table or in a buffet. Therefore, the time and temperature of holding food must be kept in mind while serving food. Koftas should be put in gravy only at the time of serving and cheese soufflé should be served straight from the oven. Food science must be kept in mind while serving food.
6. **Dish Washing:** This is often the most neglected part of food service industry. Hard water contains calcium and magnesium ions, which when used repeatedly for washing dishes form scales on their surface. Such coating prevents heat transfer and also leaves deposits for the growth of microorganisms. Further, the steam obtained from such water is alkaline and it can corrode utensils containing tin and aluminium. The composition of the detergent which is used for cleaning glassware

is different from the one which is used for cleaning metal utensils. Food science is essential for cleaning and maintenance of hygiene and sanitation.

Thus, food science is required in all day-to-day activities of food service industry. In addition, the knowledge of food science also comes in hand when a consumer demands change and new products have to be developed to survive competition. Variations in skills, ingredients, cooking equipments, budget and legislations also necessitate change in menus and recipes of dishes (Figure 1.3).

The careers associated with food service industry, which require the knowledge of food science, include the following:

1. **Chefs:** Specialist chef, saucier, pastry chef, master chef, head chef, gardé manager chef, executive sous chef, executive chef, corporate chef, chef de partié, chef de cuisine and chef.
2. **Cooks:** Cook, dietary cook, apprentice cook, grill cook, institutional cook, licensed cook and short-order cook.
3. **Food Service Personnel:** Restaurant and food service managers, F&B supervisors, counter attendants, waiters and kitchen helpers.
4. **Bakers:** Head baker, bakery supervisor and baker apprentice.
5. **Butchers and Meat Cutters:** Head butcher, meat cutter, butcher apprentice and retailers.
6. **Retailers and Wholesalers:** Managers and supervisors for food-supermarket chains, sales representatives of food manufacturing companies and retailers of food products (fresh or processed).
7. **Marketing Experts:** Advertising and marketing managers and specialists and public relation officers of food manufacturing companies, hotels, restaurants and fast food outlets.
8. **Educators:** Teachers, consultants, journalists and writers.
9. **Inventors:** Quality control scientist/technologist/technician and experts–research and development (new product development).

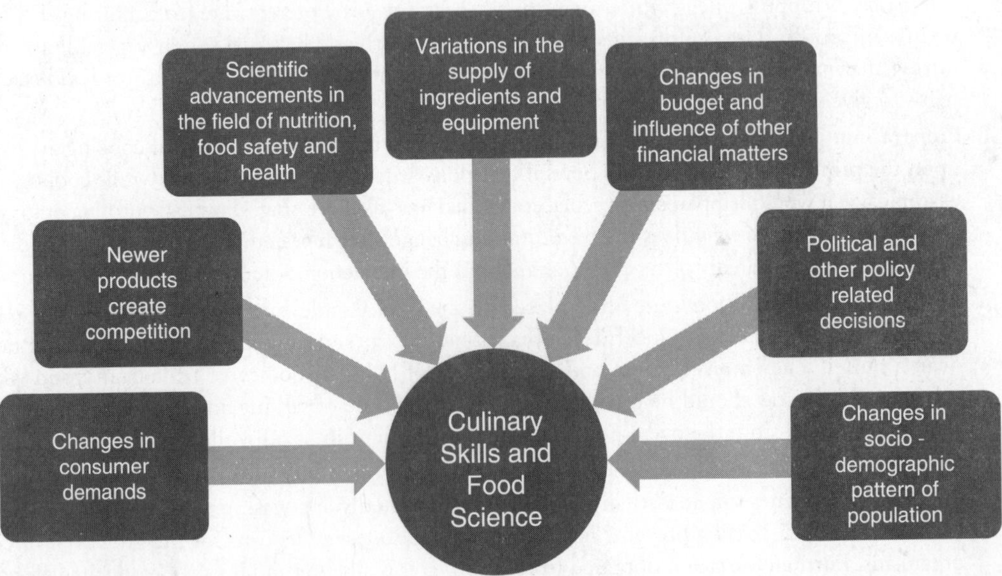

Figure 1.3 Culinary skills and food science: an integrated approach to meet the challenges of hospitality industry

In addition to the above-mentioned personnel, all personnel involved with store management, purchase of supplies, food graders, packers, dishwashers and plant sanitation helpers must have the knowledge of this subject. Such jobs would be available at the following industries:

Hotels	Catering Companies	Airlines
Restaurants	Hostels	Railways
Fast Food Outlets	Ice Cream Parlours	Hospitals
Resorts	Dhabas/Kiosks	Health Care Establishments
Canteens	Food Industry	Publishers
Mess	Research Organizations	Media Companies
Cafeterias	Schools and Colleges	Food Packaging Companies

People involved in the above-mentioned industries become professional experts when they are able to deliver high-quality services to the customers through their respective organizations. The knowledge of food science helps them to perform better in several ways. It helps them to design the process required for handling, processing and storing foods while utilizing the available resources most judiciously. They are more innovative and like to experiment with food, which results in the improvement of existing products and the development of newer products. Information on food science facilitates wiser selection of packaging material, which promotes the enhancement of 'quality shelf life' (refer to Chapter 5). In fact, sensory evaluation performed by trained panel of experts is more reliable. The results obtained can be interpreted based on their molecular origin and more palatable dishes can be prepared. To summarize, it can be said that tasty, nutritious, safe and convenient foods can be prepared with greater success if the culinary skills are based on food science.

Know Your Progress

True or False Questions

1. Food chemistry deals with molecular composition of food.
2. Time, money and workforce are the three major resources required for food production.
3. Space foods processed for space shuttles have a longer shelf life than those manufactured for International Space Stations.
4. RFID technology helps to identify the presence of micro-organisms in food.
5. Packaging materials which enhance/maintain the shelf life of a product should be used.

Very Short Answer Questions

1. Define the term food science.
2. Name any two food products/dishes which require the knowledge of food microbiology for their preparation.
3. What is food engineering?
4. Describe the term food biochemistry.
5. What is food analysis?

Short Answer Questions

1. State the objectives of studying food science.
2. What is the association between food science and agriculture?
3. How can food science help in reducing the incidence of under nutrition?
4. Discuss the role of food science in cooking food.
5. Why should food science be kept in mind while serving food?

Long Answer Questions

1. Food science is an interdisciplinary field. Justify this statement.
2. Enlist any 10 milestones achieved by food science.
3. What challenges can be encountered by a culinary expert working in a hotel? What is the role of food science in overcoming these challenges? Give suitable examples.
4. Enlist at least five ways in which the following professionals can benefit from the knowledge of food science:
   (a) Baker
   (b) Meat cutter
   (c) Executive sous chef

Beyond the Text

- A student of hospitality management is presented with six food products (canned pineapple slices, ice cream, instant coffee, tomato ketchup, olive pickle and soya granules). Can you help them to identify the major processing technique employed to prepare each of these products? (Refer to Chapter 5.)
- Responding to the consumer demand for healthier options, a food processor has developed a trans-fat-free, sugar-free and high-fibre cake. Before launching the product, they wish to test the acceptability of their new product. What would you advise them? (Refer to Chapters 2, 3 and 6.)
- An apprentice cook often creates errors while preparing mayonnaise, white sauce and egg drop soup. Give them some useful tips. Justify the tips on the principles of food science. (Refer to Chapters 2, 4, 7 and 8.)

Chapter 2

Carbohydrates

Chapter Objectives

After reading this chapter, the reader should be able to:

- learn the various properties of different carbohydrates.
- understand the carbohydrate related changes occuring in food during cooking/processing.
- identify/select the carbohydrate which is most suitable for a specific use.
- have a bird's-eye view on the myriad uses of carbohydrates.

Have you wondered why rice grains swell upon cooking, a teaspoon of custard powder (cornstarch) thickens milk sufficiently to form a custard, semolina porridge tends to thicken when allowed to stand for some time, jams develop a smooth texture when pectin is added to them, psyllium seed husks (isabgol) remain dispersed in water while stirring but swell and settle down when allowed to stand and breadcrumbs do not dissolve in water but instead they give a crisp texture to cutlets? Carbohydrate is a common food constituent which is responsible for such changes while cooking/processing.

The word 'carbohydrates' literally means *hydrates of carbon*, i.e., they are composed of carbon and water; or in simple words, carbohydrates are composed of carbon, hydrogen and oxygen. These elements are arranged in various combinations, and the simplest unit of carbohydrate is $C_n(H_2O)_n$ or $(CH_2O)_n$. The differences in the arrangement of carbon, hydrogen and oxygen result in differences in the properties of different carbohydrates, such as solubility, sweetness and rate of fermentation. Carbohydrates are broadly categorized into three types viz. monosaccharides, disaccharides and polysaccharides.

Monosaccharides: They are the smallest/single units of carbohydrates, also known as simple sugars. The most common monosaccharides are the hexoses. Their structure comprises 6 carbon, 12 hydrogen and 6 oxygen atoms. Glucose, fructose and galactose are examples of hexoses (Figure 2.1). However, if the carbon chain contains 5 carbon atoms, the resultant monosaccharides are known as pentoses, e.g., arabinose and xylose.

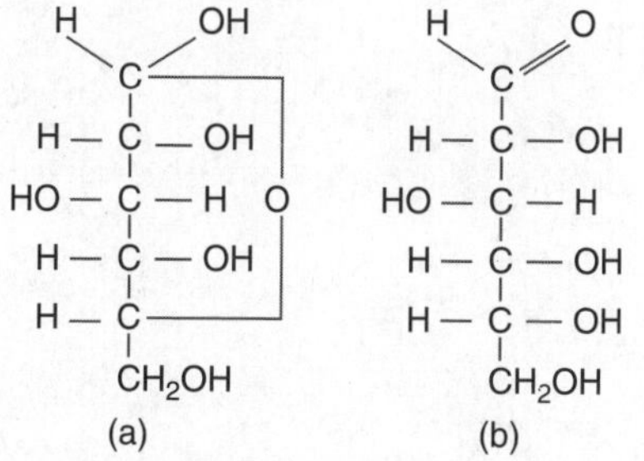

Figure 2.1 Glucose: (a) pyranose structure and (b) Fischer structure

Disaccharides and Oligosaccharides: Disaccharides are composed of two units of monosaccharides. Carbohydrates composed of 2–10 units of monosaccharides are known as oligosaccharides. Disaccharides are the most abundant oligosaccharides. The linkages between any two units of monosaccharides are known as glycosidic bonds. The most significant disaccharides present in foods include sucrose, lactose and maltose (Figure 2.2). The composition of these disaccharides includes the following:

1. Sucrose comprises of one unit each of glucose and fructose.
2. Lactose contains one unit each of glucose and galactose.
3. Maltose is composed of two units of glucose.

Polysaccharides: They are composed of more than 10 repeat units of monosaccharides. The generic name for polysaccharides is glycan and they can be classified into two types viz. homoglycans and heteroglycans. Starch, dextrin, glycogen, cellulose, hemicellulose, pectin, gums, inulin and chitin are the examples of polysaccharides present in food (with exception to glycogen). The term 'resistant polysaccharides' or 'starches' refers to those carbohydrates which remain intact throughout the cooking process or recrystallize after cooling and also resist enzymatic breakdown in human body.

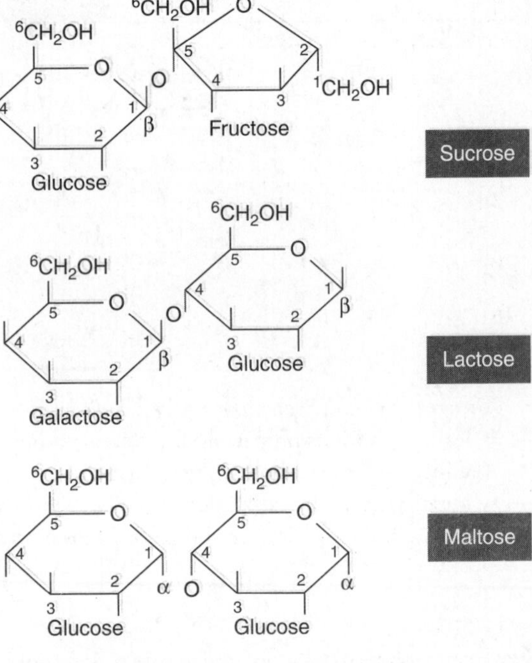

Figure 2.2 Structure of disaccharides: sucrose, lactose and maltose

The salient characteristics and properties of some important carbohydrates are given in Figure 2.3.

Sugars (glucose, fructose, sucrose, maltose and lactose)

- Soluble in water
- Sweet taste
- Digestible mono, di and oligosaccharides
- Variable moisture retention (fructose highest and lactose lowest)
- Improve mouth feel
- Form crystals
- Provide body
- Readily fermented
- Caramalize on heating
- Contribute to maillard reaction
- Preservatives in concentrated solutions
- Act as humectants
- Uses: Dry beverage mixes chewing, gums, toffees, chocolates, snacks, sports, drinks.

Starches

- Not soluble in water
- Bland taste
- Digestible polysaccharides
- Form pastes/gels in hot water
- Gels can be modified by acid/sugar
- Paste/gels retrograde upon freezing/ageing
- Dextrins formed by partial breakdown
- Used as thickeners
- Used on food surface as drying agents
- Pre-cooked starches used in commercial instant foods
- Uses: Dry beverage mixes, soup mixes, bakery products, sweet meats, snacks etc.

Pectins and Carbohydrate Gums

- Soluble in water
- Bland taste
- Non digestible plant polysaccharides
- Pectin forms gels
- Improve texture
- Thickeners
- Stabilizers
- Lubricants
- Rehydrating agents
- Clarifying agents
- Flavour encapsulators
- Film formers
- Foam enhancers
- Suspension formers
- Uses: Beverages, health drinks, jam, jelly, marmalade, soup mixes, processed products.

Cellulose and Hemicelluloses

- Insoluble in water
- Usually bland taste
- Non digestible plant polysaccharides
- Enhance viscosity
- Increase melting point
- Improve texture
- Dispersing agents
- Fillers
- Prevent formation of large ice crystals
- Uses: Ice creams, frozen desserts, health drinks etc.

Figure 2.3 Characteristics and properties of some important carbohydrates

2.1 Properties

Carbohydrates display a wide range of properties (Figures 2.3) which make them suitable for use in a host of dishes and processed food products. The difference in properties is primarily due to variations in their structure, i.e., molecular arrangement. Salient properties of carbohydrates include the following:

1. **Solubility:** Carbohydrates are hydrophilic to different degrees. Glucose, sucrose, fructose, gums and pectin are highly soluble in water but sparingly soluble in alcohol and organic solvents. Fructose is the most soluble carbohydrate followed by sucrose, whereas lactose is the least soluble. Celluloses and hemicelluloses are insoluble in hot and cold water. The solubility process can be enhanced and a more concentrated solution can be prepared by raising the temperature of water. For instance, in 110 grams of water, 181 grams of sucrose can be dissolved at 10°C, but if the temperature is raised to 90°C, 426 grams of sucrose can be dissolved in the same quantity of water. Starches are not readily soluble in cold water but form gels in the presence of heat. Thus, highly concentrated and viscous solutions can be prepared by the application of heat.
2. **Absorption of Moisture:** Sugars being hygroscopic in nature absorb and retain moisture. It is for this reason that the sugar (sucrose) crystals tend to stick together and form a lump when left exposed to atmospheric moisture such as during rainy season. Therefore, sugars should be stored in air-tight containers so as to maintain their quality and shelf life. Fructose has higher hygroscopicity (quicker to absorb moisture and slower to release it to the environment) than sucrose and dextrose; therefore, it is an excellent humectant, i.e., it retains moisture for a long period of time, even at low relative humidity and hence a preferred alternative to other sweeteners in candies and cakes. However, lactose and maltose are the least hygroscopic sugars and, therefore, they are used as bulking agents and viscosity inducers. Starches and celluloses have lower hygroscopic properties and are used in reconstitutable powders/granules which must not stick together during storage and transportation.
3. **Sweetness or Taste:** The sweetness of different carbohydrates varies. According to the sweetness index (SI), the value of 1.0 is given to sucrose which is taken as a standard or reference value for comparison with other carbohydrates. Among the commonly consumed carbohydrates, fructose (SI = 1.7) is sweeter than sucrose (SI = 1.0) while glucose (SI = 0.7), galactose (SI = 0.32), maltose (SI = 0.3–0.6) and lactose (SI = 0.16) are having lesser sweetness (in descending order). Carbohydrates which are bland or have less sweetness are often used as sweetness moderators. For example, dextrose is often used in cookie fillings as a partial or complete substitute for sucrose to reduce sweetness. Therefore, maltose, lactose, starches, dextrin, cellulose, hemicellulose, gums and pectin are not used with the objective of providing sweet taste to a dish.
4. **Hydrolysis by Acid:** Disaccharides and certain polysaccharides can be hydrolysed (broken down) in the presence of an acid. This process can be hastened in the presence of heat. Sucrose is easier to hydrolyse as compared to maltose and lactose. Sucrose gets hydrolysed to glucose and fructose and the resultant product is known as invert syrup. The invert syrup does not crystallize at low or high temperatures and is sweeter than sucrose. It has a low water activity (refer to Chapter 5) and therefore is reasonably resistant to microbial spoilage.
5. **Hydrolysis by Enzyme:** Selective enzymes can be employed to hydrolyse carbohydrates. The enzymatic action gets hastened in the presence of an acid. Enzyme invertase or sucrase is used to hydrolyse sucrose to fructose and glucose, β-amylase produce maltose from cornstarch, and α-amylase helps in splitting germinated cereal starch to prepare diastatic malt products.
6. **Decomposition by Alkali:** Polysaccharides such as cellulose, hemicellulose and mucilages rapidly soften and lose their shape in the presence of an alkali, especially in the presence of heat. Monosaccharides get

converted to bitter-tasting strong-flavoured compounds which are not acceptable to the consumers. Sucrose is least affected by alkali, if used in small amounts (such as the use of baking powder in cake).

7. **Fermentation:** Carbohydrates such as sucrose, fructose, maltose and starch can be fermented by yeast to produce carbon dioxide and water. This principle is used in the bakery industry for raising dough of breads, buns and pizza bases. The carbon dioxide gas leavens the product, making it soft and spongy, whereas alcohol volatilizes (evaporates) during baking. Several malt-based syrups are also produced by fermentation of cereal starch and maltose.
8. **Effect of Heat:** Carbohydrates respond to high temperatures in a variety of ways. Sugars such as sucrose, maltose, fructose and galactose melt upon application of heat, and at higher temperatures, they tend to caramelize. Fructose caramelizes at a lower temperature as compared to sucrose, maltose and galactose. Starches get converted to dextrins. Glucose, sucrose and fructose solutions form highly viscous syrups when cooked. Cooking of such thick syrups helps in the formation of crystals such as those of household table sugar. Starches form gels which are used in a variety of dishes such as kheer, puddings, porridges, pastas, gravies and sauces. Gums and pectins form thick stable dispersions in the presence of heat. All carbohydrates get decomposed to carbon dioxide and water when heated for prolonged duration.
9. **Texture and Other Sensory Attributes:** Carbohydrates promote the development of variety of textures ranging from liquid (syrups), semi-solid (sauces, gravies, soups, porridges), firm (jellies, marmalades, murabbas), solid (cakes, biscuits, confectionaries, sweetmeats, pastas) to hard (brittles, caramels). While hygroscopic carbohydrates help to retain moisture in products (candies, biscuit/cream roll fillings, cakes, muffins), those with low hygroscopic properties are used as films, coatings, fillers and dispersing agents in chewing gums, candies, health supplement powders and snacks. While sucrose and glucose do not have flavour on their own, maltose and its products are distinct. Carbohydrates delay protein coagulation, thereby giving adequate time to a dish for being cooked, prevent protein denaturation and create a soft product. Certain carbohydrates such as dextrose produce a cooling sensation in mouth especially when consumed in dried form (negative latent heat of crystallization).
10. **Enhancement of Shelf Life:** Carbohydrate solutions with low water activity (high Brix), i.e., highly saturated or concentrated solutions, have a prolonged shelf life. For instance, the high osmotic pressure created by sucrose (70 per cent and above concentration) as in case of preserves creates unfavourable conditions for growth and multiplication of bacteria, yeasts and moulds. At high concentrations, sucrose prevents normal diffusion of water into microbial cells or concentration of water from these cells. The preservative effect of sugar can be observed in syrups, confectionary items, honey, canned fruits, fruit drinks, jams, jellies, marmalades, fruit butter and conserves.

GOOD TO KNOW

What Is the Association Between Brix and Concentration of a Sugar Solution?

Balling hydrometer measures Brix, i.e., the concentration of a sugar by weight in a solution at 20°C. To prepare a 20° Brix solution, 250 grams of sugar needs to be added to one litre of water. The resultant volume of solution would be 1.157 litre and thus the weight of sugar in one litre of syrup would be 216 grams. Thus, to make a 30° Brix and a 40° Brix solution, 430 grams and 669 grams of sugar need to be dissolved in one litre of water.

Based on the properties discussed, carbohydrates which display similar characteristics can be grouped as follows:

1. Sugars (glucose, sucrose, fructose, maltose and dextrose)
2. Starches

3. Gums and pectin
4. Cellulose, hemicellulose and mucilages

The properties of these four groups of carbohydrates have been represented in Figures 2.4 to 2.7.

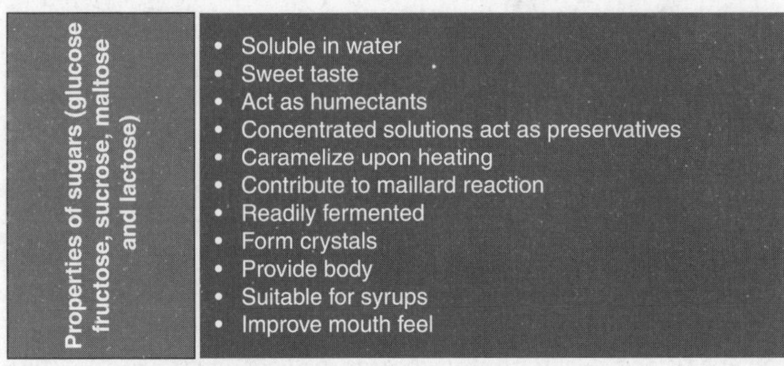

Figure 2.4 Properties of sugars

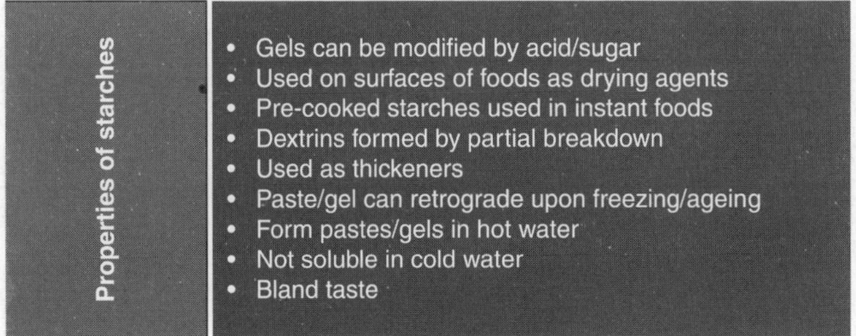

Figure 2.5 Properties of starches

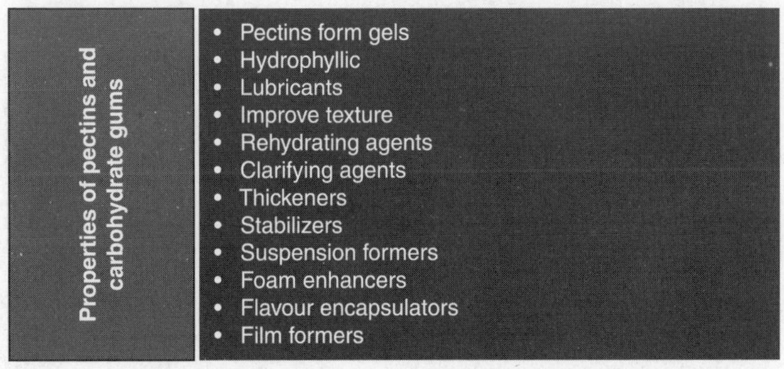

Figure 2.6 Properties of pectins and carbohydrate gums

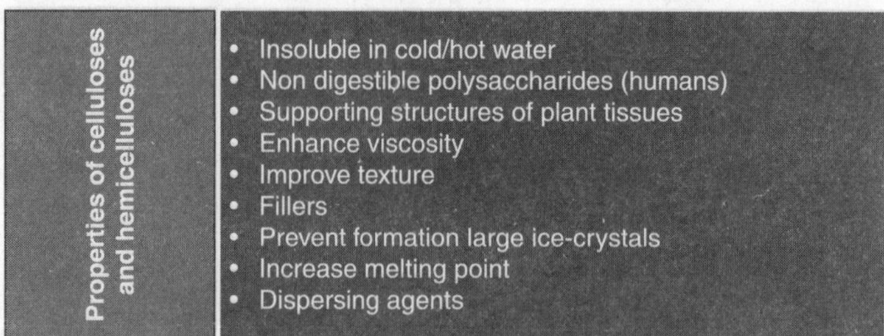

Figure 2.7 Properties of celluloses and hemicelluloses

2.2 Effect of Cooking on Carbohydrates

A host of changes are observed when carbohydrates are cooked. These changes are influenced by the temperature/duration of heat, ingredients present, cooking methods and storage/packaging/transportation/distribution conditions. The commonly observed changes include gelatinization, retrogradation and syneresis, dextrinization, crystallization, inversion and caramelization. While these changes are carbohydrate specific, many a times a combination of these may occur in a dish as a variety of carbohydrates are often present in the ingredients of a dish.

2.2.1 Gelatinization

Starch is composed of two types of semi-crystalline granules (also called glucans) *viz.* amylose and amylopectin (Figure 2.8). Amylose is a linear polymer of α(1→4) linked α-D-glucopyranose and constitutes 20–25 per cent of most starches. Amylose molecules contain in the region of 10^3 glucose units and form helical structures which help to entrap other molecules (water, fat and alcohol). Amylopectin on the other hand is a randomly branched polymer of α-D-glucopyranose comprising of linear chains of α(1→4) linkages having branches of α(1→6) linkage. The molecule of amylopectin is larger than that of amylose and contains about 10^6 units of glucose. Gelatinization is the most significant change in food starch and is associated with molecular interactions of amylose and amylopectin with water in the presence of heat. This process occurs in three major steps (Figure 2.9):

Figure 2.8 Structure of starch: Amylose and amylopectin

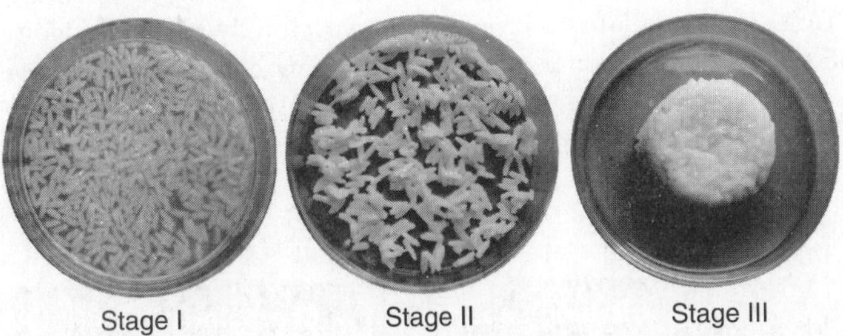

Figure 2.9 Stages of gelatinization for rice starch

1. Starch granules come in contact with water in the presence of heat, imbibe water and swell.
2. The granules become increasingly permeable to water and solutes (sugar and salt), swell further and enhance the viscosity of the dish sharply.
3. Swollen starch granules leach amylose, viscosity increases further and a paste is formed. The hydrogen bonding interactions between amylose and amylopectin continue to occur and ultimately a gel-like structure is formed (if the mixture is left undisturbed). The developments associated with cooling of starch gel are referred to as 'gelation'. Thus, gelation is the conversion of a sol into a gel which does not flow.

Gelatinization of starch depends upon the following factors:

1. **Temperature of Cooking:** In order to ensure complete gelatinization and eliminate raw taste of starch, the starch mixture must be brought to its boiling or gelatinization temperature. The temperature at which maximum swelling of starch granules occurs is known as gelatinization temperature. This temperature varies for different starches. For instance, in case of starches obtained from roots, swelling of granules begins at 65–70°C and gelatinization is usually complete at temperatures below 90°C (Table 2.1). However, starches of cereals gelatinize at higher temperatures near 95–100°C. Starch mixtures cooked below the optimum temperature tend to remain thin. If the temperature is increased rapidly, the resultant paste is thicker than the paste which is cooked with gradual increase in temperature. Further, starches which are subjected to prolonged heating or boiling temperatures tend to fragment or disintegrate.
2. **Initial Concentration of Starch:** The amount of starch present in a mixture determines its final viscosity. Higher the concentration of starch, the greater is the viscosity. However, a very high concentration results in the development of a very tight gel which is usually inedible. The amount of starch required to form a paste of desired thickness would vary with the type of starch being used. Cornstarch would be required in lesser amount as compared to potato starch for preparing a paste of same thickness (with same amount of water/liquid).
3. **Type of Starch:** The shape and composition of starch granules influence the gelatinization of starch and the ultimate gel formed. The amylose and amylopectin content of starch determines the type of gel formed. Starches with high amylose content form viscous translucent gels, whereas starches with high amylopectin content (waxy starches) form viscous clear pastes which do not form a gel upon cooling. Thus, waxy starches form a soft thick mass instead of stiff gels. It is for this reason that waxy corn and sorghum starches are found useful in pie-fillings and other cream-like fillings in bakery products. Cereal starches are more effective thickeners as

compared to root starches. Cornstarch produces a distinctly stiff gel as compared to rice starch which produces a tender gel. Potato starch and arrowroot produce an even more tender gel (Table 2.1).

Table 2.1 Commonly used starches and their gelatinization characteristics

Starch	Thickening Capacity	Characteristics of Hot Paste
Potato starch	Moderate	Less viscous; gummy translucent paste.
Tapioca	Moderate to low	Less viscous; gummy translucent paste.
Arrowroot starch	High	Optimum thickening upon cooking; firm moderately translucent gel.
Cornstarch	High	Optimum thickening upon cooking; stiff translucent gel.
Rice starch	Moderate	Optimum thickening upon cooking; firm moderately translucent gel.
Wheat flour	Moderate to low	Optimum thickening upon cooking; firm opaque gel.

4. **Agitation of Starch:** Stirring of starch in water or any other liquid in which it is cooked ensures thorough or even dispersion of the granules. If adequate stirring is not carried out, clumping occurs as a result of which, the granules on the outer surface of lump gelatinize, whereas those in the centre remain dry. This not only reduces the final viscosity of the product but also spoils the overall texture. If the starch mixture is stirred too aggressively, the starch granules burst fast and result in the development of a sticky paste of reduced viscosity. Stirring the starch in cold water before cooking is ideal for facilitating the development of a paste with homogenous texture.
5. **Other Ingredients in Dish:** Seldom is a starch cooked alone in water, usually a number of ingredients are present along with it such as sugar, salt, milk, fruits/vegetables and eggs. Custard would contain milk, eggs, sugar, fruits and essence; macaroni would contain salt, vegetables and cheese or chicken. The effects of such ingredients on the gelatinization process and the properties of starch are enlisted in the following paragraphs.

   Sugar: It competes with starch for water and hence the final viscosity reduces, because there is a decrease in the imbibition of water by starch granules (refer to Chapter 8). The resultant gel is less rigid but more translucent as the amount of sugar increases.

THINK TANK

Why do cornstarch puddings at times become gummy and heavy?

Cornstarch puddings should have a smooth texture and a delicate light mouth feel. Very brisk or vigorous stirring during cooking causes rapid breaking of swollen starch granules which results in a gummy texture of the pudding. Inadequate or careless mixing of ingredients in cold water/milk causes lumps to form and gives a heavy pudding.

Salt: Salt, i.e., sodium chloride, is hygroscopic and similar to sugar, it also competes for water. However, salt is added in small amounts to a dish and the resultant effect is not very evident on the gel.

Acid: It may be added to the dish in the form of vinegar, lemon or tamarind or may be present as part of an ingredient such as citrus fruits (custards) or vegetables, particularly tomatoes (cream of tomato soup, macaroni, *dum-aloo*). Acid reduces the thickness of the starch paste and forms a less firm gel, because it causes hydrolysis of the starch molecules which gets catalysed by heat. Thus, as the product is cooked (duration) and higher temperatures are attained, the rate of hydrolysis also increases. Therefore, acid should be added only when the gelatinization process is over, e.g., in the preparation of lemon meringue pie. Sugar limits the hydrolytic effect of acid, e.g.,in fruit custards.

Fat: It tends to form a layer or film around the starch granules, thereby reducing their water absorption capacity. This also limits the bursting of swollen starch granules. The paste thus formed is less viscous, and the gel is less firm but shiny. This principle is applied during the preparation of *pulao*, where we do not want the rice grains to stick to each other.

Protein: It raises the temperature of initial swelling and final gelatinization point. Therefore, the cooking time increases, which is useful in the preparation of products with a dense texture such as cake. The gel formed is less firm than that formed from a starch–water mixture.

THINK TANK

Why does cream of tomato soup curdle?

The tomato soup should be moderately thick with smooth texture, free from lumps and fat film. Curdling occurs when the white sauce is poured into the tomato puree during which the first portion of milk in white sauce comes in contact with acid medium (tomatoes). Milk curdles in acid medium. To prevent this, hot tomato puree should be added to the white sauce while stirring so that an alkaline medium is maintained. This should be done just prior to serving. Further, very fresh milk should be used in the preparation of white sauce.

Thus, the gelatinization process needs to be carefully monitored especially while preparing dishes for commercial reasons.

2.2.2 Retrogradation

It is also known as shrinkage of gel. After gelatinization, when the starch paste is allowed to cool undisturbed, the free amylose molecules leach out and tend to associate together to form crystalline aggregates. This causes shrinkage of gel which is followed by syneresis. Thus, retrogradation is the reversal of a dispersed amorphous state of starch to an insoluble aggregated or crystalline condition. It can occur in any cooked starch–water mixture. Dry texture in stale bread is an

THINK TANK

Why does fat at times separate out from the white sauce?

Formation of a fat film or separation of fat from the white sauce is indicative of incomplete gelatinization or inadequate cooking/cooking at low temperature. In a thin sauce, this defect can be improved by reheating the sauce while carefully stirring. In a thick sauce, a small amount of water/milk may be added and the sauce should be stirred while reheating till the fat and liquid recombine with the sauce and give it a smooth texture.

example of retrogradation. It can be removed by heating the bread without the loss of moisture. Addition of surface-active agents and salt can prevent retrogradation such as in the case of savoury crackers.

2.2.3 Syneresis

Syneresis or weeping of gel usually (but not always) follows retrogradation. It is characterized by the release of moisture when a cut is made in the gel. The leakage of moisture from the cut gel would increase with time. Syneresis is particularly a problem in frozen puddings and pie fillings, because the ice crystals disrupt the gel network and also rupture the swollen starch granules.

2.2.4 Dextrinization

It is a process which is observed when the starch is cooked by the application of dry heat and is characterized with the formation of dextrins. Dextrins are formed during toasting of bread, roasting of semolina or on the surface of biscuits facing the tray. Dextrins are polysaccharides with molecular size intermediate between starch and sugar, i.e., the structure of dextrins is shorter than that of the parent starch from which they have been formed. They are brown in colour, mildly sweet with distinct flavour, more soluble in water and possess lower thickening ability than starch. Depending upon the duration of heating and the temperature used, dextrins of variable molecular weight are formed. The starch of flour which has been browned only lightly gets converted to dextrins with comparatively larger molecular weight which are slightly soluble in water but exhibit good thickening ability. However, starch flour, which has been heated for a prolonged time, forms dark brown dextrins which have enhanced solubility but low thickening ability.

2.2.5 Maillard Reaction

Maillard reaction, also known as non-enzymatic browning, is primarily a reaction between carbohydrate and protein in the presence of heat. It involves a series of steps involving reducing sugars (a form of carbohydrate) and amino group of protein which results in the formation of brown-coloured melanoidin compounds such as amadori compound. Such compounds can be observed on the base of utensil in which milk is being heated. The brown colouration on the surface of biscuits or chapatti is also to a great extent due to Maillard reaction (Figure 2.10). Thus, Maillard reaction compounds very often help in improving the appearance, taste and flavour of certain food products. However, in milk and milk-based products such as khoa, rabri, ice cream and custard, such non-enzymatic browning should be avoided as it lowers customer acceptability.

Figure 2.10 Maillard compounds at the bottom of a pan wherein milk and sugar have been cooked for a prolonged duration

GOOD TO KNOW

Toxic Sugar Derivatives–Maillard Reaction

Research has indicated that maillard reaction in certain foods cooked by specific methods can result in the development of mutagenic compounds, predominantly, heterocyclic amines and acrylamide. Heterocyclic amines have been found in meat grilled at high temperature or cooked for long duration. Acrylamide has been found in certain potato-based snack foods.

2.2.6 Crystallization

When supersaturated solutions of glucose or sucrose are heated under controlled conditions, aggregation of sucrose molecules to form sugar crystals occurs. These crystals may be so fine that they can hardly be perceived by the tongue or as big as a grain size. Crystals of sugar are formed when sugarcane juice is concentrated by the application of heat. When sugar is dissolved in water and heated to form a supersaturated solution (>112°C), it once again forms crystals, and this is known as recrystallization. The crystals thus formed would vary in texture, size, colour and other sensory properties. Even, the weight to volume ratio would vary depending upon the following cooking conditions:

> **GOOD TO KNOW**
>
> **ICUMSA Numbers and the Purity of Sugars**
>
> The International Commission for Uniform Methods of Sugar Analysis gives ICUMSA numbers and lower numbers indicate higher level of purity of refined sugar. The specifications and standards can be seen online at www.icumsa.org

1. Higher the concentration of sucrose, the more rapid is the crystal formation. Crystals generally form when sugar concentration is >85 per cent.
2. Cooking at higher temperatures would result in finer crystals and vice-versa.
3. Stirring/agitation of solution while cooking promotes finer crystal formation. Too much stirring is also not advisable, as it will break the formed crystals.
4. Addition of ingredients such as milk, eggs, nuts, fat and flour to the sugar solution would also hinder crystal formation.
5. Addition of acid such as cream of tartar, lemon juice, vinegar or citrus fruits would also hinder crystal formation such as during the manufacture of jam, fruit custard and murabbas.

2.2.7 Inversion

The use of acid in the presence of heat to convert sucrose (a disaccharide) into fructose and glucose (monosaccharide) is known as inversion. The formation of glucose and fructose enhances the tendency of resultant product to non-enzymatic browning. Presence of invert sugar in a product prevents the formation of crystals and hence renders a smooth, creamy texture. Toffees with a liquid centre, chocolate-coated fruits and soft-centred chocolates generally contain invert syrup. Invert syrup is also used in combination with sucrose in packaged cakes to keep them moist and soft, even if kept at low temperatures.

2.2.8 Caramelization

Certain carbohydrates such as sucrose, fructose and maltose, when subjected to dry heat, melt, dehydrate, darken and become bitter. This thermal decomposition of carbohydrates is known as caramelization or non-enzymatic browning. Different carbohydrates caramelize at different temperatures. Sucrose and fructose caramelize at about 170°C and 110°C, respectively. During the formation of caramel, several volatile and non-volatile substances (2-furaldehyde, 5-hydroxymethyl-2-furaldehyde, caramelin, caramelan, caramelen, acids, aldehydes and ketones) are formed which render a distinct strong flavour and taste. Caramelized sugar can be light brown as in the preparation of peanut brittles or 'burnt-brown' and is used as an additive in the manufacture of vinegar, whisky, rum, wines, soft drinks, brown bread and biscuits.

2.3 Uses of Carbohydrates in Food Preparation

Carbohydrates have been used extensively in the food processing industry, because of the wide range of characteristics they exhibit. Very often, more than one carbohydrate is used in a dish or a product with contradictory properties, so as to create a product with new sensory qualities. For instance, the dough of extruded snacks prepared from cornstarch would contain good amount of moisture so that the starch can gelatinize, but when the product is obtained from the extruder, it is often coated with maltose/dextrin or cellulose to prevent absorption of moisture from the atmosphere. Carbohydrates have been used as follows:

Binders	Texture promoters	Film formers
Crystal formers	Crystal inhibitors	Flavour carriers
Sweeteners	Sweetness moderators	Bloom inhibitors
Drying agents	Humectants	Preservative
Emulsifiers	Stabilizers	Colouring agents

Keeping in mind the discussed capacities of carbohydrates, they have been used in a variety of food preparations. The salient ones are discussed below.

1. **Confectionary and Sweetmeats:** Sucrose is the most popular carbohydrate used in this industry, although fructose, maltose and dextrin are also increasingly being used. Some salient uses of sucrose and its related products include the following:

   (a) They are used as sweetening agents, e.g., in porridges,ice creams, meringues, crystalline and non-crystalline candies as well as several sweetmeats such as *burfi, ladoos, kulfi* and *kheer* to name a few.

   (b) Sugar syrups are often used as an adjunct to several sweetmeats such as rasgullas, gulab jamuns and jalebis, wherein the sugar molecules enter the product through osmosis. Flavoured and artificially coloured sugar syrups are used as dressings/coatings to attractively decorate meringues, soufflés, sundaes, ice creams and puddings. The uses of sugar syrups of varying thickness are briefed in Table 2.2.

   (c) Use of fructose such as in high fructose corn syrup (HFCS) helps in providing characteristic citrus flavour to confectionary items. Corn syrup, HFCS and invert syrup inhibit crystallization of sucrose and therefore they are used in small amounts in non-crystalline candies (brittles, chikkis).

   (d) Sucrose is hygroscopic and therefore tends to absorb water/moisture when added to a product. This improves the consistency (thickness) of the product such as *aam papad.* On the other hand, starch, dextrin, maltose and lactose have low hygroscopicity and therefore are used as coatings on certain confectionaries to prevent them from developing a sticky texture, if the relative humidity of storage temperature is high.

   (e) Supersaturated sugar solutions tend to thicken when cooled. This property is most prominent in case of jaggery. Therefore, jaggery and sugar also act as binding agents in certain products such as boondi ladoo.

   (f) Different types of sweeteners influence the texture and structure of the products in different ways. Several ladoos, burfi and other sweetmeats (especially those which contain equal amounts of sugar and flour), toffees, chocolates, marshmallows, brittle and glace owe their texture to the type of carbohydrate (sweetener) present in them.

Table 2.2 Uses of sugar syrups of different concentrations

Temperature (°C)	Stage of Concentration	Characteristics	Product
110–112	Thread	Shape: none (liquid) Colour: white/creamish white Taste: sweet	Syrup (*chachni*)
112–115	Soft ball	Shape: easy to modify Colour: creamish white Density: high Taste: sweet	Fondant, Fudge
118–120	Firm ball	Shape: non-modifiable Colour: straw Density: lesser than soft ball, less translucent Taste: sweet	Caramels
121–130	Hard ball	Shape: non-modifiable Colour: darker than firm ball Density: lighter and slightly transparent Taste: sweet	Nougat, Marshmallows
132–143	Soft crack	Shape: non-modifiable, syrups spread into threads which need to bind together instantly. Density: lower than hard ball, transparent Colour: light brown Taste: sweet	Butterscotch Taffies
149–154	Hard crack	Shape: same as soft crack Density: lighter than soft crack, transparent, brittle. Colour: medium brown Taste: sweet	Brittle, Glace
160	Clear liquid	Shape: none (liquid) Density: high Colour: medium brown Taste: mildly bitter	Barley sugar
170	Brown liquid	Shape: none (liquid) or as desired if put in cold water immediately. Density: high (syrup) low (caramel) Colour: medium to dark brown Taste: bitter	Caramel

2. **Baked Products:** The use of carbohydrates in flours for bakery products is most extensive. It provides body and texture to breads, cakes, pastries, cakes, biscuits, crackers and rolls. The use of a little amount of sugar in the dough facilitates fermentation which helps to produce a spongy texture and provide a characteristic flavour typical to certain bakery products such as bread, buns and pizza base. The ability of carbohydrates to form films, inhibit crystallization, aid in moisture retention and create an

optimum level of sweetness, make them suitable for the preparation of products such as fruit leather glazes and frostings. Carbohydrates are also being used as an ingredient in cream-type fillings in biscuits, wafers and cream rolls as they provide a creamy smooth texture to the filling and form a coating on the layer which does not allow it to flow out of the product.

3. **Beverages:** Carbohydrates such as lactose and maltose are often added to flavoured beverages as they provide body to the drink and at the same time optimize sweetness of the product when used alone or in combination with sucrose. Low osmolality, complete solubility, bland flavour and easy, fast digestibility make glucose, sucrose and fructose an ideal ingredient in sports drinks and other food products for sports persons. Sucrose, of course, has its ubiquitous presence in squashes, fruit cordials, fruit drinks and soy milk to name a few.
4. **Soups:** Starch, present in refined wheat flour, corn flour and potatoes, works as a key ingredient in several cream soups such as cream of tomato soup, cream of vegetable or chicken soup. It provides a smooth thick creamy consistency. Being bland in flavour and white in colour, it does not hinder with the mild flavour of vegetables, chicken and seasonings.
5. **Sauces:** Starch is used as a prime ingredient in the preparation of white sauces. Gelatinization of starch in milk provides body to the sauce. Since this polysaccharide is bland in flavour, not sweet and creates a smooth mouth feel, it is considered optimum for use in white and cheese sauce (Figure 2.11). In tomato sauce, it provides body and intensifies red colour while adding a modest shine to the product.
6. **Salad Dressings:** Several processed salad dressings contain carbohydrates as they provide body to the product, promote clinging property and decrease excessive gumming/stringiness in the product. This is particularly relevant for eggless mayonnaise.
7. **Snacks:** Carbohydrates are used for varied functions in the preparation of snacks. On one hand, carbohydrates such as starches (upon gelatinization) act as binders in snacks with high moisture content such as potato cutlets, bondas and vegetable Scotch egg. On the other hand, in low-moisture products such as breakfast cereals, nuts and extruded products, carbohydrates such as dextrin, maltose, and cellulose are used to inhibit water absorption and act as coating agents and flavour carriers.
8. **Preserves:** Sucrose is the principal carbohydrate which has been used for withdrawing water from fruits through the process of osmosis. It not only helps to reduce the water content but also provides a firmer texture to the fruit such as in the case of preserves. It provides body, texture and taste to the products such as jam, marmalade and fruit butter.
9. **Dry Mixes:** When carbohydrates with low hydroscopicity are used in dry mixes, reconstitutable powders or granules, they do not become sticky. The applications and functions of carbohydrates in such products include the following:

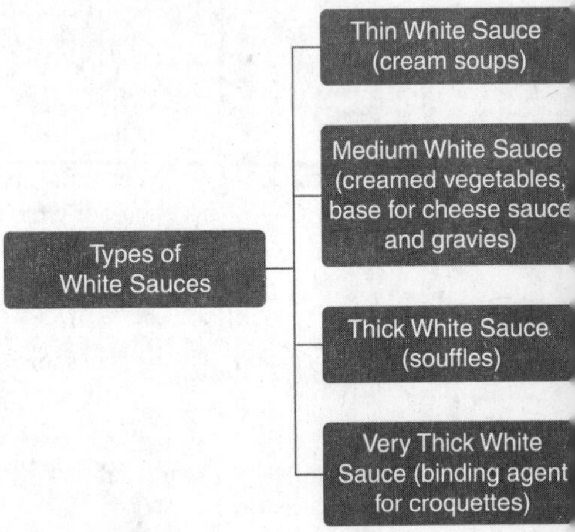

Types of White Sauces
Thin White Sauce (cream soups)
Medium White Sauce (creamed vegetables, base for cheese sauce and gravies)
Thick White Sauce (souffles)
Very Thick White Sauce (binding agent for croquettes)

Figure 2.11 Types of white sauces

(a) Powdered drinks: act as bulking agents, promote rapid dispersion of powder in water/milk/any other liquid and being bland themselves, they do not influence taste.
(b) Soup powders and sauce mixes: provide body, protect from moisture, encapsulate flavour and quickly dissolve/gelatinize when cooked.
(c) Spice powders: act as bulking agents, non-sweet carriers of taste and flavour and prevent moisture absorption from environment.
(d) Cheese powders and flakes: encapsulates flavours and protects proteins when used during spray drying. It improves sprinkling or flow properties of the products.
(e) Dairy whiteners: promote even fat dispersion and improved mouth feel.

10. **Infant Formulas:** Most carbohydrates (except gums, pectin, cellulose and mucilages) are bland in flavour and easier to digest. Therefore, they are used in infant formulas. Pre-cooked starches, particularly rice and cornstarch, are used in infant foods as they are easy to digest and form a smooth consistency when mixed in warm water. Starches in combination with glucose, sucrose and dextrin provide mild sweetness to such products.

Box 2.1 What Is a Modified Starch?

It is a starch that has been chemically modified to make it functional under the conditions of processing and storage such as high temperatures, high shear, low pH, freezing and thawing. Modified starches are E coded according to International Numbering System (INS), i.e., 1400–1415. The E numbers 1401, 1402, 1403 and 1405 are commonly used as food ingredients.

11. **Ready-to-cook/Ready-to-eat:** Modified starches and pre-cooked starches are also being used extensively in ready-to-cook, ready-to-eat cereal-based products such as pastas, breakfast cereals, extruded products, infant foods, mashed potatoes, papads, frozen parathas, frozen puddings/pie fillings and sponge base for crunchy candies (Box 2.1).
12. **Fat Replacers:** Certain carbohydrate products such as maltodextrin, polydextrose, and modified food starch have been physically and chemically processed to produce fat replacers which are useful for preparing low-colorie food products. The applications and functions of such fat replacers include the following:

(a) Bakery products: help to retain moisture and retard staling.
(b) Frozen desserts: enhance viscosity and improve texture.
(c) Dairy products: enhance viscosity, stabilize emulsions and aid in producing gels.
(d) Confectionary: improve and add variety to mouth feel and texture.
(e) Soups, sauces and gravies: improve mouth feel/texture and increase viscosity.
(f) Salad dressings: help in stabilizing emulsions, enhance viscosity and improve mouth feel.
(g) Margarines, shortenings and table spreads: improve mouth feel.
(h) Processed meat products: increase water holding capacity and improve/add variety to texture and mouth feel.

However, carbohydrate-based fat replacers cannot be used for frying purpose as they are not stable at high temperatures. The shelf life of products containing such fat replacers is comparatively shorter than those containing fat.

13. **Edible Packaging:** The possible use of carbohydrates as edible packaging material is being researched. For instance, a variety of corn has been developed, the starch of which contains about 80 per cent amylose and less than 20 per cent amylopectin. Such edible films can be used on products such as ready-to-cook pastas, because they would dissolve upon contact with hot water and form part of the dish. The use of such films is also being explored for wrapping meat products (during ageing), cheese (during ripening) and candies (for marketing).

It is evident from these discussions on the properties, effect of cooking and uses of carbohydrates that this nutrient can be employed in endless ways to create dishes and products with a variety of sensory characteristics. Natural foods usually contain a combination of carbohydrates and therefore behave differently to various cooking methods. Therefore, isolated or pure carbohydrates can be used alone or in combination with natural foods to achieve desired characteristics in a dish. It must also be remembered that the effect of cooking discussed in this chapter are for pure carbohydrates, but usually a dish contains other ingredients wherein this nutrient interacts with fats, proteins and additives. Thus, while cooking with a specific objective in mind, one must evaluate the effects of other ingredients on carbohydrates present in the dish. Exploiting the properties of this nutrient judiciously can make the cooking experience successful and satisfying.

Know Your Progress

True or False Questions

1. Sucrose is a disaccharide.
2. Fructose is insoluble in water.
3. Gums can be used as thickeners.
4. Soft crack stage can be achieved by boiling sugar solution at 120°C.
5. Carbohydrate-based fat replacers spoil the texture of frozen desserts.

Very Short Answer Questions

1. Give five food sources of glucose.
2. Describe syneresis.
3. Enlist five uses of gums in food.
4. Name a few functional fibres.
5. Give the sweetness index of sucrose, maltose, fructose and lactose.

Short Answer Questions

1. Why does dry dextrin give cooling sensation when placed in mouth?
2. What is inversion?
3. Why do lumps form at times while making a starch paste?
4. Why does fat separate out at times in white sauce?
5. What is the difference between dextrinization and Maillard reaction?

Long Answer Questions

1. Give the properties of carbohydrates.
2. What is gelatinization? What are the factors which influence gelatinization of starch?
3. What is crystallization? How is it different from inversion?
4. Discuss the uses of sugar in food processing.
5. Discuss in detail the various uses of carbohydrates.

Beyond the Text

- Read the labels of the following products and identify the carbohydrates present in them. Also find out the potential role of carbohydrate(s) in each product.

Chewing Gum
Eclairs
Sponge Cake
Instant Soup Powder
Toffees (non-milk)
Health Supplement
Soft-drink (lemon flavoured)
Dairy whitener

Chapter 3

Fats and Oils

Chapter Objectives

After reading this chapter, the reader should be able to:

- identify and select fat or oil on the basis of their composition.
- know how fats and oils are processed to meet the 'end-use-specifications'.
- understand the heat related changes in fat.
- keep in mind the factors which can prevent and promote fat spoilage (rancidity) so as to use fat most judiciously.

The use of fats and oils has forever been expanding in the food industry. Fats and oils continue to enjoy a prestigious (though controversial) space when it comes to food processing; at both small and large scale. Thei use is not just restricted to transfer of heat (frying, baking and roasting), modification of texture, developmen of strong or delicate flavours and as release agents, binders, emulsifiers or moisture retainers but also in th production of sugar substitutes. However, despite such vast uses fats and oils have always been under th scrutiny by health professionals, primarily because of the physical and chemical changes they undergo durin food processing in addition to their capacity of making the dish calorie dense. This chapter shall discuss th composition, processing, spoilage and uses of fat. In this chapter, the word 'fat' has been used interchangeabl to describe 'fat as a nutrient' and 'fat as food'.

Fats are composed of carbon, hydrogen and oxygen. Carbon atoms are attached to each other in the form of chains to which hydrogen and oxygen are attached in varied combinations with the help of bonds. This combination of atoms forms fat molecules. Each molecule of fat is composed of a molecule each of triglyceride and glycerol. The word 'triglyceride' refers to three molecules of fatty acids. Naturally occurring fatty acids have even number of carbon atoms in their chain, generally between 4 and 28. The fatty acid composition/structure determines the physical and chemical properties of a fat molecule and hence of the fat/oil in which it is present. Box 3.1 gives the definition of fat and Table 3.1 gives some characteristic features of fatty acids.

Box 3.1 What Are Fats?

Fats are scientifically termed as *lipids*. According to standard definitions, lipids are

1. chemically heterogeneous group of compounds having in common the property of being insoluble in water, but soluble in non-polar solvents such as chloroform, methanol, benzene, hexane and alcohols.
2. molecules with long chain hydrocarbons.
3. present in or derived from living organisms.

However, this definition does not include the following:

1. Lipids with one to four carbon atoms (C_1–C_4) in their fatty acid chain, because such very short chain fatty acids are soluble in water.
2. Trans fatty acids which are produced usually during food processing operations such as frying and hydrogenation of fat and hence, not always obtained from living organisms.

Table 3.1 Characteristics of fatty acids

Fatty Acid	No. of Carbon Atoms	No. of Double Bonds	Melting Point (°C)	Typical Food Source
Saturated fatty acids				
Butyric	4	0	−8	Butterfat
Caproic	6	0	−3	Butterfat
Lauric	12	0	44	Coconut oil
Myristic	14	0	53	Butterfat,coconut oil
Palmitic	16	0	62	Animal fat
Stearic	18	0	70	Animal fat
Arachidic	20	0	75	Groundnut oil
Unsaturated fatty acids				
Myristoleic	14	1	18	Butterfat
Oleic	18	1	16	Olive/canola oil
Linoleic	18	2	−6	Refined oils
Vaccenic*	18	1	44	Butter, hydrogenated fat
Elaidic*	18	1	44	Butter, hydrogenated fat
Linolenic	18	3	−13	Soybean/canola/flaxseed oil
Arachidonic	20	4	−50	Lard, goat

Note: Vaccenic and Elaidic acids are trans forms of fatty acids.

3.1 Classification of Fats

Fats can be classified on the basis of their composition, that is, the type of fatty acids (or degree of saturation) present within or on the basis of the source from which they are obtained. Classification of fats helps in selecting a fat/oil most suitable for

1. preparing the dish or product.
2. maintaining/adhering to food safety standards.
3. meeting the health claims or specifications as mentioned on the label of the product or menu card.

3.1.1 Degree of Saturation

Degree of saturation refers to the number of binding sites which are free on a carbon atom present in a carbon atom chain of a fatty acid. A carbon atom has four binding sites to which either hydrogen or carbon is attached. If the binding site is free, the bond is free and susceptible to attach any other atom. Fatty acids, wherein, all available carbon binding sites not linked to another carbon, are attached to only hydrogen atoms and are therefore saturated are known as **saturated fatty acids (SFAs)**. SFAs do not have a double bond. However, if one carbon binding site is free because of the absence of a hydrogen atom, the resultant fatty acid has a single double bond and is known as **monounsaturated fatty acid (MUFA)**. If more than one carbon binding sites are free, the fatty acid is known as **polyunsaturated fatty acid (PUFA)**. Figure 3.1 gives the structure of fatty acids. MUFA and PUFA are also known as **unsaturated fatty acids (USFA)**.

Butyric, caproic, lauric, myristic, stearic, arachidic and behenic acids are the examples of SFAs. Caproleic, lauroleic, palmitoleic, oleic, vaccenic and elaidic acids are the examples of MUFA; whereas, linoleic, linolenic, arachidonic, eicosapentaenoic and docosahexaenoic acids are the examples of PUFA. Table 3.1 gives examples of the typical food sources of each of these fatty acids.

Each fat has its own unique fatty acid composition and are categorized on the basis of their SFA and USFA content.

Saturated Fats: If the SFA content exceeds the amount of USFA present, it is known as a saturated fat. Such fats contain high amount of butyric, caproic, lauric, myristic, stearic, arachidic and behenic fatty acids. Fats rich in SFA tend to become solid at low temperatures, semi-solid at room temperature and liquid when heated. They have low smoking point and tend to burn easily upon heating. These primarily include butter, cream, malai, pure ghee, lard, tallow and hydrogenated fat such as margarine and vanaspati. They play an important role in the preparation of bakery products because of their commendable shortening and plasticizing properties.

Unsaturated Fats: They are generally obtained from plant sources and contain higher amount of USFA as compared to SFA. Fats rich in USFA are called oils. Certain oils contain higher amount of MUFAs, whereas others contain higher amount of PUFAs. Olive, canola and fish oils are rich sources of MUFAs, whereas safflower, sunflower, cottonseed, corn, soybean, mustard, linseed and marine oils are good sources of PUFAs. Unsaturated fats, that is, oils (except coconut oil) tend to remain liquid at room temperature as well as at high and low temperatures. Unsaturated fats have a lower melting point as compared to saturated fats. Oils are primarily used for transferring heat, that is, as cooking and frying medium.

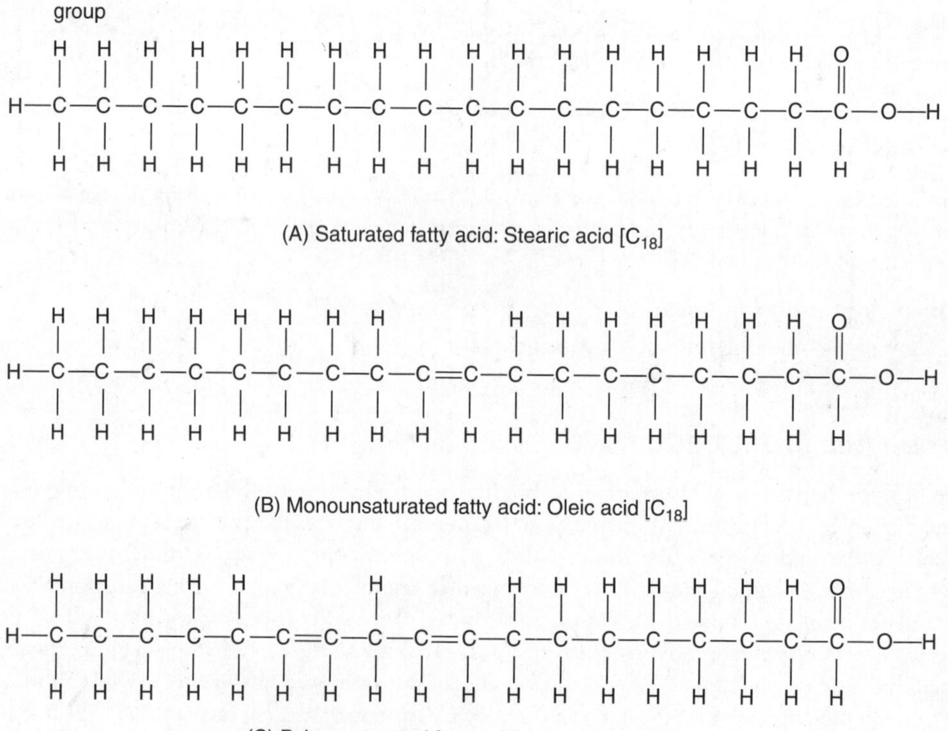

Figure 3.1 Structure of saturated and unsaturated fatty acids

> ***Melting point:*** *The temperature at which a sample of fat reaches a homogenous liquid consistency is known as melting point of fat. It varies for each fatty acid and hence each fat/oil. Further, greater the degree of unsaturation, lower is the melting point.*

3.1.2 Origin: Animal and Plant Fats

Fats are being extracted from endless number of sources. Today, even unconventional foods are being explored to extract fats/oils and create blends so as to meet end-use specifications in the culinary industry.

Animal Fat: It includes fat obtained from milk and adipose tissues of dead animals. Non-dairy animal fat is obtained from animal offal, bones and trimmings from meat processors, fish species unsuitable for marketing as fillets and other fish products (menhaden, pilchard, herring) and fish cannery wastes. Generally, the fat of buffalo, cow, pig, lamb and birds is extracted, although any animal tissue can be processed. Dairy fats which are commonly consumed include butter, cream, malai and various types of cream. Since animal fats are good sources of SFAs, they are semisolid at room temperature, liquid when heated and solid when refrigerated. Their smoking temperature is lower than plant fats.

Plant Fats: The annual field crops and the by-products obtained during milling, especially, of cereals are being used to extract oils. The major sources are as follows:

Seeds – mustard, linseed, cottonseed, grapeseed, mango seed.
Fruits – coconut, olive, almond, palm, avocado.
Flowers – sunflower, safflower.
Others – soybean, wheat germ, cocoa bean, rice bran, corn husk/fibres.

Being rich sources of USFAs, oils are liquid at room, high and low temperatures. They have high smoking point and are therefore more desirable for frying.

3.2 Processing of Fats and Oils

Lipids are present along with the plant and animal tissues in their natural form. They need to be extracted and processed to yield oils/fats that can be used for cooking purpose. Processing plant or animal tissues to obtain edible fat involves four major steps, that is, recovery of crude oil, refining, conversion and stabilization.

> *The defatted by-product obtained after extraction of fat is known as meal. Defatted meals are rich sources of protein and are often used to make low-cost nutritious products and as livestock feeds.*

Crude Oil Recovery: Animal fats and marine oils are obtained through a process known as *rendering* wherein fat is extracted from animal tissues. Seeds/legumes are subjected to cleaning, dehulling or decortication (if required) followed by grinding or flaking, pressing, expelling and solvent extraction to facilitate maximum recovery of oil. In case of fruits such as palms and olives, the oil extraction process involves sterilization, digestion, pressing, centrifuging and vacuum drying.

Refining: The crude oil obtained is viscous, translucent, contains impurities and is suceptible to spoilage. Therefore, it is subjected to a series of refining operations which include filtering, degumming, caustic refining, bleaching, de-waxing and deodorization. The refined oil thus obtained is bland, light coloured, transparent and stable.

Conversion: Natural fats and oils may not always meet the specifications desired by a chef for creating a dish. Therefore, food technologists have devised several methods which help to produce a wide range of fats/fat products with optimized properties (physical, chemical and nutritional). Some of the commonly employed methods include:

1. **Winterization:** It is a process which helps to produce an oil that would remain transparent even when it is refrigerated; a property desired for salad oils especially for preparing cold salads.
2. **Fractional Crystallization:** It helps to separate the olein and stearin fraction from each other. For instance, in case of palm oil, this process helps to produce *palm olein, palm stearin, super palm olein* and *palm mid-fraction.* Palm olein and palm stearin are used for producing shortenings and margarine, super palm olein is used for cooking/frying, whereas palm mid-fraction is used for producing cocoa butter substitutes.
3. **Hydrogenation:** Addition of hydrogen atoms at the site of double bonds in a carbon chain in the presence of hydrogen gas and a catalyst (generally nickel) is known as hydrogenation. It helps to convert liquid oils into fats which are solid and brittle at room temperature. *Vanaspati, margarine* and *shortenings* are examples of hydrogenated fats. They are valued for their high melting point, plasticity, stability, characteristic mouthfeel and taste. Hydrogenated fats are becoming unpopular because they contain high amount of trans fats which increase the risk of developing coronary heart diseases. The trans fat content of a few foods has been given in Table 3.2.
4. **Interesterification:** This process helps in converting fats rich in saturated fatty acids into fats rich in unsaturated fatty acids and vice versa such as in the production of *lard, margarines, palm oil, milk fat and fat substitutes* (olestra, salatrim).

Table 3.2 Trans fat content of certain foods

Food Product	Trans Fat (%)	Food Product	Trans Fat (%)
Vanaspati	27–50	Stick margarine	17–36
Milk and milk products	1.9–7.9	Soft margarine	10–18
Meat (ruminants)	2.0–10.6	Tub margarine	3.8–13.2
Pork fat	≤0.5	Foods with hydrogenated fat	0–34.9

Stabilization: It helps to provide characteristic texture and consistency to shortenings and margarines. This can be done by either plasticizing or stehling which is followed by tempering.

1. **Plasticizing** process helps to develop whiter shortenings and also improves their leavening properties. It is used in the production of *soft and firm table margarines.*
2. **Stehling** is used for producing *fluid shortenings* (liquid) which are pourable and pumpable; essential to bread and cake shortenings and non-dairy products.
3. **Tempering** improves plasticity, creaming properties and performance of bakery margarines and shortenings. Untempered fats are grainy and brittle and lack proper spreading qualities.

In order to maintain the quality of processed fats and oils, they are packed in the presence of nitrogen gas and stored in a cool dark place. This helps to ensure prevention of any kind of deterioration in the quality of the finished product.

3.3 Effect of Heating on Fats and Oils

Fats are frequently heated during food preparation which helps to provide desired sensory characteristics. However, fats tend to thermally and oxidatively decompose due to which several volatile and non-volatile compounds are formed that alter the functional, nutritional and sensory qualities of fat. The interaction of fat, food and environment in the presence of heat brings about several changes that can be grouped as physical and chemical.

3.3.1 Physical Changes

When food comes in contact with fat/oil, heat gets transferred from oil to the food, water gets evaporated and fat/oil is absorbed by the food. Increased viscosity, darkening of colour, foaming along with decrease in smoke point are the principal physical changes observed due to prolonged/repeated heating. These changes develop due to the formation of non-volatile decomposition products (discussed as chemical changes in the next subsection).

3.3.2 Chemical Changes

Oil on exposure to heat gets decomposed to several volatile products and non-volatile monomeric and polymeric compounds through three major chemical processes viz. hydrolysis, oxidation and polymerization (Figure 3.2).

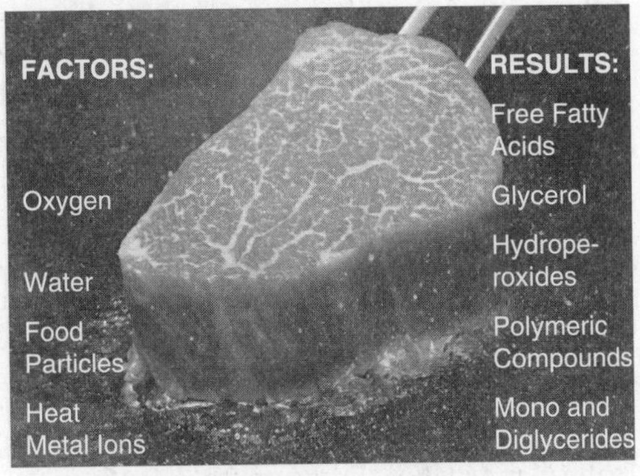

Figure 3.2 Deleterious changes in oil during frying

Hydrolysis: The interaction between oil, air and water is known as hydrolysis. Water and steam present in the food hydrolyse triglycerides to produce mono and diglycerides and subsequently free fatty acids and glycerol. Glycerol evaporates when oil is heated at ≥150°C. Factors such as temperature of oil, surface area of food, water/steam in food influence the rate of hydrolysis. This process reduces the stability of oil.

$$\text{Triglyceride} + \text{Water} \xrightarrow{\Delta} \text{Fatty Acids} + \text{Diglyceride}$$

$$\downarrow \quad \Delta(\geq 150^\circ\text{C}) \quad \downarrow$$

$$\text{Fatty Acids} + \text{Monoglyceride} \xrightarrow{\Delta(\geq 150^\circ\text{C})} \text{Glycerol}$$

Oxidation: Oxygen, such as that present in fresh oil or introduced into the oil from the environment and food, activates a series of reactions resulting in the formation of free radicals, hydroperoxides and conjugated dienoic acids. If oil is heated further at temperatures ≥190°C, the unstable hydroperoxides get converted to aldehydes and ketones. The unsaturated aldehydes such as 2,4 decadienal, 2-octenal and 2-heptenal contribute to the characteristic fried flavour of food. However, saturated and certain unsaturated aldehydes such as hexanal, octanal, nonanal contribute to off-odour (fruity or plastic).

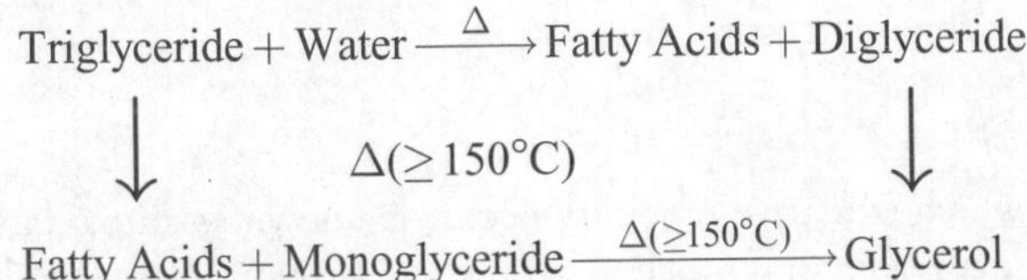

Polymerization: It generally occurs during frying or burning of fat. The formation of polymers occurs through the Diels-Alder reaction. Polymers increase the viscosity of oil.

The discussed changes in oil are influenced by several factors and are given in Figure 3.2.

USEFUL ACTIVITY

Take 100 ml of oil and heat it (below smoking temperature) in a pan for 15, 30, 45, 60 and 90 minutes respectively. Take out a teaspoon of sample after each heating period is over. Observe and compare the samples for the following characteristics:

- Colour
- Odour
- Viscosity
- Taste

3.4 Spoilage of Fat

Under-processing, repeated exposures to high temperatures, air and water, and improper packaging and storage can cause spoilage of fats and oils. The deterioration of fats and fat-containing foods due to lipid oxidation is known as *rancidity*. It results in the development of unpleasant flavours and loss of organoleptic, functional as well as nutritional qualities of fat. Rancidity is of two types: hydrolytic rancidity and oxidative rancidity.

Hydrolytic Rancidity: It is also known as *lipolytic rancidity*, because it involves breakdown (lipolysis/hydrolysis) of fats into free fatty acids and glycerol by the action of enzymes, principally lipase. The presence of free fatty acids lowers the smoke point of fats. The products fried in such fats tend to absorb greater amount of fat and develop cracks on their outer surface. Hydrolytic rancidity is of great significance to the shelf life of milk and milk products, especially raw milk and its products which have not undergone pasteurization or other forms of adequate heat treatment. This type of rancidity can be prevented by heating fat during the time of processing so as to destroy the enzyme lipase. For instance, pasteurizing butter at the time of production. The peculiar flavour of sour butter is due to the breakdown of fatty acids to form short-chain fatty acids (such as butyric and caproic acids) which are volatile and impart off-flavour to butter especially the ones made at home.

Oxidative Rancidity: The process of oxidative rancidity is a self-perpetuating reaction and is therefore termed as *autoxidation*. It refers to deterioration of fats, especially oils rich in unsaturated fatty acids (particularly PUFA). In the presence of oxygen (air, water and food) hydroperoxides are formed which are highly volatile substances and give rise to the development of off-flavour and rancid taste. Heat, light, ionizing radiations, catalysts and certain enzymes such as lipoxygenase promote oxidative rancidity. It can occur in oils, and foods containing even small amount of oil/fat such as animal tissues, wheat germ, corn flour, biscuits, chocolates and several confectionary items to name a few.

3.4.1 Autoxidation

The process or the mechanism by which oxidative rancidity occurs is known as autoxidation. The term 'autoxidation' has been given because it is a natural process that takes place between molecular oxygen and unsaturated fatty acids, and the rate of oxidation increases as the reaction proceeds. The autoxidation occurs through a free radical chain mechanism involving three steps: initiation, propagation and termination.

Initiation: The oxidation reaction begins when a hydrogen (H) atom adjacent to a double bond in the fatty acid (RH) molecule/moiety gets removed to form alkyl free radical ($R^{\bullet}$). The hydrogen is replaced by oxygen in the parent fatty acid molecule.

$$RH \xrightarrow{\text{heat/light/metal ion}} R^{\bullet} + H^{\bullet}$$

$$RH + O_2 \xrightarrow{\text{heat/light/metal ion}} ROOH$$

Propogation: The alkyl free radicals formed during the initiation reaction react with atmospheric oxygen to form an unstable peroxy free radical, which in turn promote removal of hydrogen from another fatty acid molecule to form hydroperoxide (ROOH) and a new alkyl free radical ($R^{\bullet}$). The new alkyl free radicals initiate further oxidation and contribute to chain reaction.

$$R^{\bullet} + O_2 \xrightarrow{\text{heat/light/metal ion}} ROO$$

$$ROO + RH \xrightarrow{\text{heat/light/metal ion}} ROOH + R^{\bullet}$$

$$R^{\bullet} + O_2 \xrightarrow{\text{heat/light/metal ion}} ROO$$

Termination: During the chain reaction when two alkyl free radicals (R.) react to form a non-radical product (RR), the process of autoxidation gets terminated.

$$R^{\bullet} + R^{\bullet} \longrightarrow RR$$

The hydrocarbons and alkyl free radicals formed during autoxidation undergo a series of reactions to result in the formation of alcohols, aldehydes, ketones and hydrocarbons. It is these breakdown products that contribute to the off-flavour of oils spoilt by oxidative rancidity. The rate of autoxidation is influenced by the fatty acid composition of oils; the higher the degree of unsaturation, the higher is the rate of oxidation. Further, the process of autoxidation gets catalysed by light, heat, metal ions and certain radiations. The form in which oxygen molecules are present around the fat molecules also exerts an influence on the rate of reaction; singlet oxygen reacts faster than triplet oxygen. For instance, singlet oxygen has been found to react with linoleic fatty acid 1,450 times faster than triplet oxygen.

3.4.2 Flavour Reversion

It refers to the change in the flavour of a fat/oil or a fat-containing product and is an indication for the onset of rancidity. There is development of an off-flavour, and the fat/product has poor sensory attributes. Factors associated with flavour reversion include heat, ultraviolet rays, oxygen and metal ions such as iron and copper. Hydrogenated fats and oils containing high amount of linolenic fatty acid (such as soybean oil) are more susceptible to flavour reversion. The flavour of reverted oil has been described as fishy, grassy, deep-fried or similar to paint and cardboard.

3.4.3 Prevention of Fat Spoilage

Spoilage of fat and fat-containing foods is a challenge for the food industry, because it leads to the loss of shelf life, palatability, functionality and nutritional quality. While hydrolytic rancidity is easier to control through the application of heat to fat, oxidative rancidity is difficult to manage due to the ubiquitous presence of oxygen. Therefore, antioxidants are added to fats/oils and fat-rich products to prevent their spoilage. Antioxidants are substances which delay the onset of oxidation or slow down the rate at which it proceeds, because they bind with free oxygen radicals and hence do not allow them to get attached to fatty acid chains. They can be present as natural constituents in food or may be added as an additive during processing of fat/food. However, an antioxidant must have the following properties:

1. Non-toxic
2. Effective at low concentrations
3. Stable during processing
4. No significant colour/flavour/odour
5. Inexpensive

Antioxidants not only extend the shelf life but also widen the choice of oils that can be used in specific products. Based upon their mechanism of action, antioxidants can be classified as primary and secondary antioxidants.

Primary Antioxidants: They inhibit the initiation or interrupt the propagation step of autoxidation by acting as acceptors of free oxygen radicals. The most commonly used antioxidants (synthetic) include:

1. Butylated hydroxyanisole (BHA)
2. Butylated hydroxytoluene (BHT)
3. Propyl gallate (PG)
4. Tertiary butylhydroquinone (TBHQ)

Tocopherols and carotenoids are natural primary antioxidants.

Secondary Antioxidants: They are also referred to as *synergists*, because they promote the activity of primary antioxidants. They help in slowing down the rate of autoxidation but cannot convert free radicals to more stable compounds. Some commonly used secondary antioxidants include:

1. Chelators–citric acid, phosphoric acid and ethylenediaminetetraacetic acid (EDTA).
2. Oxygen scavangers–ascorbic acid, ascorbyl palmitate and sodium erythorbate.
3. Singlet oxygen scavangers–Lycopene and lutein.

3.5 Care of Fat

Spoilage and deterioration of fat can be prevented by keeping the following points in mind:

1. Fats and oils should be stored in a cool dark place.
2. Processed fats should be kept in dark-coloured or yellow-/green-coloured bottles so that ultraviolet rays do not initiate fatty acid breakdown.
3. Presence of antioxidants should be checked on the label.
4. Unprocessed fat (such as home-made butter/cream) should be used quickly and always kept refrigerated.
5. Fat/oil should never be heated above their smoke point (greyish-blue smoke on the surface of oil on excessive heating), that is, never allow the fat to burn. The smoking temperature of certain fats/oils has been given in Box 3.2.
6. Avoid using the same oil for frying again and again; instead, sieve the oil, store in a translucent air tight container which is kept in a cool place (preferably refrigerator). Use this oil for cooking purpose.
7. Use the fat most appropriate to the dish and read the label for quality marks.

Box 3.2 Smoke, Flash and Fire Point of Fats and Oils

Smoke Point: temperature at which a cooking fat or oil begins to breakdown into glycerol and free fatty acids and produce a bluish smoke.

Flash Point: temperature (~320°C for most oils) when tiny wisps of fire begin to leap from the surface of fat/oil.

Fire Point: temperature of heating when oil/fat surface gets ablaze.

Smoking Point of Various Fats and Oils

Fat/Oil	Temperature (°C)	Fat/Oil	Temperature (°C)
Butter	208	Canola	238
Lard	183–205	Corn	236
Hydrogenated fat	223–230	Safflower	265
Shortenings	190–195 (174*)	Sunflower	246
Coconut oil	138	Soybean	241
Sesame	215	Olive (virgin)	234

Note: * refers to the temperature at which the emulsifier (not the fat) burns and smoke comes out.

While these points can help in preventing or minimizing deterioration in the quality of fat, it is always good to ascertain the quality of fat/oil before using, especially in case of large-scale food production. This can be done by checking the iodine value, acid value, refractive index, saponification value, Reichert-Meissel value and through sensory evaluation.

3.6 Uses of Fat in Food Preparation

Fats have unique functional properties which make them an inevitable ingredient in a variety of dishes and products. These properties include the following:

1. Moisture retention
2. Leavening
3. Shortening
4. Plasticizing
5. Emulsification
6. Heat transfer

The use of fats is not just limited to day-to-day cookery, bakery, confectionary and dairy products but also processed fat products such as margarine, shortening agents, coating agents and emulsifiers. Some important applications of fats/oils include:

1. They are used as ***cooking medium*** in almost every household. Several vegetable, pulse and meat preparations are made by using oil/fat as a means to transfer heat; roast onion/tomatoes/spices to prepare gravy or roast semolina while preparing halwa to name a few.
2. ***Frying*** is the most popular use of fat. Shallow and deep fat frying involve the use of fat above 165°C. It is an efficient method of heat transfer that facilitates quick cooking. It not only adds flavour but also changes the texture and colour of food. Oils obtained from plants are preferred over animal fats for frying purpose. Oils which are used for frying purpose should have the following characteristics (Figure 3.3):
   (a) It should have high oxidative stability, that is, resistance to spoilage due to exposure of fatty acids to air and water. Water is present in the foodstuffs being fried while air/oxygen is in the environment surrounding the fat.
   (b) The frying medium should have a high smoke point. It should never be heated to reach flash or fire point (Box 3.3).
   (c) The oil should not darken in colour upon frying and subsequent use.
   (d) It should be able to provide the desired (no or distinctive) flavour to food. Mustard oil provides a distinctive flavour to pakoras. The same oil may not be suitable for frying *jalebis.*
   (e) The saturated and trans fatty acid content of the oil should be low.
   (f) It should preferably be a good source of MUFA.
   (g) The oil should be free from contamination/adulteration.

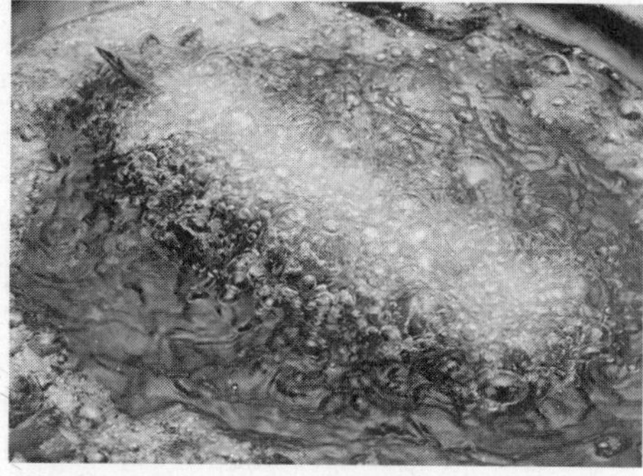

Figure 3.3 Frying: The most popular use of fats

In addition to the given points, it is necessary to use oils with low turnover rate, that is, absorption of fat by the food should be minimum. Factors influencing absorption of fat are given in Figure 3.4.

3. A wide range of properties of fat are used in the preparation of ***bakery products*** such as breads, cakes, biscuits, cookies, pie crusts, short pastry and flaky pastry. In the preparation of batter for cake, the prime function of fat is aeration, whereas that in the preparation of dough for cookies it is plasticity.

In the case of puff pastry (croissants), fat acts as a barrier, that is, it separates layers of dough from each other. In the case of short pastry, fat needs to be mixed finely throughout the dough as a thin film so that layers are not formed. Bakery fats include butter, lard, tallow, refined oils and their processed fat products, that is, shortenings and margarine.

4. Fats and oils are also used for the production of ***salad oils, mayonnaise, salad creams and French dressings.*** Winterized soy, canola, rapeseed, corn and sunflower oil are used as salad oils which enhance the flavour of raw vegetables/seasonings added to salad. Mayonnaise, salad creams and French dressings employ the property of emulsification of fats. Basic mayonnaise recipe contains 75–80 per cent vegetable oil, 9.4–10.8 per cent vinegar, 7–9 per cent egg yolk and small amounts of seasonings (salt, mustard powder, pepper). It is a stable emulsion. French dressing, on the other hand, is a temporary emulsion. It contains oil, vinegar/lemon juice and seasonings. Therefore, it needs to be shaken before use. Salad creams are similar to mayonnaise but contain lesser amount of oil (30–40 per cent) and some amount of starch/gums to give thickening.
5. Fats with typical melting characteristics such as cocoa butter have been extensively used in the manufacture of ***chocolates and confectionary items.*** The high saturated fatty acid content gives cocoa butter a characteristic texture and melting range. It is hard at 20–25°C, and at 30–35°C it melts completely. Because of this reason, chocolate remains hard/brittle at room temperature but melts quickly giving a smooth creamy texture when kept in mouth. Cocoa butter alternatives are also being produced nowadays as a healthier alternative.
6. Dairy fats and, to some extent, vegetable oils are being used for the production of ***ice creams and frozen desserts***. The fat content of ice creams usually varies between 5 per cent and 12 per cent and it provides structure to the product. It helps to retain the entrapped air, improves melting characteristics, provides creamy texture and improves taste.
7. Vegetable oils such as soybean oil are used as ***coating agents***. These oils are sprayed finely on the products such as crackers, buns and nuts to improve appearance, provide gloss, maintain crispness, retain flavour, prevent oxidative damage or staling and caking of the product. At times antioxidants (vitamin A/E) are added to the oils so as to improve their oxidative stability.
8. A variety of fats and oils are being used for the manufacture of ***filled milk, non-dairy coffee whiteners, non-dairy whipping creams and emulsifiers.***

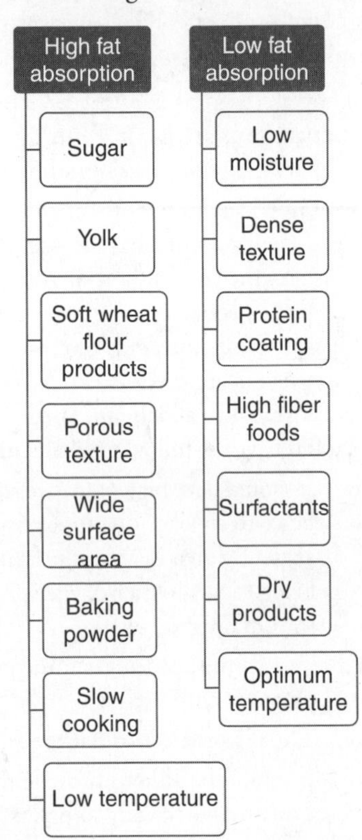

Figure 3.4 Factors influencing absorption of fat

The uses of fats/oils and their products are endless and their interactions with other ingredients can help to develop a variety of dishes and processed products. Research and technological advancements are facilitating modifications in natural fats so as to modify their properties to meet the end-use specifications. This can be seen from the attempts being made for the development of low trans-fat hydrogenated fats and cocoa butter alternatives such as cocoa butter improvers, cocoa butter replacers and cocoa butter substitutes. Therefore a chef must have a clear understanding regarding the composition, processing and characteristics of various fats, their interaction(s) with other nutrients as well as the storage and care of fats; so as to be able to prepare safe, nutritious and palatable dishes.

Know Your Progress

True or False Questions

1. Butyric acid is a monounsaturated fatty acid.
2. Rendering is a processing method used for extracting fat from seeds.
3. Salad oils must be winterized.
4. BHA is a secondary antioxidant.
5. Fat acts as a plasticizer in cookie dough.

Very Short Answer Questions

1. Give examples of monounsaturated fatty acids.
2. Why are PUFAs more susceptible to autoxidation?
3. Which property of fat is required in the preparation of flaky pastry?
4. What is the difference between smoke point and flash point?
5. Why do chocolates remain hard/brittle at room temperature but melt quickly giving a smooth creamy texture when kept in mouth?

Short Answer Questions

1. Classify fats on the basis of saturation.
2. Describe hydrogenation of fats.
3. What are synergists?
4. Enlist the points to be kept in mind while selecting oil for frying.
5. What is flavour reversion?

Long Answer Questions

1. Describe briefly, with the help of a flow diagram, the processing of oils.
2. What is the effect of heat on fats/oils?
3. Describe the process of autoxidation.
4. Enumerate giving suitable examples, the various uses of fats/oils.
5. Identify two recipes each wherein fat acts as:

   (a) leavening agent. (b) shortening agent. (c) emulsifier.

Beyond the Text

- Food applications of fats have no boundaries and, therefore, the market is always flooded with newer types/blends of fats/fat products. Find out the recent advancements in the following:

  1. Margarine
  2. Cocoa butter replacers
  3. Salad creams
  4. Frying oils

Chapter 4

Proteins

Chapter Objectives

After reading this chapter, the reader should be able to:

- learn the structure and composition of proteins.
- associate the properties of proteins with their composition and function.
- correlate the effect of ingredients and cooking procedures with the behaviour of food proteins.
- identify food proteins which are most suitable for preparing a dish.
- learn about the various food applications of proteins.

Food proteins are currently under extensive research and review. While much is known about their physicochemical properties, predictions regarding their functional use often go wrong based on their structure or molecular arrangement. Considering their nutritional benefits over carbohydrates and fat, proteins are the preferred ingredients by the chefs and consumers. However, great attention needs to be made to ensure safe incorporation of this food component because chemical alterations during certain food processing methods can render them potentially toxic.

4.1 Structure of Proteins

Proteins are complex macromolecules composed of carbon, hydrogen, oxygen, nitrogen and sometimes sulphur. These elements combine to form different types of amino acids (Figure 4.1). In a single protein, several *amino acids* are linked by *peptide bonds* in the form of long filaments often referred to as *polypeptide chains*. These polypeptide chains wrap themselves in a virtually infinite number of spherical or helical forms, which explains the wide variety of functions performed by proteins.

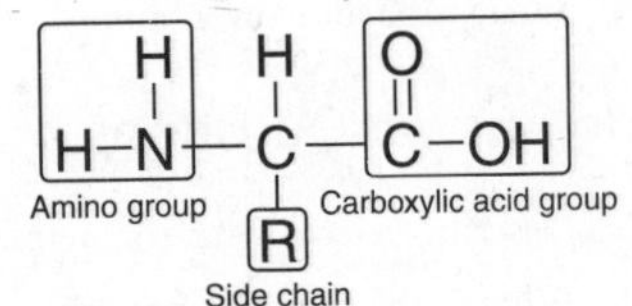

Figure 4.1 Structure of an amino acid

Based upon the molecular arrangement, the structures of proteins are classified as follows (Figure 4.2):

Primary: It refers to the linear chain of amino acids connected to each other with the help of peptide bonds. It is resistant to chemical changes except enzymatic breakdown.

Secondary: It refers to the hydrogen bonding between the amino acid residues, that is, the R groups of amino acids present on the primary structure creates α-helix and β-pleated structures. These structures are usually found to repeat at regular intervals. The α-helix is a coiled rod-like structure that involves a single polypeptide chain. β-pleated sheets such as zig-zag structures are formed when the hydrogen bonding occurs perpendicular to the direction of the polypeptide chain.

Tertiary: It is a three-dimensional structure wherein the helixes and pleated sheets get folded to form a compact arrangement. Based on the tertiary structures, proteins are categorized as fibrous and globular.

1. Fibrous proteins (such as meat proteins) have a long rod-like structure, are physically strong and hence contribute towards the structure of foodstuffs. They are quite resistant to heat and mild changes in the cooking media (acid/alkali).
2. Globular proteins (such as milk proteins) on the other hand have multiple folding on the helical and pleated structures. They are highly reactive to cooking environment (temperature, acid, alkali). Their structure is not stable to cooking/processing.

Quaternary: It is present only in very large proteins which have several polypeptide chains. Haemoglobin has two α-helix and two β-pleated chains attached to each other; it thus has a quaternary structure.

Most natural foodstuffs contain a combination of primary, secondary and tertiary structures. Quaternary structures are present usually in meat products.

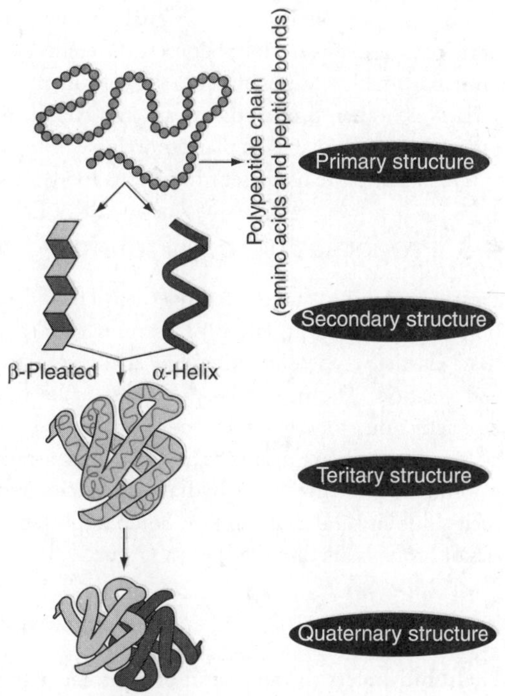

Figure 4.2 Structures of proteins

4.2 Animal and Plant Proteins

Almost all natural foodstuffs (except fat and sugar) contain protein, though in variable amounts. Among animal products; eggs, milk, cheese, curd, meat, aquatic animals and poultry are good sources of protein. Among plant food sources, pulses and legumes contain good amount of protein. Cereal proteins, especially,

Table 4.1 Proteins present in various foodstuffs

Foodstuff	Amount of protein /100 Gram Edible Portion*	Major Proteins Present
Egg (whole, hen)	13.3	Ovalbumin, Conalbumin, Ovomucoid, Ovomucin, Lysozyme, Avidin, Ovoglobulin, Ovoinhibitor, Lipovitellins, Phosvitin and Livetin.
Meat	8.9–26.6	Myosin, Actin, Actomyosin, Collagen, Elastin, Troponin and Tropomyosin.
Milk	1.1–4.3	Casein, Lactalbumin, Lactoglobulin, Serum albumin Immunoglobulins and Proteose.
Pulses	9.6–43.2	Globulins and Albumins.
Cereals	4.7–12.5	Albumins, Globulins, Prolamine, Glutenin and Gliadin.

**Note:* Values for non-processed natural forms – ICMR, 2007.

that of wheat, though present in small amount, have some unique properties. The quantity and quality of protein varies considerably between different foodstuffs. For instance, not only the composition of milk proteins would be different from meat proteins but milk obtained from different animals would also have variations in their protein composition. Overall, animal proteins have been found to be more useful in food processing as compared to plant proteins.

Details of proteins present in various foodstuffs are given in the Table 4.1.

4.3 Properties of Food Proteins

Proteins exhibit a variety of properties. Their rheological properties provide a variety of textures such as chewiness, elasticity, viscosity, adhesion, stickiness, stringiness and gelation. The interfacial properties make them surface active due to which they act as excellent foam formers and emulsifying agents. The molecular arrangement of proteins influences their hydration capacity and hence their solubility and dispersibility. Some important properties of proteins are discussed below (Figure 4.3).

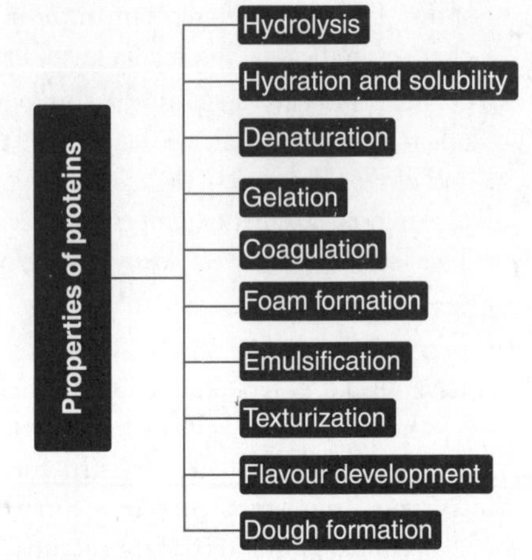

Figure 4.3 Salient properties of food proteins

4.3.1 Hydrolysis

Hydrolysis refers to the partial or complete breakdown of proteins and can be carried out with the help of acid, alkali or enzymes. The most common example is curdling or flocculation of milk when it becomes sufficiently acidic (sour taste) to approach the isoelectric pH of casein (milk protein). While acid and alkali induced hydrolysis causes some loss of amino acids, the same is not true in case of enzymes.

Enzymatic hydrolysis of food proteins is carried out by using pepsin, papain, trypsin, chymotrypsin and thermolysin. Tenderization of meat by using papaya leaves or papain powder is a common example of protein hydrolysis. Whey and soy protein hydrolysates (products of enzyme hydrolysis) are increasingly becoming popular in the food processing industry because they have high dispersibility, quick wettability, high foaming power and good emulsifying properties. The nutritional quality of protein is currently being improved by a selective proteolysis process which is commonly referred to as the 'plastein reaction'.

> **GOOD TO KNOW**
>
> **What Is Isoelectric Point?**
>
> Proteins (like amino acids) are amphoteric in nature; that is, they carry either a positive or a negative charge. When a protein does not carry either a positive or a negative charge, it is known as its isoelectric point (pH). Since the amino acid composition of each protein varies, so does the isoelectric point. The functional properties of proteins are influenced by their isoelectric charge.

4.3.2 Hydration and Solubility

Several rheological and textural properties of proteins are influenced by their interaction with water. Proteins can bind (water-binding capacity) and hold water (water-holding capacity). The protein–protein and protein–water interactions determine the *solubility* of proteins. Hydrophobic interactions promote protein–protein interactions and result in decreased solubility. On the other hand, hydrophilic interactions promote protein–water interactions and result in increased solubility. Solubility of proteins is influenced by pH, temperature and minerals. Majority of food proteins (exceptions such as egg albumin) are acidic and hence exhibit maximum solubility at alkaline pH. Further, most food proteins are soluble between 0°C and 40°C. Magnesium, calcium and sodium decrease solubility of proteins. These discussed properties are critical in determining the functional properties of proteins such as dispersability, wettability, solubility, juiciness, tenderness and their use as thickening or viscosity agents, foam/gel formers and emulsifiers.

4.3.3 Denaturation

Alterations in the secondary, tertiary and quaternary structures of protein due to changes in its environment such as those observed during cooking and processing of food is known as denaturation. The change in environment refers to the temperature, pH (acid/alkali), solvent/ionic strength (sugar/salt) and physical pressure employed during cooking of food. Denaturation of proteins has several advantages and disadvantages.

Advantages

1. Partial denaturation improves foaming and emulsifying properties.
2. It is necessary for gelation of food proteins.
3. Thermal denaturation improves digestibility of proteins.

Disadvantages

1. Excessive thermal denaturation diminishes foaming and emulsifying properties.
2. It reduces solubility and dispersibility of protein powders.
3. Even partial denaturation can cause flocculation and precipitation on storage.

Several physical and chemical factors influence denaturation of proteins such as:

1. **High temperature:** Thermal denaturation of protein (Figure 4.4) is witnessed in every day cooking (40–80°C); though each protein has its own critical temperature of denaturation (Box 4.1). Heat-induced denaturation helps to manipulate foaming and emulsifying properties, promotes Maillard reaction, and improves digestibility and sensory characteristics of the dish.
2. **Low temperature:** Certain proteins get denatured when stored at cold or freezing temperatures. Factors such as formation of ice crystals, rupturing of cellular membranes and dehydration have been found to be associated with denaturation. For example, glycine–a soy protein–aggregates and precipitates when stored at 2°C. The ability of casein to interact with renin gets modified when skim milk is stored at 4°C. Most proteins get denatured at temperatures between –15°C and 0°C.

Box 4.1 Thermal Denaturation Temperature of Certain Proteins

Protein	T_d (°C)
Haemoglobin	67
Egg Albumin	76
Myoglobin	79
Avidin	85
Soy glycine	92
Broad beans	94
Oat globulin	108

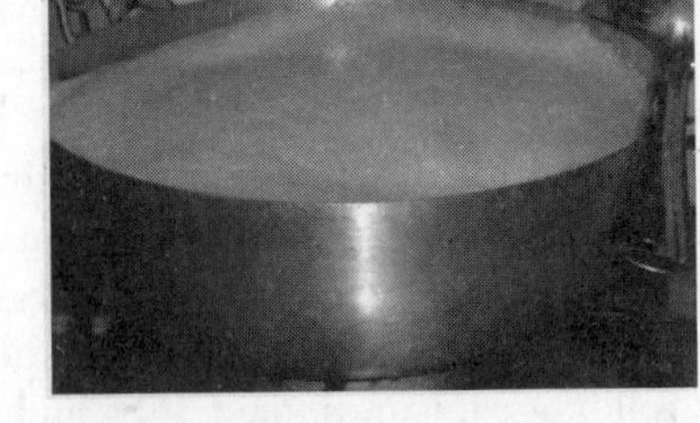

Figure 4.4 Scum formation while heating milk: thermal denaturation of proteins

3. **Pressure and Shear:** In the presence of pressure, proteins get thermally denatured at a lower temperature. Soy protein and egg white gels are softer when prepared in the presence of pressure along with heat rather than heat alone. Mechanical agitation such as whipping, beating, shaking, kneading and mincing cause denaturation of proteins. Such denaturation of proteins is often observed during egg white foam formation when the white is excessively whipped.
4. **pH:** Both acid (lemon, vinegar, cream of tartar and tomatoes) and alkali (sodium bicarbonate) cause denaturation of proteins which is usually partial and reversible. The effect of alkali is usually more pronounced as compared to that of acid.
5. **Sugar, Salt and Additives:** Sugars such as sucrose, lactose, glucose and glycerol tend to stabilize the protein structure and prevent its denaturation. Salt elevates the thermal denaturation temperature (T_d) and hence at low concentrations prevents protein denaturation. At high concentration, salt acts as a dehydrating agent and tends to remove water from protein molecules thereby destabilizing their structure. Additives such as urea, guanidine hydrochloride and sodium dodecyl sulphate also cause denaturation of proteins.
6. **Water:** It decreases the thermal denaturation temperature (T_d) but its concentration influences protein denaturation. T_d decreases sharply at water concentration of 0–0.35 gram/gram of protein and marginally at water concentration of 0.35–0.75 gram/gram of protein. Concentration of water above 0.75 gram/gram of protein has no influence on T_d and hence protein denaturation.

4.3.4 Gelation

Gels can be formed from carbohydrates and proteins. A protein gel is a protein–protein and protein–water interaction characterized by association, aggregation, polymerization, precipitation, flocculation and coagulation (Figure 4.5). The process of protein gelation involves heating a moderately concentrated protein solution (sol). During heating, the protein in the 'sol state' gets converted into the 'pro-gel state'. The pro-gel state when allowed to cool at ambient or refrigeration temperature gets converted into a protein gel.

Proteins form two types of gels, viz. *coagulum gels and translucent gels*. In coagulum type gels, the rate of aggregation and network formation is faster than denaturation; as a result, the gel network begins to set even while being heated, for example, during the preparation of scrambled egg. Formation of coagulum type gels is commonly

THINK TANK

1. Why does the texture of *paneer* maintained at holding temperature become rubbery?

When paneer, meat or any protein is kept at holding temperature to keep it warm while being served for a long duration, it gets excessively denatured. Excessively denatured protein is difficult to chew, has a rubber-like texture and is difficult to digest.

2. Why is raw egg white added to clarify heated soup stock?

When raw egg white is added to hot soup stock, the egg proteins coagulate and while doing so they trap tiny particles (impurities) from the liquid that can later be strained and separated out.

referred to as *coagulation of protein* (Figure 4.5). Factors influencing coagulation of protein are given in Figure 4.6. In the case of translucent gels, the gel network sets only when cooled after cooking, such as in the case of gelatin-based desserts. Translucent gels hold more water than coagulum type gels and hence are less prone to syneresis. The gelation property of proteins is used for preparing eggs, meat, cheese, tofu and yogurt. This characteristic of protein also makes them useful as a thickening agent and stabilizer for emulsions and foams.

Figure 4.5 Coagulation of egg proteins

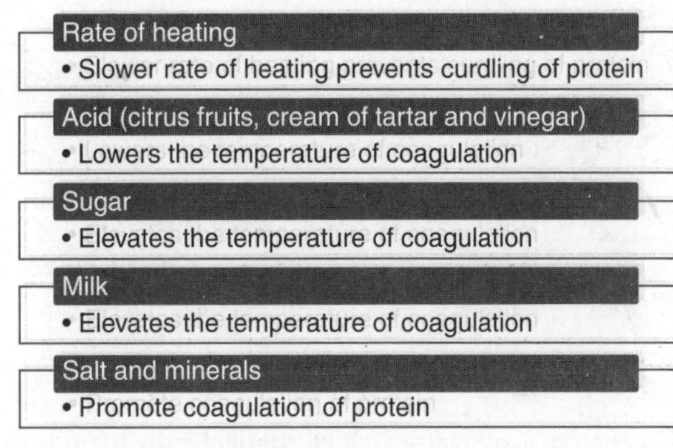

Figure 4.6 Factors influencing coagulation of proteins

4.3.5 Foam Formation

Mechanical agitation such as whipping, beating or shaking results in the formation of thin, tenacious protein films which can hold large quantities of air in the form of bubbles or air cells.

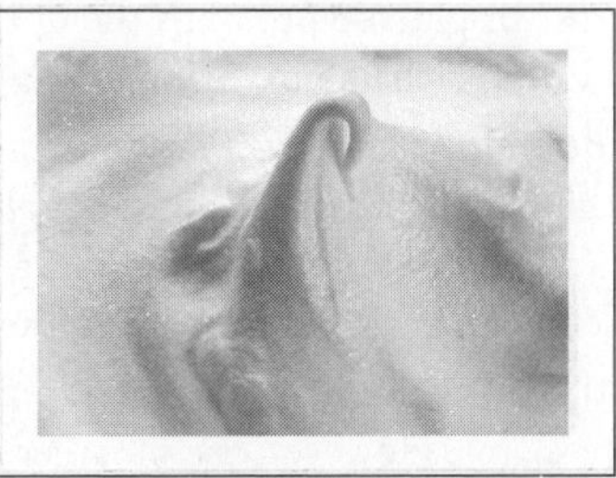

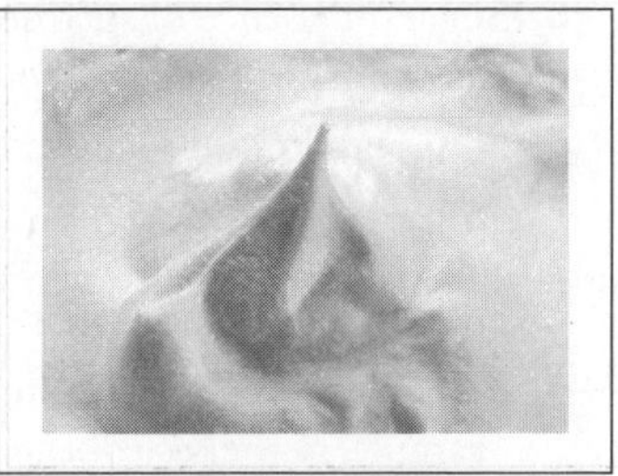

Figure 4.7 (a) Foamy stage (b) Soft peak stage (c) Stiff peak stage

The process of foam formation occurs in three distinct stages viz.

1. **Foamy Stage:** When a protein solution is whipped, coarse bubbles form on the surface and some liquid remains at the bottom of the bowl. The product sticks to the surface of the bowl (Figure 4.7a).
2. **Soft Peak Stage:** If whipping is continued further, more air gets incorporated and there is no liquid left at the bottom of the bowl. The air cells increase in number and the size gradually reduces. When the beater is slowly pulled out of the foam, peaks are formed which bend over their tip. Such peaks are known as soft-peaks and if made from egg white, they can be used to prepare soft meringues (Figure 4.7b).

3. **Stiff Peak Stage:** Further whipping results in the development of stiffer stable peaks. They are known as stiff peaks. Stiff peak stage as obtained from egg whites is used in several bakery and confectionary products such as chiffon cakes (Figure 4.7c).

GOOD TO KNOW

What Is Foaming Capacity and Foam Stability?

Foaming capacity refers to the surface area or volume that can be created by whipping a protein solution. Higher the volume of foam, greater is the air entrapped within the air-cells of the foam. It is expressed as overrun or foaming power.

$$\text{Overrun} = \text{Volume of foam} \div \text{Volume of initial liquid} \times 100$$

$$\text{Foaming power} = \text{Volume of air incorporated} \div \text{Volume of liquid} \times 100$$

Foam stability refers to the ability of a protein to maintain the foam formed despite mechanical pressure or gravitational pull. It is expressed as time required for 50 per cent reduction in the foam volume.

There are several factors which influence the foam formation and its stability. They are as follows:

1. **Type and Concentration of Protein:** The foaming power of isolated proteins (gelatin, whey protein isolate, soy protein) is greater than natural proteins (egg albumin). The higher the concentration of protein, the stiffer and more stable is the foam.
2. **Fat:** Presence of fat is detrimental to the foam formation. Since fat molecules do not exhibit viscoelastic properties, if they are present along with protein during whipping, small bubbles form rapidly and collapse immediately.
3. **Sugars:** Presence of sucrose, lactose or glucose prolongs the duration of whipping required to reach the end point of desired foam. However, the foam formed has a fine texture and is very stable.
4. **Salt:** The effect of sodium chloride on foaming power and stability varies with the type of protein. The foaming power and stability of globular proteins such as egg albumin, lactalbumin, gluten and soy protein increases with increasing concentration of salt. However, the effect of salt on whey proteins is entirely opposite. Presence of calcium and magnesium ions improves foamability and foam stability.
5. **pH:** Addition of small amount of acid such as cream of tartar or lemon enhances foam stability, for example, in the case of egg white foams. However, use of larger amount of acid can cause hydrolysis or breakdown of protein which reduces foam volume and stability.
6. **Water:** Presence of water increases the volume of foam considerably but makes the foam highly unstable.
7. **Temperature:** Proteins maintained at room temperature form larger volumes than those maintained at refrigeration temperature, for example, in the case of egg white. Unstable foams are formed from proteins maintained at warm temperature prior to whipping.
8. **Method Employed:** Use of an electric beater generally creates small air bubbles and hence a fine textured foam as compared to a mechanical beater. Excessive beating or beating at a very rapid pace produces dry foam which breaks while blending it with other ingredients.
9. **Container:** Bowls with round edges and slanted walls are more suitable for foam formation as compared to those with vertical walls and sharp edges.

USEFUL ACTIVITY

Take seven eggs. Separate egg white of one egg in a steel bowl and whip with a hand beater till a stiff peak stage is formed. Transfer the foam lightly into a funnel held on a measuring cylinder. Note the drainage volume (liquid) in the measuring cylinder after 20 minutes. Repeat the exercise by adding or using:

1. 10ml water
2. 2 teaspoon sugar
3. 1/2 teaspoon vinegar
4. 1 teaspoon oil/yolk
5. mechanical beater
6. a different shaped bowl

4.3.6 Emulsification

The interfacial property of protein which facilitates its migration at the oil–water interface promotes the emulsification property of proteins. The ability of a protein to form an emulsion is measured as emulsion capacity. The *emulsion capacity* of a protein refers to the volume of oil (ml) that can be emulsified per gram of protein before phase inversion (change of oil-in-water emulsion to water-in-oil emulsion) occurs. Animal proteins such as ovalbumin and serum albumin have higher emulsifying capacity as compared to soy proteins and other plant proteins. Protein-based emulsions are often more stable than those formed with the help of small-molecule surfactants alone. The physical and sensory features of protein-based emulsions are influenced by the type of protein, amount of protein and oil, melting point of oil, temperature, pH, surfactant used, type of equipment and method of forming emulsion (refer to Section 7.1 of Chapter 7). The emulsification property of proteins can be observed in milk wherein the fat globules are surrounded by a thin film of lipoproteins. Egg yolk, coconut milk, soymilk are also the examples of protein-based emulsions.

4.3.7 Other Properties

Proteins have the ability to bind or entrap flavour compounds. Proteins, which are able to tightly bind with flavour, retain them during processing and release them during mastication of food, are used as flavour carriers. The interaction of proteins with carbohydrates during heating (Maillard reaction) and the subsequent breakdown of amino acids (Strecker degradation) also result in the development of various distinctive aromas. For instance, breakdown of amino acids phenyalanine, leucine, lysine, cysteine and arginine produces caramel, bread-like, buttery, smoky and popcorn-like flavours respectively. Proteins can also be texturized to form meat analogues, and certain proteins (gluten: glutenin and gliadin) exhibit dough-forming property. Figure 4.8 provides a detailed view of gluten strands, which provide wheat flour dough, the characteristic feature of being rolled to form a chappati or a naan; a property not exhibited by rice flour.

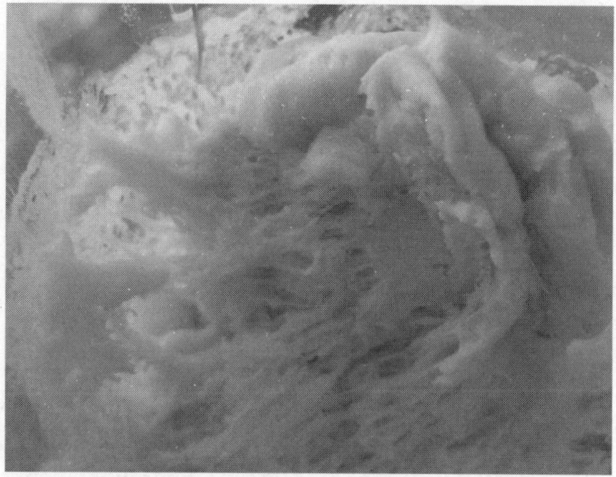

Figure 4.8 Gluten strands in a fermented dough of wheat flour

4.4 Uses of Proteins in Food Preparation

Each food contains a combination of proteins. The performance of a food protein is thus influenced by its inherent composition, composition of the food and its interaction with the cooking environment. When a food is cooked, each protein performs more than one function and the net effect on the final product involves protein–protein interaction and the interaction of protein with other ingredients, cooking media and process.

The salient functions performed by food proteins are enlisted below:

Denaturation	Hydration	Texturization
Coagulation	Dispersion	Extrusion
Emulsification	Gelation	Flavour/colour development
Fat adsorption	Cohesion	Viscosity enhancer
Foam formation/leavening	Aggregation	Dough development

Although proteins are present in several foodstuffs, milk and egg proteins have been found to be the most functional, followed by meat and plant proteins. Some of the uses of food proteins are discussed as follows.

Bakery Products: Egg, milk, whey and cereal (especially, wheat) proteins are used extensively in bakery products as they possess several functional properties ideal for their preparation. Milk proteins enhance moisture absorption capacity of dough, control fermentation, facilitate moisture transfer during gelatinization of starch, improve baking quality of weak flours, enhance flavour and crust formation, promote moisture retention and retard staling of bakery products. In the case of egg proteins, the foaming, water-binding, coagulation, emulsification and flavour-binding capacities make them ideal for bakery products. Wheat proteins help in the development of an elastic dough which provides structure to several bakery products (breads, cookies, breakfast cereals, cakes, pastries and pastry glazes). Caseinate is particularly being used in pastry glazes, frozen cakes, cake mixes and chewy cookies. Whey protein concentrates provide a good texture to protein-enriched breads (≥22% protein), cakes, muffins and croissants.

Dairy Products: The property of casein (milk protein) to coagulate in the presence of acid/renin along with the heat makes it most suitable for the production of cottage cheese and several other forms of processed cheese. Rennet casein is used in the manufacture of cheese analogues (cheese-like products made from milk products and vegetable oil). The presence of protein in ice cream helps to prevent the formation of large ice crystals and facilitates incorporation of air in the ice cream mix. This helps to improve texture and mouthfeel of ice creams.

Desserts: Egg proteins find a special place in several dessert preparations such as stirred custard, baked custard, custard pies, cream pies and puddings as well as hard and soft meringues. While eggs act as a thickener, structure provider and emulsifier in custards, they act as a foaming agent in the case of meringues. Casein, caseinates and the co-precipitates of casein are used in ice creams, frozen desserts, mousses, instant puddings and whipped toppings on various desserts.

Beverages and Related Foods: Whey protein concentrates, being soluble in acid medium, are being added to fruit-based beverages, soft drinks, fruit juices, powdered or frozen orange beverages and milk-based flavoured beverages. Casein is usually added to health drinks, chocolate drinks, cream liqueurs, wine aperitifs and certain types of beer. Protein hydrolysates are being used in geriatric foods, infant formulas, sports drinks and therapeutic diets/tube feed formulas.

Confectionary and Sweet Meats: Several toffees, caramels, fudges, nougat and marshmallows contain milk proteins, particularly 'casein'. Since whey protein concentrates exhibit good foaming qualities, they are used in the manufacture of candy floss and aerated candy mixes. The coagulation property of milk protein casein is used in the preparation of paneer-based sweet meats such as *rasgullas* and *chennamurki*, whereas the same property of egg proteins is employed during the cooking of *egg halwa*. The characteristic flavour of *besan halwa/ladoo* is due to Maillard reaction and Strecker degradation of amino acids.

Meat Products: Meat muscle itself is a concentrated source of protein. The property of protein hydrolysis, hydration, flavour binding and coagulation is used in preparing dishes from meat, fish, sea food and poultry. The comminuted meat products often contain casein to improve texture and flavour. Whey proteins are used in fortified brine solutions, frankfurters and luncheon rolls.

Texturized Vegetable Protein: The ability of globular proteins to get transformed into fibrous proteins in the presence of an alkali is used in the production of texturized protein products. When vegetable proteins, particularly soy protein, are texturized, the product obtained is chewy yet soft, elastic and juicy. Such properties are appreciated by consumers. Texturized soy protein (TSP) is available as soy granules and soy chunks in the market. TSP powder is often added to the meals provided to children, pregnant women and lactating mothers under supplementary feeding programmes. It is being added to soup/sauce mixes and for the manufacture of comminuted meat products, meat analogues and imitation meat products.

Fat Replacers: Several protein-based fat replacers have been developed with the objective of reducing fat intake and at the same time improving the nutritional attributes of food products. They help to improve texture and mouthfeel when used as a partial fat substitute in baked products, processed snacks, confectionary, margarines, shortenings, spreads, soups, sauces and gravies. The water-holding capacity of certain protein-based fat replacers makes them suitable for processed meat products. They are also used as stabilizers and emulsifiers in certain dairy products. However, such fat replacers are not suitable for frying and baking.

Emulsifiers: Proteins are used in the production of salad dressings, mayonnaise, bread spreads, table spreads, margarine, butter, sausages and frozen desserts because of their emulsification capacity.

Flavours: Certain proteins such as serum albumin, β-lactoglobulin and soy protein have been used as flavour carriers. They have been used successfully in meat analogues containing plant proteins which has successfully improved consumer acceptance of such products. They have also been used as flavour modifiers in fabricated foods.

The use of food proteins is not just limited to the applications discussed till now. Newer technologies are resulting in the development of several unique protein-rich foods such as nutrition bars and high-protein chocolates. Soy protein is being used for preventing bloom formation in fruit juices and for preparing edible packaging films (especially active packagings) and dyes. However, several food safety issues are being raised regarding the food safety aspects of processing proteins under adverse conditions (Figure 4.9).

Thus, it can be concluded that the structure of proteins influence their physical and chemical properties which in turn affects their functionality required during food preparation. When proteins are cooked, it is a complex interplay of several functions that results in the development of a palatable dish. For instance, while preparing stirred custard, the functions of

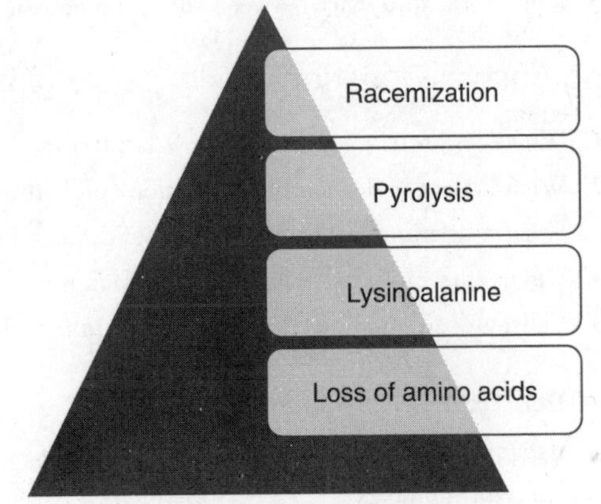

Figure 4.9 Food safety issues: Processing of food proteins

water-binding, gelation, emulsification and coagulation are performed by egg and milk proteins. Functions performed by each protein are unique and cannot be performed by another. Therefore, advancements in food technology today have resulted in isolation and modification of food proteins to meet the tailor-made requirements of a dish/product. These proteins can be used alone or in combination with natural foodstuffs to prepare your favourite dish!

Know Your Progress

True or False Questions

1. Globular proteins have quaternary structure.
2. Actin is a meat protein.
3. Milk lowers the temperature of coagulation.
4. Sugar stabilizes egg white foams.
5. Emulsifying capacity of animal proteins is lower than that of plant proteins.

Very Short Answer Questions

1. What is isoelectric point?
2. Enlist the different proteins present in egg.
3. What is T_d?
4. Define foaming capacity of protein.
5. What do you understand by water-holding capacity of protein?

Short Answer Questions

1. Describe the structure of proteins.
2. What factors influence the coagulation of protein?
3. What are the benefits of incorporating milk proteins in bakery products?
4. Give any five uses of the following in food preparation:
   (a) Egg foam formation (b) Coagulation of egg proteins
5. What is the difference between coagulum gels and translucent gels?

Long Answer Questions

1. Why does protein get denatured? What factors influence protein denaturation?
2. What do you understand by hydration and hydrolysis of proteins?
3. What is gelation? Give five examples of dishes which involve this property of protein.
4. Discuss the application of proteins in food preparation.
5. Enlist any 10 functions of protein. Explain the principle by giving suitable examples.

Beyond the Text

- Read the labels or recipes of the following products and identify the key proteins present in them
  Nutrition bars Meringue Soup powders Cheese souffle

Chapter 5

Food Processing

Chapter Objectives

After reading this chapter, the reader should be able to:

- understand the significance of processing food.
- learn the common pre-preparation techniques.
- describe the important techniques by which food is processed.
- know the effects of processing on food quality.

Food processing is not new to humans; they have been engaged in finding different ways to preserve the food since time immemorial. Early humans knew some of the processing techniques such as sun-drying, salting and fermentation, but others such as pasteurization, canning and concentration are more recent. Quick freezing, reverse osmosis and irradiation represent some of the modern developments in this field. Currently, food processing has become such an integral part of food supply that it is difficult to imagine a world where the populations would have to rely on unprocessed indigenous and in-season foods. In present times, a chef employs host of pre-preparation techniques such as peeling, chopping, mincing and dicing; as well as processing techniques such as steaming, stewing, baking, refrigerating and freezing to prepare and store various dishes. Empowering him/her with sound knowledge of various processing techniques and their effects on processing can go a long way in effective and safe execution of these tasks. This chapter discusses the objectives and techniques of food processing along with their effects on food quality.

5.1 Objectives of Processing Food

Food processing can be defined as the process of transforming fresh food into wholesome, nutritious and acceptable products which are safe for human consumption. The objectives of food processing are to

1. preserve and enhance the sensory attributes colour, flavour, texture and nutritive value of food.
2. prevent or delay spoilage caused by microbial growth and chemical reactions.
3. extend the shelf life of food while maintaining the highest quality standards.
4. make the food available throughout the year and in times of scarcities.

5. ensure that the food is wholesome, nutritious and safe for human consumption.
6. facilitate easy and successful storage, long distance transportation and distribution of food.
7. create employment and generate income for the processor.

5.2 Common Pre-preparation Techniques

Processing transforms natural food through a series of operations to high-quality products. Prior to processing, the raw materials undergo a number of procedures such as cleaning, sorting, grading, peeling and size reduction.

5.2.1 Cleaning

The process used for removing undesirable materials from food is called cleaning. Removal of damaged, insect-infested or deteriorated food ensures that the spoilage is not spread to the remaining bulk. It also helps to prevent subsequent damage to equipment by materials such as stones or metal pieces. Cleaning includes washing (Figure 5.1) which accomplishes the following objectives in fruits, vegetables, cereals, pulses and meat products:

1. Removal of soil, insects and undesirable materials (e.g. husk particles and stones from rice, wheat and pulses, blood spots from meat).
2. Getting rid of pesticides, fungicides and insecticides.
3. Reducing the number of surface microorganisms.
4. Improving the appearance and acceptability.

Further, washing lowers the temperature and reduces the rate of respiration and transpiration among the fruits and vegetables. A detergent or 1.5 per cent solution of dilute hydrochloric acid is usually used to remove insecticide and pesticide residues. Chlorine, potassium permanganate or salt solution is used for disinfection.

Figure 5.1 Washing: A type of cleaning process

5.2.2 Sorting and Grading

Though the terms 'sorting' and 'grading' can be used interchangeably, sorting implies categorization of food based on a measurable physical property such as shape, size, colour and weight. On the other hand, grading is the process of separating the foods into categories based on their quality. These days, sophisticated equipment have been developed which sort and grade the foods simultaneously. Similar to cleaning, sorting and grading should also be carried out as early as possible to save time, labour and expenditure on materials which are subsequently discarded due to poor quality.

Sorting by size and shape is carried out using flatbed screens (for flour, sugar and spices) or drum screens (for nuts, pea and pulses). Grading is done by employing various types of machines such as screen grader, roller grader and length graders.

5.2.3 Peeling

It removes the protective covering from fruits and vegetables by knives, peelers or following methods:

1. **Scalding:** It is a process of dipping fruits and vegetables such as potatoes and tomatoes in boiling water for a short time so as to ease the peeling process.
2. **Abrasion Peeling:** Large-scale peeling employs special machines such as potato peeler which is a cylindrical equipment with a rough surface. When potatoes are rotated in this cylinder, they rub on the surface and get peeled. The peels are washed away by a continuous supply of water.
3. **Flame Peeling:** Used for onions, this method employs a high temperature to burn off outer hair and peels which are subsequently removed by high-pressure water sprays.
4. **Lye Peeling:** Boiling vegetables in a solution of 1–2 per cent sodium hydroxide decomposes the pectin substances present below the epidermis and loosens the skin. This procedure, known as lye peeling, is used for vegetables such as potatoes, tomatoes, carrots and sweet potato.

5.2.4 Size Reduction

Suitable knives, slicers, choppers, dicers and grinders are employed for reducing the size of food ingredients so that they are easy and quick to cook (Table 5.1).

1. It makes cooking easier and faster.
2. It adds variety to recipes and garnishing.
3. It improves the appearance of the product.
4. It ensures uniform blending and cooking of food.

Table 5.1 Some terms relating to size reduction

Term	Description	Food Application
Cut	Divide into smaller pieces usually with a knife	Fruits and vegetables
Chop	Cut haphazardly to divide into fine pieces	Onions, ginger
Dice	Cut into uniformly sized cubes	Pineapple
Slice	Cut into thin vertical shapes	Cucumber, onions
Macedoine	Cut into 5 mm dice	Carrots
Julienne	Cut into 3 × 2.5 mm thin strips (e.g. matchsticks)	Celery, beets
Brunoise	Cut into 2 mm dice (can be diced from julienne)	Carrots
Puree	Blend to a soft creamy paste	Mango
Mince	Convert to a very fine and soft paste	Meat
Flake	Shred into thin shavings	Cheese
Grinding	Crush into powder	Sugar
Milling	Grinding and removing unacceptable parts	Cereals and pulses

Certain pre-preparation techniques such as marinating, bleeding/singeing/flaying, trussing, de-scaling/filleting/skinning and ageing are typical to the processing of non-vegetarian foods. Other pre-preparation techniques include pumping, mixing (blending, beating, whipping and folding), filtration, centrifugation, coating (dredging, breading and battering) or lamination to name a few (Figure 5.2).

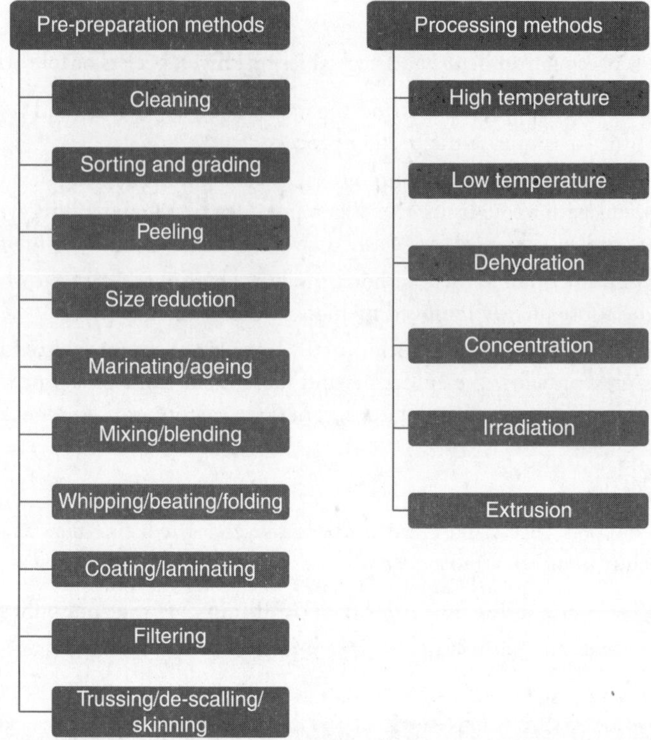

Figure 5.2 Important pre-processing and processing techniques

5.3 Important Processing Techniques

After pre-processing, a raw material undergoes suitable processing techniques to get transformed into a final product which is acceptable to the consumer. Some of the important techniques include high or low temperature processing, removal of moisture (dehydration and concentration), irradiation, fermentation and extrusion (Figure 5.3).

5.3.1 High Temperature

Heating can be regarded as the foundation of food processing and preservation. At high temperature, enzymes get inactivated and most microorganisms and their spores are killed; thereby, the shelf life of food is prolonged. Heat processing is carried out using moist heat, dry heat or hot oil (Figure 5.4). Some modern techniques employ direct and radiated energy to produce heat. Figure 5.3 illustrates the different methods of heat treatment. The temperature, duration and process of heat application depend upon the food composition, microbial load and desired characteristics of the finished product in a particular environment.

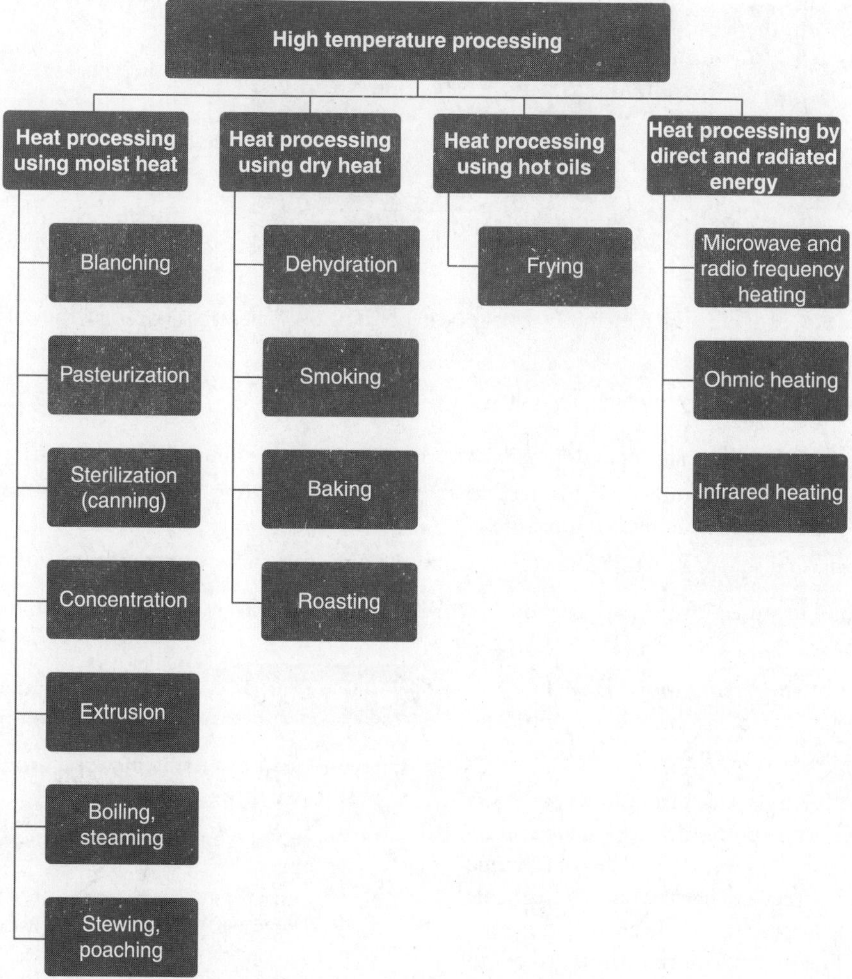

Figure 5.3 Methods of processing by use of high temperature

The major advantages of heat treatment include the following:

1. Heating above 55°C denatures enzymes, halts ripening, enzymatic browning and spoilage of food e.g souring of milk/cream.
2. Selective heat treatments such as pasteurization help to kill pathogenic microorganisms e.g tuberculosis-causing bacterium (*Mycobacterium tuberculosis*), *Brucella abortus* and *Streptococcus pyogenes.*
3. Heating above 100°C with modified pressure (steam under pressure) helps in sterilization of food products and equipment. Sterilization kills all microbes and their spores and denatures the toxins.
4. Certain methods such as blanching help to retain the natural shape, colour and flavour of foods.
5. The storage, distribution and transportation costs are lesser for several (not all) heat-processed foods (jams, jellies, marmalades, canned food) as compared to cold-processed foods.

Figure 5.4 Heat processing method: (a) Moist heat (boiling), (b) dry heat (baking) and (c) hot fat (frying)

The disadvantages of heat processing include the following:

1. Prolonged heating causes development of burnt flavour, dark colour and loss of nutrients.
2. Excessive heat denatures proteins, reduces their solubility, destroys vitamins, breaks emulsions anc dries out foods by removal of moisture.
3. The natural shape, structure and colour cannot be preserved in most methods (except blanching).

Some commonly used processing methods involving the use of heat are discussed in the followin paragraphs.

I Heat Processing Using Moist Heat

Varying degree of moist heat is employed by blanching, pasteurization and sterilization.

Blanching: It is a pre-treatment carried out at less than 100°C before processing (e.g canning, freezing and dehydrating) vegetables. The contact time depends on vegetable characteristics of vegetable such as size, shape, heat conductivity and natural levels of their enzymes. Blanching is carried out to achieve the following objectives:

1. Inactivating enzymes, particularly, polyphenol oxidase which causes browning.
2. Preventing changes in colour and flavour.
3. Removing the astringent taste of some plant foods such as gooseberry.
4. Removing air from the intercellular spaces, preheating the food and reducing the volume to facilitate more filling during packaging such as canning.

THINK TANK

Can you match the following food products with suitable processing technique?

Food products	Processing technique
* Biscuits, cakes, breads, tarts and pies	• Roasting
* Coffee beans, nuts	• Frying
* Condensed milk	• Baking
* Packaged milk (Amul/ Mother Dairy)	• Dehydration
* Dried milk powder	• Extrusion
* French fries/doughnuts	• Concentration
* Soya granules	• Pasteurization

Blanching is of several types viz. hot water blanching, cold water blanching, oil blanching and microwav blanching. Hot water blanching results in the loss of water-soluble flavours and nutrients, particularl ascorbic acid through leaching. On the other hand, steam blanching prevents the loss of water soluble cor stituents and also saves the effort of changing hot water repeatedly. Microwave blanching is a new techniqu which further reduces the nutrient loss.

Pasteurization: It employs high temperature for a short duration to kill the undesirable microbes present in food (*Mycobacterium tuberculosis*, *Brucella abortus* and *Streptococcus pyogenes*) and inactivate most enzymes, followed by sharp cooling, e.g., milk. It does not affect the chemical composition or sensory attributes of the product. Only a small amount of ascorbic acid present naturally in milk is lost. Discovered by Louis Pasteur, it is a technique of partial preservation and needs to be followed by other methods of preservation such as refrigeration and/or aseptic packaging of milk, fruit juices, cream, vinegar and beer (Box 5.1).

Box 5.1 Methods of Pasteurization

Pasteurization can be carried out by employing three processes:

1. *Low Temperature Long Time (LTLT) Pasteurization*: Milk is held at 63°C for 30 minutes and rapidly cooled to 10°C. The flavour and nutritive value of milk/juice are adversely affected.
2. *High Temperature Short Time (HTST) Pasteurization*: Milk is heated at a temperature of 72°C and held for 15 seconds. The nutritive value and flavour of milk is better retained.
3. *Ultra-heat treatment (UHT)*: Temperature of 135°C or more is employed for 1–2 seconds. This recent method reduces the number of microorganisms substantially and prolongs the shelf-life from four to nine months without bringing any changes in the nutritional and sensory quality.

Sterilization: This heat treatment employs a high temperature of about 121°C (temperature at which food fluids get converted completely or partially into vapours/steam) resulting in complete destruction of microbes and their spores including *Clostridium botulinum* and putrefactive anaerobes. However, such a harsh treatment affects adversely the organoleptic and nutritive quality of food. It is advisable to commercially sterilize the food to retain its quality. *Commercial sterilization* is referred to as the degree of sterilization which destroys all pathogenic, toxin-forming and spoilage-causing organisms without harming the heat-resistant bacterial spores. Growth of these microbes is prevented by vacuum and high acidity as in canning.

Canning: Many fruits and vegetables are available as whole (cherries, peas), sliced (pineapple), diced (fruit tidbits) or pureed (tomatoes) in the market. Numerous fruit and vegetable products as well as meat, fish and their products are also canned. It is a complete method of preservation, which destroys all pathogenic, toxin-forming and spoilage-causing organisms and extends the shelf life up to one year. Canned foods can develop a cooked flavour and texture; there is loss of heat labile nutrients especially vitamin B and C; and acidic foods can react with cans to produce metallic flavour.

For canning, the food is graded, washed and pre-prepared before being filled in cans along with a medium–either sugar syrup for fruits or brine (salt solution) for vegetables. After exhausting air, the can is sealed hermetically and then heat processed at a high temperature. Acidic fruits are processed at a boiling temperature (100°C), whereas vegetables, being alkaline, need a higher temperature (121°C) of processing. Heat processing destroys *Clostridium botulinum* and its spores. Aseptic canning is a modern HTST technique in which the food is sterilized separately outside the container; is placed in previously sterilized package and then sealed in a sterile environment.

Cooking by various methods such as **boiling (Figure 5.4a), steaming, stewing** and **poaching** also employs moist heat. The temperature used is about 100°C at which most of the microorganisms are killed, but bacterial spores remain unharmed. Cooked food should be served hot or stored at low temperature to extend its shelf life. **Concentration** and **extrusion,** discussed later in subsections 5.3.4 and 5.3.7 of this chapter, are other heat-processing techniques which use moist heat.

II Heat Processing Using Dry Heat

Baking, roasting, smoking and dehydration employ dry heat. Dehydration will be described in subsection 5.3.3 of this chapter. Baking and roasting both use heated air to alter the sensory quality of food.

Baking: It is a process that employs dry heat to flour-based foods or fruits to alter their eating quality. Baking preserves the baked products by using high temperature to destroy the microorganisms and to remove moisture. The wide variety of baked products produced by the hotel industry can be grouped as low moisture (biscuits, crackers), intermediate moisture (cakes, pastries, soft cookies) and high moisture (breads, rolls, pizza, cheese cakes, fruit tarts, pies).

Roasting: The term 'roasting' refers to a dry heat method used for meat, cocoa, coffee beans, nuts and vegetables to alter their eating quality and enhancing flavour. The high temperature destroys microorganisms and also results in loss of moisture. As a result, the shelf life of roasted foods is increased.

Smoking: It is a method used to preserve protein-rich foods such as meat, fish and cheese by a combination of dry heat and antimicrobial substances in the smoke. Heat kills microbes, inactivates enzymes and decreases moisture, whereas the chemicals prevent oxidation and other deteriorative changes. Consequently, the keeping quality of smoked food is enhanced.

III Heat Processing Using Hot Oil

Frying: It alters the eating quality of food by heating it in oil (Figure 5.4c). Fried foods acquire a characteristic golden brown colour, a crisp texture, a distinct mouthfeel and a characteristic flavour. Preservative effect of frying results from thermal destruction of microorganisms and enzymes, as well as from the lowered moisture content. Foods dried more thoroughly by frying (e.g. potato chips) have a longer keeping life at room temperature as compared to fried foods which retain a moist interior (e.g. doughnuts).

IV Heat Processing by Direct and Radiated Heat

Microwave, radio frequency and **ohmic heating** are direct methods in which heat is generated within the product which helps to preserve it. On the other hand, **infrared heating** is an indirect method.

5.3.2 Low Temperature

Low temperature storage delays or prevents food spoilage because it:

1. Prolongs shelf life without adversely affecting the nutritive value and sensory quality of food.
2. Controls the chemical and enzymatic changes in food.
3. Retards the growth and activity of microbes.
4. Minimizes the loss of flavour such as in juice extracted from citrus fruits.
5. Precipitates waxes from edible oils.
6. Facilitates certain unit operations such as slicing of bread and meat cutting.
7. Increases the solubility of carbon dioxide in water as in soft drinks.

The disadvantages of low temperature are as follows:

1. Repeated freezing and thawing are detrimental to food quality.
2. Maintenance of an efficient cold chain during processing, storage, distribution and sale proves to be expensive.

3. Microbial spoilage due to a break in cold chain is very common.
4. Cryogenic liquids which leak out from the food during thawing result in loss of nutrients and food constituents.

efrigeration and freezing are two different methods which employ low temperature for slowing down hemical and microbial changes (Table 5.2).

able 5.2 Difference between refrigeration and freezing

Refrigeration	Freezing
Temperature for refrigeration: 4.5 to 7.0°C Temperature for cold storage: −2 to 16°C	Temperature for freezing: <0°C Temperature for frozen storage: < −18°C
Most microorganisms stop growing at refrigeration temperature	Nearly all microorganisms cease to grow and some are killed on freezing
Shelf life is increased from few days to weeks	Shelf life is prolonged up to months or even years

efrigeration and cold storage: Low temperature of refrigeration prevents the growth of microorganisms but oes not usually destroy them. In other words, their slow metabolic activity may still continue. Commercial and ousehold refrigerators keep a temperature of 4.5 to 7°C; whereas, in cool storage, the temperature ranges from 2°C to 16°C. Ideally, refrigeration of perishable foods such as milk, meat, fish poultry, fruits and vegetables nould start immediately after slaughter or harvest and be maintained throughout transportation and storage. However, chilling (e.g in bananas, potatoes and bread) can result in undesirable quality changes, i.e., bananas now adverse darkening of skin, potatoes become sweet and rate of staling in bread is increased.

reezing and frozen storage: Freezing is a rocess which converts water into ice and nakes it unavailable for the microorganisms and the enzymes. As a result, it prevents microbial growth entirely and slows own the enzymatic activity. Freezing also ontributes in making a large variety of onvenience foods available to the present ay consumer. If the process is carried out roperly, it prolongs the shelf life of foods without adversely changing their sensory and nutritional attributes. Nowadays, it is possible to prepare a ood item or a meal completely prior to freezing and serve it just after a single thawing-heating operation. Jse of microwaves can ensure faster thawing. Commercial freezing can be carried out by three methods: eezing in air, freezing by indirect contact with the refrigerant and freezing by direct immersion in a efrigerating medium.

USEFUL TIPS

- Quick freezing helps to maintain sensory attributes, especially shape and structure, much more than slow freezing. Therefore, it is employed in freezing of fish.
- Slow freezing enhances shelf life more than quick freezing. Therefore, it is useful in preserving unstructured foods such as tomato puree.

Frozen storage is an integral part of freezing. It refers to the storage of frozen food at a temperature of −18°C r less so as to maintain it in a frozen condition. Frozen food in optimum frozen storage conditions can be reserved for many months or few years, if provided with proper packaging.

5.3.3 Dehydration

Dehydration or drying, one of the oldest methods of food processing, literally refers to the removal or reduction in the moisture content of food. Generally, meat, fish and dairy products are dehydrated to a moisture content of 3 per cent or less, vegetable/fruit products to about 5 per cent and cereal products to about 12 per cent. Certain products (spices) may contain moisture content as low as 1 per cent. However, the moisture level of dehydrated foods is determined by its end-use, sensory features and expected shelf life.

Removal of moisture helps to lower the water activity of the dehydrated products. Lowering the water activity (free water available to microorganisms) below 0.85 inhibits the growth and multiplication of microorganisms and bacterial spore germination. A good dehydrated product can be re-hydrated or re-constituted (if required) to resemble closely its natural counterpart (milk powder can b easily reconstituted with water to resemble natural milk). However, not all products can be re-structure or re-constituted, because the cellular structure gets destroyed bringing vivid changes in colour, texture taste and mouthfeel.

GOOD TO KNOW

Water Activity (a_w)

'Water activity' is the quotient of water vapour pressure of the food divided by the vapour pressure of pure water at the same temperature. The a_w of pure water is 1.00, whereas that of most bacteria, yeast and mould is 0.91, 0.88 and 0.80. An a_w below 0.85 inhibits the growth and multiplication of most microbes. Based on their water activity, foods can be divided into following three categories:

1. High Moisture Foods (HMF): a_w= 1.0–0.90 (squashes, cordials)
2. Intermediate Moisture Foods (IMF): a_w = 0.90–0.60 (jam, jellies, marmalade)
3. Low Moisture Foods (LMF):a_w = 0.60–0.00 (chips, milk powder, turmeric powder)

Dehydration has several advantages which are as follows:

1. Dried products are lighter, lesser volumetric/bulky, easy and economical to store, pack and transport.
2. Dehydrated products have a very low microbial content.
3. Enzymatic activities get reduced or stopped due to loss of moisture and application of heat.
4. Dehydration simplifies food processing and prolongs the shelf life of products.
5. They add variety to the range of convenience, ready-to-use and ready-to-cook products.

The disadvantages of this process include the following:

1. The cellular structure gets destroyed resulting in major alterations in shape, colour, texture, taste and mouthfeel.
2. Certain nutrients such as vitamin C, riboflavin, thiamin and folic acid get destroyed.
3. Discolouration due to Maillard reaction or enzymatic browning may occur if proper pre-preparation techniques (such as sulphiting in case of potato chips) are not carried out prior to dehydration.
4. Re-hydration or re-constitution may not always result in production of foods which are close to their natural counterparts.

Methods of Drying: Many methods have been developed for dehydrating foods.

1. **Sun Drying:** It is the most economical, easy and frequently used method. Apples, mango candy/powder, raisins, papads and cereal grains are still sun dried in many parts of our country. However, the moisture content cannot be reduced to below 15 per cent; nutrient losses are high and sensory characteristics are adversely affected.

2. **Drying with Dehydrators:** Use of dehydrators for drying preserves the nutrients and quality of products. The process involves the passage of heated air with controlled relative humidity over the food to be dried or the passage of food through such air. It includes air convection drying such as kiln drying, tray or cabinet drying, tunnel or continuous belt drying, puff drying, fluidized bed drying and spray drying. Drum or roller drying and vacuum drying result in good quality products.
3. **Freeze Drying:** It uses low temperature in the presence of vacuum so that sublimation of ice can be achieved to result in water loss. The moisture content of freeze dried foods can be as low as 5 per cent.
4. **Other Methods:** Electronic and ohmic heating are being employed for reducing water content of dried or concentrated products. It is being used to prepare egg white powder.

Pre-preparation of foods before drying is important and includes procedures such as selection/sorting/grading for size/maturity and desired composition, appropriate washing and cleaning, blanching/scalding, sulphiting and use of alkali (if required). Careful assessment of moisture loss during dehydration, handling 'sweating' of food products immediately upon removal from dehydrator and appropriate storage/packaging help in devloping high quality product.

5.3.4 Concentration

Concentration is a process of converting liquid foods into foods of viscous consistency or those which have a semi-solid or paste-like texture such as fruit and vegetable juice concentrates, sugar syrups, squashes, jams, jellies and marmalades, tomato ketchup and sweetened condensed milk. Commercially prepared concentrated foods which have moisture content between 15 per cent and 40 per cent are known as **Intermediate Moisture Foods**, e.g., soft candies, jams, jellies, marmalades, honey and raisins. Concentrated products have a low a_w and hence a longer shelf life as compared to their liquid counterparts. However, they would have a shorter shelf life as compared to their dehydrated counterparts. Box 5.2 gives an outline of the process/principle of processing concentrated products.

Concentrated food products have the following advantages:

1. Loss of moisture and increase in osmotic pressure (addition of solutes) result in a lowered water activity and thus a reduced microbial load and an extended shelf life.
2. Removal of water results in reduction in weight and volume of the product. For example, 5 litres of tomato juice would get reduced to about 1.73 kg after removal of water.
3. They are economical to handle, pack, store and transport.

Box 5.2 Steps in the Manufacture of Concentrated Food Products

The manufacture of concentrated food products is generally a two-step process involving

1. an initial reduction in water content so as to result in an increased total solid content and
2. binding of the remaining water with a solute such as sugar (70 per cent solution) or salt (18–25 per cent solution).

The second step may be optional. For example, in order to convert tomato juice into tomato paste, only the first step, i.e., evaporation of water is required. However, if tomato juice is to be converted into tomato ketchup, water from the juice would be initially evaporated followed by addition of salt and/or sugar.

4. Concentration is a cost-effective method of utilizing perishable crops during peak harvest period and improved availability of food throughout the year.
5. They add variety to the existing range of semi-processed and processed products.

Concentrated food products have the following disadvantages:

1. Organoleptic properties of the product change such as development of cooked flavour or darkening of colour (the colour of milk changes from white to off-white).
2. Over concentration may cause crystallization of carbohydrates, giving a gritty texture to the product.
3. Excessive use of salt may cause precipitation of proteins in the presence of heat such as in case of milk. Excessive use of sugar may cause syneresis of gel as in case of custard.
4. The products may contain some microbes or their spores and hence should be refrigerated once the seal has been opened by the customer. Use of additives may be required to preserve concentrated food products.

Methods of Concentration: Foods may be concentrated by three techniques viz. evaporation, freeze concentration and reverse osmosis. These methods may be used alone or in combination with other food processing techniques.

Evaporation: Removal of water through water vapours by the application of heat is known as evaporation. It is the most commonly employed technique and is of several types such as solar evaporation.

Freeze Concentration: It is based on the principle that aqueous solutions can be concentrated by freezing part of the water as ice and mechanically separating the ice by draining, pressing or centrifugation. All liquids and solids have different freezing points. Water freezes at 0°C, whereas fluids in food freeze at temperatures lower than 0°C. Thus, if, e.g., orange juice is kept for freezing, the initial ice crystals formed would contain only water and not the flavour/colour components or nutrients. If these ice crystals are removed, the resultant juice would be more concentrated. This process gives good quality products but is very expensive.

Reverse Osmosis and Ultrafiltration: These techniques are being used for fruit and vegetable juice concentration (~60°Brix). It is based on the principle of osmosis (movement of molecules from higher concentration to lower concentration) and employs the use of selective permeable membranes. Since it is a non-thermal process, retention of nutrients and sensory attributes is the highest.

5.3.5 Fermentation

Curd, yogurt, cultured milk, cheese, miso, tempeh, poi, pidan, soy sauce, sauerkraut, vinegar, beer, whisky, bread and certain bakery products as well as *idli, dosa, dhokla, uttapam* and *kulcha* are some of the food products which can be prepared by employing the principles of fermentation (refer to Section 23.2 of Chapter 23). Fermentation can be described as a process involving breakdown of carbohydrates, carbohydrate-like materials, proteins, fats and other structural/molecular modifications by microorganisms such as bacteria, yeast and fungi (refer to Box 5.3 for details).

Box 5.3 Types of Fermentation Processes

Fermentative – involves the action of microbes on carbohydrate and carbohydrate-like materials.

Proteolytic/putrefactive – involves microbial breakdown of food proteins.

Lipolytic – involves the breakdown of fatty acids by microbes.

The spongy texture of bread and idli or the bubbling effect observed in wine/beer is due to the liberation of gases (principally carbon dioxide) produced by microbes (Figure 5.5). While certain fermentations occur in aerobic (presence of air) conditions, others can occur in anaerobic (absence of air) conditions. Fermentation has the following advantages:

USEFUL ACTIVITY

Ferment your own dough!

Take 50 gm of refined wheat flour, one–fourth teaspoon of frozen baker's yeast (*saccharomyces cerevisae*) and make a soft dough using tepid water. Keep the dough ball in a beaker and note its height. Allow the dough to stand in sun or proover. Note the rise and changes in dough after 15 minutes, 30 minutes and 45 minutes (Figure 5.5).

Figure 5.5 Fermentation-related rise in dough

1. Selective and careful fermentation can help in creating endless possibilities of new dishes and variations in existing recipes.
2. It helps in preservation of food and improving shelf life. The metabolic by-products of several useful microorganisms act as inhibitory substances for many pathogenic organisms. For example, the acidic environment created by lactobacilli during the preparation of curd hinders the growth of *Clostridium botulinum* (a harmful microbe).
3. Fermented foods are usually more nutritious and are good sources of vitamin B and C.
4. Microorganisms make the structure of food more permeable to heat/water used for cooking and digestive juices. Therefore, fermented foods are easier to digest and faster to cook.
5. It adds variety to our diet and is used in the manufacture of nutrients such as riboflavin, vitamin B_{12}, certain fatty acids and amino acids.

Like any other process, fermentation has certain disadvantages also, such as the following:

1. The energy value of fermented foods is little lower than their unfermented counterparts.
2. Large-scale fermentation needs to be handled very carefully. Stringent monitoring of culture composition and microbial load of the product being manufactured are essential to avoid episodes of food poisoning.

In order to produce fermented products, controlled starter cultures containing a single or a combination of microbes must be used. The composition of culture should be the same every time a particular product is made so that all sensory characteristics of the product remain same, batch after batch. Selection, preparation and maintenance of culture are often carried out by a microbiologist along with a food technologist. At household or small-scale level, the culture is often a readily available fermented food such as sour curd or a microbe in easy-to-use form such as baker's yeast. They involve a single fermentation process. Table 5.4 gives details of microbes used in the manufacture of certain fermented products.

Table 5.4 Cultures used in fermented foods

Food Product	Microorganism(s) Used
Italian cheese/Swiss cheese	*Lactobacillus bulgaricus/lactis/helveticus, Streptococcus faecalis*
Sour cream	*Streptococcus lactis*
Yogurt	*Streptococcus thermophilus, Leuconostoc bulgaricus*
Cultured buttermilk	*Streptococcus cremoris/lactis, Leuconostoc cremoris*
San Francisco dough bread	*Torulopsis holmii, Lactobacillus species*
Rye bread	*Lactobacillus plantarum, Lactobacillus brevis, Lactobacillus bulgaricus*
Other breads	*Saccharomyces cerevisiae, Lactobacillus species*
Soy sauce	*Aspergillus oryzae, Pediococcus halophilus, Lactobacillus species, saccharomyces rouxii*
Tamari sauce	*Aspergillus tamarii*
Angkhak	*Monascus purpureus*
Tempeh	*Rhizopus stolonifer/oryzae/oligosporus*
Kefir	*Streptococcus lactis/cremoris* and *yeast*
Miso	*Saccharomyces rouxii, Zygosaccharomyces, Lactacid bacilli, Aspergillus oryzae*
Idli	*Leuconostoc mesenteroides, Streptococcus faecalis, Pediococcus cerevisiae*
Beer	*Saccharomyces carlsbergensis*

5.3.6 Irradiation

Food irradiation is a comparatively new technology (Box 5.4). It works upon the principle that ionizing radiations interact with food material by transferring energy to electrons and ionizing molecules, by creating positive and negative charge (ions).

The unit of measuring ionizing radiations has recently been changed from 'rad' to 'gray'. One gray is equal to 100 rad. Food irradiation doses are generally characterized as low (<1 kGy), medium (<1–10 kGy) and high (>10 kGy).

Food irradiation has several advantages such as the following:

1. No significant changes develop in the sensory characteristics (colour, texture, taste and flavour) of food.
2. It helps in destroying microorganisms, worms and insects thereby, improving shelf life of the product.
3. No harmful effects such as that of radio activity have been observed, if radiation doses as recommended by FDA/PFA are used.
4. No significant rise in temperature occurs during irradiation; therefore, heat sensitive foods such as milk, fruits and vegetables are most suitable for irradiation.
5. Change in the structure and the amount of nutrients is quite low.
6. No residual changes or substances have been observed in foods as a result of irradiation.

Box 5.4 Definition of Irradiation

According to the Statutory Regulations under part VII, rule 73 of PFA (Prevention of Food Adulteration Act), 1955, the definitions for irradiation of food are as under:

1. 'Irradiation' means any physical procedure, involving the intentional exposure of food to ionizing radiations.
2. 'Irradiated food' means articles of food subjected to radiation by:
   (a) Gamma rays.
   (b) X-rays generated from machine sources operated at or below an energy level of 5 million volts.
   (c) Sub-atomic particles, namely, electrons generated from machine sources operated at or below an energy level of 10 million electron volts, to dose levels as specified in Schedule I of the atomic energy (Control of Irradiation of Food Rules, 1991).

7. Irradiation can be done even on packaged and sealed containers.
8. It can facilitate effective quarantine procedure for infested and infected foods, thereby promoting international trade.

Irradiation has the following disadvantages:

1. Excessive irradiation can adversely affect nutrients, flavour, pigments and enzymes.
2. Foods irradiated in plastic containers are at a risk of developing off-flavour due to the development of volatile compounds.
3. It is difficult to differentiate between batches of products subjected to irradiation or not because this process does not result in any physical or significant molecular changes in food.
4. The dosage given to food is difficult to determine.
5. It inhibits germination and hence the reproduction ability of seeds is lost. Radiations can stunt, alter or kill plants, animals and microorganisms.
6. The capital cost of infrastructure development (irradiation plant) is very high.
7. It needs to be carried out carefully under expert supervision to prevent exposure to radiations.

Figure 5.6 Radura symbol used for irradiated food

According to joint FAO/IAEA/WHO Expert Committee on Food Irradiation (IJECFI), irradiation of food up to an average dose of 10 kGy causes no toxicological hazards and introduces no special nutritional or microbiological problem. Food irradiation has been endorsed as a safe and effective tool to increase food safety and reduce the incidence of food-borne illness by the American Medical Association's Council on Scientific Affairs. Currently, 37 countries, including India, have approved the use of irradiation for processing or preservation of food. Figure 5.6 shows the symbol which must be present on the label of irradiated food.

5.3.7 Extrusion

Hand-extrusion is a household method in India and has been used for preparing *jalebis, imratis* and *namkeens*. However, its large-scale application in food industry has opened endless possibilities of new products and newer textures and shapes (Figure 5.7). It is based on the principle of mass and flow transfer in the presence of pressure and at times high temperature. Extrusion helps to convert raw material (one or more ingredients) into a product with desired shape and form by forcing the material through a small opening (having the die to give desired shape) using pressure. It includes a series of steps such as mixing, kneading, shearing, cooling, shaping and forming. Extrusion is as two types viz., cold and hot extrusion.

Cold extrusion is a non-thermal process used to mix, knead, disperse, texturize, dissolve and form a product or ingredient. Typical food products include ready-to-use pastry dough, individual pieces of candies or biscuits (before baking), confections, pasta, hot dogs, Indian namkeens and selected pet foods. Such products generally have a dense texture/composition.

Hot extrusion or extrusion cooking involves the use of heat. The ingredients get mixed and cooked as they are subjected to high temperature and pressure while flowing through the extruder to form extrudate at the exit end. Texturized soy protein (granules/chunks), breakfast cereals, corn grits, pre-gelatinized soy flour, pretzels, corn puffs and namkeens such as mad angles/kurkure are manufactured by this process. They do not require further cooking. The products are generally porous because of the release of steam due to sharp temperature gradient within the extruder and the atmosphere outside.

Figure 5.7 Desired shapes can be produced by extrusion technique

Extrusion has the following advantages:

1. It is the most energy-efficient processing method for manufacturing ready-to-eat foods.
2. It can result in endless variations in sensory characteristics, especially texture, shape and size.
3. Hot extrusion process involves the use of moist heat under pressure which results in destruction of most pathogenic microorganisms and inactivation of enzymes.
4. Moist heat helps in cooking of the ingredients within the extruder itself.
5. Extruded products have a very long shelf life.

Extrusion has the following disadvantages:

1. The natural shape, structure, colour such as that of fruits and vegetables cannot be retained.
2. Large-scale extrusion involves the use of commercial extruders which require big capital cost.
3. Packaging of extruded foods with low moisture content needs to be done immediately to avoid loss of texture.

The food processing methods discussed here are rarely used alone. Most food products/dishes are prepared by combining a number of methods.

5.4 Effects of Processing on Food

When food is processed or cooked, several desirable and undesirable changes occur in its constituents. Food constituents include nutrients (carbohydrates, fats, proteins, vitamins and minerals), water, pigments, enzymes, emulsifiers, organic acids, oxidants, antioxidants and flavour compounds. The effect of food processing methods and the complex interactions between these constituents results in the development of a spectrum of food products with varied characteristic features. The effect of cooking on nutritive value of food has been discussed in Chapter 18 of Nutrition section in this book. The effects of food processing on other constituents is discussed in the following paragraphs.

5.4.1 Effect on Pigments

Why is it that carrots are orange-red, certain varieties of capsicum are green whereas others are orange, white or yellow; eggplant may be white or purple; grapes are either green or purple? Why is it that the outer skin or peel at times is green but the vegetable inside is off-white, the colour of peach and apricot near the core is different from that near the outer skin? Why does lentil (dehusked masoor dal) loose its pink colour when cooked? This is so, because all raw foods inherently contain pigments which provide them with different colours/shades. Each food may contain one or a combination of several pigments to result in a particular colour. Food processing can result in change in the structure of these pigments and hence the colour of dish prepared (Table 5.6). Major pigments present in food include chlorophyll, carotenoids, anthocyanins, anthoxanthins, myoglobin and haemoglobin (Figure 5.8). Their salient food sources and characteristics are briefed in Table 5.6.

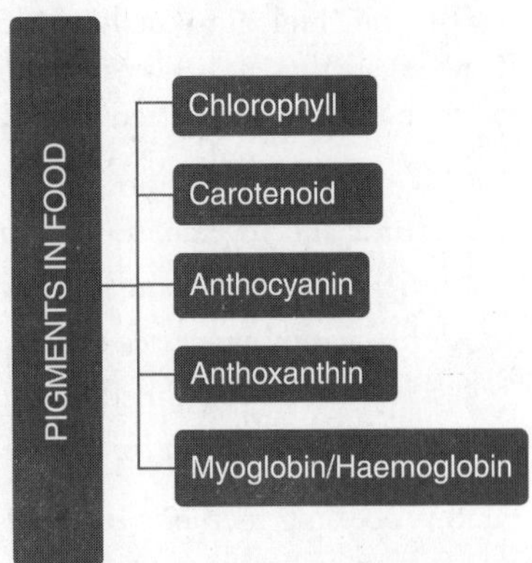

Figure 5.8 Pigments in food

Table 5.6 Effect of processing on different types of pigments

Pigment	Colour	Present In	Solubility	Effect of Acid	Effect of Alkali	Effect of Prolonged Heat
Chlorophyll	Green	Bitter gourd, green leafy vegetables, cucumber, outer covering of bottle gourd, green apple	Fat soluble	Dull olive green (pheophytin)	Intense green (chlorophyllin)	Dull green
Carotenoids	Orange-red	Papaya, mango, carrot, capsicum, tomato, annato	Fat soluble	Stable	Stable	Heat stable
Anthocyanins	Red-purple	Cherries, beetroot, pomegranate, red cabbage, berries, currants, outer covering of egg plant	Water soluble	Bright/deep-red	Blue or bluish-purple, brown	None/darker
Anthoxanthins	Colourless to pale yellow	Potatoes, litchi, colocasia, bottle-gourd, apple, wheat flour	Water soluble	Whiter	Dark-dull-yellow	Deeper pale-yellow
Myoglobin	Red	Meat, poultry and marine foods	Not soluble	None	None	Dark brown colour (met-myoglobin)

5.4.2 Effect on Enzymes

Enzymes are biological catalysts which promote several biochemical reactions without becoming part of it. Enzymes are present in all natural unprocessed foods. They are most active at about 37°C and at a pH close to neutrality. Enzymatic activities are of significance to a food processor because

1. enzymes facilitate both ripening and, if not controlled, over-ripening of natural foods, especially fruits and vegetables (pre- and post-harvest).
2. they can bring changes in food flavour, colour, texture, mouthfeel and nutritive value.
3. the heating processes employed during food manufacture can inactivate or destroy the enzymes.
4. they can be controlled during fermentative processes so as to bring desirable changes.
5. their activities can be initiated by using commercial enzyme preparations.

Food processing methods involving the use of temperature between 50°C and 55°C result in complete inactivation, whereas temperatures above 55°C result in denaturation/destruction of enzymes as in case of blanching, steaming, boiling, pressure cooking, and pasteurization. Use of low temperature can also influence enzyme activity. Partially frozen products do exhibit some enzyme activities. Rapid freezing results in rupturing of plant and animal tissues which leads to release of enzymes. Thus, even at 0°C, enzymatic breakdown of protein may become the principal cause of quality loss in meat, fish, and poultry. Use of acid and, in some cases, alkali can result in denaturation of the protein structure in enzymes and hence loss of their activity. This principle is used in the denaturation of phenolase enzyme which is responsible for browning of cut fruits and vegetables. Dehydration and concentration methods of food processing reduce enzyme activities selectively.

Commercially available enzymes such as carbohydrases, proteases, lipases, oxidoreductases and renin can be used efficiently in the manufacture of bread, cheese, milk chocolate, carbonated beverages, dried eggs, invert sugar and certain wines. Enzymes such as papain, ficin and bromelain are frequently used to tenderize large cuts of meat. Enzymes are also used to remove haze from fruit juices and wines and to give sparkling property to jellies.

5.4.3 Effect on Natural Toxicants

Certain foods contain natural toxins which can be destroyed if food processing methods are employed judiciously. Some of the toxic constituents (Figure 5.9) present include the following:

1. *Trypsin inhibitors* which are present in soybean and kidney beans. Autoclaving or pressure cooking at 120°C or above can result in inactivating the trypsin inhibitor.
2. *β-N-oxalyl-L-α,β diamino propionic acid* (beta-ODAP) is a neurotoxin present in Lathyrus sativus (kesari dal). Steeping and parboiling processes remove 80–90 per cent of this toxin. It is also known as *β-N-oxalyl amino-L-alanine acid* (BOAA).
3. *Vicine, Covicine* and *DOPA* (*dihydroxy phenyl alanine*) are present in the cotyledons of beans. Sprouting followed by boiling results in the reduction of these toxins.
4. *Haemagglutinins* are present in field beans, white beans, double beans and horse gram. They are heat labile.
5. *Saponins* are found in soybeans, and they are eliminated by a simple process of soaking in water before cooking.
6. *Goitrogens* are heat sensitive, and they are present in small amounts in vegetables of crucifera family such as cabbage and cauliflower, and soybeans, lentils and groundnuts.

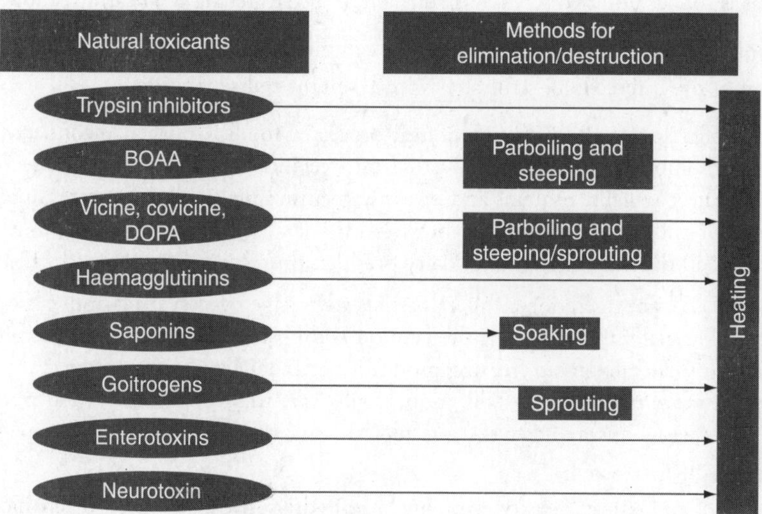

Figure 5.9 Natural toxicants in food and methods of their elimination or destruction

The above toxins are naturally present in foods. However, certain microorganisms also produce toxins (refer to the 'food safety' section of this book for details) such as the following:

1. *Enterotoxin* produced by *Staphylococcus aureus*. It is found in under-processed milk and milk products, meat and meat products, salads, puddings, cream sauces and custard-filled bakery products. Pasteurization or heating at 190.6–218.3°C, maintaining an acid pH, prompt refrigeration and use of additives (bacteriostatic agents) can help to denature or prevent the release of this toxin.
2. *Neurotoxin* produced by *Clostridium botulinum*. It is present in improperly processed foods, particularly canned products and fish. Heating at 90°C for 15 minutes destroys this toxin but the spores of this bacterium get destroyed only if heated at 120°C for four minutes.

Aflatoxins, patulin, ochratoxin, luteoskyrin and *penicillic acid* are some other examples of toxins. Most of these are heat sensitive. Addition of additives to food during processing helps in preventing their production thereby ensuring the safety of food. Food safety reports should be obtained for susceptible foods especially in case of large-scale production.

5.4.4 Effect on Sensory Attributes

The food processing options available today have changed the magnitude of the food service units. An endless variety of organoleptic properties (colour, texture, taste, flavour and mouthfeel) can be created while cooking/processing food. For instance, milk can be converted to curd by fermentation, milk powder by dehydration and condensed milk by concentration techniques. Combinations of several food processing techniques are frequently applied such as during the preparation of ice cream (high and low temperature along with concentration). While it is beyond the scope of this chapter to discuss all sensory changes and their reasons, a few important ones are being discussed.

High temperature: The structure of food (such as that of tomato and eggplant) changes during the application of heat. This is so, because two types of tissue damage occur during the heating of food:

1. Destruction or damage to the semi-permeable cell membrane.
2. Disruption of the intercellular structures with resultant cell separation.

This results in loss of crispness and softening of heat-processed foods. Collagen, a connective tissue protein of meat, which is insoluble in water gets converted into gelatin, which is water soluble. The viscosity of fluids containing heat-coagulable proteins and/or easily gelatinizable starches increases upon application of heat such as in case of milk and eggs (refer Chapters 2–4 for details). HTST heating helps to preserve colour and is therefore used in the manufacturing of dairy products, purees, pea-soups, sauces and tomato paste.

Low temperature: Use of refrigeration temperature has advantage over freezing in preserving firmness and crispness of vegetables. Careful pre-freezing preparation and post-freezing storage of the product is vital to final product quality. Generally, rapid freezing produces better quality frozen products than slow freezing. It is so because, the more rapid the crystallization, smaller are the ice crystals (1–2 μm) and lesser is the damage caused to tissues. Thus, rapid freezing results in lesser cell dehydration, breakage of cell walls and less textural damage.

Removal of water: The texture, density, wettability, re-hydration capacity and mechanical properties of dehydrated foods depend upon the drying process. For example, freeze drying produces products with porous crust and superior re-hydration capacity; whereas, hot air drying results in a dense product with an impermeable crust. Dehydrated vegetables in instant noodles and breakfast cereal flakes are examples of freeze drying. During drying, the surface layers loose more moisture and at an earlier stage as compared to the inner core. Therefore, the surface layers shrink which causes checking, cracking and warping.

GOOD TO KNOW

Recrystallization: Large fluctuations (>3°C) in the temperature of frozen foods such as during retail display or carry home packages or repeated thawing causes ice crystals to melt slightly, refreeze (ripen or grow), coalesce and move to the product surface. The size of ice crystals thus increases each time there is a fluctuation of storage temperature. This process is known as **re-crystallization** (Figure 5.10).

Figure 5.10 Recrystallization of ice on frozen (a) cereals, (b) vegetables and (c) meat

Freezer burn: At times, unpackaged products/foods are blast freezed or kept directly on ice which results in the development of **freezer burn** due to moisture loss by evaporation from the surface of the product. Freezer burn appear as brownish grainy spots or patches of dry tough tissues and is often mistaken for mould growth. If freezer burn is mild, it gets resolved by re-hydration. It can be prevented by increasing relative humidity and/or temperature of storage area and packaging products in heavyweight, moisture-resistant wraps.

Despite keeping these and several other points in mind, variations in sensory features are inevitable, because nature cannot produce identical raw material. For example, tomato ketchup produced in summer months would have different sensory qualities as compared to that manufactured in winters, unless, suitable additives are added to each batch. Effects of food processing should be used to our advantage so as to result in greater customer satisfaction and brand loyalty.

5.4.5 Effect on Shelf Life

Shelf life refers to the period of storage after harvesting or manufacture during which the product retains its characteristic features and remains suitable for consumption or other intended processes. It is also known as practical storage life or useful life. International Institute of Refrigeration describes it as high-quality life (Box 5.5).

A combination of food processing methods positively influence shelf life to a greater extent as compared to a single method. For example, psychrophilic bacteria survive/multiply at cold temperatures, mesophilic at room temperature and thermophilic at high temperature. Thus, if in the preparation of a dish (such as ice cream), both low and high temperatures are used, the microbial load would be least. Use of extreme temperatures, even for a shorter duration, results in prolonged shelf life. The lower the cold storage temperature, the longer is the shelf life. Poultry can be stored only for 2–3 days in a refrigerator, but it can be stored for 3–6 months in a freezer. High acid foods require lesser heat for sterilization as compared to low acid foods. For example, *Bacillus subtilis* is able to survive heat treatment only for 4 minutes at 4.4 pH but for 11 minutes at pH 6.8 to 7.8. Moist heat is more effective in killing spoilage organisms as compared to dry heat. For example, spores of *B. subtilis* are killed in less than 10 minutes in steam at 120°C, but in the presence of dry heat, it requires 30 minutes at 170°C. Packaging in the absence of oxygen or in the presence of alternative gases such as nitrogen helps to eliminate aerobes and other organisms, thereby prolonging shelf life. Drying, concentration, irradiation and the use of additives, helps in delaying deterioration of products.

Box 5.5 What Is High-Quality Life

High-quality life is the period elapsed from processing (such as freezing) of an initially high-quality product and the moment when, by sensory assessment, a statistically significant difference ($p < 0.01$) from the initial high quality (such as immediately after freezing) can be established.

The study of food processing methods is very expansive and it is beyond the scope of this chapter to discuss all techniques. However, it should be remembered that the raw material used for food processing should be handled with the objective of preventing contamination, minimizing physical loss, maintaining physical appearance and other sensory characteristics, minimizing nutrient losses, preventing development of undesirable substances and minimizing handling time and frequency. Seldom is a food processed or dish cooked by employing any single method of food processing. Combinations of several food processing techniques are frequently applied such as during the preparation of puddings (high and low temperature along with concentration). Using a combination of methods helps to selectively adopt the features of each method suitable to the properties of a particular dish. Continuous research and development to find out methods of product and process improvement can help in making the entire exercise of food processing more satisfying and profitable to a chef.

Know Your Progress

True or False Questions

1. Scalding is a process of boiling vegetables in a solution of 1–2 per cent sodium hydroxide to decompose the pectin substances present below the epidermis and loosen the skin.

2. *High Temperature Short Time (HTST) Pasteurization* employs a temperature of 72°C for heating the milk and holding it for 15 seconds.
3. Commercially prepared concentrated foods which have moisture content between 10 per cent and 20 per cent are known as *Intermediate Moisture Foods*, e.g., soft candies, jams, jellies, marmalades, honey and raisins.
4. Fermentation can be described as a process involving breakdown of carbohydrates, carbohydrate-like materials, proteins, fats and other structural/molecular modifications by microorganisms such as bacteria, yeast and fungi.
5. 'Irradiation' means any physical procedure, involving the unintentional exposure of food to ionizing radiations.

Very Short Answer Questions

1. What do you mean by 'irradiated food'?
2. Define blanching.
3. What is myoglobin?
4. How is 'freezer burn' caused?
5. What is pasteurization?

Short Answer Questions

1. What are the objectives of food processing?
2. Differentiate between freezing and refrigeration.
3. Define the following terms:
   (a) Commercial sterilization (b) Freeze drying
   (c) Chlorophyll (d) Enzyme
   (e) Reverse osmosis
4. What are the advantages and disadvantages of concentration?
5. How is cold extrusion different from hot extrusion?

Long Answer Questions

1. What are the common pre-processing techniques? Discuss in detail.
2. Justify the statement 'Heating can be regarded as the foundation of food processing and preservation'.
3. Discuss the various methods of concentration. List the advantages and disadvantages of this processing technique.
4. Describe the benefits and limitations of using dehydration for food processing. What are the different ways of drying?
5. What is the effect of processing on pigments, enzymes and sensory attributes?

Beyond the Text

- Pick up any five processed foods such as milk, parsley leaves, jam, soy sauce, peas and read the packet/label carefully. Identify the salient food processing techniques employed for manufacturing and/or packaging these products. Repeat the same exercise for any two recipes (ingredients and the final dish).

Chapter 6

Evaluation of Food

Chapter Objectives

After reading this chapter, the reader should be able to:

- differentiate between sensory and objective evaluation.
- describe the quality of food in terms of colour, appearance, flavour and texture.
- learn different methods of sensory evaluation and their application.
- understand the fundamentals of proximate analysis of food constituents.
- associate how rheology affects the sensory properties of food.

The choice of food and its amount eaten by people largely depends on its quality. The food may be a culinary dish prepared by a chef or a food product marketed by a multinational company but quality in terms of what it looks like, what its aroma is and how it tastes would decide its acceptability. One of the major challenges facing culinary professionals in food and hospitality business is to ensure the ultimate satisfaction of consumer. Acquainting themselves with the ways of food evaluation in general and sensory methods in particular would enable them to carry out the task more effectively.

6.1 Sensory Versus Objective Evaluation

'Quality' is a complex term which can be defined simply as the degree of excellence. Food quality includes sensory attributes and non-sensory characteristics such as nutritive value, microbial content and storage life. The process of measuring, investigating and interpreting the attributes of quality is known as evaluation.

Generally food evaluation is carried out to

1. develop and market new food products.
2. compare quality of food with competitor's products.
3. assess consumer preferences for food products.
4. bring about changes in menus/food products for better consumer acceptance.
5. judge routine quality and predict storage life of foods.
6. ensure better consumer acceptance of foods and improve their cost effectiveness.

THINK TANK

A chef makes a sugar-free version of his special mocktail by replacing sugar with an artificial sweetener. He has the following questions in his mind:

- Are the two beverages different in their quality attributes?
- How are the two beverages different in their quality?
- What is the acceptability for the new beverage?

These are some examples of the situations requiring food evaluation.

Food can be evaluated either by sense organs or by use of instruments. Accordingly, the process is classified as sensory evaluation and objective evaluation.

Sensory Evaluation: The science of assessing excellence of food by senses of sight, taste, touch, smell and hearing is known as sensory evaluation; it measures perception of stimuli through the sense organs (Figure 6.1). The definition emphasizes on 'what is perceived' which may be different from the physical properties of the food.

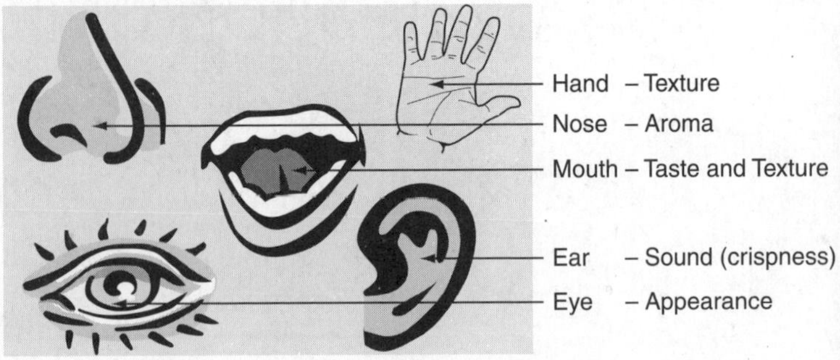

Hand – Texture
Nose – Aroma
Mouth – Taste and Texture
Ear – Sound (crispness)
Eye – Appearance

Figure 6.1 Organs involved in sensory evaluation

Sensory information obtained from trained/untrained individuals is unique since it cannot be adequately furnished by instruments or chemical techniques. Evaluation by sensory organs is also referred to as subjective or organoleptic evaluation. Although the results tend to be less consistent and less accurate, better consumer acceptability and cost effectiveness of food product are ensured. Moreover, sensory evaluation reduces uncertainty and lowers the risk in decision making.

Objective Evaluation: Evaluation of food quality using instruments, physical methods and/or chemical techniques is known as objective evaluation. In contrast to subjective judgement, it assesses one attribute of food quality at a time and the results are more consistent and accurate. One of the major drawbacks with objective evaluation is that it cannot measure consumer preferences. It is generally employed for routine quality control and includes the following methods:

1. **Chemical Methods** estimate nutrient content, adulterants and other constituents of food through chemical tests. In addition, high-pressure liquid chromatography (HPLC) is used for separating individual sugars and organic acids, whereas gas chromatography is employed for separating volatile aromatic acids.

Table 6.1 Comparison of sensory and objective evaluation of food

Sensory (Subjective) Evaluation of Food	Objective Evaluation of Food
Carried out primarily by individuals	Carried out by using instruments and equipment
Uses human senses of sight, touch, hearing, taste and smell	Uses physical or chemical techniques
Can be biased (subjective)	Is more objective
Can judge the overall acceptability of food	Can judge one aspect of food quality at a time
Results are less accurate and less consistent	Results are more accurate and consistent
Affected by illness (particularly cold) of the judge	Not affected by illness of the judge
Standards are difficult to establish	Standards are easy to establish
Can determine consumer acceptance	Cannot determine consumer acceptance
Requires more time and finances	Requires less time and finances, but initial cost of instruments can be high
Employed for new product development and marketing	Employed for routine quality control and for predicting storage life

2. **Physiochemical Methods** use appropriate instruments for measurement for example, pH meter and refractometer are used for assessing pH and sugar concentration, respectively.
3. **Microbial Methods** examine the number and type of microorganisms present in food and thereby predict its safety and storage life.
4. **Physical Methods** assess physical parameters such as weight, volume, specific gravity and moisture content of food. Special instruments are employed to assess the sensory attributes of food such as texture (Table 6.2).

Sensory and objective methods are complementary to each other and both are important for judging the quality of food. A comparison of the two methods is presented in Table 6.1.

6.2 Sensory Characteristics of Food

Culinary professionals frequently judge the quality of their dishes organoleptically so as to satisfy the palate of their clients. Appearance, flavour and texture are the characteristics which are generally assessed by the senses. Some of these aspects can also be measured through different instruments available for the purpose.

6.2.1 Colour and Appearance

The appearance of food is judged by its colour, size, shape, symmetry, structure and other characteristics such as transparency, turbidity or opaqueness, dullness or gloss and wholeness or damage.

1. **Colour** is an important attribute for processed and unprocessed foods. It not only plays a pertinent role in the visual appeal of a menu but is also associated with other attributes such as ripening of fruits (green guava turns yellow on ripening). Additionally, colour can be used to assess the end point of frying (in French fries), the storage history (in chocolates) and efficiency of mixing (cake batter lightens in colour with incorporation of air).
2. **Size** is used to grade fruits and vegetables prior to processing. *Shape* of cut fruits and vegetables is an important consideration, particularly, for serving aesthetically appetizing salads.
3. **Structure** plays a crucial role in denoting quality of baked products such as cakes and breads where a fine porous appearance is desirable, but the presence of large holes or tunnels is undesirable.
4. **Transparency** is required in some juices such as apple juice, whereas *turbidity* is desirable in orange juice.
5. **Wholeness** denotes the degree of whole or broken pieces and is needed in products such as biscuits, pineapple slices and potato chips. Physical damage reduces the acceptability of these products drastically.
6. **Gloss** is indicated by the reflecting property of food such as preserves, jams and jellies.

Objective evaluation of colour can be carried out by various instruments such as Hunterlab colour and colour-difference meter, Macbeth-Munsell disc colorimeter, tintometer and spectrophotometer; whereas screen graders are used to sort fruits and vegetables based on their size.

6.2.2 Flavour

Flavour denotes the composite sensation of *taste, aroma* and *mouthfeel.* Taste is perceived when the taste buds present at the tip, sides and back of the tongue are stimulated. Aroma, on the other hand, is detected by olfactory epithelium in the nasal cavity. Mouthfeel is a sensation perceived by tongue, mouth and jaws. Temperature can also affect the intensity of overall flavour; high temperature enhances flavour by volatilizing aromatic compounds. The main flavour compounds of some foods are listed as follows:

Apple	⟶ Ethyl 2-methylbutyrate	Almond	⟶ Benzaldehyde
Banana	⟶ Isopentyl acetate	Cherries	⟶ Benzaldehyde
Grape	⟶ Methyl anthranilate	Butter	⟶ 2,3 Butanediol
Lemon	⟶ Citral	Meat	⟶ Hetcrocyclic thioles and disulphides

Taste is an important component of flavour. Most foods have a blend of two or more of five major tastes: sweet, salty, sour, bitter and umami. Some examples of foods and the compound responsible for the specific taste are as follows:

Sweet – Table sugar (sucrose), fruit sugar (fructose), artificial sweeteners (aspartame, sucralose)

Salty – Table salt (sodium chloride), low sodium salt (potassium chloride)

Sour – Grapes (tartaric acid), apple (malic acid), lemon (citric acid)

Bitter – Quinine, grapefruit (naringin), dried orange peel (hesperidin), coffee (caffeine), chocolate (theobromine),

Umami – Meat (glutamate), baked products (maltol), ajinomoto (monosodium glutamate)

Aroma contributes immensely to the pleasure of eating. The food must produce some volatile substance which must come in contact with the receptors in the epithelium of the olfactory organ for the aroma to be perceived. The smelling mechanism is highly intricate and sensitive. There are thousands of aromas and the sensitivity of olfactory organ is 10,000 times greater than that of the taste organ. Chefs use selected blends of spices/herbs to bring out the most favourable aroma of a dish.

Mouthfeel is the way a food feels in the mouth. Tongue, mouth and jaw can assess the form, shape and feel of a food. Some aspects of mouthfeel are chewiness, astringency and hotness. A food is termed 'hot' or 'burning' depending on the amount of pepper or chillies present in it. This sensation, sensed as 'pain' by nerve fibres (present in the mouth and the nose), is desirable only up to a limit.

Since humans can detect odourants at a level lower than that are detected by instruments, sensory evaluation is considered a better method for measuring flavour. Nevertheless, some objective methods are also employed; for example, evaluation of acidity is done by using a pH meter, whereas sugar concentration is measured by a refractometer.

DID YOU KNOW?

Taste buds are located in the bright red spots (known as papillae) on the surface of tongue and each bud contains nearly 100 receptor cells. Taste buds for four major types of tastes have regional distribution – sweet sensation at the tip, bitter at the back, sour at the sides and salty both at sides and tip of the tongue.

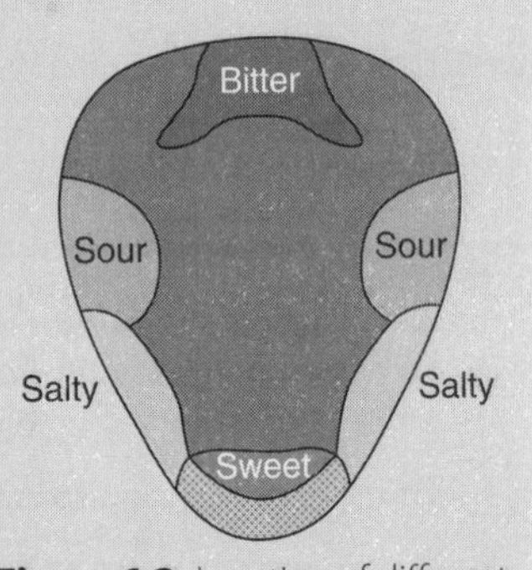

Figure 6.2 Location of different taste buds on the tongue

6.2.3 Texture

Texture of food is characterized by how it feels on touch or on the palate. In the case of cookies and chips, crispness is judged by the hearing sensation. Generally, the texture is assessed in terms of tenderness, graininess, brittleness and consistency.

1. **Tenderness** in a pastry can be judged by breaking it with hands, by cutting it with knife or by chewing it in the mouth. It is also desirable in vegetables such as peas and can be assessed by mouthfeel. Tenderness of meat is judged by the number of times it needs to be chewed before swallowing. The meat cuts with low connective tissue are tender and require less chewing.
2. **Graininess** in ice creams and fudge denotes large-sized undesirable ice crystals which can be felt by tongue. The texture of baked products such as cakes, breads and biscuit is assessed by the size of cells and thickness of cell walls in the crumb. The cells along with the cell walls are termed as 'grain' which can be judged by sight. Texture of bread and cake is porous and fine, whereas it is flaky for many types of biscuits.
3. **Brittleness** is desirable in fresh fruits such as apple and pears, as well as vegetables such as carrot and radish. When the cells of these foods are turgid with water, they offer some resistance to teeth biting into them and the mouthfeel is crunchy. On the other hand, biscuits and toasted breakfast cereals have low moisture content and are crisp in texture. The crispness of these products is judged by hearing and mouthfeel.
4. **Consistency** can be assessed by mouth, tongue and jaws, as well as by sight and touch. Culinary professionals know that the consistency of soup, gravy or sauce can range from thin to thick. In case of custard, the consistency may additionally be smooth, lumpy or curdled. It must be remembered that temperature can alter the consistency; for example, a sweet corn soup with perfect consistency becomes thick and viscous on cooling.

Objective evaluation of texture can be carried out by using suitable instruments such as consistometer, viscometer, tenderometer and succulometer. Instron Universal Testing Machine is a multipurpose machine which can determine several aspects of texture (Table 6.2).

Table 6.2 Application of some instruments used for objective evaluation of food texture

Instrument	Application	Characteristic Measured
Viscometer	Salad dressing	Viscosity
Consistometer	Ketchup, sugar syrup	Consistency
Farinograph	Bread dough	Plasticity
Succulometer	Fruits	Juiciness
Tenderometer	Peas	Tenderness
Shortometer	Pastry, cookies, crackers	Tenderness
Bloom Gelometer (Penetrometer)	Gelatin dessert	Gel strength
Squeeze tester	Bread	Softness
Shear press	Fruits	Firmness/crispness
Universal testing machine	Cookies, meat	Cohesiveness, hardness, springiness, gumminess, chewiness

6.3 Methods of Sensory Evaluation

Despite technological advancement in the field of objective evaluation, some measurements are best made by human senses. Modern sensory testing is carried out in a facility which can be simple or highly sophisticated.

Whatsoever may be the scale of facility, the certain provisions are required for the sensory testing procedure (Box 6.1).

DID YOU KNOW?

A Million-dollar, State-of-the-art Sensory Facility

Deerfield III Sensory Facility is a million-dollar, state-of-the-art facility which was opened at Illinois in 2010. It features 24 sensory testing booths with controlled lighting, temperature and pressure and provides electronic reporting with direct entry software. There is a 2,500-square-feet test kitchen with rooms for conducting sensory tests.

Source: Mermelstein (2010)

Box 6.1 Pre-requisites for Sensory Testing Procedure

Facilities and environment	– White/off-white colour – Controlled lighting – Good ventilation – Individual testing booths – Controlled noise level – No extraneous smell – Red lights to mask colour	**Samples**	– Labelled with random 3-digit code – Served at equivalent time after cooking/preparation – In same volume/size – At same temperature – Placed on a tray with a ballot – Served in random order – Cracker and water provided
Panelists	– Should have sensory skill for the required attribute – Should not have any medical problem which can affect evaluation – Bias and distraction should be minimized – Should not have eaten spicy/strong flavoured food before tasting		

Based on the type of information they provide, sensory evaluation methods can be grouped primarily as analytical methods and affective methods. *Analytical methods* provide information about analysis of products, for example, discrimination tests and descriptive analysis. *Affective or preference methods* provide information relating to preference or liking for a product, for example, consumer tests. These methods are generally not combined during evaluation.

6.3.1 Discrimination Tests

The discrimination tests, also known as difference tests, are carried out by a small group (5 to 12 in number) of trained judges employed to ascertain whether there is any perceivable difference between two food products. They can be used in many situations such as change in quality of a dish/product on using a new ingredient or a different packaging; or finding out difference between quality of a new product and that of a competitor' product. Discrimination tests are of three types (Figure 6.3). The judges are asked to describe differences in the samples presented to them in any of the following ways:

1. **Triangle Test:** Three samples, one different and two identical, are presented simultaneously to the panellist who has to identify the odd sample.

2. **Duo-trio Test:** This test also uses three samples, one different and two identical. First, the panellist is presented with one of the identical samples known as the reference sample. Subsequently, the other two samples are presented in a random order. The panellist has to match one of these samples with the reference sample.
3. **Paired Comparison Test:** Two samples in a pair are presented to the panellist who is asked to indicate if there is any difference between them.

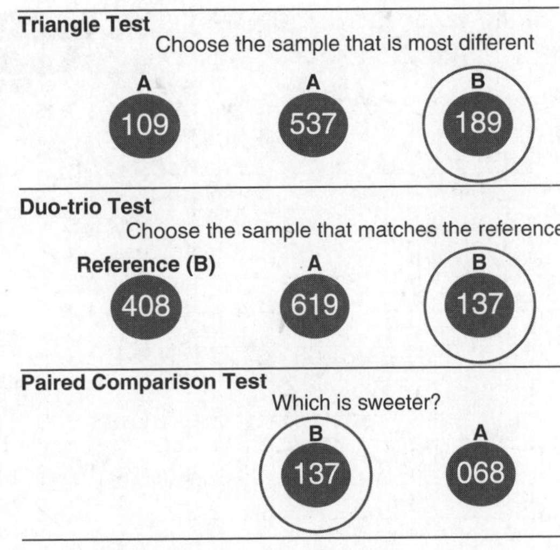

Figure 6.3 Types of discrimination tests

Note: Samples are presented to the judges only with the 3-digit code number. In this figure different samples are represented with alphabets (A and B) for better understanding by the reader.

These tests are quick and simple; they show that the difference (if any) was authentic. They form the foundation for sensory evaluation and can determine qualitatively the difference between two samples, but they do not reveal which sample is better and why. In recent years, some newer methods have been described. One such method is **R-index** which asks the panellists to indicate their confidence or sureness for their choice; whereas **n-Alternative Forced Choice (n-AFC)** asks the subjects to determine which product is stronger in a pre-identified attribute.

6.3.2 Rating Tests

In contrast to the discrimination tests which are qualitative, rating tests are quantitative and are very often used for consumer testing. They employ more than two samples and can tell which one is better than the other. Some methods of rating tests are listed as follows:

1. **Ranking Test:** The panellists are presented 2–7 samples to rank them in an increasing or decreasing order based on one or more attributes such as flavour, colour or texture.
2. **Single Sample (Monadic) Test:** Some products leave an after-taste or flavour; so it becomes impractical to taste the next product immediately after the first. Single sample test is used for such food products to find out if there is any off-flavour.
3. **Hedonic Rating Test:** This test can measure food acceptability by different groups of consumers in college hostels, army mess and school canteens. The panellists rate the food on a Hedonic scale of 0–9 ranging from 'like extremely' to 'dislike extremely' (Box 6.2). For children, a smiley scale is employed (Figure 6.4).
4. **Numerical Scoring Test:** In this test, a score system is employed to evaluate the quality of food products. For example, a score of 90 may be awarded to excellent quality, whereas a score of 20 may be assigned to poor quality product.
5. **Composite Scoring Test:** In contrast to numerical scoring test, this test rates every quality attribute such as colour, texture and flavour separately and then the scores are compounded. Total score represents the overall acceptability of the product.

Box 6.2 Score Card for Sensory Evaluation Hedonic Rating Test

Name of judge (optional) ____ Date ____________

Name of product _______ Quality attribute ____________

Degree of Preference	Sample A
Like extremely	
Like very much	
Like moderately	
Like slightly	
Neither like nor dislike	
Dislike slightly	
Dislike moderately	
Dislike very much	
Dislike extremely	

Comments ____________________

Figure 6.4 The smiley scale for kids

6.3.3 Sensitivity Tests

Unlike the tests discussed in the previous section, sensitivity tests assess the ability of individuals to identify and detect different aspects of food quality such as taste, aroma and astringency. Sensitivity tests help in training panellists for evaluating specific food products.

1. **Sensitivity Threshold Test:** The panellists are presented with food, beverages and even pure substances such as water for a specific characteristic such as sugar in increasing concentration.
2. **Dilution Test:** This test is used to detect the presence of an unknown substance added in small amount to a natural food product. The quality and acceptability of test material is represented by a *dilution number* which denotes the percentage of test material in the mixture of food product. Bigger the dilution number, better is the quality of test material.

6.3.4 Descriptive Analysis

Descriptive tests are carried out by a small group of highly trained judges with the objective of assessing both quantitative and qualitative aspects of food products. Most food companies have their own panel for evaluating each of their products. The panellists measure differences between products and identify the basis for the differences and their magnitude. They are asked to rate intensity for various sensory attributes; thereafter, t-test is employed for analysis to determine if the results are statistically different. *Flavour profile* is a method developed for measuring flavour as a composite of different tastes and odours. Similarly, *texture profile* has also been developed.

6.3.5 Affective Tests (Consumer Tests)

Affective tests, also known as **preference tests, acceptance tests or consumer tests,** are used to ascertain whether a particular group of consumers prefer or like a food product. These tests employ a large group of untrained judges numbering 75–100 selected from general public. For example, the acceptability of a new chocolate being launched can be tested by a group of children selected from a school. Since the number of panellists is large, consumer testing requires more time and finances. Very often, these panellists are not trained and are asked to indicate their liking or disliking for the product on a 9-point Hedonic scale (Box 6.1). Consumer testing is carried out in a less controlled environment and the consumers are less reliable panellists as compared to those employed for analytical sensory evaluation. These tests are employed for new food development. A summary of the characteristics of commonly used sensory tests is presented in Table 6.3. The chef mentioned in Think Tank can also find answers to his questions (about the sensory attributes of his new mocktail) from this table.

Table 6.3 Characteristics of different methods of sensory evaluation

Feature	Discrimination Tests	Descriptive Analysis	Consumer Tests
Basic question	Are the two products different from one another?	How do the products differ in all sensory attributes?	What is the acceptability of the product?
Characteristics of panellists and their task	• 5–12 • screened for sharpness of perception • asked to ascertain perceivable difference	• 8–10 • highly trained • asked to rate all sensory attributes	• 75–100 • semi-trained • asked degree of liking/preference
Advantages	• quick • simple • authentic	• give detailed analysis • supported by statistical analysis • can be presented graphically	• quick • simple • provide essential information relating to consumer liking
Limitations	• do not tell which sample is better and why	• is time consuming	• may be difficult to get representative sample of consumers • time consuming • expensive
Types of tests employed	• Paired comparison test • Triangle test • Duo-trio test • R-index • n-Alternative Forced Choice (n-AFC)	• Flavour profile • Texture profile • Sensory spectrum • Quantitative Descriptive Analysis(QDA)	• Hedonic scale • Smiley scale • Ranking scale • Numerical scoring • Composite scoring

6.4 Introduction to Proximate Analysis of Food Constituents

Besides sensory attributes, there are non-sensory nutritional and biochemical components of food which cannot be measured by the consumer. The quantitative analysis which gives information about the food constituents such as moisture, ash, acidity, crude fat, proteins, sugars and the crude fibre is known as proximate analysis. Special equipment and techniques are required for such analysis.

Providing nutrition information on the label is voluntary in our country, but the modern day consumer is becoming increasingly interested in nutrient-related information (particularly, the caloric content) about the food being purchased. Chemically, crude fat is measured by using Soxhlet apparatus, whereas Kjeldahl method is used for assessing the protein content. Ash content of a food refers to the inorganic residue remaining after burning the organic matter in a muffle furnace. It usually denotes the mineral content and in some cases suggests the presence of adulterants.

Care must be taken while sampling food for proximate analysis. The sample selected should represent all homogenous parts of a heterogeneous food. The sample needs to be prepared prior to analysis; it may be necessary to store and preserve the samples before they are analysed.

6.5 Rheological Aspects of Food

The sensory properties of food are influenced to some extent by **rheology** which is defined as the study of deformation and flow of matter and has three aspects: elasticity, viscosity and plasticity. Rheological properties are based on the flow and deformation when foods are subjected to stress (cutting, shearing and pulling).

1. **Elasticity** is a property exhibited by solids e.g. fermented bread dough which stretches in proportion to the amount of force (stress) applied and returns back to its original position when the force is removed. It breaks when the force applied exceeds beyond a point. Elasticity is also exhibited by baked products such as cake and bread as well as by gels prepared from pectin, gelatin or starch.
2. **Viscosity** is a property exhibited by liquids such as soups and salad dressings. It is defined as the resistance to flow of a liquid when it is subject to shearing force. As the thickness of liquid increases, viscosity increases. For example, tomato ketchup is more viscous than tomato soup. Viscosity is also affected by temperature; an increase in temperature lowers the viscosity in a salad dressing.
3. **Plasticity** is exhibited by foods such as cream icing, frosting and peanut butter. A plastic food resists changing position until a certain amount of force (yield stress) is applied. For example, warm butter is plastic and can be spread easily on a bread slice. Plasticity of shortening is a desirable characteristic for incorporating air in a cake batter.

We learnt that quality is an important criterion for determining the final desirability of food. Objective evaluation assesses one attribute of food quality at a time and the results are more consistent and accurate, but it cannot measure the consumer preferences. Sensory evaluation, on the other hand, measures perception of stimuli through the sense organs and is considered a better method particularly for measuring flavour. Different methods of sensory evaluation include discrimination tests, rating rests, sensitivity tests and descriptive analysis. Discrimination tests and descriptive analysis are examples of analytical methods which provide information about analysis of products, whereas consumer tests are affective methods which provide information relating to preference or liking for a product. Adequate knowledge about evaluation of food quality can go a long way in ensuring consumer satisfaction which forms the core of hospitality industry.

Know Your Progress

True or False Questions

1. Sensory evaluation is employed for new product development of food.
2. In triangle test, three different samples are presented simultaneously to the judge who has to indicate the difference in the three samples.
3. Hedonic scale is used by the panellist to rate the food on a scale of 0–9 ranging from 'like extremely' to 'dislike extremely'.
4. Acceptance tests are used to ascertain whether a particular group of consumers prefer a food product.
5. Brabender Farinograph is used for measuring plasticity of bread dough by objective evaluation.

Very Short Answer Questions

1. Define the following terms:
   (a) Flavour
   (b) Viscosity
   (c) Duo-trio test
   (d) Ranking test
   (e) Sensitivity tests

Short Answer Questions

1. How does sensory evaluation differ from objective evaluation?
2. What are the characteristics related to colour and appearance of food?
3. Describe the different methods employed for carrying out rating tests.
4. Differentiate between discrimination tests and affective (consumer) tests.
5. What are the various rheological aspects related to sensory quality of food?

Long Answer Questions

1. Describe the quality attributes associated with the appearance, texture and flavour of food.
2. Discuss in detail the methods of sensory evaluation.

Beyond the Text

- Pick up five processed food products (e.g. biscuit, chips, chocolate, cereal bar and ice cream) from a retail outlet and carry out their sensory evaluation using hedonic rating score cards for any one characteristic (appearance, texture or flavour).

Chapter 7

Emulsions

Chapter Objectives

After reading this chapter, the reader should be able to:

- identify natural and man-made food emulsions.
- define emulsions and know their properties.
- understand the theory of emulsification and factors affecting stability of emulsions.
- classify emulsions based on dispersed and continuous phase.
- recognize different emulsifiers and their role in food emulsions.

What do milk, egg yolk, ice cream, mayonnaise and margarine have in common? They are examples of food emulsions which contain water and oil; one dispersed in another. We know that water and oil do not normally mix. If they are agitated together, one liquid forms droplets which get temporarily dispersed in the other. However, in due course of time, the droplets coalesce and the two liquids separate as two distinct layers. Such emulsions are prone to undesirable physical changes during storage. Since food service professionals are involved in the preparation and handling of various emulsified products such as salad creams, cake batter, cream soups and gravies, they should acquaint themselves with the types, principles of formation and stability of emulsions.

7.1 Emulsification

When oil and water are mixed in a container, they tend to revert to a thermodynamically stable state; that is, oil layer over the water layer. By agitation, droplets of one liquid are dispersed in to another to form an emulsion which can be stabilized by adding a third ingredient known as emulsifier. The emulsion so formed consists of a dispersed phase, a continuous phase and the emulsifier; the size of droplets in the dispersed phase ranges from 0.1 to 10 μm (Box 7.1). Milk, cream, butter and egg yolk are the examples of natural food emulsions; whereas, margarine, ice cream, cake batter, mayonnaise, salad creams, vinaigrette and cream soups are food products which are man-made emulsions.

THINK TANK

Forming an Emulsion

Take a cup of water in a transparent bottle and add two tablespoons of edible oil. Shake vigorously and let the mixture stand for 10 minutes. What do you observe?

Box 7.1 Definition of Emulsion

Emulsion is defined as a heterogeneous system consisting of one immiscible liquid intimately dispersed in another one, in the form of droplets with a diameter generally over 0.1 μm (Figure 7.1).

Salad dressings

Butter/margarine

Ice creams

Figure 7.1 Some formed emulsions available at household and commercial level

7.1.1 Formation of Emulsions

The process of converting two immiscible liquids into a stable emulsion is known as *emulsification*. Breaking the droplets of one of these liquid requires energy which is provided by agitation through a beater, stirrer, mixer or homogenizer. As soon as the droplets are formed, the newly formed interfaces must be covered by the emulsifier to prevent the droplets from coalescing. Therefore, the emulsifier should be present in sufficient quantity to coat the surfaces completely. As the emulsifier gets adsorbed at the surface and the interfacial tension (tension at the interface of two immiscible liquids) decreases more droplets are formed which gradually get smaller and assume colloidal size. Among the two immiscible liquids, the one with higher surface tension forms the droplets (the dispersed phase), whereas the other one flows around the droplets forming the continuous phase. The steps in emulsification of mayonnaise are presented in Box 7.2.

Box 7.2 Steps in Emulsification of Mayonnaise

1. Egg yolk (containing natural emulsifiers) is mixed with vinegar (aqueous fluid).
2. Oil is added in drops.
3. Energy for breaking oil drops into droplets is provided by beating or homogenizing.
4. Emulsifier adsorbs at the surface of oil droplets and the interfacial tension decreases.
5. More droplets are formed which gradually get smaller in size.
6. Aqueous liquid flows around the oil droplets to form the continuous phase.
7. As numerous small droplets get surrounded by a thin continuous phase, the emulsion becomes more viscous.
8. The interfacial film of emulsifier contributes to the viscoelasticity and the stability of emulsion.

7.1.2 Stability of Emulsions

During storage, emulsions may exhibit some undesirable physical changes. These stability-related changes generally occur through the following processes (Figure 7.2):

1. *Creaming* involves sedimentation of dispersed droplets as a result of which the emulsion forms two layers (one richer in fat and the other poorer in fat).

2. *Flocculation* is agglomeration of droplets to form loose and irregular clusters.
3. *Coalescence* is an irreversible process which results in the reunion of small droplets to form large droplets.

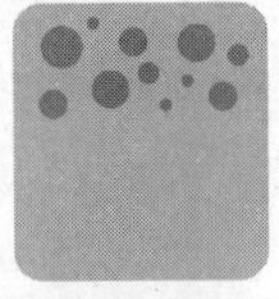

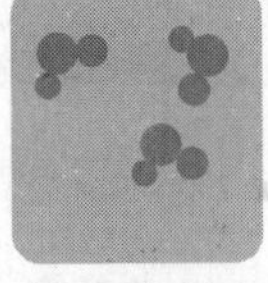

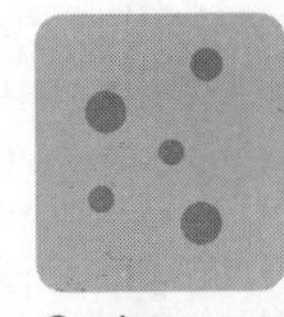

Figure 7.2 Diagrammatic representation of creaming, flocculation and coalescence

Stability of an emulsion is its ability to resist changes in properties during storage in different environmental conditions. Stability against creaming, flocculation and coalescence can be achieved through different ways such as using an emulsifier (which forms a strong interfacial film around the dispersed droplets), adding electric charge to the surface of droplet and/or increasing the viscosity of continuous phase. There are many factors that affect the stability of emulsions:

- **Type of emulsifier:** Some emulsifiers (particularly the ones containing large protein molecules) result in better stability of emulsions, since they are able to form stronger interfacial films. For example, casein helps in stabilizing the natural emulsion in milk.
- **Concentration of emulsifier:** Adequate amount of emulsifier is desirable for stability as it is more likely to form a continuous film to cover all droplets completely.
- **Size of droplets:** Large droplets tend to coalesce, whereas fine droplets make the emulsion more stable. Droplet size can be reduced by using beaters, blenders and homogenizers.
- **Change in pH:** Addition of acid (such as vinegar, lemon) can reduce the stability of interfacial film, particularly when it is made up of proteins.
- **Viscosity:** Thicker emulsions (such as salad creams) are more stable as the movement of droplets in a viscous system is less and the two phases take longer time to separate. Gums, pectin and gelatin are usually added to make these emulsions more viscous.
- **Change in temperature:** When an emulsion is warmed, the oil droplets become more fluid and tend to coalesce. Stability of an emulsion may be enhanced at refrigeration temperature. At freezing temperature, most emulsions break due to the denaturation of protein at interfacial film or physical disruption of this film by the ice crystals.
- **Agitation:** Violent shaking can break an emulsion. For example, emulsion of cream is disrupted by churning to separate the butter; use of warm or chilled water further helps in the process. On the other hand, controlled agitation helps in emulsifying fat into the cake batter.

7.1.3 Properties of Emulsions

Emulsions are integral to food preparation, whether it is the preparation of a cake batter or that of mayonnaise. Their properties depend on temperature, size and distribution of droplets in the dispersed phase and the gums added to the continuous phase.

1. *Optical properties* of emulsions vary depending on the refractive index and droplet size of the two liquids. An emulsion is transparent when refractive indices of the two liquids are same and when the size of droplets is less than or equal to 0.05 μm. Dispersion of larger droplets (0.05–1.0 μm) in the continuous phase (e.g. milk, mayonnaise) results

THINK TANK

Properties of Emulsions

Respond to the following questions in context of any of these emulsions: soft butter, cream, margarine, mayonnaise or Hollandaise sauce:

1. Is it opaque or transparent?
2. Is the mouthfeel creamy, smooth or grainy?
3. Is it plastic (changes shape on applying mild pressure)?
4. Is it pourable from a jar?

in scattering of light which makes the emulsion optically opaque. The colour can range from white and creamy to pale yellow.

2. *Interfacial properties* are determined by the interfacial region that separates the oil from the aqueous phase. They play an important role in deciding the properties of emulsions such as their rheology and stability.
3. *Rheology* (refer to Section 6.5 in Chapter 6) is a property of emulsions which influences their sensory quality in terms of the mouthfeel (creaminess and smoothness), the texture (spreadability and viscoelasticity) and flowability. The rheology (flow property) of an emulsion is generally determined by the viscosity of continuous phase.
4. *Stability* of emulsions has already been discussed (refer to Section 7.1.2).

7.2 Types of Emulsions

Though it is simpler to understand emulsions as a three-component system consisting of oil, water and mulsifier, most food emulsions are rather complex. Several water-soluble components such as sugars, salts, cids, bases, surfactants, proteins, polysaccharides and additives may be present in the aqueous phase. Many at-soluble substances such as fatty acids, vitamins, cholesterols and triacylglycerols can also be found in the il phase. The interfacial region may contain surface-active components including surfactants, proteins, hospholipids and polysaccharides. Based on their stability, source and the type of dispersed and continuous hase, the emulsions can be grouped in different ways (Box 7.3).

Box 7.3 Types of Emulsions

Based on stability

1. *Temporary emulsions* separate into different layers on standing; for example, French dressing.
2. *Permanent emulsions* are stable when they are left to stand in general environment; for example, mayonnaise.

Based on source

1. *Natural emulsions* are found in nature. Milk is one such emulsion.
2. *Man-made emulsions* are culinary inventions such as mayonnaise, ice cream and margarine.

Based on dispersed and continuous phase

1. *Oil-in-water emulsions*; for example, milk and mayonnaise.
2. *Water-in-oil emulsions*; for example, butter and margarine.

.2.1 Oil-in-Water Emulsions

mulsions having oil droplets as dispersed phase and water as the continuous phase are termed as 'oil-in-water /w) emulsions' (Figure 7.3). Most of the food emulsions such as milk, cream, mayonnaise, salad creams, naigrette, cream soups, gravies, cake batter and ice cream fall under this category.

1. **Milk** is a *natural emulsion* which is stable. It is an *o/w emulsion* where fat droplets are stabilized by a complex protein membrane. Milk is a *permanent emulsion* but non-homogenized fresh milk exhibits creaming (sedimentation of dispersed droplets as a result of which the emulsion forms two layers: one rich in fat and other poor in fat) on standing. Milk contains numerous small droplets and comparatively less large droplets of fat. Casein adsorbed on the fat globules has an important role in the stability of milk emulsion. The larger ones tend to rise and cluster together as cream but the interfacial film is still intact and so is the emulsion. **Cream** (30–45 per cent fat) is also an emulsion

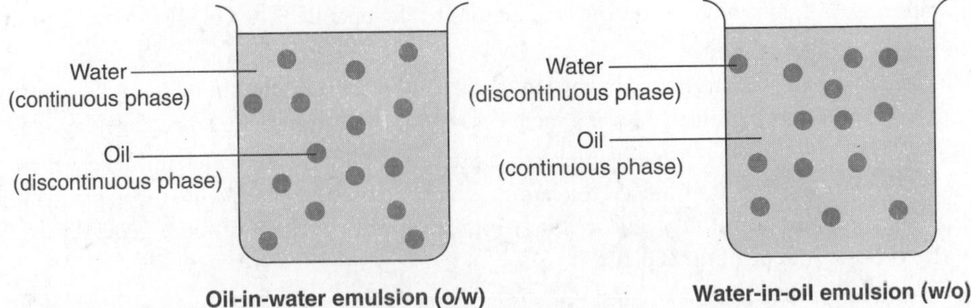

Figure 7.3 Oil-in-water and water-in-oil emulsions

which can be separated by centrifuging milk. On the other hand, homogenization breaks down the fat globules into finer droplets and prevents creaming in milk.

2. **Mayonnaise** is a *permanent emulsion* which is stable under general environmental conditions. It i a thick and creamy sauce or salad dressing. The basic ingredients used for this *o/w emulsion* are oil vinegar and egg yolk. Fat constitutes 75 per cent of mayonnaise. Egg proteins form a continuou film around fat droplets and thus act as emulsifiers. **Salad creams** are low oil (30–40 per cent fat products containing cooked starch ingredients and gums for more viscosity.
3. **French dressing** is an example of *temporary emulsion.* This o/w emulsion is formed primarily b means of oil, vinegar, dry mustard powder and paprika. As these ingredients are shaken vigorousl mustard and paprika adsorb at the interface and lower the interfacial tension facilitating emulsio formation; but they are not able to form a stable film around the oil droplets. Therefore, when shak ing is stopped, the droplets tend to coalesce and the oil layer separates from vinegar layer.

7.2.2 Water-in-Oil Emulsions

The emulsions which have water as the discontinuous (dispersed) phase and oil as the continuous phase ar known as *water-in-oil (w/o) emulsions* (Figure 7.3); for example, butter, margarine and spreads.

1. **Butter** is a w/o emulsion containing 80–82 per cent fat. This *permanent emulsion* is prepared by churning and kneading cream to break the interfacial film. As a result, there is a phase inversion from o/w to w/o emulsion. Warm or chilled water is used to facilitate the process. Butter is a *natural emulsion* known for its excellent flavour.
2. **Margarine** is a *permanent emulsion* which is a *man-made* substitute for butter to meet new nutritional demands such as lower levels of cholesterol and energy and desirable levels of vitamins. Production margarines involves emulsification of oil and water phases, crystallization of the fat phase and final its plastification. **Spreads** are prepared from liquid vegetable oils which are hardened by parti hydrogenation.

ACTIVITY

Dye-Solubility Test to Identify o/w and w/o Emulsions

Pick up samples (one tablespoon each) of any five food emulsions (e.g. milk, egg yolk, mayonnaise, margarine, butter) and stain them separately with a water-based food colouring (e.g. erythrosine). Spread each sample on a slide and observe under a magnifying glass to note whether the continuous phase or the dispersed phase is stained. Accordingly, group them as o/w or w/o emulsions.

Table 7.1 Common food emulsions and the stabilizing factors

Food Emulsion	Type of Emulsion		Emulsion Stabilizing Factors
	Natural or Man-Made	Oil-in-Water or Water-in-Oil	
Milk, cream	Natural	Oil-in-water	• Lipoprotein membrane • Phospholipids • Casein adsorbed on fat droplets
Mayonnaise	Man-made	Oil-in-water	• Egg yolk proteins • Phospholipids (lecithin)
French Dressing/Vinaigrette	Man-made	Oil-in-water	• Mustard powder • Paprika
Ice cream (emulsion-foam)	Man-made	Oil-in-water	• Semi-solid frozen continuous phase • Added surfactants control fat agglomeration • Foam stabilized by agglomerated fat globules forming surface of air cells
Egg yolk	Natural	Oil-in-water	Phospholipids and lecithin
Cake batter (emulsion-foam)	Man-made	Oil-in-water	• Egg proteins • Wheat proteins
Butter	Natural	Water-in-oil	Water droplets dispersed in semi-solid plastic fat phase
Margarine	Man-made	Water-in-oil	• Water droplets dispersed in a matrix of fat crystals • Added surfactants reduce surface tension and promote emulsification

Food emulsions are stabilized by emulsification and natural/added emulsifiers. Some of the o/w and w/o food emulsions (natural/man-made) are presented in Table 7.1 along with the factors contributing to their stability.

7.3 Emulsifying Agents and Their Role in Food Emulsions

For a colloidal scientist, 'surfactants' and 'emulsifiers' are two distinct terms, though they may be used interchangeably by a chef. *Emulsifier* (emulsifying agent) is a surface-active agent (surfactant) that confers stability on an emulsion by decreasing interfacial tension and by forming a protective film around each droplet to prevent their coalescence. *Surfactants* reduce

DID YOU KNOW?

Bancroft's Rule

Bancroft stated that the phase in which an emulsifier is most soluble forms the continuous phase of an emulsion.

USEFUL TIP

Disperse emulsifier in the preferred continuous phase before the other liquid is added. For example, while preparing mayonnaise, egg yolk should be mixed in vinegar so that the continuous phase is aqueous.

interfacial tension and get adsorbed at the surface, but many of them are not able to form a stable film at the interface to prevent coalescence of dispersed droplets. Therefore, all surfactants do not make good emulsifiers. Emulsifying agents play a crucial role in stabilizing emulsions. They perform the many functions such as decreasing the interfacial tension, adsorbing at the oil/water interface, forming a stable, continuous and viscoelastic film around each droplet, and preventing coalescence of dispersed droplets.

Based on their function, emulsifiers can be divided in to three groups: emulsifiers, finely divided particles and hydrocolloids. Stability of an emulsion is optimal when all the following three groups of emulsifiers are used.

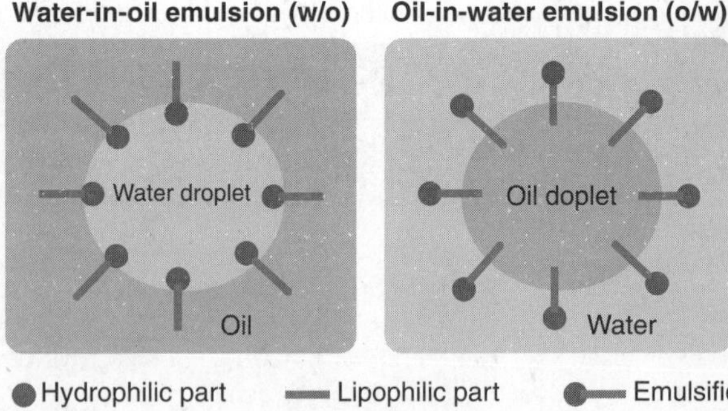

Figure 7.4 Diagrammatic representation of food emulsifier stabilizing water-in-oil and oil-in-water emulsions

Food emulsifiers are surface-active agents which contain both hydrophilic (water loving) and lipophilic (fat loving) ends (Figure 7.4). Natural emulsifier such as phospholipids (lecithin present in egg yolk) is used for stabilizing the emulsion in mayonnaise.

Finely divided particles do not usually exhibit any surface-active property. They are adsorbed at the interface and thus prevent coalescence of dispersed droplets; for example, mustard powder, pepper powder, magnesium hydroxide, aluminium hydroxide and calcium hydroxide.

Hydrocolloids such as plant gums, pectin, gelatin and cellulose derivatives increase viscosity of the continuous phase and thereby help in stabilizing o/w emulsions (refer to Chapter 8 on Colloids).

GOOD TO KNOW

Hydrophile-Lipophile Balance of Emulsifiers

Hydrophile-lipophile balance (HLB) is a numerical expression for the relative simultaneous attraction of an emulsifier for water (hydrophilic) and for oil (lipophilic). It indicates the solubility of an emulsifier in the oil and water phases.

1. *Low HLB emulsifiers*; for example, glycerine monostearate tend to form w/o emulsions.
2. *Intermediate HLB emulsifiers*; for example, gelatin tends to form o/w emulsions.
3. *High HLB emulsifiers*; for example, gum acacia, gum tragacanth tend to form o/w emulsions

A number of natural and synthetic emulsifiers are used in different food emulsions.

1. *Natural emulsifiers* such as proteins (present in egg, milk, soy and meat) are among the best emulsifiers. Proteins are large molecules which uncoil or denature and get adsorbed at the interface to form a stable interfacial film. They have both hydrophilic and lipophilic ends. When they uncoil, their hydrophilic ends tend to orient towards water, whereas the lipophilic ends are directed to the oil phase.
2. *Synthetic emulsifiers* act differently from natural emulsifiers. They should be more aptly named as 'surfactants' since they are smaller molecules as compared to proteins, and they aid in emulsification by helping in fat dispersion but they do not stabilize the emulsion. Glycerol monostearate (GMS) is an example of monoglyceride added to the bread dough for dispersing fat droplets. SPANS and TWEENS are other two synthetic emulsifiers which form w/o and o/w emulsions, respectively.

We learnt about the formulation and stability of emulsions in this chapter. An emulsion is a dispersion of one immiscible liquid in another (usually oil and water) in the form of droplets. It consists of a dispersed phase, a continuous phase and an emulsifier. Nature of emulsifier and its concentration are crucial in determining the stability of an emulsion. Emulsifier adsorbs at interface, reduces surface tension and forms a continuous and stable viscoelastic interfacial layer to prevent the droplets from coalescing. Stability of an emulsion can be enhanced by increasing the viscosity of continuous phase by adding gums. Most food emulsions are o/w emulsions, e.g., milk, mayonnaise, cream and ice cream; whereas, some are w/o emulsions, e.g., butter and margarine. Both natural and synthetic emulsifiers can be used for stabilizing the emulsions.

Know Your Progress

True or False Questions

1. Viscosity of an emulsion is usually decided by the viscosity of the continuous phase.
2. An emulsion is in a thermodynamically stable state.
3. During emulsification, the emulsifier gets adsorbed at the surface of droplets and the interfacial tension increases.
4. Coalescence is a reversible process resulting in reunion of small droplets to form large droplets.
5. Dispersion of larger droplets in the continuous phase makes the emulsion optically opaque.

Very Short Answer Questions

1. Define the following terms:
   (a) Emulsion
   (b) Flocculation
   (c) Emulsifier
   (d) Creaming
   (e) Stability of an emulsion

Short Answer Questions

1. Differentiate between a surfactant and an emulsifier.
2. Classify emulsifiers based on their function.
3. Describe briefly the theory of emulsification for an oil-in-water emulsion.
4. Discuss the factors affecting the stability of emulsions.
5. State the properties of emulsions.

Long Answer Questions

1. What are the different types of emulsions? Describe the theory of emulsification with special reference to mayonnaise preparation.
2. Discuss the role of emulsifying agents in food preparation.

Beyond the Text

- Pick up five salad dressing (e.g. Thousand Island, Caeser, Ranch, Vinaigrett and Italian dressing). Read the labels to note down all the ingredients. Identify the substances which stabilize these emulsions.

Chapter 8

Colloids

Chapter Objectives

After reading this chapter, the reader should be able to:

- define and differentiate between solution, suspension and colloid.
- identify the colloidal systems applicable to foods.
- understand the properties of colloidal systems.
- learn the application of sols, gels and foams in food preparation.

Foods are complex in nature, and they contain many constituents (including nutrients) as particles of varying size. The particle size of sugar dispersed in a solution is so small that it is not visible. On the other hand, seeds suspended in tomato juice are large enough to be seen through the naked eye. Between the solution and the suspension lies a third category of mixture known as colloid which has particles of intermediate size. Examples of colloidal mixtures can be found among natural foods (egg white and milk) and prepared foods (custard and jam). Knowledge about the properties and applicatio of different colloidal systems is necessary to improve and maintain their quality in food processing.

THINK TANK

Size of dispersed particles in foods

Consider the following foods:
(i) Sugar solution, (ii) Milk and (iii) Tomoto juice

How are these food mixtures different in terms of their particle size?

8.1 Definitions: Solution, Suspension and Colloid

Nutrients (proteins, carbohydrates, fats, minerals and vitamins) are present in foods along with oth constituents (phytochemicals, enzymes, organic acids, pigments and flavouring substances). In most case water forms the continuous phase in which these substances are dispersed as particles. Based on the size dispersed particles, the dispersions can be classified as solution, colloid or suspension (Figure 8.1).

True Solution: Solutions are homogenous and transparent in nature and are present in nearly all food Compounds such as sugars and amino acids have low molecular weight and form solutions. In a true sol tion (e.g. sugar syrup), the solute or the dissolved substance (sugar) is uniformly distributed throughout th solvent or the medium (water) in which it is dissolved. The particles in a solution are not visible as their si is less than 1 mμ (10^{-6} mm).

Colloid: The size of colloidal particles is intermediate; they are small enough to remain dispersed but are lar enough to scatter light and not get dissolved. Their size lies between 1 mμ and 100 mμ (10^{-6}mm to 10^{-4} mm

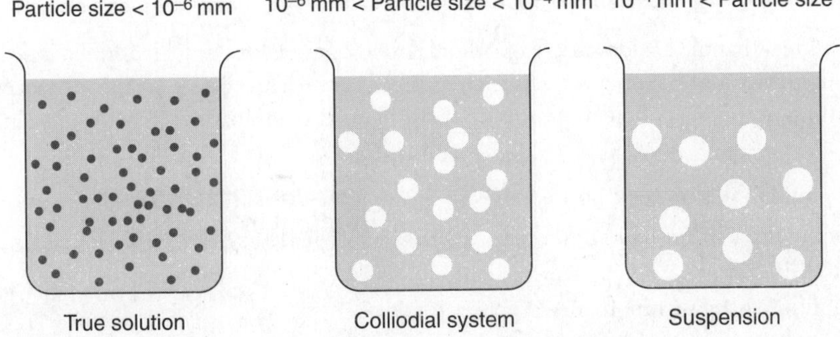

Figure 8.1 Diagrammatic representation of particle size in solution, colloid and suspension

Table 8.1 Difference between solution, colloid and suspension

Property	True Solution	Colloidal System	Suspension
Particle size	<1 mμ (10^{-6} mm)	Between 1 mμ and 100 mμ (10^{-6} mm and 10^{4} mm)	>100 mμ (10^{-4} mm)
Particle subdivision	Molecules and ions	Molecular aggregates	Molecular aggregates
Nature	Homogenous	Heterogeneous	Heterogeneous
Visibility of particles	Particles are not visible to naked eye	Particles are not visible to naked eye but can be seen through an ultramicroscope	Particles are visible to naked eye
Tyndall effect	Do not exhibit Tyndall effect	Exhibit Tyndall effect	May or may not exhibit Tyndall effect
Optical property	Transparent	Translucent	Opaque
Filterability	Particles diffuse through filter paper and parchment paper	Particles diffuse through filter paper but not through parchment membrane	Particles diffuse neither through filter paper nor parchment membrane

Compounds such as starch, gum and protein have high molecular weight and form colloids. Major differences between true solution, colloid and suspension are summarized in Table 8.1.

Suspension: It is a heterogeneous mixture in which the particles measure more than 100 mμ (10^{-4} mm) in size, are visible to the eye and tend to settle down on standing. For example, seeds suspended in tomato juice gradually settle when they are left to stand.

A colloidal system is heterogeneous and has two separate phases: dispersed and continuous. The *dispersed phase* (discontinuous phase) is the material which exists in the colloidal condition, whereas the *continuous phase* (dispersion medium) forms the base of the colloidal system. Usually, water or oil forms the continuous phase, whereas dispersed phase can be a solid, liquid or gas. Skimmed milk is a colloidal system in which water forms the continuous phase, whereas milk protein (casein) exists as the dispersed phase.

8.2 Classification of Colloidal Systems

Culinary professionals handle a wide variety of colloidal foods and beverages such as milk, curds, eggs, salad dressings, soups, gravies and bakery products such as breads and cakes. Based on the physical state of dispersed and continuous phase, colloidal systems can exist in eight combinations of which five are applicable to foods.

1. Solid in liquid (e.g. skimmed milk)
2. Liquid in another immiscible liquid (e.g. French dressing)
3. Gas in liquid (e.g. egg white foam)
4. Gas in solid (e.g. bread)
5. Solid in gas (e.g. smoking of fish)

DID YOU KNOW?

The word 'colloid' is derived from the Greek word *kolla* meaning 'glue'. Thomas Graham is known as father of colloidal chemistry as he recognized the existence of colloidal state in 1861. He discovered that colloids do not pass through the parchment membrane; whereas crystalloids readily do so.

Table 8.2 presents five types of colloidal systems found in foods. Skimmed milk and starch suspension are the examples of diphasic colloids which have only one phase dispersed in water. Some foods consist of two or three phases dispersed in a continuous phase. Mayonnaise and whipped cream have protein particles, oil droplets and air bubbles dispersed in water and are the examples of tetraphasic systems (also be termed as 'emulsion-sol-foam').

Table 8.2 Types of colloidal systems applicable to food

Dispersed Phase	Continuous Phase	Name of Dispersion	Examples
Solid	Liquid	Sol	Whey, skimmed milk, starch suspension, gravies, soups, kheer
		Gel	Jam, jelly, gelatin gels, starch gels
Liquid	Liquid	Emulsion	Whole milk, mayonnaise, French dressing, margarine, butter, cream
Gas	Liquid	Foam	Whipped cream, whipped egg white
Gas	Solid	Solid foam	Cake, bread, baked meringue, idli, candy floss
Solid	Gas	Solid aerosol	Smoked fish or meat

Colloids may be water-loving or water-hating. Egg white is a water-loving colloid which transforms to water-hating colloid on coagulation.

1. *Hydrophilic colloids* are water-loving colloids which have high affinity for water; for example, milk, egg white, pectin, starch and gelatin.
2. *Hydrophobic colloids* are water-hating colloids with low affinity for water. They are mainly aqueous dispersions of inorganic substances which are not common in foods.

A colloidal system in which the particles are dispersed in water is called a *hydrocolloid,* e.g. agar-agar and carrageenan derived from seaweed, gelatin from bovine animals and pectin from apple pomace. They find many uses in food preparation such as improving viscosity in sauces and fat replacement in emulsions.

GOOD TO KNOW

Uses of hydrocolloids for the chef

Generally, a chef uses kitchen ingredients such as flour, starch, cream and/or egg yolk to achieve the desired texture in a sauce. However, in recent years, food science has made way in to some of the distinguished kitchens of the world. One example is the use of hydrocolloids (modified starch, xanthun gum, cellulose gum, pectin and gellan gum) by innovative chefs to manipulate the traditional texture of sauces. Since hydrocolloids can mimic fatty mouthfeel and perception, they can also be used for fat replacement.

8.3 Properties of Colloidal Systems

Colloids exhibit some distinct properties which distinguish them from solutions. Many of these properties are useful in food preparation.

1. **Tyndall Effect:** When a strong beam of light is passed through a colloidal dispersion, the particles are not visible but the beam leaves a bright distinct path. This phenomenon is known as Tyndall effect and can be attributed to scattering or diffusing of light rays from the surface of colloidal particles due to deflection (Figure 8.2). Tyndall effect is more obvious in hydrophilic colloids.

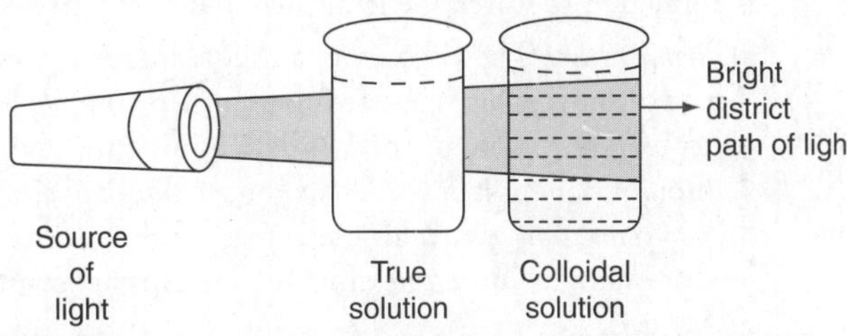

Figure 8.2 Tyndall effect observed on passing a strong beam of light through a colloidal solution

2. **Brownian Movement:** Particles in a colloid exhibit continuous and random zigzag motion which is known as Brownian movement. It results from the collision of dispersed particles with the molecules of dispersion medium. Smaller the size of the particle, more rigorous is this movement.
3. **Electric Charge:** The particles in a colloid selectively absorb ions and become electrically charged. All these particles acquire the same charge which is either negative (–) or positive (+). These similar-charged particles repel each other, do not clump together and remain in suspension. Addition of acid can neutralize the electric charge on the dispersed particles and lead to the precipitation of colloidal dispersion. For example, addition of acid to milk neutralizes the electric charge on dispersed casein and leads to curdling.
4. **Adsorption:** Colloidal particles exhibit adsorption, that is, they grip on their surface molecules of gases, vapour and other matter which they come in contact with (Figure 8.3). This property finds use in many food preparations. In ice creams prepared with added fruit, addition of gelatin to the milk ensures a smooth product. Similarly, the use of white sauce instead of milk alone in tomato soup helps in preventing curdling of casein.

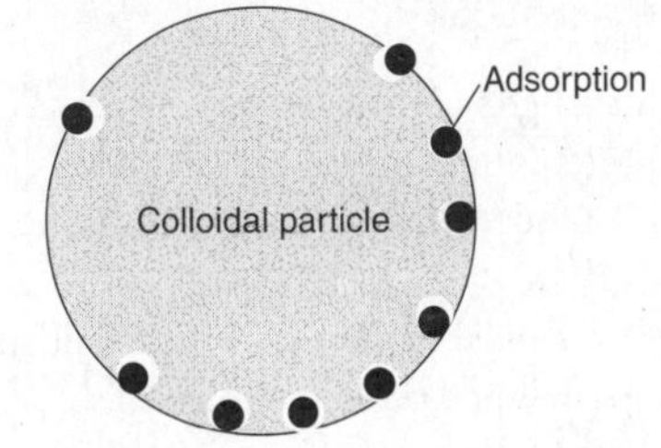

Figure 8.3 Diagrammatic representation of adsorption of matter by colloidal particles

5. **Osmotic Pressure:** Since colloidal particles do not pass through plant or animal membranes, they do not participate in osmosis. In other words, they do not exert any osmotic pressure.
6. **Imbibition:** Colloids can pick up water and swell when they come in contact with liquids. This property, known as imbibition, usually results in swelling of the colloid. When a starch sol is heated, imbibition of water into the granules takes place and they swell. However, a gel which has a fairly rigid structure may not swell after the uptake of water. For example, boiled rice can imbibe liquid from gravy without swelling.
7. **Flow Properties:** Sols are free-flowing at room temperature, whereas gels are firm (refer to Section 8.4). The viscosity (resistance to flow) of hydrophilic sols such as starch solution is greater as compared to water. Generally, colloidal viscosity is affected by the following factors:
   (a) *Temperature:* The viscosity of a colloid increases when temperature is lowered. A jam boiling at the end point is still a free-flowing sol; it sets to form a solid gel only when it is cooled down to room temperature. One method of determining the end point of jam is through cold plate test where a drop of boiling jam is dropped on a cold plate to ascertain sol-gel transformation.
   (b) *Concentration of colloid:* Increasing the amount of colloidal solids can also increase the viscosity. For example, more egg white is added to the formulation to achieve thicker custard.
8. **Effect of Colloidal Particles on Freezing and Boiling Point:** In contrast to the solute particles of solution, colloidal particles have negligible effect on the freezing or boiling points of the food. In other words, they do not affect the boiling or freezing temperature of a colloidal system.

> **GOOD TO KNOW**
>
> **Useful Tip**
>
> If a soup tastes too salty, you can make it palatable by adding some egg white which on heating will adsorb salt on to its surface. Allow the egg white with adsorbed salt to settle at the bottom of the container from where you can remove it subsequently.

These properties are specific to colloidal dispersions and are not exhibited by solutions.

8.4 Application of Colloidal Systems in Food Preparation

Among the colloidal systems, the emulsions have already been discussed in Chapter 7. We will now learn about the properties of sols, gels and foams along with their application in food preparation.

8.4.1 Sols

Sols and gels are usually reversible colloidal dispersions of proteins or complex carbohydrates (polysaccharides). Sol is a fluid, whereas gel is a semi-solid colloidal system (Box 8.1).

A colloidal system in which solid particles are dispersed in liquid phase is termed as 'sol', e.g. gelatin mixed in water and starch suspension. Skimmed milk is also a sol, because the milk protein is dispersed in water. Being colloids, sols exhibit some characteristic properties.

> **THINK TANK**
>
> **Why do some free flowing liquids transform to a semi-solid state?**
>
> 1. Egg, sugar and milk when baked form a semi-solid custard.
> 2. Fruit pulp, pectin, acid and sugar when heated and cooled form a semi-solid jam.

Box 8.1 Definition of Sol and Gel

Sol is a colloidal system in which solid particles are dispersed in liquid phase. It is a free-flowing liquid which has clear or opaque appearance and exhibits Tyndall effect and Brownian motion.

Gel is a colloidal system in which solid forms a three-dimensional cross-linked network which holds the liquid. It contains mostly water, yet it is firm and does not flow.

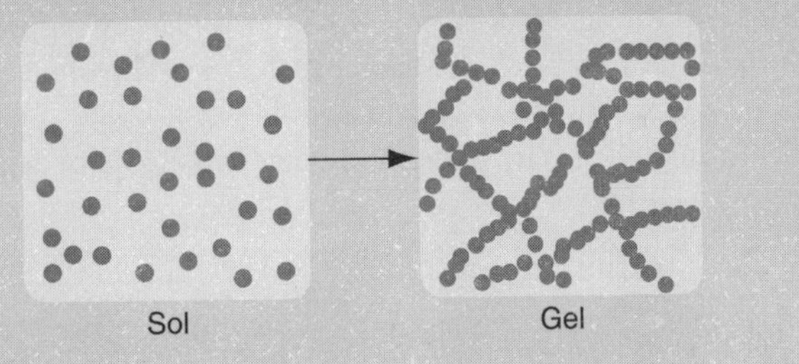

Figure 8.5 Diagrammatic representation of sol–gel transformation

1. Sols are liquids which are pourable at room temperature.
2. They have an opaque or clear appearance.
3. Since they do not exhibit rigidity of form, they are free flowing.
4. The dispersed particles in a sol scatter and polarize light (Tyndall effect).
5. Colloidal particles in a sol also exhibit rapid irregular movement (Brownian movement).
6. Sol particles have similar charge on their surface, so they do not coalesce and they repel each other.

The viscosity of a sol can be increased by increasing the concentration of dispersed solid, by lowering the temperature or by aging. Sols having higher concentration of dispersed solids get transformed spontaneously to a gel which is firm at room temperature and does not flow from a spoon.

8.4.2 Gels

During the sol-gel transformation, the solid (dispersed phase) forms a three-dimensional network in whose interstices the liquid (continuous medium) is retained (Figure 8.5). As a result, the sol loses its fluidity and becomes firm at room temperature.

Based on the nature of gelling agent present in them, gels can broadly be grouped into two types: protein gels and polysaccharide gels (Box 8.2).

Box 8.2 Custard (a protein gel) and Jam (a polysaccharide gel)

Custard is prepared by using egg, sugar and milk as the main ingredients; salt and vanilla are added for flavour. Egg, sugar and salt are blended and milk (scalded at 85°C) is added. This mixture is then baked into a moderate oven at 177°C. During baking, egg proteins (mainly ovalbumin) are denatured by heat and form a network in which milk fluid is held. As a result, the custard sets to form a gel.

Jam is prepared by heating fruit pulp, pectin, acid and sugar to 105°C so that the pectin–water equilibrium becomes unstable. As a result, pectin forms a network of fibrils in the interspaces of which sugar solution is held. The resultant product sets as a gel.

Protein Gels contain gelling substances which are protein in nature such as casein in curd, egg albumin in custards and gluten in flour-based batters and doughs (refer to sub-section 4.3.4 in Chapter 4). There are diverse methods by which protein present in different foods results in gel formation. The milk protein (casein) gels on prolonged heating or by the action of acid or enzymes. The egg white gets denatured by heat and subsequently leads to gelation; whereas, the gluten forms a gel in the dough when flour is kneaded.

Polysaccharide Gels comprise of gelling substances which fall under the category of polysaccharides such as starch in white sauce and pectin in jams and jellies. Starch, pectin and algal polysaccharides such as agar, alginate and carrageenan form gels with a three-dimensional network in which water molecules are held (refer to sub-section 2.2.1 in Chapter 2). Special skills are required to form pectin gels which require a fragile balance of pectin, water, sugar and acid.

Gel formation and its strength, rigidity and elasticity depend on the following factors:

Type and concentration of gelling agent: Proteins form gel at a concentration of 10 per cent or less. On the other hand, pectin and gelatin form a gel at a concentration of 1 per cent or lower.

Sugar content: Sugar is hygroscopic; so it competes with water in the continuous phase. If it is added to a starch sol at the beginning of cooking, gel formation is delayed and the firmness of the cooked and cooled product is lowered.

Salt content: Depending on its concentration, salt can affect gel formation either favourably or unfavourably.

pH: During sol-gel formation, pH plays an important role. Addition of acid to starch results in the formation of dextrins. Consequently, the viscosity of starch paste and the firmness of cooled product are decreased. In the case of jams and jellies, pectin leads to gel formation at a pH of 3.0.

Temperature: A decrease in temperature facilitates gel formation by increasing the viscosity of a sol. For example, gelatin gel is kept in a freezer to set.

Sometimes, weeping is observed in baked custards, frozen baked products and fruit jellies. This spontaneous exudation of fluid from a gel on aging which results in contraction of the gel and affects its rigidity adversely is termed as 'Syneresis'. Faulty gelation or frequent freeze-thaw cycles can aggravate syneresis (refer to seb-section 2.2.3 in Chapter 2).

8.4.3 Foam

A colloidal system where gas bubbles are dispersed in a liquid is known as foam (Box 8.3). It is formed by mechanical agitation of a liquid so that atmospheric air is incorporated in the form of small bubbles which are separated from each other by liquid walls. Examples of food foams include whipped cream, ice cream, egg white; cake batter and froth on the beer (Box 8.4).

Box 8.3 Definition of Foam

Foam is a colloid in which a large proportion of gas in the form of bubbles is dispersed in a liquid, solid or gel. The diameter of gas bubbles ranges from 1μm to few centimetres, but the thickness of foam walls (also known as film or lamellae) is usually in colloidal size range.

Food foams have the following characteristics:

1. Food foams are opaque in appearance.
2. They contain large amount of entrapped gas as bubbles.
3. The diameter of gas bubbles ranges from 1μm to few centimetres.
4. They have large surface area between the dispersed gaseous phase and the continuous liquid phase.
5. Walls of foams (known as film or lamellae) are rigid and elastic and they reflect light.
6. Depending on the size of gas bubbles and the thickness of liquid walls, foams are categorized as light or dense.

Foam formation can be facilitated if the liquid used has low surface tension, low vapour pressure and is elastic enough to facilitate the formation of a continuous film around the air bubbles (refer to sub-section 4.3.5

Box 8.4 Egg Foams and Meringues

Egg foams are used for preparing products such as meringues, puffy omelette and soufflé. When egg whites are beaten, they incorporate air and become foamy. On further beating, the foam gets thicker, finer and whiter and begins to form soft peaks when the beater is withdrawn. On more beating, the peaks become stiff (refer to Figure 4.7 in Chapter 4). Care should be taken to incorporate other ingredients as soon as it is formed; otherwise the foam loses its elasticity.

Meringues are egg white foams with sugar incorporated into them. They are of two types: soft meringues which are used as toppings for lemon and cream pies, and hard meringues which are used as confectionery. For preparing hard meringues, egg whites are beaten to soft peak stage and sugar is added; beating is carried on till the stiff peak stage. The mixture is then dropped on a greased baking tray with the help of a spoon. It is baked at 120°C. Hard meringue is delicate brown in colour, puffy in appearance and crisp, but tender in texture.

in Chapter 4). Prior to incorporation of gas, the continuous phase must contain a *foaming agent* such as a surface-active lipids, proteins or cellulose derivatives which facilitate foam formation or enhance its stability by preventing the coalescence of bubbles (refer to sub-section 4.3.5 in Chapter 4). Factors affecting foam formation are as follows:

1. *Salt* decreases film strength and thereby reduces foam stability.
2. *Sugar* increases viscosity and thereby enhances foam stability. This property of sugar is applied in the preparation of meringues.
3. *Acids* such as tartaric acid and citric acid (present in lemon) decrease the time of foam formation and enhance foam stability.
4. *Water* makes foams with large volume.

Foam stability can be enhanced by different ways such as increasing the elasticity of liquid forming the continuous phase or by adding hydrophobic particulate matter (spices, cocoa). Sugar increases viscosity, whereas gums and proteins have high water-binding capacity and thereby contribute to foam stability. Addition of foam stabilizers is another option used in food industry for stabilizing ice cream foams.

Heat stability of foams is of considerable importance for achieving good volume bakery products such as cakes, breads and meringues as well as Indian products such as idli, bhatura and dhokla. These products have a porous solid structure and are examples of solid foams formed by evaporation of water on heating liquid/semi-solid foams.

Antifoaming agents are sometimes used to prevent undesirable foam formation which can otherwise decrease the processing rate and also leads to product loss. For example, edible oils are added to prevent foaming when jam/jelly is boiled.

In this chapter, we studied about food colloidal systems particularly sols, gels and foams. We learnt that particles in a colloid are larger than those in a solution but smaller than those present in a suspension. Colloidal systems can be distinguished from other dispersions on the basis of their characteristic properties such as Tyndall effect, Brownian motion, electric charge, adsorption, imbibition and flow properties. A colloid in which solid particles are dispersed in liquid phase is known as sol; when it forms a three-dimensional cross-linked network that hold the liquid, it is termed as 'gel'; whereas foam is a colloid in which gas bubbles are dispersed in a liquid, solid or gel. Formation of sols, gels and foams as well as their properties, stability and application in food preparation were deliberated upon with suitable examples.

Know Your Progress

True or False Questions

1. The spontaneous exudation of fluid from baked custard is an example of *syneresis* of gel.
2. Absorption is a property of colloids where particles grip on their surface the molecules of other matter they come in contact with.
3. Colloidal particles increase the freezing or boiling points of the food.
4. The viscosity of a colloid increases with increase in amount of colloidal solids.
5. Colloidal particles participate in osmosis and exert osmotic pressure.

Very Short Answer Questions

1. Define the following terms
   (a) Colloid
   (b) True solution
   (c) Tyndall effect
   (d) Imbibition
   (e) Brownian movement

Short Answer Questions

1. Differentiate between solution, colloid and suspension.
2. How is a sol different from a gel?
3. What are the factors affecting foam formation and stability?
4. Discuss the factors affecting strength, rigidity and elasticity of a gel.
5. What are sols? Discuss their application in food preparation.

Long Answer Questions

1. Describe the different types of colloidal systems applicable to foods.
2. Discuss the properties of colloids with suitable examples.

Beyond the Text

- Identify 15 foods which are colloids and group them as sols, gel or foams. State the dispersed and continuous phase for each food with particular emphasis on the number of dispersed phases.

Chapter 9

Flavours

Chapter Objectives

After reading this chapter, the reader should be able to:

- define and understand different aspects of flavour.
- know the basics of flavour development in tea, coffee and wine.
- understand the flavour-related changes occurring in meat, fish and poultry during cooking/processing and storage.
- identify major volatile compounds present in spices.

We learnt in Chapter 6 that flavour is an important attribute for food quality. Whatsoever may be the colour, appearance, nutritive value or wholesomeness of food, its acceptance/rejection depends predominantly on its flavour. Being a non-static characteristic, flavour undergoes changes during post-harvest and post-mortem handling of plant and animal foods, and also during their cooking, processing and storage. Various adjectives are used to describe flavours ('fruity', 'floral', 'spicy', 'nutty', 'caramelized' and 'roasted') and off-flavours ('rotten', 'rancid', 'stale' and 'fishy'). An understanding of flavour of various foods and the related changes during processing would go a long way to equip a chef to satisfy the palate of consumers. Although this chapter discusses only the flavours of tea, coffee, wine, meat, fish, poultry and spices, flavours are an important constituent of fruits and vegetables. Cereals and pulses are comparatively bland in terms of flavour.

9.1 Definition of Flavour

Each food is characterized by its own flavour which includes taste, odour and mouthfeel. The definition of flavour emphasizes on the integrated psychological response produced when taste and aroma receptors are stimulated (Box 9.1).

Box 9.1 Definition of Flavour

Flavour is defined as the sensation produced by a material taken in the mouth, perceived principally by the senses of taste and smell, and also by the general pain, tactile and temperature receptors in the mouth. Flavour also denotes the sum of the characteristics of the material which produces that sensation.

Source: Hall 1968.

Taste: An important component of flavour is the taste. Most foods have a blend of two or more of the five major tastes: sweet, salty, sour, bitter and umami. These tastes are perceived when water soluble substances stimulate the taste buds (refer to sub-section 6.2.2 of Chapter 6).

Umami is a savoury taste which is important for foods such as meat, cheese and soya sauce. Glutamic acid present in monosodium glutamate has long been recognized as a flavour enhancer but is now included as a primary taste of umami. It enhances the flavour of many food products such as poultry, soups and vegetables.

Odour: The aroma of a food is perceived when volatile odorous compounds come in contact with receptors in the epithelium of the olfactory organ. It is a vital attribute in beverages such as tea, coffee and wines; foods such as meat, fish and poultry; and spices.

Mouthfeel: In the composite sensation of flavour, mouthfeel has an important contribution. Tongue, mouth and jaw are the organs which assess the form, shape and feel of a food. Chewiness is important for baked meringues, whereas astringency is exhibited by red wines and tea.

Among other factors, temperature and texture in terms of smoothness, granularity, roughness and viscosity affect the perception of flavour.

9.2 Descriptions of Food Flavours

Food flavours can be categorized as *primary flavours* present naturally in food, e.g., fruits, vegetables and spices or *secondary flavours* developed during processing–roasting, boiling, frying, grilling, baking and fermentation in foods and beverages such as tea, coffee and wine. Often, chemical flavourings are intentionally added to enhance the aroma of food (refer to Chapter 25). Before discussing flavour, let us understand what hinders with good aroma.

Foods' taint and off-flavours hinder or interfere with the natural food flavour. A food taint generally arises due to chemical contamination, whereas an off-flavour results due to the products of chemical or microbial spoilage of food (Table 9.1).

Table 9.1 Possible causes and compounds responsible for off-flavours

Off-flavour	Possible Cause	Major Compound
Acrid	Microbial growth	Acrolein
Cardboard	Autoxidation of fats/oils	2,4-Nonadienal
Fruity	Degradation product of PET packaging	Acetaldehyde
Kerosene	Degradation of sorbate by fungi	1,3-Pentadiene
Woody	Drain cleaners/moth proofing agents	1,4-Dichlorobenzene
Plastic	Migration from polystyrene containers	Styrene
Smoky	Microbial degradation of vanillin/lignin	Guaiacol

Recent advances in analytical techniques such as chromatography, nuclear magnetic resonance and infrared spectroscopy have led to better understanding of chemical nature of food flavours. Majority of these are minor non-nutritive constituents, usually present in low concentration and are generally heat labile and unstable. Fundamental knowledge about these components can help food processors and chefs to optimize flavour development.

9.2.1 Tea

Tea is known for its pleasant flavour and stimulating effect. Tea and tea blends are obtained by processing hand-picked young tender shoots of tea shrubs: the white-haired bud and two adjacent young leaves. Younger the leaves, better is the flavour of tea. The country of origin (India, China, Sri Lanka and Japan)

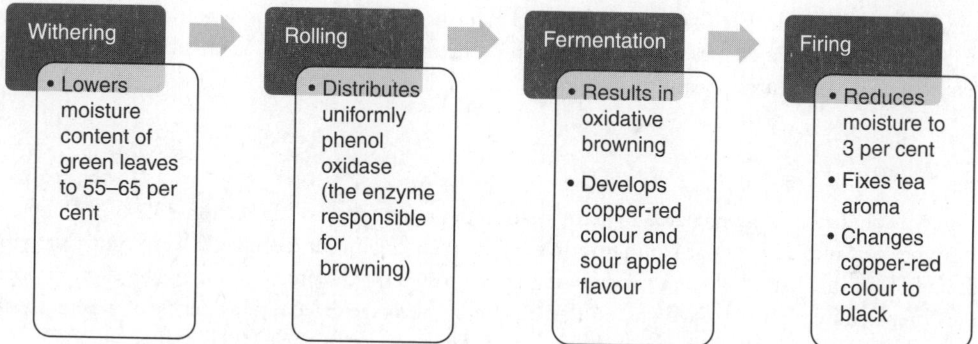

Figure 9.1 Stages in processing of black tea

and processing also influence its aroma. More than 450 flavour constituents have been identified in tea which owe its distinct flavour to lactones, aldehydes, alcohols, acids, pyridines, caffeine and tannins.

Processing for black tea involves withering, rolling, fermentation and firing (Figure 9.1). During fermentation, Strecker degradation reactions of amino acids with oxidized flavonols lead to the development of aroma. Linalool contributes to 'floral' aroma, whereas furanones exhibit 'caramel' or 'seasoning-like' aroma in black tea.

Green tea is a very light, clear and slightly bitter beverage. Its processing involves steaming of plucked leaves to inactivate the oxidative enzymes but withering and fermentation are omitted. In case of yellow tea, only fermentation is omitted; whereas red tea is partially fermented. Aroma substances produced as a result of oxidation of unsaturated fatty acids in green tea contribute to its 'hay-like' flavour. Green tea contains more catechins which are antioxidants and have health benefits.

DID YOU KNOW?

Consumers' power gets back the original flavour of tea blend!

Earl Grey of Twinings is a tea blend with characteristic flavour and aroma of oil extract of bergamot orange. When its flavour was changed in mid-2011, the Earl Grey lovers in UK revolted and forced Twinings to start the sale of the original tea blend.

Source: Lusher, 2011

9.2.2 Coffee

Coffee is a beverage prepared from coffee beans which are the seeds of crimson fruits. These seeds may be raw, roasted, whole or ground. The process of roasting green coffee beans results in brown colour and delightful flavour associated with roasted coffee. Maltol, a flavour enhancer, is formed when coffee is roasted. The brown compounds present in the soluble fraction of roasted coffee result from Maillard reaction or caramelization of carbohydrates (refer to Chapter 10).

Caffeine, present in coffee ranging from 0.8 to 4 per cent, is mildly bitter in taste. Secondary products formed due to the decomposition of carbohydrate and protein mixture also contribute to the bitter flavour of roasted coffee. Volatile fraction of this beverage has a rather complex composition, and more than 650 flavour compounds have been identified. These include hydrocarbons, aldehydes, ketones, phenols, furans, pyrazines, pyrrols and thiophenes. Presence of large amount of furans and pyrrols coupled with oxazoles, thiazoles and phenols contribute to the distinct flavour of coffee.

The flavour of coffee should be 'rich' and 'exotic' but it may turn 'beany' or 'baggy' due to insufficient roasting; or it may be 'baked', 'bready' or 'bitter' due to over roasting of coffee beans. Coffee rapidly loses its fresh flavour and tends to become 'lifeless', 'bland' or 'flat' due to its unstable aroma. Optimum flavour is achieved at mildly acidic pH (4.9–5.2). If the pH is lower, coffee tastes sour; whereas the flavour is flat at higher pH. In case of canned coffee drinks, pH should be maintained as low as possible to prevent the loss of fresh 'roasty' flavour during heat processing.

9.2.3 Wine

Wine is an alcoholic beverage prepared by fermentation of grapes (refer to sub-section 23.2.6 in Chapter 23). Various types of wines derive their name from regions of their origin; for example, Champagne originated in France and Sherry in Spain. Different varieties of grapes are used for the production of wines; for example, red wines are prepared from purple or red grapes. These alcoholic beverages have varying degree of sweetness, sugar–acid ratio, flavour, colour, astringency and effervescence .The proportion of alcohol is also different, i.e., 10–14 per cent in table wines and sparkling wine; 14 per cent in dry wine and 14–21 per cent in sweet dessert wine.

Around 650 volatile compounds have been identified in wines. Wine aroma can be attributed to a number of chemical substances such as esters, aldehydes, terpenes, aldehydes, alcohols, norisoprenoids and sulphur compounds. Some of these originate from grapes, whereas others develop during fermentation or during ageing. In contrast to the still wines, the sparkling wines such as Champagne undergo secondary fermentation after bottling and therefore contain excess carbon dioxide produced during fermentation. Wine aroma, an extremely important attribute in sensory evaluation, may be tested by using a Wine Aroma Wheel.

9.2.4 Meat, Fish and Poultry

Muscle foods have mild flavour in the raw fresh state. Their aroma actually develops during cooking and processing when the proteins, lipids and carbohydrates produce significant flavour precursors. Meat flavour is also affected by many other factors such as the species of animal and its diet; enzymatic activity and ageing after slaughter; addition of ingredients (for marination and flavour enhancement); and pre- and post-processing storage conditions (oxygen availability, aging time, temperature, humidity and gas atmosphere). Microbial growth during storage can also lead to the development of undesirable volatile compounds which generate off-odours signalling the end of shelf life (refer to sub-section 22.4.3 in Chapter 22).

Meat: Raw meat has mild aroma but it contains a pool of compounds which act as aroma and flavour precursors. For example, glycoprotein and inosinic acid (inosine-5-monophosphate) develop characteristic meat flavour on heating. A large number of non-volatile compounds such as amino acids, peptides, nucleotides, organic acids and sugar are also present in meat. During cooking, reactions involving amino acids and sugar form volatile compounds which enhance flavour of the meat. When lipid components get involved in Maillard reaction, the species-specific flavours of meat develop (refer to Chapter 10).

> **GOOD TO KNOW**
>
> **Marination of meat and flavour improvement**
>
> Marinades tenderize meat and also add to its flavour. They consist of three main components: acid (lemon juice, vinegar or curds), oil and herbs/spices. The acid component denatures the meat proteins partially to open up tunnels through which the flavour can enter into the meat structure. The meat proteins also have flavour-binding property.

Lean beef, pork and lamb have similar flavours due to the presence of similar amino acids and sugars; nevertheless, the fats differ in composition and impart slight variations. Volatile compounds identified in meat extract include acetic acid, propionic acid, dimethyl sulphide, ammonia, acetone, methyl mercaptan and carbonyls.

GOOD TO KNOW

Warmed-over Flavour in Meat

Lipid oxidation of fresh, cooked, stored (refrigerated or frozen) or reheated meat can adversely affect their odour and flavour to produce warmed-over flavour (WOF) which is also referred to as 'stale', 'cardboard-like' or 'rancid' flavour. Such oxidation reactions can be triggered by light, heat, metals (iron and copper) and myoglobin (meat pigment which contains iron) as well as some ingredients (salt).

Flavour of Cured Meats

Nitrites prevent oxidative rancidity of fat and the production of warmed-over flavour (WOF). They play an important role in the development of distinct flavour for each of the cured meat products such as ham, salami and bacon. Added flavouring agents and processes such as smoking also contribute to the flavour.

Fish: The sweetness of fish may be due to proline–an amino acid present in it in abundance; whereas ribose and glucose contribute to its flavour. Hydrogen sulphide, methyl mercaptan, dimethyl sulphide, pyridine, piperidine and pyrrolidine contribute to fish aroma, whereas inosine-5-monophosphate acts as a flavour enhancer. Umami component arises from the synergism between inosine monophosphate and free amino acid. Tuna and mackerel species of fish contain a large amount of free histidine which possibly contributes to their distinct flavour. Sugars and amino acids present in fish may be involved in Maillard-type browning in canned fish. Bacterial spoilage can result in rapid loss of fresh flavour of fish. As spoilage advances, pH increases and ammonia is released from fish resulting in off-flavour. The off-flavour in canned salmon may be related to the presence of dimethyl sulphide.

Poultry: Lipid oxidation plays an important role in the development of a characteristic flavour of poultry. The carbonyls derived from linoleic and arachidonic acids contribute to the distinctive flavour of stewed chicken. The amount of accumulated α-tocopherol is much higher in chicken as compared to turkey. Therefore, carbonyls are formed to a higher extent in turkey than in chicken when they are roasted.

9.2.5 Spices

Spices are highly flavoured natural products of vegetable origin which are added in small amount to foods for zestiness, tanginess and savouriness, and for enhancing their flavour and aroma Figure 9.2. Presence of aromatic constituents coupled with hotness and pungency impart the characteristic flavour. Essential oils (containing volatile compounds) and oleoresins (also include non-volatile fats or oils) present in the spices contribute to their flavour and pungency. Many volatile compounds are present in spices: 113 in cinnamon, 95 in cloves, 146 in ginger and 122 in pepper. Major volatile compounds present in spices are listed in Table 9.2.

Figure 9.2 Spices used in Indian curries

Diverse cuisines all over the world are characterized by specific combination of spices, condiments and herbs.

Table 9.2 Major flavour-related volatile compounds present in culinary spices and their application in food preparation

Spices, Herbs and Condiments	Major Flavour-related Volatile Compounds	Application in Food Preparation
Allspice	Eugenol	Cakes, pies and puddings
Aniseed	Anethole	Indian breads
Asafoetida	Ferulic ester	Indian curries
Caraway seeds	Carvone	Sausages, meats and canned foods
Cardamom	Cineole, terpinyl acetate	Indian sweets
Cinnamon	Cinnamaldehyde, eugenol	Cakes, cookies
Clove	Eugenol	Sausages, meat gravies, pickles and desserts
Coriander seeds	Coriandrol	Curry powder and Indian curries
Cumin seeds	Cuminol	Curry powder and Indian curries
Dill	Carvone	Pickles
Fenugreek seeds	Sotolon	Sour pickles and soup powder
Ginger	Zingiberene, gingerol	Pickles, syrups and meat preparations
Nutmeg	Safrole, myristicin	Pudding and fruit pies
Oregano	Carvacrol	Pizza
Pepper	Piperine	Sandwiches, salads and curries
Peppermint	Menthol	Candies
Rosemary	Verbenone, linalool	Mediterranean cuisine
Saffron	Safranal	Indian desserts
Thyme	Thymol	Mediterranean cuisine
Turmeric	Turmerone	Curry powder
Vanilla	Vanillin	Cakes and ice creams

European cuisine uses spices sparingly, and Mediterranean dishes predominantly use thyme, sage, bay, oregano and rosemary. Thai curries have flavour of lemon grass, lime and fresh coriander leaves, whereas a wide range of spices including turmeric, coriander seeds, cumin seeds and asafoetida are added to Indian curries.

In this chapter, we learnt that flavour encompasses taste, aroma and mouthfeel of a food taken in mouth. Though flavour is present naturally in spices, it develops in tea, coffee and wine as well as in meat, fish and poultry during cooking/processing. The role of various compounds in flavour development of these foods and beverages was discussed. While planning menus or launching new products, the growing interest of consumers in different cuisines of the world should be taken into consideration. In this context, knowledge about flavour trends is necessary. Bell Flavours and Fragrances has predicted top 10 flavour trends for 2012 which include Coconut Crème, Honey Vanilla, Cupuacu, Rich Umami, Demi Glace and Paneer Cheese. While using food flavours/flavourings, due attention must be paid to food safety related issues so as to prevent any possible harm to the consumer.

Know Your Progress

True or False Questions

1. Flavour includes the senses of taste and smell but not the general pain, tactile and temperature sensation when food is taken in the mouth.
2. Glutamic acid present in monosodium glutamate has long been recognized as a flavour enhancer but it is not included as a primary taste.
3. Wine aroma can be attributed to a number of chemical substances such as esters, aldehydes, terpenes, aldehydes, alcohols, norisoprenoids and sulphur compounds.
4. Hydrogen sulphide, methyl mercaptan and dimethyl sulphide, pyridine, piperidine and pyrrolidine contribute to fish aroma, whereas inosine-5-monophosphate acts as a flavour enhancer.
5. European dishes predominantly use thyme, sage, bay, oregano and rosemary; whereas, Thai curries use lemon grass, lime and fresh coriander leaves as spices and herbs.

Very Short Answer Questions

1. Define the following terms:
   (a) Umami taste
   (b) Warmed-over flavour
2. What are the possible causes of off-flavour in fish?
3. How does pH affect the flavour of coffee?
4. What is the role of fermentation in the development of flavour in black tea?
5. In contrast to still wine, sparkling wine such as Champagne contains carbon dioxide. Why?

Short Answer Questions

1. List five off-flavours along with the possible causes.
2. How do different steps of tea processing contribute to its flavour?
3. What are the different flavour constituents of meat?
4. List five chemical compounds that contribute to the flavour of wine.
5. Write the major volatile constituents of five Indian spices.

Long Answer Questions

1. What is the role of related constituents in the flavour of meat, fish and poultry?
2. Discuss the flavour of tea, coffee and wines.

Beyond the Text

- Carry out a survey to find out the different flavoured tea and coffee available in the market.

Chapter 10

Browning

Chapter Objectives

After reading this chapter, the reader should be able to:

- differentiate between enzymatic and non-enzymatic browning.
- learn the mechanism of enzymatic browning.
- know the different pathways of non-enzymatic browning (Maillard reaction, reaction of oxidation products and caramelization).
- understand the role of metallic contamination in discolouration of food.
- associate the development of brown colour in foods with enzymatic and non-enzymatic reactions.
- apply appropriate measures for preventing undesirable browning and optimizing desirable browning during processing and storage of food.

Browning of food takes place during pre-preparation, processing and storage of foods. In some cases, it is undesirable, whereas it is beneficial in other foods. Darkening of evaporated milk or the cut surface of an apple is a change that degrades the visual appeal. On the other hand, browning of crust in bakery products enhances their delicate sensory pleasure. Catering professionals need to have adequate knowledge about the different mechanisms of browning so that they can minimize its adverse consequences and optimize the favourable ones.

Browning reactions in foods are extremely complex and follow intricate chemical pathways. They can be broadly categorized as enzymatic and non-enzymatic browning. The major difference between the two is that enzymatic browning takes place in the presence of an enzyme, whereas non-enzymatic browning does not require any enzyme for the reaction (Table 10.1).

THINK TANK

Browning Reactions in Foods

Consider the following changes in food colour:

- The cut surfaces of potatoes turn brown on coming in contact with air.
- Milk develops light brown colour during the preparation of *kheer*.
- Packaged orange juice becomes brownish on storage.
- Muscle of white fish develops brown colour during frozen storage.
- Sugar melts and turns brown when heated above 160°C.
- The use of rusted knife for cutting leads to further browning in apples.

Do you know the pathways for these changes?

Table 10.1 Examples of enzymatic and non-enzymatic browning in foods

Type of Browning	Examples of Foods Exhibiting Browning	
	Undesirable	**Desirable**
Enzymatic browning	Cut vegetables (potato, eggplant), fruits (apple, avocado) and sea food (shrimp and lobster)	Tea and coffee. Raisins, prunes, dates and figs.
Non-enzymatic browning	Evaporated milk, Orange juice	Beer, coffee, chocolate, Maple syrup, caramel, brittle. Crust of baked (bread), fried (potato chips) and roasted (meat, nuts) foods.

10.1 Enzymatic Browning

The cut surfaces of fruits, such as apples, and vegetables, such as potatoes, on coming in contact with air turn brown; this is an example of *enzymatic browning* which significantly impairs the visual appeal of these plant foods. But it is highly beneficial in the processing of tea and coffee, and also for developing colour and flavour in dried fruits such as dates, figs and raisins (Box 10.1).

Box 10.1 Enzymatic Browning, Phenolic Compounds and Polyphenoloxidase

Enzymatic browning is an oxidative reaction which has three requisites: phenolic compounds, polyphenoloxidase enzyme and oxygen.

Phenolic compounds are present in different plant foods and in some sea foods. Varying in their composition, they include chlorogenic acid (apple, pear, potato, lettuce and coffee beans), caffeic acid (apple, coffee beans, avocado and grapes), catechins (tea and cocoa), tyrosine (mushrooms, lettuce, lobster and shrimps), flavanoids (tea) and tannins (tea and grapes).

Polyphenoloxidase enzyme (also known as polyphenolase, tyrosinase or phenolase) is found in almost all plants. Its levels are particularly high in potatoes, apples, peaches, pears, cherries, tea leaves and coffee beans but it is not present in citrus fruits and melons. This enzyme contains two copper atoms and is responsible for the oxidation of phenolic compounds to dark-coloured polymers known as melanins.

In intact tissues, the cellular structure prevents the phenolase enzyme from coming in contact with the phenolic compounds. However, when the tissue is cut and exposed to air, the enzyme gets released from the cells, is activated and reacts with the phenolic compounds in the presence of oxygen to form ortho-quinones which further polymerize to brown-coloured **melanins** (Figure 10.1). The reaction is favoured

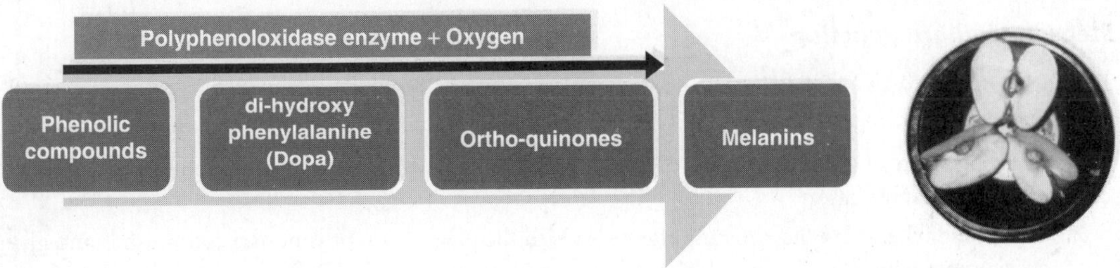

Figure 10.1 Simplified process of enzymatic browning

by warm temperature and slightly alkaline pH ranging from 5 to 7; it is accelerated in the presence of iron (e.g. cutting with a rusted knife) or copper (container or equipment). However, the presence of ascorbic acid delays enzymatic browning.

10.2 Non-enzymatic Browning

In contrast to enzyme-catalysed browning, non-enzymatic browning is generally seen in cooked foods such as roasted coffee beans and baked breads, biscuits and cakes. It is also observed in dried milk powder or packed citrus juices as an undesirable change. Three distinct mechanisms have been identified for non-enzymatic browning: Maillard reaction, reaction of oxidation products and caramelization of sugars.

10.2.1 Maillard Reaction

One of most important reaction of browning in cooked foods is the Maillard reaction which additionally contributes to flavour enhancement. It was first reported by French chemist Louis Camille Maillard in 1912.

The *Maillard reaction* is a protein–sugar interaction that follows a sequence of reactions to form a series of products (Box 10.2). It terminates with the formation of insoluble brown nitrogenous polymers known as **melanoidins**. These melanoidins have different structures, molecular weights and nitrogen content, and they contribute to diverse shades of brown colour in foods. Besides hundreds of volatile compounds are produced which contribute to the production of distinct flavours in different foods (Box 10.3).

Many cooked, fried, roasted and baked foods exhibit this type of non-enzymatic browning. Some of the Maillard reaction products exhibit antioxidant properties and thus have protective health benefits; others may lead to nutritive losses (essential amino acids) and decreased digestibility (by inhibiting enzymes), whereas products such as acrylamide are possibly carcinogenic (Figure 10.2).

Box 10.2 Maillard Reaction and Reducing Sugars

Maillard reaction is a reaction between the amino group of a protein (or peptide or amino acid) and the aldehyde group of a reducing sugar at high temperature.

Reducing sugars are simple carbohydrates that react in basic solution with oxidizing agents.

Examples of reducing sugars include fructose, glucose, maltose, galactose and lactose. Though table sugar (sucrose) is not a reducing sugar, it can hydrolyse to glucose and fructose (at high temperature and low pH) and then participate in Maillard reaction.

Box 10.3 Cooking of Meat, Maillard Reaction and Flavour Development

While cooking meat, sulphur-containing amino acids (cysteine) and reducing sugars undergo Maillard reaction to form heterocyclic compounds (furan thiols and disulphides) which are responsible for the 'meaty' flavour. Formation of some of these compounds depends on pH; this possibly explains the improved flavour of meat when pH is lowered by incorporating lemon or curd in the recipe. Artificial meat flavourings also use these compounds.

Steps in Maillard Reaction

The complex process of Maillard reaction was explained in 1952 by Hodge through a scheme which still forms the basis for its understanding. The scheme presented in Figure 10.3 includes the following reactions:

1. **Condensation** of carbonyl group of the reducing sugar with the amino compound of protein produces a Schiff base.
2. The **Amodari and Heyns rearrangement** leads to the production of Amodari compounds and glycosylamines which are the unstable intermediate products. This reaction requires high temperature (above 100°C) and is hastened at low moisture and high pH.

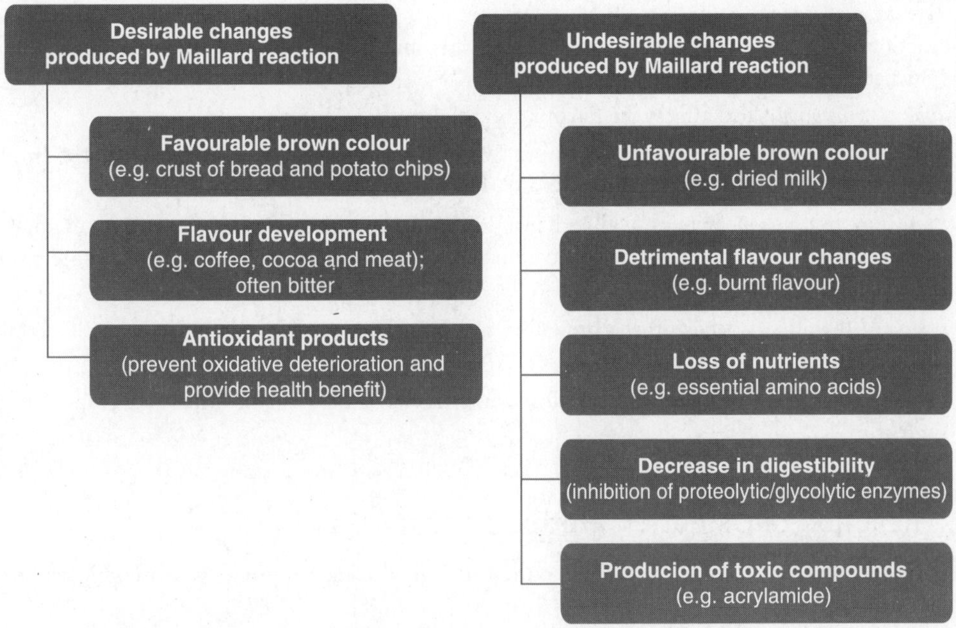

Figure 10.2 Desirable and undesirable changes during Maillard reaction

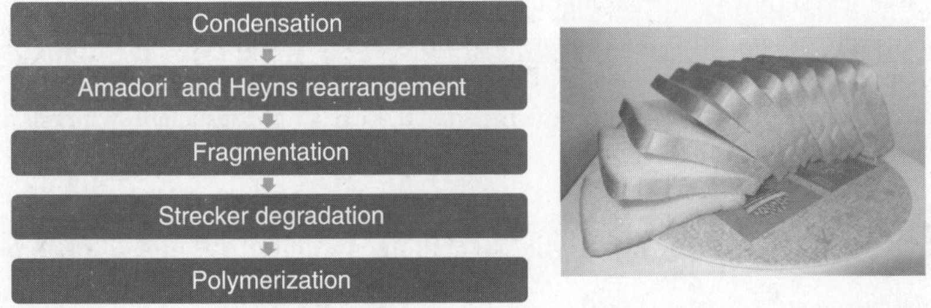

Figure 10.3 Steps in Maillard reaction

3. **Fragmentation** and **Strecker degradation** of these products lead to the formation of aldehydes, furfurals, furanones and other carbonyls which play a significant role in developing flavour associated with the Maillard reaction. Strecker degradation is a reaction between α-dicarbonyls and amino acids, and produces *Strecker aldehydes* contributing to fruity, sweet aroma (ethanol), malty aroma (methyl propanal) and flowery/honey-like aroma (2-phenyl ethanol).
4. **Polymerization** results in the formation of nitrogen-containing pigments (melanoidins) contributing to the dark brown colour. In addition, flavour compounds such as pyrazines, pyrroles, furans, oxazoles, thiazoles and thiophenes are produced.

Factors affecting Maillard reaction

Depending on the type of sugar and amino acid involved in the reaction, diverse range of brown colour (yellowish to brown) and flavour are produced. The following factors affect the rate of this reaction:

1. *Type of sugar* decides the rate of Maillard reaction. Only reducing sugars can take part in the reaction and among them, pentoses (ribose, xylose and arabinose) are more reactive than hexoses (glucose, fructose and galactose).
2. *Type of amino acid* also affects the flavour and the brown colour obtained.
3. *Temperature* increases the rate of browning two to three times for each 10°C rise. For example, the cake baked at high temperature obtains a darker brown colour.
4. *Protein concentration* favours Maillard reaction and the rate of reaction increases at high sugar concentration. *Kheer* develops brown colour due to higher concentration of milk protein and the added sugar.
5. *High pH* (alkaline) speeds up the browning reaction. For example, adding soda to chocolate cake further darkens its colour.
6. *Low moisture* promotes Maillard reaction which is more apparent in dried and evaporated foods such as milk.

10.2.2 Reaction of Oxidation Products

Non-enzymatic browning can also result due to the reaction of oxidation products of ascorbic acid or those of polyunsaturated fatty acids.

Reaction between oxidation products of ascorbic acid and proteins (or amino acids): Oxidation of ascorbic acid is an enzymatic reaction catalysed by the enzyme ascorbic oxidase. It leads to the formation of dehydro ascorbic acid (DHAA) which is highly unstable and spontaneously breaks down to form furfurals. In addition to furfurals, other oxidation products of ascorbic acid are reductones and 3-deoxy-pentosone. In the presence of amino acids, the ascorbic acid, DHAA and their oxidation products form brown pigments and decrease the nutritive value of the food products (Box 10.4). Example of such browning is the undesirable darkening of the colour in citrus juices.

Box 10.4 Browning Due to Reaction of Oxidation Products

Oxidation products of ascorbic acid (dehydro ascorbic acid, furfurals, reductones and 3-deoxy pentosone)	+	**Amino acids**	=	**Brown pigments**
Oxidation products of polyunsaturated fatty acids (aldehydes and ketones)	+	**Amino acids**	=	**Brown pigments**

Reaction between oxidation products of polyunsaturated fatty acids and proteins (or amino acids): Lipid oxidation results in the formation of hydroperoxides and its secondary products such as aldehydes and ketones (refer to Chapter 3). Many fatty foods during processing and storage exhibit browning which can be attributed to the reaction of lipid oxidation products with amino acids, amines and proteins. The cross-linkage of aldehyde/ketone with protein is probably the first step which is followed by pyrrole polymerization to produce brown-coloured polymers (Box 10.4). For example, white fish containing polyunsaturated fatty acids and high concentration of protein shows discolouration in muscle during frozen storage.

10.2.3 Caramelization

When sugar is heated at a high temperature (160°C or above) in the absence of water and amino acids, it turns brown and this sugar–sugar interaction is known as *caramelization*. This is also a type of non-enzymatic browning. The non-crystalline and water-soluble caramel so formed is bitter, pungent and less sweet in taste, whereas the colour is deep amber or golden brown. Development of caramel colour and flavour in caramel custard is a desirable change, whereas uncontrolled heating of sugar can lead to objectionable burnt colour and flavour.

Though the caramelization temperature varies for different sugars–fructose 110°C; sucrose, glucose and galactose nearly 160°C; and maltose 180°C–it is beyond their melting point. Caramelization is supposed to be a reaction similar to the one when carbohydrates react with strong alkali or acid to produce brown colour. It involves isomerization and dehydration of sugar at high temperature. The first step is the reversible isomerization of aldoses (D-glucose) or ketoses (D-fructose) to produce an enediol intermediate which further dehydrates to form dicarbonyls. Enediol and dicarbonyl form a series of degradation products which produce caramel flavour and brown colour in products such as peanut brittle (Figure 10.4).

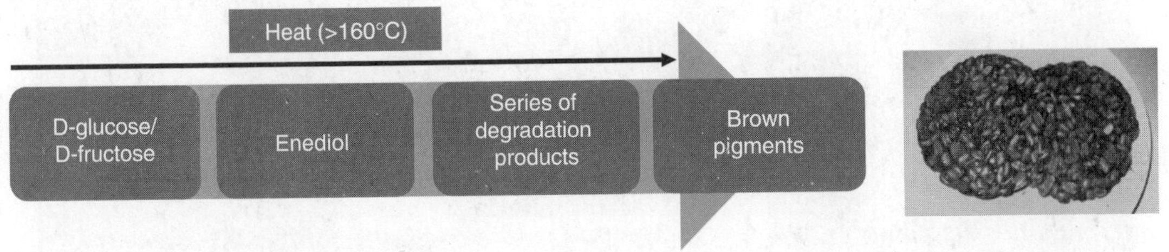

Figure 10.4 Simplified mechanism of browning due to caramelization

10.3 Discolouration Due to Metallic Contamination

Metal contamination may result from the food, the pre-harvesting sprays or the water and equipment used for processing. Besides accelerating enzymatic browning, they can result in spoiling food colour without the aid of enzymes. The following are some examples that illustrate undesirable food discolouration due to metallic contamination:

Reaction with sulphur: Metal ions such as iron and copper can react with sulphur-containing foods (e.g. corn) and result in grey discolouration. They can also act with sulphur dioxide (present in the sugar added to the canned fruits) to form the respective sulphides which are black in colour. The green ring formation on the surface of yolk in hard boiled eggs is also due to the production of ferrous sulphide (iron from the yolk and sulphur from egg white).

Reaction with phenolic compounds: Iron can combine with phenolic compounds to form complexes. The bluish grey discolouration of boiled potato is an example of such interaction. In addition to this, black discolouration (ferric tannate) in canned fruits usually results from the reaction between the iron (present in tin-plated steel cans) and tannins (present in fruits).

Reaction with lycopene: Tomato ketchup sometimes exhibits brown discolouration. It is due to the reaction between lycopene (the pigment responsible for the red colour of tomatoes) and iron (which finds a route to tomato ketchup through the processing equipment, the can body or the bottle lid).

Reaction with anthocyanin: Metal ions combine with anthocyanin pigment (present in some fruits) to form purple, blue, green or dark-coloured complexes. For example, cherry pie darkens on being cut with an iron knife and 'inking' of skin is observed in fresh peaches (Box 10.5).

Box 10.5 What Causes 'Inking' on Peaches?

Brown and black spots on the skin of peaches, often termed as 'inking', are caused by physical damage combined with metal contamination from fungicide and insecticide pre-harvest sprays. The physical injury releases anthocyanin/phenolic pigments which form complexes with metal ions such as iron, metal and aluminium. Some safe pre-harvest treatments have been developed to minimize the problem.

Source: Crisosto et al. 1999.

10.4 Role in Food Preparation

Irrespective of their pathways, enzymatic and non-enzymatic browning plays a crucial role in developing the distinct colour and flavour of a variety of dishes and products. Some of its applications during food preparation are presented in Figure 10.5 and discussed in the following paragraphs.

Figure 10.5 Desirable browning in food preparation

1. **Crust Formation in Baked, Fried and Roasted Products:** Maillard browning contributes to the crust colour and flavour of bakery products (bread, cakes, biscuits and muffins); fried foods (potato chips, samosa, pakora and tikki) and roasted foods (meat and nuts). The brown and crisp crust of these products is a characteristic that is important for their acceptance by the consumer. The colour starts developing only when these products are dehydrated to some extent and have reached a particular temperature. It is important for chefs, bakers, cooks and food processors to know that formation of acrylamide in a wide range of fried and oven-cooked foods is a matter of concern due to its possible role in causing cancer among humans. Acrylamide forms only when the food is heated to a temperature above 180°C.

2. **Brown Colour of Tea, Coffee and Cocoa:** Tea processing includes withering, rolling, fermentation and firing. It is during fermentation that the tea leaves undergo oxidative enzymatic browning to develop copper-red colour and sour apple flavour. However, the brown colour of roasted cocoa and coffee beans results from Maillard reaction and caramelization.
3. **Colour Development in Beer:** Colour is the first thing a beer drinker notices when the beer is poured into a glass; some like it bright golden, whereas others prefer the reddish black colour. During beer production, when malted and un-malted grains are dried, the brewers and winemakers control the length and degree of temperature to optimize the Maillard reaction so that the correct hue is created.
4. **Dark Colour and Flavour of Maple Syrup:** Culinary chefs use maple syrup as a sweetening and flavouring ingredient in a wide variety of dishes. The distinct dark colour and flavour of maple syrup is acquired during its processing when high concentration of glucose and thermal evaporation lead to Maillard reaction and caramelization.
5. **Chocolate Colour and Flavour:** Maillard reaction is also responsible for converting the musty smelling cocoa beans to dark-coloured and nice-smelling beans during the roasting process. Thereafter, these beans are used to prepare chocolate which is an integral part of confectionery, bakery and cooking.
6. **Preparation of Caramel:** Chefs and bakers caramelize sugar by heating it to a temperature above 160°C and use it to top desserts and cakes. Caramel is also used in sugar confectionery (refer to Section 2.3 in Chapter 2). However, care is needed during the process to avoid over-browning or burnt flavour.

10.5 Prevention of Undesirable Browning

Undesirable browning can result in food spoilage during processing and storage. It is necessary to prevent these reactions not only to preserve the sensory attributes of food but also to ensure nutritional quality and food safety.

10.5.1 Prevention of Enzymatic Browning

Enzymatic browning of fruits and vegetables is generally undesirable as it can ruin the visual appeal and lower the consumer acceptability. It can be prevented primarily by eliminating one of the requisites, that is, phenolic compounds, phenoloxidase enzyme, copper or oxygen. Different measures adopted to prevent enzymatic browning in cut fruits and vegetables (Figure 10.6) are described as follows:

1. **Exclusion of Oxygen:** Since enzymatic browning is an oxidative reaction, it cannot take place in the absence of oxygen. Immersing the cut fruit or vegetable in water, syrup or brine excludes oxygen partially. Vacuum treatment also removes the oxygen.
2. **Inactivation of Enzyme by Application of Heat:** Being proteinaceous in nature, enzymes are inactivated at high temperature. Blanching is a mild heat treatment that inactivates polyphenoloxidase and thereby prevents enzymatic browning (refer to sub-section 5.3.1 in Chapter 5). It is employed as a pre-treatment in dehydration of potato chips to prevent darkening of the cut surfaces during processing and storage. Effective pasteurization of fruit juices also destroys the enzymes.
3. **Storing at Low Temperature:** Enzymatic activity is delayed at low temperature but is not completely halted. Keeping salads at refrigerating temperature controls enzymatic browning partially.
4. **Lowering the pH:** The polyphenoloxidase enzyme needs slightly alkaline pH (ranging between 5 and 7) for its activity. Adding citric, malic, phosphoric or ascorbic acid inhibits its activity. Lemon juice used by salad bar operators to prevent browning of cut fruits, naturally contains citric acid.

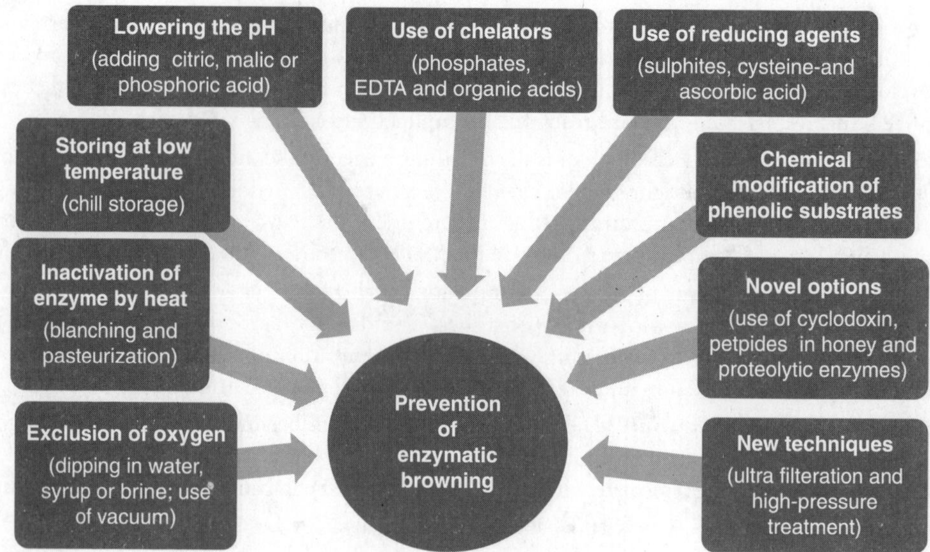

Figure 10.6 Measures for preventing enzymatic browning

5. **Use of Chelators:** They bind with metal catalysts (such as copper and iron) and prevent them from accelerating the enzymatic reaction. Phosphates, ethylene diamine tetra acetic acid (EDTA) and organic acids are examples of chelators.
6. **Use of Reducing Agents:** Substances that prevent oxidation reactions are termed as 'reducing agents', for example, sulphites and cysteine. Sulphites are the cheapest and most efficient chemicals for controlling browning. Potassium metabisulphite (KMS) is a commonly used sulphiting agent in the dehydration of vegetables. However, its use has some limitations such as bleaching the colour of vegetables, imparting undesirable odour to the product and destroying thiamin. Use of cysteine is a better alternative. Ascorbic acid also reverses the reaction of enzymatic browning by reducing ortho-quinones back to dihydroxy polyphenols.
7. **Chemical Modification of Substrates:** It is possible to chemically modify phenolic substances such as chlorogenic acid, caffeic acid and tyrosine to prevent enzymatic browning.

Among the novel options of inhibiting enzymatic browning are the use of cyclodextrins, peptides (present in honey), carbohydrate derivatives and proteolytic enzymes (ficin, papain and bromelain). Processing techniques such as dehydration and freezing, and newer techniques such as ultrafiltration, high-pressure treatment and irradiation can also be used to prevent the enzymatic browning.

10.5.2 Prevention of Non-enzymatic Browning

The products of Maillard reaction can result in the loss of nutrients (such as essential amino acids), decreased digestibility and formation of mutagenic and carcinogenic compounds (acrylamide). Therefore, it is necessary to control/prevent Maillard browning by the following measures:

1. **Lowering the Heat of Cooking:** Being temperature dependent, Maillard reaction can be controlled through temperature. For example, keeping the temperature of deep fryer under control can help in preventing the potato chips from getting over-brown and developing burnt flavour.

2. **Storing at Low Temperature:** Storage at low temperature (0–5°C) helps in delaying browning in products such as evaporated milk.
3. **Lowering the pH:** Maillard reaction is slowed down when pH is decreased below 6. Various acids (lemon and vinegar) can be used to reduce pH and control browning.
4. **Lowering the Water Activity (a_w):** The optimum a_w for Maillard reaction is intermediate (0.5–0.8). Therefore, this reaction is most common in intermediate moisture foods. Lowering the a_w below 0.5 prevents browning despite the increase in the concentration of solutes.

Oxygen should be excluded to prevent non-enzymatic browning due to the reaction of oxidation products of ascorbic acid or those of lipids. Oxidation of lipids can be controlled by excluding air; preventing the oils from light; use of antioxidants; use of sequestrants; and storing oils and fried foods at low temperature (refer to Chapter 3). It is crucial to control the oxidation of ascorbic acid during extraction of fruit juices by the use of low temperature, decrease in holding time, deaeration (removal of air) and pasteurization. The decomposition of ascorbic acid can be slowed down by making pH acidic (<7) and removing the metal catalysts. Non-enzymatic browning of fruit juices can be prevented by process optimization, ion-exchange treatment and chemical inhibitors, whereas synthetic absorbers, nanofilteration and activated charcoal are employed as restorative methods. Metallic discolouration can be minimized by preventing the food from coming in contact with metals such as iron and copper or by using chelators.

We learnt that food can turn brown with or without the aid of enzymes during pre-preparation, cooking, processing and storage. Some browning reactions are beneficial, whereas others can result in unfavourable colour, undesirable flavour, nutrient loss, decrease in digestibility or production of toxic compounds. By optimizing desirable browning reactions, culinary chefs and food technologists can create products with the most acceptable hue. We also learnt the different ways of controlling undesirable browning not only to preserve the appearance of food but also to provide foods that are nutritious and safe for our esteemed customers.

Know Your Progress

True or False Questions

1. Enzymatic browning is an oxidative reaction which has three main requisites: phenolic compounds, oxygen and phenolase enzyme.
2. Browning due to Maillard reaction is exhibited by baked, roasted and fried foods.
3. Darkening of colour in citrus juices is an example of non-enzymatic browning due to the reaction between oxidation products of fatty acids and proteins.
4. When sugar is heated at a high temperature (160°C or above) in the absence of water and amino acids, it turns brown and this sugar-sugar interaction is known as Strecker degradation.
5. Undesirable changes produced by Maillard reaction include loss of nutrients, lowering of digestibility and production of toxic compounds (e.g. acrylamide).

Very Short Answer Questions

1. Define the following terms
   (a) Amodari compounds (b) Acrylamide (c) Strecker degradation (d) Melanoidins (e) Chelators

Short Answer Questions

1. Describe the process of enzymatic browning.
2. What is Maillard reaction? Discuss its mechanism and the consequences.

3. Discuss the role of browning in food preparation.
4. How can enzymatic browning be prevented in cut fruits and vegetables?
5. Describe the role of oxidation products in non-enzymatic browning.

Long Answer Questions

1. What are the pathways of different types of non-enzymatic browning?
2. Discuss the steps in Maillard reaction along with the factors that affect this reaction.

Beyond the Text

- Prepare the following solutions in small beakers. Peel a potato and slice it to seven pieces of approximately same size. Keep one slice exposed to air (control) and quickly put one slice each in the prepared solutions. Follow the instructions and observe browning after 30 minutes.

1. Plain water (100 ml)
2. Salt solution (100 ml water + 1/2 tsp salt)
3 KMS solution (100 ml water + 1/4 tsp KMS)
4. Lemon juice (100 ml)
5. Plain water (100 ml) and keep in refrigerator
6. Plain water (200 ml) and blanch for 5 minutes

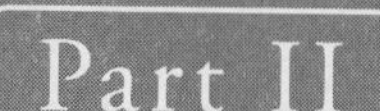

Part II

Nutrition

Chapter 11

Nutrition: Basic Aspects

Chapter Objectives

After reading this chapter, the reader should be able to:

- understand the concept of health, nutrition and nutrients.
- appreciate the diverse roles of food in maintaining health.
- associate the states of optimum nutrition and malnutrition with health and disease.
- classify nutrients on the basis of their requirements.

Consider the following situations:

Situation 1: A five-star hotel is assigned the task of hosting a buffet lunch for the national and international participants of a conference. The theme is low glycemic index and antioxidant-rich foods. The chef in-charge seeks the help of a new recruit who has knowledge about such foods to develop menu as per the desired specifications.

Situation 2: A confectionery manufacturing company wanted its product development officer to prepare a chocolate bar (25 gm) containing 10 gm of protein and 5 gm of carbohydrate. Foods which are high in protein but low in carbohydrate were selected for developing the chocolate of desired nutrient content.

Situation 3: According to the Food Safety and Standards Act 2011, it is now mandatory to provide information regarding the nutrient content on the label of every packaged product. A ***namkeen*** manufacturer who has no knowledge about computing nutritive value of foods had to hire an expert. His profit reduced considerably.

Situation 4: High-protein cholesterol-free burger buns were required to be prepared for a chain of fast food restaurants. Since milk and eggs could not be used, the baker decided to add soy flour and whey protein powder instead.

Knowledge about nutrition becomes even more important when food is being prepared and served for the same customer several times a day and for several months such as in case of hostels, offices/schools/colleges canteens and mess. This information becomes critical to the health of a patient when a kitchen manager, who is not acquainted with basic aspects of nutrition, wrongly substitutes an ingredient. Customers can often be quite demanding when perceiving the ingredients present in a finely crafted dish. Above all, understanding our food and how it can affect our body, helps in maintaining good health.

11.1 Introduction to Health, Nutrition and Nutrients

Issues related to health and nutrition have always been pivotal in modulating the direction of food processing and service industry. Over the past few decades, nutrition has become the fundamental factor in determining the choice of ingredients. Advancements in the field of food, nutrition and medical science as well as an ever demanding aware customer make it necessary for a food handler to have basic understanding about this field.

Health: The World Health Organization (WHO) defined health in 1946 as 'a state of complete physical, mental and social well being and not merely the absence of disease or infirmity'. This definition is still widely accepted. The word 'state' refers to health as a resource for everyday life, and not an objective of living. Therefore, health should be referred to as a positive concept which emphasizes social and personal resources as well as physical capacities. In order to identify whether an individual is healthy or suffering from a disease, WHO has given several elaborative classifications such as the International Classification of Functioning, Disability and Health (ICF) and the International Classification of Diseases (ICD).

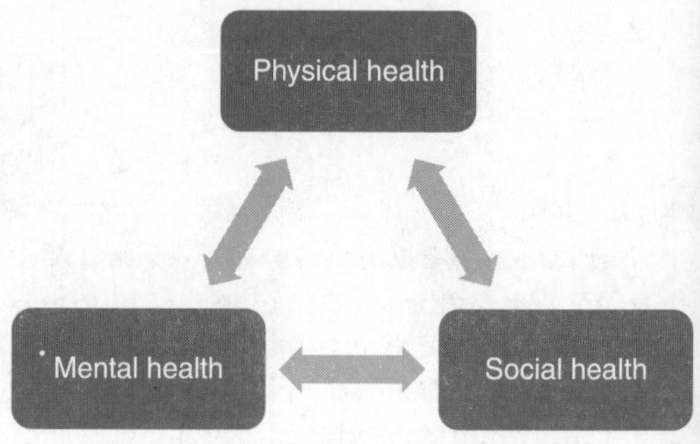

Figure 11.1 Health triangle

Maintenance and promotion of health can be achieved by a combination of physical (also referred to as physiological), mental and social well-being. All these three aspects of health are inter-related to each other and this concept is often referred to as the health triangle (Figure 11.1). Also the determinants of health can be seen in Figure 11.2.

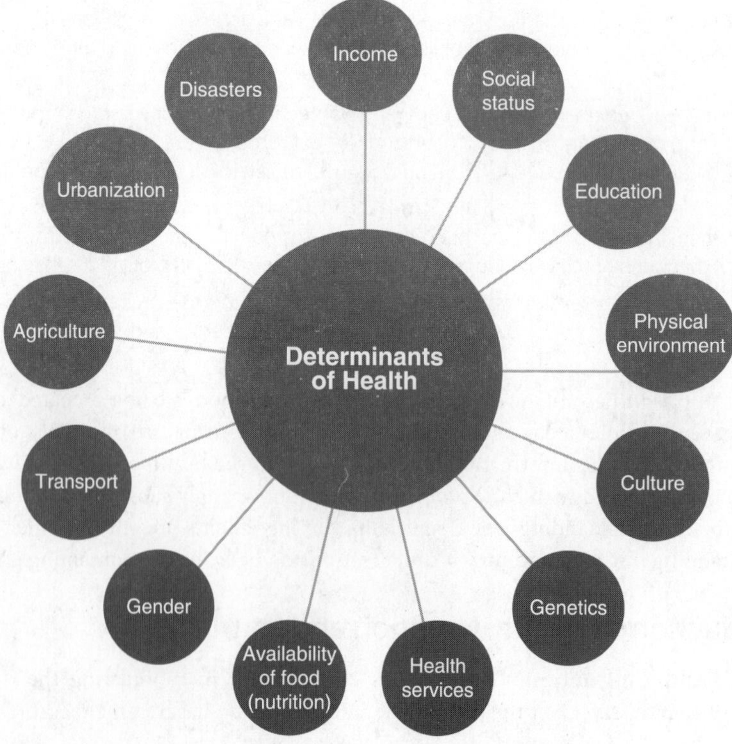

Figure 11.2 Determinants of health

hysical health: It is also called physiological health. Physical health refers to the physiological status ncluding the work efficiency of various organs which should be such that all metabolic processes in the ody can be carried out towards achieving and maintaining optimum health.

ocial health: The positive interaction of individuals with their family members, peers and the society uring all phases of life and their usual stress and strain, is known as social health.

Mental health: WHO has defined mental health as 'a state of well being in which the individual realizes his r her own abilities, can cope with the normal stresses of life, can work productively and fruitfully, and is ble to make a contribution to his or her community'. The word 'mental health' is being replaced by mental vell-being. It refers to emotional well-being, including the capacity to live a full and creative life and the exibility to deal with the day-to-day inevitable challenges of life.

Optimum health can be achieved (to a great extent) by consuming a balanced diet and following a rudent lifestyle. However, health is not always within an individual's control and is often influenced y several modifiable and non-modifiable factors. These factors are also called determinants of health Figure 11.2).

GOOD TO KNOW

The Millennium Development Goals

In order to improve health and well-being of a population, it is essential to improve the environment and the factors associated with morbidity and mortality. Thus, in view of the current global situation (particularly of developing and under-developed nations), the United Nations Millennium Declaration (2000) has put forth eight Millennium Development Goals which need to be addressed by the 191 member countries including India by 2015. These include the following:

1. Eradication of extreme poverty and hunger.
2. Accomplishing universal primary education.
3. Promoting gender equality and empowering women.
4. Reducing child mortality.
5. Improving maternal health.
6. Combating HIV/AIDS, malaria and other diseases.
7. Ensuring environmental sustainability and,
8. Developing a global partnership for development.

utrition: It refers to the science of food and the components nutrients and other substances) present therein with relevance to ealth and disease. Thus, nutrition refers to the action, interac-on and balance of nutrients; their digestion, absorption, metabo-sm and excretion as well as their functions and significance in aintaining optimum health (physiological, social and mental) roughout life.

utrients: They are the substances present in food which can e utilized by the body to obtain energy, facilitate multiplication f cells, regulate various metabolic processes and protect from everal diseases, if consumed in optimum quantities. A nutrient an perform any or a number of the above-mentioned functions. utrients present in food include proteins, carbohydrates (includ-g fibre), fats, vitamins, minerals and water (Figure 11.3).

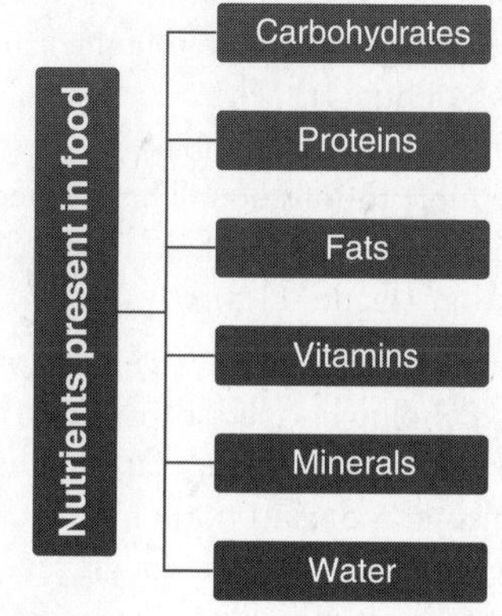

Figure 11.3 Nutrients present in food

Food, and hence nutrients, should be such as to maintain an optimum nutritional status and a positiv state of health. It is not necessary that all nutrients should be provided in every meal or every day in exactl the amount as desirable for the body. While the intake of certain nutrients may be insufficient, that o others may be excess. Thus, while optimum nutritional status may be an ideal situation, that of under- an over-nutrition can be frequently observed.

Optimum Nutrition: It is a state during which all nutrients are provided in the amount, form an proportions as required by the body to perform daily activities, store reserves that can be used during period of low intake and maintain physical, mental and social health during all phases of life. Practically, a state o optimum nutrition is difficult to achieve.

Malnutrition: It is a medical condition associated with inadequate, high or disproportionate intake of nutrient(s). Malnutrition can impair physical, mental and social health. It is commonly associated with the

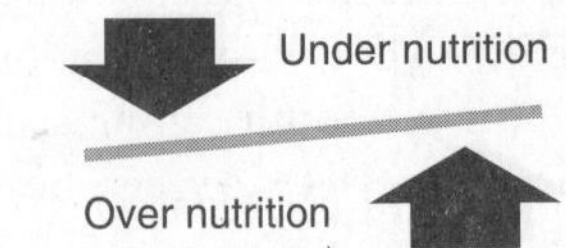

Figure 11.4 Types of malnutritior

1. consumption of energy and protein below the requirement for maintenance and growth (as the case may be).
2. consumption of excessive calories (such as obesity).
3. excessive loss of nutrients (diarrhoea/vomiting/chronic illness) or increased energy expenditure.
4. over consumption of certain nutrients (high/prolonged intake of health supplements and crash diets).

Thus, malnutrition can exist in two forms *viz.* under-nutrition and over-nutrition (Figure 11.4).

Under-nutrition: It refers to poor health resulting from depletion of nutrients in the body due to inadequate nutrient(s) intake or loss of nutrient(s) from the body for a considerable period of time. It can result from, or result in, increased frequency of illness due to infections, injuries and metabolic diseases. For instance, inadequate intake of nutrients such as proteins and vitamins lowers resistance to infections. During infection, there are losses of nutrients from body (vomiting/diarrhoea), increased breakdown of body tissues and lowered intake of food due to anorexia (reduced feeling of hunger). These further enhance the state of deficiency and make the person more susceptible to infection. This is often termed as the 'viscous cycle of under-nutrition and infection' (Figure 11.5).

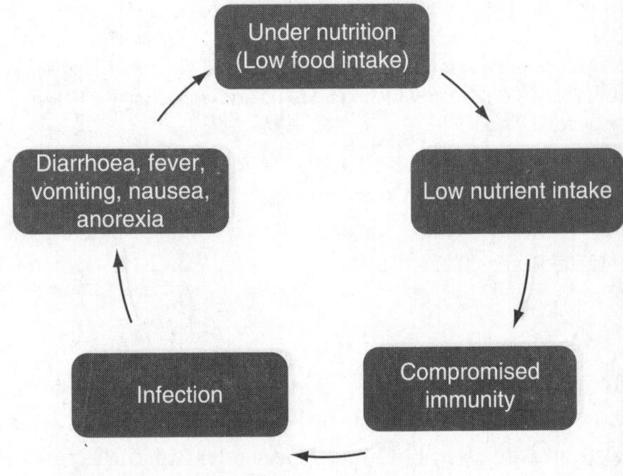

Figure 11.5 The viscous cycle of under nutrition and infectio

Over-nutrition: Excessive intake (above the daily requirement and fulfilment of body reserves) of nutrien over a prolonged period of time is known as over-nutrition. It is most often characterized by excessiv intake of calories, dietary fat and cholesterol, for example, during obesity. Excessive consumption of protei supplements during workouts can result in toxicity–a state of over nutrition.

It has been found through several research studies that both under-nutrition and over-nutrition ca co-exist in an individual. Several overweight or obese individuals are often anaemic or have low reserves (lean tissues/micro nutrients.

11.2 Food and Its Functions

Food is vital to our lives and is often consumed to satisfy hunger and nourish our body. Food also influences our social life, emotions and association with nature. The functions of food are discussed in the following paragraphs (Figure 11.6).

Figure 11.6 Functions of food

Physiological Function

The most important function of food is to maintain the normal structure and function of our body. This can be done by providing energy, building and maintaining body structures, regulating various metabolic processes and protecting from diseases. The physiological functions of food are discussed below.

Energy Giving: Food acts as fuel for human body just like petrol for the car. If food is not consumed in adequate amounts such as when meals are missed or during fasts, the person feels lethargic. Food provides us energy to perform various voluntary and involuntary activities throughout the day. Voluntary activities are those which are performed by our will and can be controlled such as brushing teeth, bathing, walking, cooking, writing and speaking. Beating of heart, functioning of kidneys/lungs, blinking of eyes, perspiration etc. are the involuntary activities as we do not have any control over them. Carbohydrates, proteins and fats present in our meals are source of energy for our body.

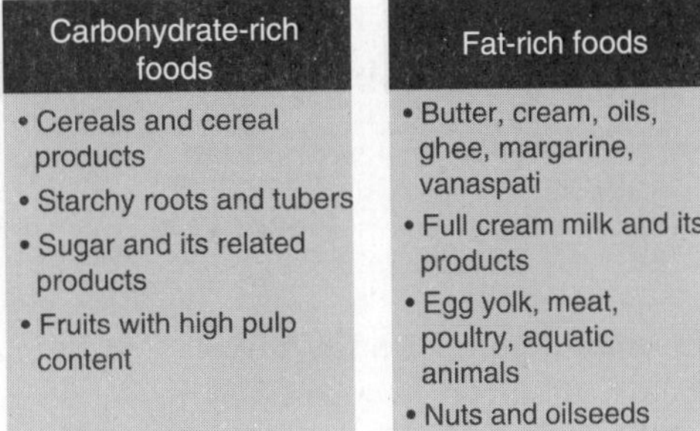

Figure 11.7 Energy giving foods

While each gram of dietary carbohydrate and protein provide 4 kcal (kilocalories) of energy, one gram of fat provides 9 kcals (refer to Chapter 12 for details). However, proteins are not the preferred source of energy as they need to perform other vital functions. Figure 11.7 gives examples for foods which are rich in carbohydrates and fat. Adipose tissues (body fat) are the reservoirs of energy in our body.

Body Building and Growth: At birth, the weight and length of an average Indian infant is 2.5–3.5 kg and 50 cm, respectively. However, when the same child reaches 18 years, the average weight and height are 49.7–58.3 kg and 156–172 cm. This increase in body mass is the result of cell multiplication which is at its peak during infancy and adolescence. The nutrient important for this function is protein (amino acids). Foods rich in protein include eggs, milk, cheese, yogurt, meat, poultry, aquatic animals, pulses, legumes and certain nuts (Figure 11.8). If protein rich foods are not consumed in adequate amounts during the 'growing years' of an individual, full growth potential is not achieved. These foods also provide the protein necessary for repairing the usual wear and tear occurring in our body. For example, the average life span of red blood cells is 120 days; they need to be replaced with new such cells continuously, and protein is necessary for the synthesis of red blood cells. Foods rich in protein also promote healing of wounds. A cut on the finger while chopping vegetables gets healed faster, if protein-rich foods are consumed.

Regulation of Body Processes: Blood clots at the site of injury but not within the blood vessels during circulation, gastric juice is released in the stomach when we feel hungry or eat food, air is inhaled and exhaled continuously in a rhythmic manner, co-ordination of various body movements are the examples of some well-regulated body processes. Fruits and vegetables play a significant role in regulating body processes because they are rich sources of vitamins, minerals, phytochemicals and fibres (Figure 11.8). Foods such as bael, pineapple, guava, cabbage, lotus stem, green leafy vegetables, beans and legumes are rich sources of dietary fibre which helps to regulate normal bowel movements, maintain normal blood glucose levels and lower elevated blood cholesterol levels. Orange-red coloured foods are rich in vitamin A which helps in maintaining normal vision. Animal foods are good sources of vitamin B_{12} which maintains integrity of red blood cells. Fish oils promote cardio-vascular health because they contain good amount of n-3 fatty acids (Chapter 13).

Body building foods	Regulatory foods	Protective foods
• Eggs, meat, poultry, water animals • Milk, cheese, curd, yogurt and ice cream • Legumes and pulses • Nuts and certain oilseeds	• All fruits and vegetables especially orange red-coloured and citrus • High fibre foods including husk/bran of cereals and pulses, edible peels of fruits and vegetables.	• All fruits and vegetables (except starch-rich roots and tubers) • Yeast • Liver • Wheat germ • Eggs, meat, milk, pulses, legumes and nuts.

Figure 11.8 Body building, regulatory and protective foods

Protection from Diseases/Adverse Conditions: Certain individuals suffer from common cough and cold more frequently than others. The frequency of illness due to infections (communicable diseases) is closely related to food choices. Certain foods help to enhance our immune system by promoting the synthesis and function of macrophages, lymphocytes and antibodies such as citrus fruits and vegetables (gooseberry, orange, lime and lemon). Protein-rich foods promote synthesis of certain immunoglobulins which strengthen our immune system. Orange-, red- and purple-coloured foods (papaya, carrots, tomato, berries, grapes and currants) provide protection against cancer, prevent cell destruction and promote precise cell multiplication. Foods such as almonds, sunflower seeds and wheat germ oil which are good sources of vitamin E help to protect tissues of the heart, lungs, eyes, liver, breasts and muscles against oxidative damage and certain types of cancer. If variety of food is consumed in adequate amounts for a considerable duration, it is easier to pass through periods of lean food intake such as missing of meals, fasts, surgery and natural disasters (flood/drought/famine).

Social Function: Food facilitates social interactions. The role of food in the hospitality industry can best justify this function. Food is served and shared by the hosts with their guests. Specific foods are associated with different types of social interactions. For example, sharing ice cream with friends, coffee with subordinates or boss, cake during birthday parties, and ladoos and sweet meats during house warming/religious functions. Sharing of food helps to develop positive/cordial relations with family, colleagues, friends and the society at large including strangers.

Psychological Function: People tend to develop positive and negative associations with food which is based on their experiences in life. Food is often used as a reward or punishment by parents for their children which later on influences their likes and dislikes. A mother usually prepares favourite food for her child who scores well in exams or makes an achievement. Food provides emotional comfort to people; many tend to eat more when anxious, depressed or bored. Snacking is frequently observed among children preparing for exams. Sharing of food during bereavement provides comfort and an opportunity to share sorrow with each other.

Spiritual Function: Certain foods are used as a token of respect to God and nature. Holy water, coconut, puffed rice, sugar balls and sweet meats are a few examples of foods associated with prayer and religious rituals. Following vegetarian dietary habits, avoidance of non-vegetarian foods during Navratra's by Hindus and on Fridays during Lent (period prior to Easter) by Christians is a mark of respect for life. Food is also distributed free of cost through community kitchens to minimize discrimination based on caste, income, gender or religion.

Thus, food influences our lives through a number of ways, the most significant being, providing the much needed nourishment to our body. It is through food that we get a variety of nutrients, and it is ultimately these nutrients what our body is composed of.

11.3 Classification of Nutrients

Food provides us with six essential nutrients *viz.* carbohydrates (including fibre), proteins, fats, vitamins, minerals and water. Almost all foods (except fats and sugars) contain these nutrients but in varying amounts. The nutrients which our body cannot synthesize or can synthesize but in inadequate amounts are known as **essential nutrients**. Therefore, essential nutrients must be supplied by our daily diets in order to remain healthy. Amino acids such as methionine, lysine, valine, leucine isoleucine and linoleic fatty acid are classic examples of essential nutrients. The six classes of nutrients present in food perform three major functions:

1. provide energy to the body.
2. contribute to body structure and facilitate growth.
3. regulate body processes and protect from diseases.

To perform these functions, all nutrients are not required in the same amount. Some are required in larger quantities, whereas others are needed in small amounts only. Based on the requirement of these nutrients by our body, they can be classified as macronutrients and micronutrients (Figure 11.9). Although macro and micronutrients have different functions to perform, many would be overlapping. For example, the synthesis of red blood cells requires protein but its multiplication, growth and maturity are dependent on several vitamins (A, C, E, folate and cyanocobalamin) and minerals (iron, zinc and copper).

Nutrients can also be classified based on their chemical composition. Nutrients which contain carbon in their structure are known as **organic** substances such as carbohydrates, fats, proteins and vitamins. Nutrients which do not contain carbon in their structure are known as **inorganic** substances such as minerals and water. In addition to nutrients, a class of substances has been identified which have the potential of promoting health. These substances are known as phytochemicals. *Phytochemicals* exhibit antioxidant activities and hence confer protection from cancer, heart diseases and diabetes. They are not classified as nutrients because they are not essential for life.

Thus, to summarize, it can be said that health does not merely refer to a disease-free life. It refers to physical, social and mental well being. Food plays an integral role in maintaining optimum health as it not only satisfies hunger and keeps us alive but it also helps in growth, maintaining body structures, regulating body processes and protecting from diseases. The social, psychological and spiritual roles of food promote social and mental well-being. There are six major classes of nutrients *viz.* carbohydrates, proteins, fats, vitamins, minerals and water. Macronutrients are required in larger amount than micronutrients. Each nutrient is

Macronutrients	Micronutrients
required in large quantities by the body	required in comparatively smaller quantities by our body
source of energy to the body	contribute to body structures and regulate body processes
examples: carbohydrates, proteins, fats	examples: vitamins (A, B, C, D, E, K), minerals (iron, iodine, sodium, calcium)
present in larger amounts in food	Present in smaller amounts in food
unit of measurement: grams	unit of measurement: milligrams, micrograms, international units, picograms
Deficiency (examples): underweight, stunting, wasting	Deficiency (examples): xeropthalmia, scurvy, beri-beri, anaemia, pellagra, anorexia
Excess (examples): overweight, obesity, kidney-related diseases	Excess (examples): toxicity, organ damage, respiratory distress, nephrolithiasis

Figure 11.9 Macronutrients and micronutrients present in our food and body

responsible for a particular function. It is also important to remember that each function is performed by the combined effort of more than one nutrient. Each food (except pure fats and sugars) contains a variety of nutrients though the quantity of each nutrient may vary.

Know Your Progress

True or False Questions

1. Health should be considered as a resource for everyday life, and not an objective of living.
2. Malnutrition is a synonym for under-nutrition.
3. Synthesis of red blood cells requires only proteins.
4. Phytochemicals are nutrient-like substances.
5. Vitamin D is a micronutrient.

Very Short Answer Questions

1. Define the word 'Health'.
2. How is mental health different from mental well-being?
3. Describe the word 'nutrients'.
4. Name the six classes of nutrients.
5. Enlist any 10 body-building foods.

Short Answer Questions

1. What functions should food perform in a college canteen?
2. Enlist the determinants of health, as applicable to your family.
3. Differentiate between macronutrients and micronutrients.

Long Answer Questions

1. Discuss the functions of food with special relevance to hotel industry.
2. What are the current millennium development goals? Are they relevant for India?
3. How is optimum nutrition different from under and over nutrition?
4. Nutrition is the backbone of healthy living–justify this statement.

Beyond the Text

- Study the food labels of the following food stuffs and identify the macro and micro nutrients present in them

Health supplement	Glucose based drink
Noodles	Spice Mix (masala mix)
Salt	Milk

Chapter 12

Energy

Chapter Objectives

After reading this chapter, the student should be able to:

- understand the concept of energy and its units of measurement.
- learn the components which comprise energy expenditure.
- reason out the factors which are responsible for variations in energy requirements at individual level.
- associate energy balance with weight management.
- know the fundamental aspects of underweight and overweight/obesity.

Consider the following situations:

Situation 1: A and B visit a super market near their college to buy breakfast cereals. After comparing the energy content (density) of various products, they purchase the cereal.

Situation 2: Knowing that there would be no time to have lunch before 3:00 p.m., 'A' selects egg sandwich with extra cheese, milk and banana from the buffet for breakfast.

Situation 3: A has kept his energy intake the same for past 20 years but still he is gradually gaining weight.

The food and hospitality industry is often encountered with challenging situations involving development of mouth-watering recipes which are at the same time meeting the objectives of specific calorie counts. Increasingly, the customers of food service industry are becoming health conscious. At an individual level, many of us lay greater stress on energy (calories) consumed through food without paying much attention to energy (calories) expended by the body. The word 'calorie' is often synonymously used for the word 'energy', but do they mean the same? What determines our energy requirement? Why do some of us gain or lose weight, whereas others are able to maintain it? Such and several other questions shall be answered through this chapter.

12.1 Definition and Unit(s) of Measurement

Energy, as we all know, is a scalar quantity which was hypothesized for the first time by Newton. Energy is present in different forms such as potential, chemical, kinetic, magnetic or electrical. In the field of nutrition, it is primarily the chemical and potential energy which is obtained by the human body from food. It is this energy which helps the human body to survive, perform various physical activities, physiological functions and several cellular/metabolic processes within. Thus, energy may be referred to as 'the capacity to do work' by the human body; both within and outside.

The unit of measuring energy with relevance to nutrition is calorie or Kilocalorie (kcal). A **calorie** refers to the amount of energy (heat) required to raise the temperature of 1 g water from 14.5°C to 15.5°C. Therefore, one kcal is equal to 1,000 calories. Thus, **kilocalorie** refers to the heat required to raise the temperature of 1 kg water from 14.5°C to 15.5°C. Certain international organizations use the unit **Joule** to measure energy in terms of mechanical work and is defined as the energy required for moving 1 kg of mass by 1 m by a force of one Newton acting on it. (One Newton is the force needed to accelerate 1 kg mass by 1 m/sec^2.) The association between these units is given in Box 12.1.

Box 12.1 Units for measuring energy

1 kcal	1,000 calories
1 kcal	4.184 kJ (Kilo Joule)
1 kJ	0.239 kcal (Kilocalorie)
1,000 kcal	4184 kJ or 4.18 MJ (Mega Joule)
1 MJ	239 kcal

12.2 Components of Energy Requirements

According to Indian Council of Medical Research (ICMR, 2010), the energy requirement of an individual is defined as *the level of energy intake from food that balances energy expenditure when the individual has a body size and composition and level of physical activity, consistent with long-term good health, also allowing for maintenance of economically essential and socially desirable activities. In children, pregnant women and lactating mothers, it includes the energy needs associated with deposition of tissues or secretion of milk at rates consistent with good health.* Thus, energy required in a day refers to the energy expended by an individual to perform the above-mentioned functions. In other words, it can be said that total energy expenditure (Figure 12.1) is the sum of basal metabolic rate (basal energy expenditure/resting energy expenditure), thermic effect of food and activity thermogenesis (physical activity).

12.2.1 Basal or Resting Energy Expenditure

Energy is required for circulation of blood, respiration, movement of organs, synthesis of cells/hormones/enzymes, functioning of central nervous system, maintenance of body temperature, contraction and relaxation of muscles and other such involuntary functions which are necessary for survival and maintenance of human body. The energy expended on these metabolic processes over a 24 hours period when a person is in a state of complete physical and mental rest in a thermo-neutral environment (no shivering/sweating) is known as **Basal Energy Expenditure** (BEE). If the measurement of such energy expenditure is made early in the morning when a person is in a 10–12 hours fasting state (no food, water, alcohol, nicotine) and has not made any physical activity, it is known as **Basal Metabolic Rate** (BMR). If any of the aforesaid conditions are not met, the energy expended is known as **Resting Energy Expenditure** (REE). Therefore, the REE is 10–18 per cent higher than the BEE. Both get influenced by body size, body composition, age, gender, physiological state (pregnancy and infancy), illness (surgery and fever), medications, environment and so on.

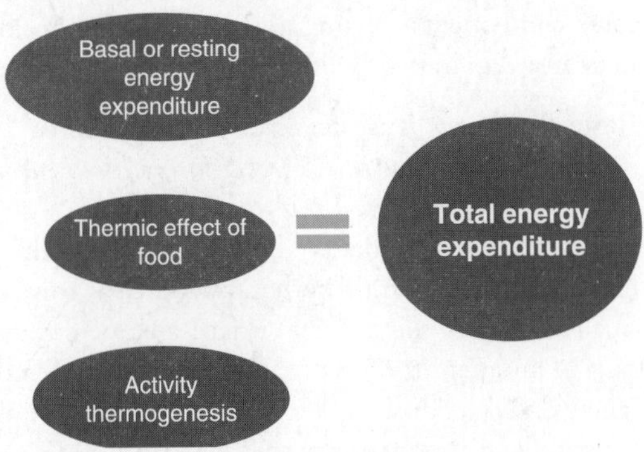

Figure 12.1 Components of energy expenditure

12.2.2 Thermic Effect of Food or Specific Dynamic Action of Food

While food is a source of energy, its consumption, digestion, absorption and utilization itself requires energy. Thus, before we receive energy from food, our body spends some energy during chewing, swallowing, mixing of gastric/pancreatic/bile juice with food in stomach, movement of small and large intestine, transportation/oxidation and inter-conversion of nutrients (carbohydrates, proteins and fats) to release and store energy.

This expenditure of energy is known as **Thermic Effect of Food** (TEF) or **Specific Dynamic Action of Food** (SDA) or **Specific Effect of Food**. Dietary fats are more efficiently metabolized as compared to carbohydrates and proteins. The size, composition and nutritive value of meal, physical activity before/during and after meal consumption, meal pattern (timing), cooking method and physiological age of digestive organs influence thermic effect of food. Thus, no two individuals can have same TEF even if they are consuming same quantity and quality of meal. In normal individuals, TEF contributes about 10 per cent to the total energy expenditure for a day.

12.2.3 Activity Thermogenesis

It is a significant and yet most variable component of energy expenditure. It refers to the energy expended daily on physical or obligatory activities (such as brushing, bathing, involvement in occupation and household chores) and discretionary activities (such as walking, exercising and playing games). The energy spent on activities which do not involve sports/exercise is known as **Non-Exercise Activity Thermogenesis (NEAT)** such as occupational/social/leisure activities. The energy expended in such activities may be as low as 500 kcal (sedentary office worker, patient) to as high as 3000 kcal (stone cutter, road maker, construction worker). No two individuals of same age, gender, body size/composition performing same activity would be spending equal calories. For example, a student sitting in a classroom who has the habit of constantly changing position in his/her chair would be spending more energy as compared to another student sitting without changing his/her body posture/position in the chair. Individuals who follow an active lifestyle and regular exercise schedule tend to have a higher activity thermogenesis and BEE/REE due to greater lean body mass and hence a lower body weight as compared to individuals who follow a sedentary lifestyle.

12.3 Factors Affecting Energy Requirements

While REE (or BMR), TEF and Activity Thermogenesis are the three components of energy requirement, there are several factors which influence each of these components (Figure 12.2) and are responsible for individual variations in energy requirements and hence energy homeostasis.

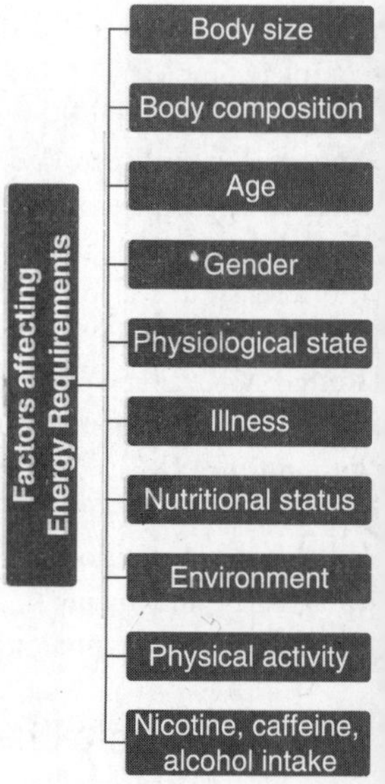

Figure 12.2 Factors affecting energy requirements

Body size: *A and B are staying in a hostel and studying the same course. A is 5 feet 3 inches tall, whereas B is 5 feet 5 inches. Both eat same quantity and quality of food. B is thin, whereas A is a little overweight.*

Body size determines the body surface area. Tall, large-build and overweight individuals have a larger body surface area as compared to those who are short, small built and thin. Body surface area determines energy lost through the skin in an attempt to maintain normal body temperature. Thus, greater heat (energy) will get dissipated to the surrounding atmosphere in case of individuals having a larger body surface area. The BEE/BMR and hence the energy requirement would be higher for large body size than small body size.

Body composition: *A and B are of same age and gender. They have similar activity schedules and food intake. However, A tends to gain weight, whereas B is able to maintain a constant body weight.*

Our body weight is primarily composed of fat mass (adipose tissues) and fat-free mass or lean body mass (muscles, water, bone and other tissues). Fat-free mass is metabolically active and requires energy for its survival and functioning. Adipose tissues are themselves a source of energy (1 g = ~7.7 kcal). Thus, an individual with higher fat-free mass would have a higher BMR and higher energy requirements as compared to an individual who is having adipose tissues above normal levels (8–24 per cent: males and 21–35 per cent: females). Therefore, the food intake should be more of an individual having a higher fat-free mass as compared to the other with lower fat-free mass despite there being of same age, gender and following similar lifestyle. Table 12.1 gives the normal body composition of males and females. Quality of diet also influences body composition and hence REE (Figure 12.3).

Table 12.1 Behnke's theoretic body composition model

Parameter	Males	Females
Total Fat	8–24 per cent	21–35 per cent
– Storage Fat	5–21 per cent	9–23 per cent
– Essential Fat	3 per cent	12 per cent
Muscle	44.8 per cent	38 per cent
Bone	14.9 per cent	12 per cent
Remainder	16.3–32.3 per cent	15–29 per cent

Source: American Journal of Clinical Nutrition 72:694, 2000.

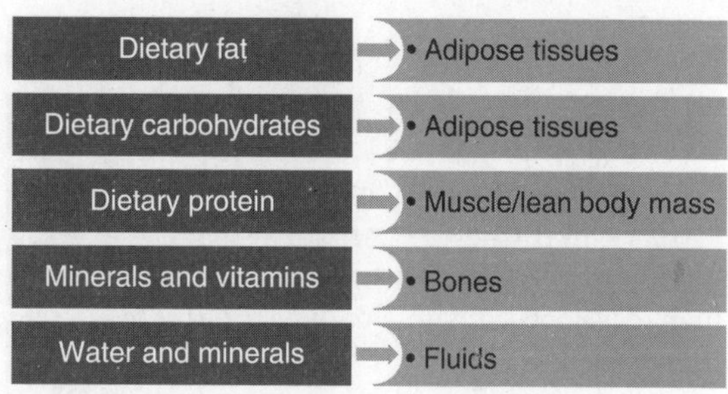

Figure 12.3 Association of dietary components with body composition

Age: *A is a 15-year-old boy who is ever hungry and eats more than his cousin brother who is 32 years old. B is a 60-year-old woman who has kept her lifestyle and food intake constant for the past 25 years but is still gaining weight.*

The energy requirements per kg body weight are higher during infancy, childhood and adolescence and tend to reduce after early adulthood. Infancy, childhood and adolescence are the periods of rapid growth and development in an individual's life span. Additional energy is required to facilitate multiplication of cells and deposition of body tissues. Therefore, the BEE per kg body weight is the highest during the first year of life followed by adolescence and thereafter childhood. In contrast to this, after adulthood and especially during old age, the BEE (BMR) drops significantly (due to the reduction in functional capacity of organs by 1 per cent each year after the age of 30 years) and there is also reduction in physical activity. This results in a subsequent reduction in fat-free mass accompanied at times by an increase in adipose tissues. Thus, individuals in mid-adulthood and elderly years tend to gain weight gradually over a period of time despite keeping their food intake same. Regular exercise can help in maintaining the amount of fat-free mass and hence normal body weight after early adulthood.

Gender: *A is a 5 feet 4 inches tall 16-year-old girl. B is a 5 feet 4 inches tall boy of same age. They are having similar lifestyle and food intake. However, B weighs lower than A.*

The energy requirement per kg body weight is higher for men as compared to women. The BMR of men is 5–10 per cent higher as compared to that of women of the same weight, height and age. This is so because men have higher (76–92 per cent) fat-free mass than women (65–79 per cent). On the other hand, the adipose tissue content of women (21–35 per cent) is higher than that of men (8–24 per cent). Thus, despite having same lifestyle and food intake, men and women of same age will have different energy requirements due to differences in body composition.

Physiological State: *Women are required to consume greater amount of food during pregnancy and lactation.*

During certain physiological states such as pregnancy and lactation, BMR increases and so do the energy requirements. For example, the BMR increases by 5.3 per cent, 11.4 per cent and 25.3 per cent during the first, second and third trimesters of pregnancy to support gestational weight gain of about 12 kg during the nine-month period. The energy requirements thus increase by +350 kcal during pregnancy and +600 kcal during lactation (as per the recommendations of ICMR, 2010).

Illness: *A and B suffered from fever. Both are of same age and gender. They had similar activity pattern, dietary habits and health status prior to illness. A recovered faster than B. B also lost weight.*
The energy requirements increase during most common illness. The basal metabolic rate increases during fever, infection, diarrhoea, sepsis, injury and surgery. For example, for every 1°C rise in body temperature above 37°C, the BMR increases by about 13 per cent. Accident, injury, burn and surgery result in breakdown of cells (catabolism). Recovery from such states requires synthesis of new cells (anabolism). If the body is supplied with adequate energy, the anabolic process is faster and recovery is quick. However, many patients eat less during sickness and therefore tend to recover late and at times lose weight during such periods.

Nutritional Status: *A is having a busy office schedule and is missing at least one meal 3–4 times a week. However, no change has been observed in his body weight.*
During prolonged periods of inadequate food intake, the energy requirements also reduce. When the body is deprived of adequate energy, such as during states of starvation, inadequate food intake and fasting, the BMR adapts to conserve energy by decreasing the metabolic activities of the body tissues. This process can reduce the energy requirement by up to 20 per cent. Such situations are observed in certain populations of underdeveloped and developing nations wherein there is low availability of food or deliberate low consumption of food.

Environment: *A eats more during winters than summers but still manages to maintain same body weight.*
Adverse environmental conditions can influence energy requirements. During winters, some body heat gets lost in the surrounding cold atmosphere. Therefore, the body has to generate greater heat (energy) to maintain normal body temperature. In summers, no significant body heat is lost because the surrounding atmosphere is at times higher than the normal body temperature. The REE increases by 5–20 per cent in cold climates as compared to warm areas. Thus, the energy requirements are higher during winters than summers.

Physical Activity: *A is a marketing executive and B is an accountant. Both are of same age, gender and height. They consume same quantity and quality of food. A is lean, and B is a little overweight.*
Physical activity influences energy expenditure; therefore, energy requirements are higher for individuals who follow an active lifestyle as compared to those who have a sedentary activity pattern. As discussed before, active lifestyle and regular exercise increase fat-free mass and simultaneously reduce fat mass. Fat-free mass increases energy expenditure and hence the requirement for energy also increases. Another factor influencing energy requirement is the choice of activities. For example, standing with weight (bag) requires more energy as compared to standing without carrying load. The energy requirement of men and women leading a sedentary lifestyle is 2320 kcal and 1900 kcal per day, respectively. However, the energy requirement of heavy workers is much more, that is, 3490 kcal and 2850 kcal per day for man and women, respectively (as per recommendations given by ICMR, 2010).

Other Factors: *A consumes eight cups of coffee a day and is a habitual smoker. When he left smoking and reduced his coffee intake, he gained weight slightly.*
The REE has been found to increase by 5 per cent, 9 per cent and 15 per cent among habitual consumers of nicotine, alcohol and caffeine, respectively. The energy requirements of such individuals per se would be higher than non-consumers. However, it must be remembered that coffee and alcoholic beverages itself contain energy which may partially/wholly compensate for the increased energy requirements. Consumption of very spicy food may also raise REE to some extent for as long as three hours.

Thus, the energy required by each individual would be different and would vary on day-to-day basis and during the various phases of life such as pregnancy, lactation, infancy, adolescence and adulthood. Since it is not possible for us to compute our energy requirement without using expensive state-of-the art scientific equipment, ICMR has given recommendations for energy intake for Indian population. These recommendations are age and gender specific. Further, the recommendations are based on reference body weight (such as of reference man and woman; refer to Box 12.2) and need to be modified if the body weight (for height) is more or less (Table 12.2).

Box 12.2 Reference Indian Man and Woman

Parameter	Reference Man	Reference Woman
Age	18 to 29 years	18 to 29 years
Weight	60 kilograms	50 kilograms
Height	1.73 meters	1.61 meters
Body Mass Index	20.3	21.2
Physiological status	Disease free, physically fit	Non-pregnant, disease free, physically fit
Occupation	8 hours	8 hours
Bed rest	8 hours	8 hours
Light work	4-6 hours	4-6 hours
Active recreation	2 hours	2 hours

Table 12.2 Recommendations for daily energy intake (Indian Council of Medical Research, 2010)

Group	Age/Category	Reference body weight (kg)	Energy (kcal)	kcal per kg body weight per day
Men (Adult)	Sedentary work	60.0	2320	39
	Moderate work		2370	46
	Heavy work		3490	58
Women (Adult)	Sedentary work	55	1900	35
	Moderate work		2230	41
	Heavy work		2850	52
	Pregnant	55+GWG*	+350	N/A
	Lactating 0–6 months		+600	N/A
	6–12 months		+520	N/A
Infants	0–6 months	5.4	92 kcal/kg/d	92
	6–12 months	8.4	80 kcal/kg/d	80
Children	1–3 years	12.9	1060	82
	4–6 years	18.0	1350	75
	7–9 years	25.1	1690	67
Boys	10–12 years	34.3	2190	64
	13–15 years	47.6	2750	58
	16–17 years	55.4	3020	55
Girls	10–12 years	35.0	2010	57
	13–15 years	46.6	2330	50
	16–17 years	52.1	2440	47

Note: *GWG: Gestational weight gain.

12.4 Nutrients as Source of Energy

Box 12.3 Approximate Energy Content of Macronutrients		
Nutrients		**kcal/g**
Carbohydrates	:	4 kcal/g
Proteins	:	4 kcal/g
Fats	:	9 kcal/g
Dietary fibre	:	2 kcal/g

The energy requirements of our body are met by the food we eat. Food contains both macronutrients and micronutrients. While macronutrients (carbohydrates, proteins and fats) are sources of energy, micronutrients (vitamins and minerals) help in releasing the energy captured within the macronutrients and also in utilizing the energy for various metabolic functions. The amount of energy obtained upon the metabolism of macronutrients given in Box 12.3. Dietary fibres were earlier not considered to be a source of energy, as they were said to be indigestible. However, recent research has indicated that certain forms of fibre get fermented in the colon to yield short-chain fatty acids such as butyric, propionic and acetic acid. These short-chain fatty acids are utilized as a source of energy by the cells of the colon and to some extent by liver (box 12.3). These nutrients are present in varying amounts in food and therefore no two foodstuffs would contain exactly the same amount of dietary energy.

12.5 Food as Source of Energy

As we all know, food is the fuel for our body. Macronutrients present in food provide energy. However, not all foods provide equal amount of energy because all foods do not contain same amount of all macronutrients. Foods which contain high amount of water and fibre are **low-calorie foods,** whereas those with high amount of carbohydrates and fat are **high-calorie foods**. For instance, fats (oil, ghee, butter and cream) and fat-rich foods such as egg yolk, red meat and full-cream milk are good sources of energy. If fat is removed, partially or wholly from these foods such as in case of toned/skimmed milk, the calorie content reduces. Cereals are also rich source of energy which comes primarily from carbohydrates. On an average, 20 g of cereal provides 70 kcals. A similar amount of sugar, honey and jaggery would provide a little higher amount of energy, that is, 80 kcals (average). Protein-rich foods are moderate sources of energy. One serving (30 g) of pulse provides around 100 kcal, whereas one serving of meat (40–50 g) would provide on an average 70 kcal of energy. Serving of 100 g of mixed fruit/vegetables would provide about 40 kcals of energy. Comparing the energy content of food (raw) or dishes for variable portion sizes is very difficult. However, it is easier to compare foods based on their energy density, that is, amount of energy per gram of food (raw or cooked). For example, the energy content of one medium-sized potato weighing 60 g is 58 kcal, whereas that of 50 g French fries is 165 kcal. Their respective energy density would be:

$$\text{Potato (raw) } 58\ kcal \div 60\ g = 0.96\ kcal/g$$

$$\text{French fries } 165\ kcal \div 50\ g = 3.3\ kcal/g$$

Calorie content of all foods and dishes can be compared by calculating their energy density. Dishes made by incorporating different foods in different amounts would contain varied amount of calories. For example, calories present in 50 ml strawberry ice cream would be lesser than that present in the same amount of mango ice cream, because the energy density of mango (0.74 kcal/g) is higher than that of strawberry (0.44 kcal/g). Judicious selection of all foods in appropriate amounts can help in incorporating just the right amount of calories in our daily diets so as to meet our daily energy requirements. Refer Figure 12.4 for energy density of some foods.

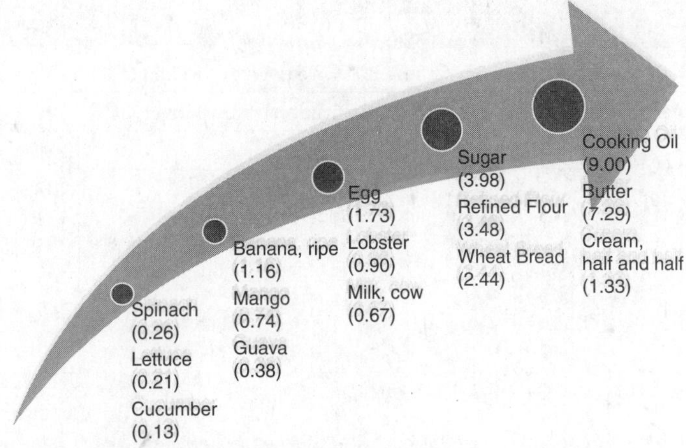

Figure 12.4 Energy density of some foods (kcal per g)

12.6 Energy Balance and Health

We expend energy in the form of several voluntary and involuntary activities. These activities determine our energy requirements which in turn are met through the food we eat. Energy balance or energy homeostasis is said to be maintained when the dietary energy intake through food is equal to the energy spent by the body on various activities (REE + TEF + Activity Thermogenesis). While it is not possible to keep an account of the energy expenditure or intake on a day-to-day basis, it is always desirable to maintain the balance over a period of time (7–15 days). Body weight is a good indicator of energy adequacy or inadequacy. Consuming too much or too little energy even for a few weeks can result in body weight changes. When energy intake is equal to that spent over a period of time, the body weight remains constant. If the intake is more than the output (energy spent), a **positive energy balance** is created and the excess energy gets deposited in the form of adipose tissues resulting in weight gain. On the contrary, if the intake is lesser than the energy expended, the body weight begins to reduce due to **negative energy balance** (refer to Figure 12.5) during which adipose tissues and muscle mass is lost. Thus, during weight gain and loss, the normal body composition may get disturbed. To identify the status of adipose tissues in an individual, body mass index (BMI) can be calculated. BMI refers to relative body fatness of an individual (refer to Box 12.4).

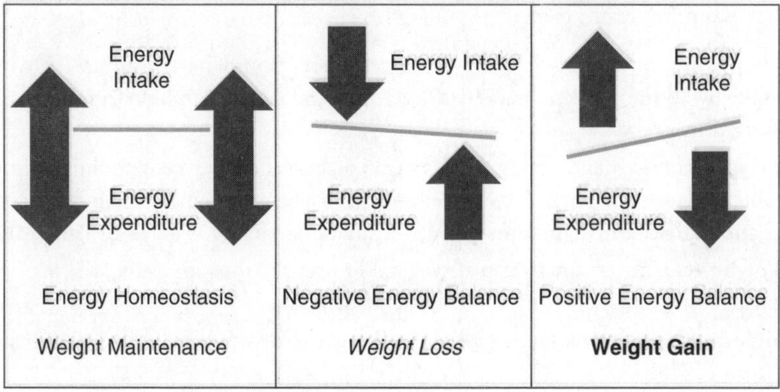

Figure 12.5 States of energy balance

Box 12.4 Body Mass Index

It is a formula used for computing relative body fatness. BMI = Weight (kg) ÷ Height (m^2)

Classification for Overweight and Obesity		**Classification for Underweight**	
Weight Status	BMI (kg/m^2)	Classification	BMI (kg/m^2)
Underweight	<18.5	Normal	>18.5
Normal range	18.5–24.9	Mild Underweight (Grade I)	17.0–18.4
Overweight	25–29.9	Moderate Underweight (Grade II)	16.0–16.9
Obesity	≥30.0	Severe Underweight (Grade III)	<16.0
Class I	30.0–34.9		
Class II	35.0–39.9		
Class III (morbid)	≥40.0		

Source: World Health Organization (1998) and FAO/WHO (1994).

In view of the high prevalence of obesity, hypertension, hyperinsulinemia, impaired glucose tolerance, dyslipidemia and other degenerative diseases among Asians, the World Health Organization (2002) has recommended slightly lower BMI cut offs, that is, 23.0, 27.5, 32.5 and 37.5 kg/m^2 for normal, overweight, class I and class II obesity.

It is important to remember that when weight is gained by an underweight individual, all the body components viz. total fat, muscle, bone and other components should increase proportionally. However, while losing weight, it is pertinent to lose only adipose tissues.

12.6.1 Underweight

When the body weight is 15–20 per cent or more below the accepted weight standards, it is referred as underweight. In other words, if the BMI (refer to Box 12.4) is less than 18.5, the person is said to be underweight. When the energy intake or that available to the body is below the requirement, a negative energy balance occur and there is weight loss or inability to gain and maintain desirable body weight. The causes of underweight are given in Figure 12.6.

Since underweight is often the symptom of a disease, the underlying cause should be medically assessed. Underweight individuals generally suffer from undernutrition, that is, deficiency of several macronutrients and micronutrients. They are associated with higher risk of morbidity and mortality (refer to Figure 12.7).

Dietary modifications: The following dietary strategies should be followed to create a positive energy balance and hence promote weight gain.

1. A high-energy, micronutrient-rich, easy-to-digest diet should be consumed. An increase in the energy intake by about 500 kcal per day above the usual intake can help in increasing about 500 g of body weight in a week.
2. Consumption of energy-dense foods such as sugar, jaggery, honey, rice, semolina, refined wheat flour, starchy roots and tubers (potato, sweet potato, colocasia and yam), whole milk and its products, probiotics and high-calorie fruits and vegetables (mango, banana, papaya and grapes) can be helpful.
3. Addition of butter and cream to soups, vegetable preparations and chapati helps to increase the calorie content without increasing the volume of dishes.
4. Choose high-fat food options such as whole milk instead of skimmed milk, meat instead of fish and butter instead of jam.

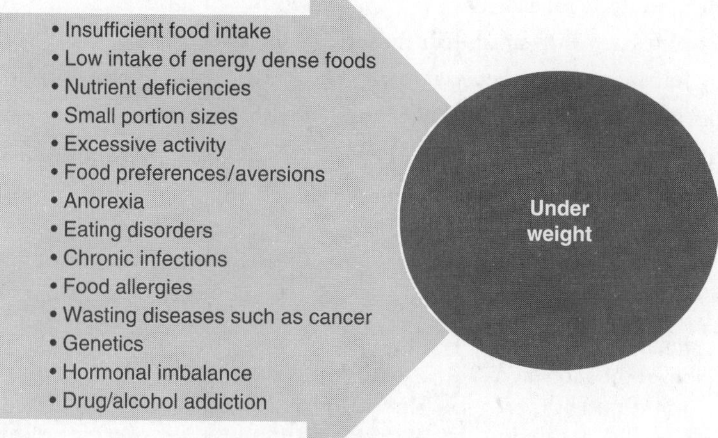

Figure 12.6 Causes of underweight

Figure 12.7 Consequences of weight imbalance: Underweight overweight obesity

5. Preferring juice or stewed fruits over their raw counterparts and creamed soups instead of salads promote higher calorie intake.
6. A regular meal pattern with small frequent meals should be followed.
7. Consuming nutritious snacks between meals can facilitate an improved nutritional status.
8. Consuming water or other beverages after and not with meals ensures adequate food intake.

Since several underweight individuals are anorexic and have multiple nutrient deficiencies, consumption of vitamin supplements and appetite stimulants can be helpful.

12.6.2 Overweight

When the body weight is 10–20 per cent or more, above the accepted weight standards, it is known as overweight. More specifically, an individual with a BMI >25 to 29.9 is overweight and those with BMI above 30 are considered to be obese (refer to Box 12.3). Excessive weight gain occurs when the body is in a state of positive energy balance, that is, energy intake is more than energy expenditure. Excess energy gets deposited in the form of adipose tissues resulting in weight gain. The causes of overweight/obesity are given in Figure 12.8.

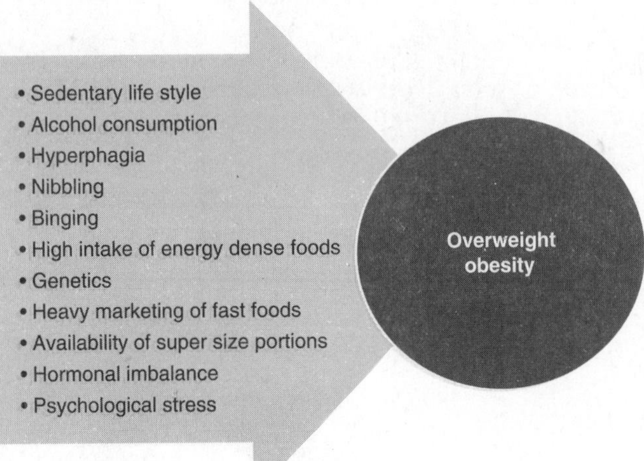

Figure 12.8 Causes of overweight and obesity

Excessive weight gain increases the risk of developing a host of diseases such as hypertension, coronary heart diseases, hyperinsulinemia, diabetes, gout, fatty liver, respiratory problems, cancer, sleep apnea, difficulties in walking, surgical complications, poor wound healing, poor antibody response to hepatitis B vaccine and lowered life expectancy (also refer to Figure 12.7).

GOOD TO KNOW

What Regulates Eating?

The following processes happen in a sequence which regulate our eating food:

Hunger: a physiological response of body to find and eat food, arising in response to deficiency of energy in the body.

Appetite: a physiological and psychological response of body which regulates amount of food consumed.

Satiation: a feeling of fullness.

Satiety: feeling of fullness after consuming a meal and regulates feeling of hunger for next meal.

Certain body proteins also play an important role in regulating our appetite. For example, Leptin, oxyntomodulin, pancreatic peptide and resistin suppress appetite, whereas Ghrelin stimulate it.

Treatment of obesity and management of weight is a challenging task and requires dietary management along with lifestyle modifications, psychological therapy, medication and surgery (if required). A low-energy diet rich in micronutrients and fibre should be given to such patients. A gradual sustained weight loss is generally more beneficial.

Dietary Modifications: The following dietary strategies should be followed to create a negative energy balance and hence facilitate weight loss.

1. A decrease in the daily energy intake by 500 kcal from the usual intake can help in reducing about 500 g of body weight per week. However, calorie-restricted diets should not provide less than 1200 kcal in case of women and 1400 kcal in case of men, unless under strict medical supervision. The energy required to lose desired weight should always be calculated by a qualified dietician.
2. High-calorie carbohydrate and fat-rich snacks, for example, should be substituted with low-calorie protein and micronutrient-rich foods; that is, emphasize on foods with low-energy density.
3. Whole cereals (cracked wheat, jowar, bajra and ragi), whole pulses (soybean, chickpeas and cowpea) and low-calorie/fibre-rich fruits and vegetables such as watermelon, muskmelon, citrus fruits, beans, peas, cabbage and leafy vegetables should be preferred.
4. Low-fibre/high-calorie refined cereals/their products (refined flour, semolina, sago and arrowroot) and high-calorie fruits/vegetables, sweets, bakery products and namkeens should be avoided.
5. Grilling, roasting, non-stick/microwave cooking, steaming and boiling should be preferred as they help to reduce fat intake.
6. Serve/buy small servings/portions of foods for yourself.
7. Breakfast should not be missed, because it retards food cravings.
8. Meals should be started with a bowl of salad wherein the edible peels of vegetables have not been removed. Addition of sprouts/low-fat cottage cheese promotes satiety.
9. Alcoholic beverages, sweetened beverages and fruit juices should be avoided.
10. Certain obese individuals suffer from eating disorders or other diseases which should be identified and treated before starting the weight loss program.

In addition to these dietary modifications, energy expenditure should be enhanced by increasing physical activity. A combination of aerobic, strength and flexibility workouts help to expend energy and improve overall fitness and appearance (refer to Figure 12.9).

Crash diets, body wraps, vibrators, belts and sauna treatments have been proved by research to have side effects. Medical treatment such as medications and surgical procedures such as bariatric surgery, gastroplasty, gastric bypass and liposuction are used for morbidly obese patients.

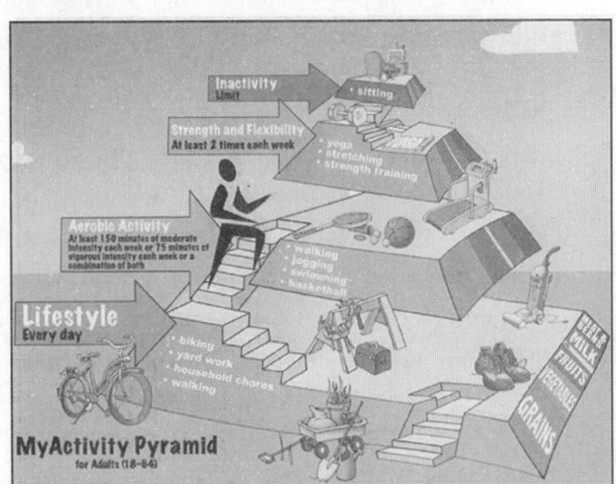

Source: U.S. Department of Agriculture. ChooseMyPlate.gov Website. Washington, DC. Food Science, Nutrition and Safety, http://www.choosemyplate.gov/print-materials-ordering/my-pyramid-archive.html accessed on August 2, 2011.

Figure 12.9 Active lifestyle essential for weight maintenance

Thus, to summarize, it can be said that energy is the fuel for our body and influences the capacity to perform. The energy required by our body is the sum total of resting energy expenditure, thermic effect of food and physical activity. These three components in turn are influenced by several factors such as body size/composition, age, gender, physiological state, illness, nutritional status, environment and alcohol/nicotine/caffeine consumption. Food fulfills the requirement of energy by our body. The amount of energy provided per gram of food is known as energy density. Consumption of high-density foods can promote a positive energy balance and hence weight gain. On the other hand, laying emphasis on low-density food can promote weight loss through negative energy balance in the body. In addition to regulating food intake, an active lifestyle and regular physical exercise is also a must.

Know Your Progress

True or False Questions

1. The REE is higher during growth phases such as infancy but not during pregnancy.
2. Increased physical exercise can raise BMR and hence REE.
3. One teaspoon of fat would provide 45 kcal of energy.
4. Thermic effect of food contributes the maximum towards energy expenditure as compared to the other two components viz. basal energy expenditure and activity thermogenesis.
5. Energy balance is best reflected by constant body weight.

Very Short Answer Questions

1. What is energy?
2. What is the difference between Kilocalorie and Joule?
3. Calculate your body mass index.
4. 'C' is 7 kg below her ideal body weight. Give an estimate of the time to gain the required weight.
5. What is negative energy balance?

Short Answer Questions

1. Describe basal metabolic rate and resting energy expenditure.
2. 'G' is a college student who is fond of outdoor games and carries out aerobic exercise every day. 'K' is a software engineer who does not get time for regular exercise. Comment upon their energy requirement.
3. Write your daily activity pattern. Identify five key areas/factors (each) which may reduce and increase your REE.
4. How does food meet our energy requirements?
5. Discuss the association of energy balance with health and disease.

Long Answer Questions

1. What are the components of energy expenditure?
2. Elaborate on the factors which influence energy requirements.

Beyond the Text

- Different foodstuffs have different calorie content. You can check that yourself. Compare the calorie content (energy density) of packaged soft (carbonated) drink with fruit drink and fruit juice.

Chapter 13

Macronutrients

Chapter Objectives

After reading this chapter, the reader should be able to:

- describe different types of carbohydrates, proteins and fats.
- have a clear understanding regarding the functions of each macronutrient.
- correlate foods with the macronutrient most abundant in them.
- know the effects of low and excessive intakes on health.
- ensure optimal intake of macronutrients.

Consider the following situations:

Situation 1: A and B were obese. To reduce weight, A decided to consume a no-fat diet, whereas B reduced the total food intake. After three months, both lost some weight; but, A had dry scaly skin.

Situation 2: C is a busy office executive and often avoids salads and fruits. He is suffering from diabetes and constipation and is increasingly becoming dependent upon medicines to manage his health.

Situation 3: D is an experienced chef. While trying a new 'healthy heart' recipe, he decided to add flaxseeds and oat flour in the buns. His attempt was highly appreciated.

Dietary errors related to choice of food and hence macronutrients can spoil health adversely as can be seen in situation 1 and 2. However, correct food choices can help in improving our health and that of our society (situation 3). Therefore, knowledge about macronutrients is essential to remain healthy and prevent the development of diseases.

Macronutrients viz. carbohydrates, fats and proteins (Chapter 11) make much of the weight of the food we incorporate in our daily diets. In fact, our body structure depends to a great extent upon the amount of each of these macronutrients consumed. They contribute significantly to normal body composition. Whether a body is lean, heavy or muscular depends on as to how much of these macronutrients are usually present in our plate.

13.1 Carbohydrates

While you are reading these words, billions of glucose molecules are being used by brain cells to remain active during this activity. Carbohydrates, which are produced by plants through the process of photosynthesis, act as fuel for our body just as petrol does for a car. They are required by the human body to perform almost all physical activities such as sleeping, brushing teeth, climbing stairs, running, talking and cooking, to name a few.

13.1.1 Classifications Based on Function

Carbohydrates have been categorized in a number of ways; classification based on structure (Chapter 2) is more relevant to food science because the structure influences interaction of this nutrient with other food components. However, this classification does not explain the variety of physiological effects exhibited by various classes of carbohydrates with relevance to health and nutrition. Therefore, carbohydrates have been classified as simple and complex, available and partially available, as well as extrinsic and intrinsic sugars. Such classifications are overlapping and therefore the terms are often used interchangeably. According to Food and Agriculture Organization (FAO), World Health Organization (WHO) and United States Department of Agriculture (USDA), several terms can be used to describe a single or a group of carbohydrates performing similar functions. The commonly used terms and classifications are discussed below.

Sugars: The term 'sugars' is conventionally used to describe the monosaccharides and disaccharides. 'Sugar', by contrast, is also used to describe purified sucrose as are the terms 'refined sugar' and 'added sugar'.

Simple and Complex Carbohydrates: Monosaccharides and/or disaccharides are known as simple carbohydrates such as glucose, fructose, galactose, sucrose, lactose and maltose. On the other hand, oligosaccharides and polysaccharides are known as complex carbohydrates such as starch, glycogen and fibre (cellulose, hemicellulose, pectins, gums, mucilages, lignins, glucans, chitin, chitosans). Simple carbohydrates get digested more efficiently and metabolized faster than the complex ones. It is important to note that, the term 'complex carbohydrate' is often used to encourage consumption of whole grains, fruits and vegetables and not starches and therefore this classification is no more popular.

Available and Partially Available Carbohydrates: Carbohydrates which can yield 4 kcal energy per gram upon digestion and absorption are known as available carbohydrates. However, those which though are indigestible by the human digestive tract but get fermented in the colon to yield short-chain fatty acids are known as partially available carbohydrates or fibre, and they yield approximately 2 kcal energy per gram. All carbohydrates other than fibre are referred to as available carbohydrates. Earlier, partially available carbohydrates (i.e. fibres) were not considered to be a source of energy and hence were termed as 'non-available' (refer to Box 13.1).

Box 13.1 What Is Fibre?

The first description of dietary fibre was given by Trowell in 1972. Dietary fibre refers to 'naturally occurring structural parts of plants i.e. non-digestible carbohydrates (non-starch polysaccharides) and lignin, which cannot be broken down by enzymes in the human digestive tract and are intact and intrinsic to plant foods'. Cellulose, hemicellulose, pectins, gums, mucilages, lignins, glucans, chitin and chitosans are some examples of fibre. Dietary fibres are of two types: soluble and insoluble.

Soluble Fibre	Insoluble Fibre
• Dissolve in water	• Do not dissolve in water, remain suspended
• Form viscous gels in food preparations and stomach	• Do not form viscous gels in food preparations and stomach
• Easily fermented by bacteria in colon	• Less readily fermented by bacteria in colon
• Useful in the prevention and treatment of obesity, heart diseases, diabetes, ulcers etc.	• Useful in the prevention and treatment of constipation, diverticulosis and obesity
• Legumes, beans, oats, barley, lady fingers and certain citrus fruits are good sources.	• Husk/bran of cereals, pulses, legumes/beans, seeds and edible peels of fruits and vegetables.

If dietary fibres are extracted from plants or parts of plants, or are manufactured and added to foods or used as supplements with the objective of improving health, then they are called *functional fibers*. Psyllium husk (*isabgol*), guar-gum powder, xanthum gum and pectin are commonly used functional fibres.

Glycemic and Non-glycemic Carbohydrates: Carbohydrates which can raise blood glucose level after consumption and/or can participate in metabolic processes of body are known as glycemic carbohydrates. On the other hand, those carbohydrates which do not have the potential to raise blood glucose level (to any extent) and/or cannot participate in metabolic processes of body are known as non-glycemic carbohydrates. These terms are often used with relevance to diabetes.

13.1.2 Functions of Carbohydrates

This macronutrient provides a major proportion of the energy we consume. It gives us the energy required for survival, work and play. It also helps to maintain the health of our digestive system and protects us from several diseases. The major functions of carbohydrates (Figure 13.1) include:

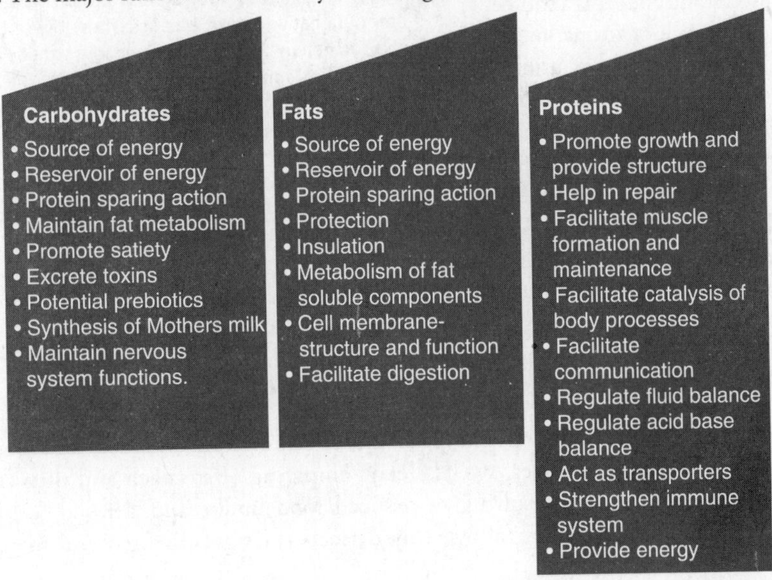

Figure 13.1 Functions of macronutrients: carbohydrates, proteins and fats

Source of Energy: Subsequent to the process of digestion and absorption, disaccharides, oligosaccharides and polysaccharides get converted into monosaccharides; the principal monosaccharide being glucose. Glucose is the primary fuel for most cells in the body and the preferred fuel for the cells of the brain, nervous system, red blood cells, placenta and foetus. Each gram of glucose upon oxidation yields 4 kcal energy.

Reservoir of Energy: Whenever we consume carbohydrates above our needs, the excess amount gets converted to glycogen in the liver and muscles. During states of negative energy balance such as during fasting, skipping meals, rigorous activity or illness, liver glycogen gets broken down readily to glucose for providing energy as needed. Muscle glycogen helps to provide energy to muscle tissues during physical activity.

Protein Sparing Action: During periods of frequent or prolonged starvation (negative energy balance), liver glycogen reserves get depleted and in such situations the body explores for alternative fuels to provide energy for survival and maintenance. In such situations, if dietary carbohydrates are not present, dietary proteins and latter body proteins are broken down to release energy. Dietary proteins should not be used for supplying energy as their primary and most important function is to form new cells. Thus, our diet must contain adequate amount of carbohydrates so as to spare the proteins for their own specialized functions.

Maintains Fat Metabolism: When carbohydrates fall deficient in the diet, adipose tissues are broken down to release energy. Even for this purpose, a small amount of carbohydrate is required. However, if carbohydrates are completely absent in the diet, the body is not able to break adipose tissues completely to release energy. It, instead, produces small compounds known as ketone bodies to release energy. Most cells can use ketone bodies as a source of energy. When the rate of formation of ketone bodies exceeds the rate of utilization, they get accumulated in the blood (ketonemia). Ketone bodies are acidic in nature and hence cause a change in pH of blood (ketoacidosis). In an attempt to maintain normal blood composition, the body tries to excrete excessive ketone bodies by increasing urination, excessive urination increases the chances of dehydration. This entire condition is known as **ketosis** and can be prevented by consuming adequate amounts of carbohydrates daily.

> **THINK TANK**
>
> **Are *Low Carb Diets* Good for Health?**
>
> Carbohydrates must provide 45–65 per cent of the total energy consumed in a day. In 'low carb diets', carbohydrates provide less than 45 per cent of the total energy intake. Such diets are low to moderate sources of fat and slightly high in proteins. Being deficient in energy, such diets create an environment of negative energy balance involving the onset of ketosis. Proteins consumed in excess of the requirement do not result in greater muscle mass; instead, they are used to provide energy during states of negative energy balance. Thus, low carb diets are not good for health unless prescribed by a dietician and taken under medical supervision.

If carbohydrates are consumed in excess of the requirements and the glycogen reserves are full, the extra amount gets converted into fat cells which are stored as adipose tissues in the body. Unlike glycogen, there is no upper limit to the amount of storage fat which can be accumulated in the body. Over a period of time, this can result in over weight or obesity.

Promote Satiety: Different carbohydrates have variable effects on hunger, satiety and food intake. Fibre absorbs water from the digestive juices, swells, delays emptying of stomach and thus creates an early and prolonged feeling of fullness resulting in reduced food intake and delayed feeling of hunger. On the other hand, monosaccharides followed by disaccharides get digested and metabolized faster than starches.

Excretion of Toxins: Glucuronic acid, a by-product of glucose metabolism, combines with several chemicals, bacterial toxins and waste products in blood, thereby, helping in their excretion. Dietary fibre also promotes excretion of harmful substances such as adulterants and pesticides.

Potential as Prebiotics: Dietary fibres particularly fructooligosaccharides (fructans) have prebiotic properties, that is, they have the ability to selectively stimulate the growth or activity of beneficial bacterial species already residing in the colon (probiotic). Insoluble fibre such as lignin and cellulose get fermented by the colonic bacteria to produce short-chain fatty acids (SCFAs). SCFAs have been found to enhance sodium and water absorption, enhance

> **THINK TANK**
>
> **Why do we feel hungry earlier after having rice than *chapati*?**
>
> *Chapati* made from whole wheat flour contains the outer husk of wheat grain and hence has a higher amount of fibre than rice (outer husk is removed during milling). Fibre delays emptying of stomach and hence gives a prolonged feeling of fullness. It is for this reason that when we eat rice or bread (low fibre) in a meal, there is an early feeling of hunger as compared to when *chapati* made from whole wheat flour or millet flour (high fibre) is consumed.

blood flow in colon, stimulate autonomic nervous system, increase production of gastrointestinal hormones and decrease risk of tumour. Thus, fibre promotes gut health.

Synthesis of Milk: Glucose and galactose combine to form lactose in the presence of mammary enzymes during lactation. Several oligosaccharides are also present in mother's milk and have been found to protect the infant from several disease-causing pathogenic microorganisms.

THINK TANK

Why do we feel hungry earlier after having rice than chapati?

Chapati made from whole wheat flour contains the outer husk of wheat grain and hence has a higher amount of fibre than rice (outer husk is removed during milling). Fibre delays emptying of stomach and hence gives a prolonged feeling of fullness. It is for this reason that when we eat rice or bread (low fibre) in a meal, there is an early feeling of hunger as compared to when *chapati* made from whole wheat flour or millet flour (high fibre) is consumed.

Nervous System Function: Glucose is an essential fuel for the central nervous system. It promotes cognitive performance especially alertness and memory processing. Adequate intake of carbohydrates, especially in the breakfast, facilitates better work performance.

13.1.3 Carbohydrates in Health and Diseases

Carbohydrates are the principal component of our diet, and they influence our health to a great extent. While dietary fibres have been found to be protective against gaining excess weight, constipation, diverticulosis, appendicitis, diabetes, heart diseases, cancer and gout; their intake needs to be reduced during undernutrition, diarrhoea, pancreatitis and hepatitis. Similarly refined sugars and starches are a boom for sports persons, underweight individuals and those suffering from digestive disturbances but their intake needs to be monitored during obesity, diabetes, certain heart diseases and constipation.

Sugars and Nutrient Intake: Foods rich in carbohydrates such as cereals, tubers and certain roots contain other nutrients as well, such as minerals, vitamins, proteins and fat but in small amounts. On the other hand, foods which are pure carbohydrates, such as sugar, molasses, honey or the products containing them as the only principal ingredient, either contain or do not contain other nutrients in traces (Table 13.1). Such foods are energy dense but are not nutrient dense; they are often termed as 'empty calories'. The intake of such foods should be kept to a minimum so as to prevent the development of other nutrient deficiencies.

Dental Caries: Sugar and starch rich foods have been found to be associated with greater risk of developing dental caries. This is so because after the consumption of sweet or carbohydrate rich foods, bacteria in the mouth ferment the sugars and, in this process, produce acid which erodes tooth enamel.

Weight Management: People who consume high fibre diets often weigh less than those who consume less fibre. Foods rich in fibre are less dense in energy as compared to those which are good sources of starches, disaccharides and monosaccharides. Such foods take a longer time to masticate and hence contribute to a lower consumption. Compare the time taken to consume an apple along with its peel and the juice extracted from it. Fibre also contributes to feeling of fullness and satiety. Therefore, overweight/obese individuals benefit by consuming a high-fibre diet whereas, those who are underweight need to consume a greater amount of sugars and starches. Several weight loss products available in the market have good amount of fibre added to them.

Table 13.1 Comparative nutrient content of certain foods

Foodstuff	Size of 100 kcal Portion	Carbohydrates (g)	Protein (g)	Calcium (mg)	Iron (mg)	Vitamin A (µg)	Vitamin C (mg)
Sugar, white	2 tbsp.	24	0	Neg.	Neg.	0	0
Molasses	2.5 tbsp	28	0	343	12.6	0	0.1
Cola type beverage	1 can	26	0	6	Neg.	0	0
Honey	1.5 tbsp	26	Neg.	2	0.2	0	Neg.
Milk, cow	150 ml	6.6	4.8	180	0.3	79.5	3.0
Soybeans	23 g	4.8	9.9	55.2	2.3	100.2	Neg.
Orange	208 g	22.6	1.4	54.1	0.6	309.1	62.4
Lettuce	476 g	11.9	9.9	238	11.4	4712.4	47.6

Diabetes: High fibers, i.e low glycemic index diets, are helpful for the prevention and management of diabetes. Fiber prolongs transit time of food from the stomach to the small intestine. This results in a gradual and sustained rise in post prandial (after meals) blood glucose. In case of low fibers, i.e high glycemic index diets, there is a sharp rise followed by a steep decline in blood glucose. A high fiber diet reduces the requirement for drugs/insulin and the risk of developing complications among diabetics. Box 13.2 gives the glycemic index of certain foods which is a measure as the glycemic index of certain foods which is a measure of the change in blood glucose after the consumption of carbohydrate-containing foods.

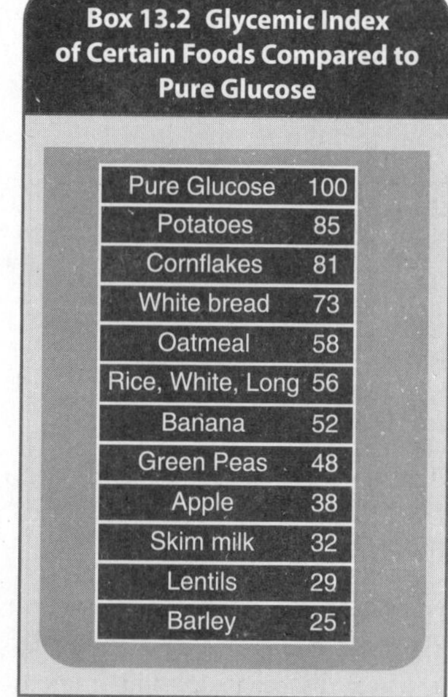

Box 13.2 Glycemic Index of Certain Foods Compared to Pure Glucose

Food	Glycemic Index
Pure Glucose	100
Potatoes	85
Cornflakes	81
White bread	73
Oatmeal	58
Rice, White, Long	56
Banana	52
Green Peas	48
Apple	38
Skim milk	32
Lentils	29
Barley	25

Heart Diseases: Soluble fibre such as those present in oats (β-glucan), cluster beans (guar-gum), fenugreek seeds, legumes particularly soybean, kidney beans, horse gram, bael fruit and papaya are particularly useful in conferring protection against heart diseases by promoting reduction in weight, blood pressure and blood lipids (cholesterol).

Gastrointestinal Disorders: Dietary fibre are helpful in preventing a number of gastrointestinal diseases. Fibre absorbs water molecules thereby making the stools soft and bulky. Such stools are easier and faster to eliminate. Thus, fibre helps to maintain normal bowel movements, alleviate constipation and prevent appendicitis/diverticulosis.

13.1.4 Food Sources

Carbohydrates are abundantly available in plant foods (Table 13.1 and Figure 13.6a). Animal products contain carbohydrates but comparatively in lesser amounts with exception to milk.

Available Carbohydrates: Cereals (rice, wheat, jowar, bajra, ragi, sago) and cereal products (semolina, refined flour, rice flakes, breads, pastas), starchy roots and tubers (arrow root, colocasia, sweet potato, tapioca, yam), legumes, sweet fruits (apricot, currants, dates, raisins, banana, mango, sapota), certain vegetables and natural sweeteners such as sugar, jaggery, honey and molasses contain good amount of available carbohydrates.

Partially Available Carbohydrates: Several foods are rich in different types of fibre such as the following:

Insoluble fibre: Whole grain cereals and pulses, their bran/husk, vegetables and fruits along with edible peels are good sources of cellulose and hemicellulose. Edible seeds, and mature vegetables, especially their edible peels, contain appreciable amounts of lignin.

Soluble fibre: Gums are present in good amount in oats, legumes, broad beans, barley, lady fingers and fenugreek seeds. Pectin is concentrated in guava, apple, strawberries, carrots and citrus fruits.

Functional fibre: Fibre is increasingly being extracted from food and used as a dietary supplement. Some salient functional fibres and their food sources include:

Chitin: shells of crab and lobsters.
Fructooligosaccharides (fructans): onion and chicory.
Beta-glucans: oats, barley, certain types of seaweed and algae.
Psyllium: husk of psyllium seeds.

13.1.5 Deficiency and Excess

Carbohydrates should comprise 45–65 per cent of the total energy intake. Thus, a 2000 kcal diet should provide around 225–325 gm of carbohydrate. When the carbohydrate **intake is deficient** for a long duration, there is resultant ketoacidosis in addition to depleted glycogen reserves and weight loss. This can cause damage to the liver and kidney. It is for this reason that 'low-carb diets' should be taken only under strict supervision of a nutritionist. When carbohydrate intake is low in the diets of infants and preschoolers it causes marasmus and kwashiorkor. Such children are stunted and wasted and have low BMI for age.

Around 25–35 gm of dietary fibre should be consumed in a day. However, nearly 50 per cent of Indian diets are deficient in fibre. The most common problem associated with low intake is constipation. Prolonged intake of low fibre diets increases the risk of developing obesity, hypertension, elevated lipids, diabetes, cancer, gout and appendicitis.

Excessive intake of carbohydrates causes hyperglycemia and resultant hyperinsulinemia and insulin resistance. Continuous exposure to elevated glucose causes oxidative damage to cell membranes, thereby reducing their lifespan. A high intake of carbohydrate is associated with obesity, diabetes, heart disease and cancer. Excessive intake of fibre or a rapid increase in the intake of fibre above habitual amounts can cause abdominal discomfort, bloating (gas) and diarrhoea. If water intake is not increased along with an increase in fibre intake, it can cause obstruction of the gastrointestinal tract due to the presence of dry and hard stools. Fibre also hinders absorption of certain minerals such as iron and calcium.

13.2 Fats

Fat-rich foods are almost always found delicious by most of us, yet we often eat them with guilt. This is so because, we frequently come across messages related to the negative aspects of fat on our health. However, fat is an essential nutrient which performs several important functions in our body and therefore it needs to be consumed everyday but in moderation, that is, as per the requirements of our body. In this chapter, the word 'fat' refers to lipids which are solid and 'oils' for lipids which are liquid at room temperature (25°C).

13.2.1 Classification of Fatty Acids

Fats (lipids) are composed of fatty acids linked together to form chains of variable lengths. Each fatty acid is composed of chains of carbon atoms. The number of carbon atoms in each chain may vary from 4 to 24. Depending upon the arrangement of hydrogen atoms with carbon atoms, the fatty acids are classified (Figure 13.2) as saturated or unsaturated (details in Chapter 3).

Saturated Fatty Acids (SFAs) are those which carry maximum possible number of hydrogen atoms in their carbon chain such as stearic acid present in chocolates and meats. They are not considered good for health, as they tend to elevate blood cholesterol (Low density lipoprotein cholesterol) levels and hence increase the risk of developing coronary heart diseases.

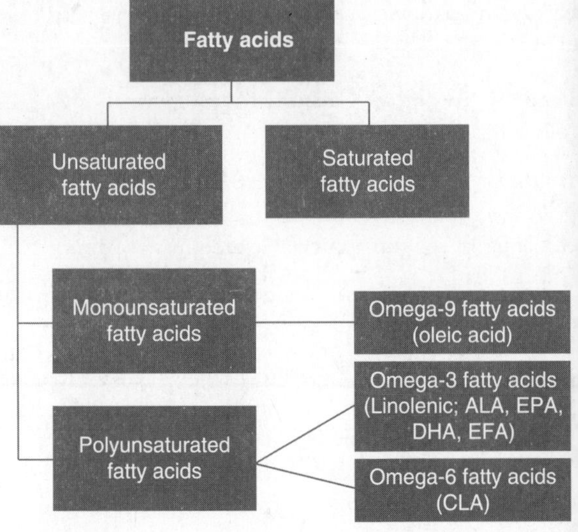

Figure 13.2 Fatty acids significant for diet and health

GOOD TO KNOW

What Is the Association of Blood Lipids with Coronary Heart Diseases (CHD)?

Lipids (cholesterol, triglycerides and phospholipids) circulate in the blood bound to proteins and are termed as 'lipoproteins', which are globular, high molecular weight particles. They can be divided into five major classes: chylomicrons, very low density lipoproteins (VLDL), intermediate density lipoproteins (IDL), low density lipoproteins (LDL) and high density lipoproteins (HDL). The sum of the cholesterol present in these lipoproteins is known as total cholesterol.

The cholesterol present in these lipoproteins (except HDL) accumulates in the arteries (fatty streaks-atherosclerosis), restricting blood flow and raising blood pressure. While LDL is most atherogenic, HDL confers protection from CHD. VLDL levels are indicative of triglyceride concentration in blood. Elevated triglycerides are a risk for developing atherosclerosis–an initial stage of coronary heart diseases (CHD) which can progress to develop fibrous plaques and myocardial infarction or stroke. The desirable values of various blood lipids as given by National Cholesterol Education Program are:

Adults

Total cholesterol:	<200 mg/dl	HDL cholesterol:	≥60 mg/dl
LDL cholesterol:	<100 mg/dl	Triglycerides:	<150 mg/dl

Children and adolescents

Total cholesterol:	<170 mg/dl	LDL cholesterol:	<110 mg/dl

Diet and physical activity influence our lipid profile.

Jnsaturated Fatty Acids (USFAs) lack hydrogen atoms in their structure and therefore have one or more ouble bonds. These include monounsaturated and polyunsaturated fatty acids.

Monounsaturated Fatty Acids (MUFAs) or omega-9 fatty acids lack two hydrogen atoms and therefore ave one double bond in their structure such as oleic acid present abundantly in olive and canola oil. 'hey lower blood cholesterol levels and confer protection against coronary heart diseases.

olyunsaturated Fatty Acids (PUFAs) lack four or more hydrogen atoms and therefore have two or more ouble bonds in their structure. Two types of PUFAs are of particular nutritional significance, that is, mega-3 and omega-6 fatty acids.

1. **Omega-3 fatty acids** such as *linolenic fatty acid* present in flaxseeds and walnuts, especially *alpha-linolenic acid (ALA)*, provide protection from inflammatory diseases such as those associated with joints and heart. Eicosapentaenoic and docosahexaenoic acids are also of great nutritional significance.
2. **Omega-6 fatty acids** such as linoleic fatty acids particularly *conjugated linoleic acid (CLA)*, present in good amounts in oils obtained from sunflower and safflower, have anti-carcinogenic properties.

'hile the discussed classification of fatty acids is based on their structure, they can also be classified based 1 their need by the body.

ssential and Non-essential Fatty Acids

ur body needs several fatty acids. Though our body can synthesize most fatty acids, it cannot synthesize few. The fatty acids which our body cannot synthesize and must therefore be present in our daily diets e known as essential fatty acids (EFAs). The fatty acids, which our body can synthesize when required, e known as non-essential fatty acids (NEFAs). Linolenic (ω-3 fatty acid) and linoleic (ω-6 fatty acid) are 'As. Arachidonic acid was earlier considered to be an EFA but has now been found to be synthesized from oleic acid in the body.

3.2.2 Functions of Fats

ts are essential to our survival and for maintenance of various body processes.

urce of Energy: Dietary fats are a concentrated source of energy. One gram of fat provides 9 kcals which a little more than the amount provided by similar amount of carbohydrates. A tablespoon of oil (pure fat) ould provide 135 kcals, whereas a tablespoon of sugar (pure carbohydrates) would provide only 50 kcal. us, inclusion of fat in a dish can help to make it more energy dense. For example, a bowl of tomato soup th cream would provide more calories as compared to the one without cream.

servoir of Energy: When we consume dietary fat and carbohydrates above our requirements, they get nverted to body fat and are stored as adipose tissues in adipocytes. During states of negative energy balance ich as fasting, starvation, skipping of meals), subsequent to depletion of liver glycogen, the adipose tissues broken down to provide energy to the body for performing various activities. Thus, fat helps people to vive during states of prolonged starvation such as natural calamities and chronic illness; a function similar that observed in animals hibernating during winters.

Protein Sparing Action: This function is similar to that of carbohydrates. During states of negative energy ba ance when dietary carbohydrate intake is also deficient, dietary and body fats are used to provide the necessa energy so that proteins can be spared for performing their own valuable functions (other than providing energy

Protection: Body fat (as visceral, adipose, subcutaneous) comprises nearly 8–35 per cent of total body weigh The visceral fat provides a protective cushion around delicate organs such as kidneys, heart, lungs, stomac gall bladder and reproductive organs. It also provides protection to bones. For example, fat pads on the pal and buttocks protect the bones from mechanical pressure; a reason why it does not hurt us while sittin walking or performing other activities. People who have very little body fat tend to develop pressure sore which during the presence of essential fatty acid deficiency gets covered with very fine hair called lanugo.

Insulation: The subcutaneous body fat helps to insulate the body. It helps to conserve body heat durir winters and prevents the body to get heated up during summers. Subcutaneous fat helps us to survi extreme climates. Individuals who have very little fat below their skin find cool temperatures intolerable ar even room temperature may be felt uncomfortably cool.

Metabolism of Fat Soluble Components: The digestion and absorption of fat soluble vitamins (A, D, and K) and certain phytochemicals such as carotenoids and lycopenes are facilitated by dietary fats becau they carry these components throughout the process of digestion and absorption. Thus, the absorption vitamin A and D would be higher through the intake of full-cream milk rather than skimmed or toned mil During states of malabsorption such as chronic diarrhoea, the digestion/absorption of fat gets reduced ai can thus result in deficiency of fat soluble vitamins and phytochemicals.

Cell Membrane: Fat is a structural component of cell membranes. This function is performed phospholipids. Cell membranes comprise of two layers of phospholipids and small amount of cholester which facilitates selective entry of certain fat and water soluble substances. Phosphatidylcholine, phospholipid, facilitates neurotransmission by nerve cells.

Facilitate Digestion: The presence of fat in the stomach stimulates biliary and pancreatic secretio It also slows down gastric emptying, thereby facilitating digestive process. In the stomach, dieta phospholipids help break fats into tiny particles for easier digestion. Phospholipids in the bile jui (secreted by gall bladder) promote emulsification of fats in the small intestine.

13.2.3 Significance of Fatty Acids in Maintaining Health

Dietary fats are often linked to obesity, heart diseases and cancer. High-fat meals create positive ener balance which results in deposition of fat cells in the body. At the same time, consuming a very low-fat d or fat-free foods has not been found to be an effective solution to reduce weight. When fat intake is neg gible, there is a subsequent increase in the intake of food or carbohydrates due to early feeling of hung Different fatty acids have variable effects on health, the salient ones are discussed.

1. **Saturated Fatty Acids:** These fatty acids increase the risk of developing heart diseases and cancer the breasts and colon. In general, SFAs tend to elevate blood cholesterol when substituted for carb hydrates or other fatty acids. For every 1 per cent increase in the total energy intake from SFAs, tl total blood cholesterol levels increase on an average by 2.7 mg/dl which increases the risk of develo ing heart diseases. Butterfat, coconut and palm kernel oils are rich in lauric, palmitic and myrist fatty acids which promote heart diseases. Therefore, the American Heart Association has suggest that SFAs should not provide more than 7 per cent of the total energy.

2. **Monounsaturated Fatty Acids (MUFAs):** MUFAs provide protection from heart diseases. When SFAs are replaced by MUFAs, there is reduction in the levels of total cholesterol, LDLc and triglycerides. MUFAs have been found to be helpful in raising HDLc (bad cholesterol) levels (good cholesterol) in blood. Therefore, consumption of olive oil, canola oil and fish is being advocated to help in reducing the incidence of hypertension and CHD.
3. **Polyunsaturated Fatty Acids (PUFAs):** Different PUFAs have variable effects on serum lipids and hence the risk of heart diseases. If SFAs are replaced by PUFAs in a low-fat diet, there is a resultant drop in total cholesterol, LDLc and HDLc levels. Increased intake of omega-6 PUFAs lowers total cholesterol. Omega-3 fatty acids (DHA and EPA) have been found to be beneficial in lowering blood pressure, reducing blood clot formation and lowering elevated serum triglycerides. These fatty acids tend to reduce inflammation which is beneficial during several diseases such as pancreatitis, hepatitis, arthritis and gout. ω-3 and ω-6 fatty acids are precursors of certain hormone-like compounds known as eicosanoids. Eicosanoids formed from ω-6 fatty acids (arachidonic) increase blood pressure, heart rate, blood clotting, immune response and inflammation. On the other hand, eicosanoids formed from ω-3 fatty acids have contrary action, that is, heart-healthy effects. Conjugated linoleic acid (CLA) is present in dairy products, and fat present in the meat of lamb, veal, calves and cow/buffalo. CLA has been found to confer protection against cancer (inhibits growth and metastasis of tumour cells) and coronary heart diseases but it adversely affects the rate of bone formation.

   Omega-3 to Omega-6 Ratio: In view of the variable effects of ω-3 and ω-6 fatty acids on health, it is important to maintain an ideal ratio of 5:1–10:1 which is almost four times lower than the current intake. It has been found that increasing the intake of ω-3 fatty acids is far more beneficial to health than reducing the intake of ω-6 fatty acids in an attempt to achieve an ideal ratio. The increased consumption of vegetable oils has led to greater intake of ω-6 fatty acids. In order to improve the intake of ω-3 fatty acids, flaxseed/canola/soybean oils, green leafy vegetables (purslane) and marine foods such as mackerel, salmon, sardines, cod liver, crab, shrimp and oysters need to be incorporated in daily dietaries.
4. **Trans Fats:** As discussed in Chapter 3, trans fats have caught the attention of health and food industry professionals. Trans fats have been found to elevate total cholesterol and LDL cholesterol even if consumed in moderate amounts, that is, 3 per cent of total calorie intake. However, when the intake of trans fats exceed 6 per cent of total calories consumed, it reduces HDL cholesterol as well. The intake of trans fatty acids should be kept to least possible by choosing 'trans fat free' products, preferring vegetable oil over hydrogenated fat, soft (liquid or tub) margarine instead of hard (stick) margarines and avoiding deep-fat fried foods made by 'halwais'.

13.2.4 Cholesterol: Basic Concepts

Cholesterol is a sterol found in cell membranes of all animal tissues. It is not present in plant food. Cholesterol is present both in our diet and in our body. The cholesterol pool of our body comprises of cholesterol synthesized in our body (endogenous) and that obtained from diet (exogenous). Cholesterol in our body performs several functions such as the following:

1. It is involved in the synthesis of sex hormones (testosterone) and adrenal hormones (cortisol).
2. Cholesterol is a structural component of cell membranes. Almost 90 per cent of cholesterol in body is present in cell membranes.
3. It is a component of vitamin D (7-dehydrocholesterol).

Desirable total cholesterol levels in blood should be <200 mg/dl. In order to maintain desirable levels of total cholesterol, the daily intake of dietary cholesterol in healthy individuals should not exceed 300 mg, whereas in case of heart patients, the intake should be kept below 200 mg. Intake of dietary cholesterol above the recommended levels has similar effects on blood lipids as SFAs. A 25 mg increase in dietary cholesterol raises serum cholesterol by about 1 mg/dl. The LDLc raising effect of cholesterol is more pronounced when coupled with a higher intake of SFAs. Food sources of dietary cholesterol are given in sub Section 13.2.5.

13.2.5 Food Sources of Fats and Cholesterol

We consume fat in two forms viz. visible and invisible (Figure 13.6b). Visible fat includes the fat/oil used as cooking medium such as oils, butter, ghee, cream and vanaspati. Invisible fat refers to the fat present inherently in natural foods such as the fat present in milk, nuts, meat, eggs, cereal grains and legumes. Pure sugars such as table sugar, jaggery, honey and molasses do not contain fat. All the visible and invisible fats present in our diet are composed of different fatty acids in variable proportions. Typical food sources of significant fatty acids present in our diet are given in Table 13.2 and fatty acid composition of certain fats/oils is given in Table 13.3.

Saturated Fatty Acids: Animal food particularly red meat (obtained from cow, buffalo, goat and swine), full-cream milk and dairy fat products (cream, butter, ghee), egg yolk, palm oil, coconut oil and cocoa butter are good sources of SFAs.

Polyunsaturated Fatty Acids: Oils obtained from corn, sunflower, soybean and cottonseed contain appreciable amounts of PUFAs.

Monounsaturated (Omega-9) Fatty Acids: Oils obtained from olive, canola, peanut and safflower, marine animals, fish oils and most nuts are rich in MUFAs.

Table 13.2 Average fatty acid composition of various food groups

Food Group	SFA (per cent)	PUFA (per cent)	MUFA (per cent)
Meats, fish, poultry	39	18	35
Fats and oils	34	68	48
Milk and milk products	20	2	8
Legumes and nuts	2	6	4
Eggs	2	2	2
Others	3	6	3

Omega-3 fatty acids: Canola oil, soybean oil, olive oil, walnuts, peanuts, pistachios, pecans, almonds, filberts, flaxseeds and certain green leafy vegetables are good sources of alpha linolenic acid. DHA and EPA are present in good amounts in fish such as mackerel, tuna, salmon, herring, trout and cod, pacific oysters and human milk. Flaxseed oil contains a very high amount (55 per cent) of omega-3 alpha-linolenic acid. A half tablespoon of canola oil provides an equal amount of alpha-linolenic acid.

Omega-6 fatty acids: Whole grain cereals and pulses, whole eggs, poultry, most vegetable oils and margarine are good sources of ω-6 fatty acids.

Trans Fatty Acids: Hydrogenated fats such as vanaspati, margarines (stick), shortenings, commercial frying fats, baked products, fast foods and high fat savoury snacks contain trans fats.

'able 13.3 Fatty acid composition of commonly consumed oils

Oil/Fat	Percentage of Total Fatty Acids		
	Total Saturates	Total Monounsaturates	Total Polyunsaturates
Coconut oil	89.5	7.8	2.0
Corn oil	12.7	29.6	57.4
Cottonseed oil	25.9	22.9	50.9
Groundnut oil	20.9	49.3	29.9
Mustard oil	10.7	56.0	32.6
Olive oil	14.8	74.5	10.0
Rice bran oil	22.1	41.0	35.7
Safflower oil	10.7	16.7	73.5
Gingelly oil	13.7	41.3	44.5
Soybean oil	13.1	28.9	57.2
Butter	69.4	28.0	2.5
Lard	46.2	45.2	11.0
Tallow	54.9	40.9	4.2

)ietary Cholesterol: Eggs, especially yolk, neat (especially organ meats such as brain, idney and liver), poultry, milk and milk roducts and to some extent shell fish contain ood amount of cholesterol. Foods obtained om plants do not contain cholesterol.

'at Replacers: Carbohydrate, protein nd fat-based (indigestible) fat replacers re available in the market such as olean, aprenin, salatrim and olestra.

THINK TANK

What are fat replacers?

Fat replacers are compounds which imitate the functional and sensory properties of fat, but contain less available energy than fats.

Are they healthy?

According to the position of American Dietetic Association (2005), majority of the fat replacers, when used in moderation by adults, can be safe and useful adjuncts to lowering the fat content of foods and may play a role in decreasing total fat and hence, energy intake.

3.2.6 Deficiency and Excess

n optimum intake of dietary fats helps to ensure maintenance of normal health (refer Box 13.3).)eficiency of fat is observed among patients suffering from malabsorption (allergies), malnutrition narasmus), hepatitis, pancreatitis and cancer or those who have been on a low-fat diet for a considerably ng time. Low-fat intake can result in weight loss if the total energy intake is low. Among children, it an result in growth faltering. Deficiency of omega-3 EFAs increases the risk of developing cardiovascular iseases, cancer, arthritis, altered immune function, learning problems, impaired vision and polydipsia. ow intake of omega-6 EFAs results in growth retardation, skin lesions, reproductive failure, fatty liver nd polydipsia.

Box 13.3 How Much Fat Should We Consume?

According to National Academy of Sciences, American Heart Association and National Cholesterol Education Program, dietary fat should meet the following guidelines:

1. Fat should provide 20–35 per cent calories of the total energy intake.
2. Saturated fats should provide <7 per cent of total calorie intake.
3. Acceptable daily intake of linoleic (ω-3) fatty acid is 17 g and 12 g per day for adult (19–50 years) men and women, respectively; 14 g and 11 g for men and women above 51 years.
4. Acceptable daily intake of linolenic (ω-6) fatty acids is 1.6 g and 1.1 g for men and women, respectively.
5. Trans fats should constitute <1 per cent of total calories.
6. Cholesterol intake should be less than 300 mg/day.

If fats are consumed in excess of the requirements, it results in increased risk for developing obesity, heart diseases, cancer and arthritis. This is particularly important while choosing an 'over-the-counter' fatty acid supplement such as those available for EPA, DHA, ALA and CLA. Excessive intake of such supplements can have deleterious impact on health. For example, high intakes of Omega-3 PUFAs can increase bleeding time, interfere with wound healing, raise LDL cholesterol and suppress immune function.

GOOD TO KNOW

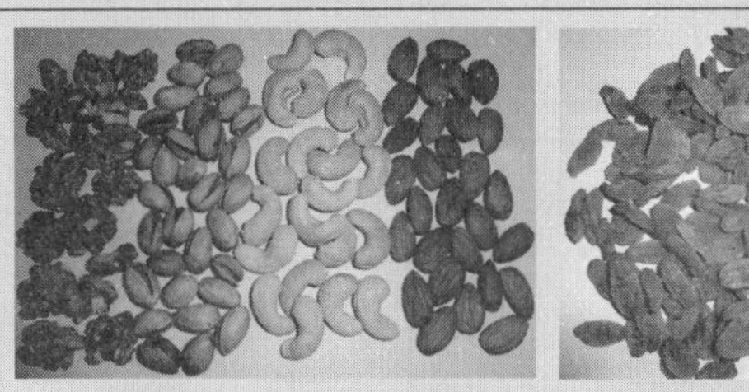

Most nuts are good source of proteins and fats

Raisins are good source of carbohydrates

13.3 Proteins

The word 'protous' which means 'of prime importance' was coined by a Dutch chemist Gerardus Mudle in 1838 to refer to the nutrient protein. Proteins are present in our body and in the food we eat. Protein constitute nearly 50 per cent of our dry weight. Our body synthesizes proteins from the amino acid made available from food and those present or synthesized by it. The ability of the body to perform thi prime function depends not only on the quantity but also on the quality of protein consumed by us i our daily diets.

13.3.1 Classification of Amino Acids and Proteins

As discussed in Chapter 4, proteins are composed of several amino acids attached to each other with the hel of peptide bonds. A total of 20 different amino acids exist in nature (food and body). While a combinatio of foods can provide all the 20 amino acids, the body can manufacture only a few. Based on the ability o our body to synthesize amino acids, they can be classified as follows:

Essential or Indispensable Amino Acids: These amino acids cannot be synthesized by our body an therefore must be present in our daily diets in adequate amount to ensure good health. In case of adults eight amino acids and for children nine amino acids are considered to be nutritionally essential.

Non-essential or Dispensable Amino Acids: These amino acids can be synthesized by our body if adequate amount of nitrogen, carbon, hydrogen and oxygen are supplied in the diet. Therefore, no state of deficiency will occur even if they are not present in adequate amounts in our diet. Eleven amino acids have been classified as non-essential.

Conditionally Essential or Conditionally Dispensable Amino Acids: During some physiological conditions, the body is not able to synthesize adequate amounts of certain non-essential amino acids and therefore, they need to be provided through food. Such amino acids are called conditionally essential; for example, at times, premature infants are not able to make certain non-essential amino acids (cysteine, arginine, glycine). The milk given to such infants needs to be fortified with conditionally essential amino acids to ensure their optimal growth and development. Similarly, the amino acids glutamine, arginine and tyrosine become indispensable during trauma/critical illness, physiological stress and phenylketonuria (a rare genetic disorder), respectively.

It is important to remember that all amino acids (whether essential, non-essential or conditionally essential) are equally important for the synthesis of body proteins (tissues) and performing various metabolic functions. Figure 13.3 lists the essential, non-essential and conditionally essential amino acids.

The amino acid composition of all food proteins is different and therefore all food proteins do not have the same nutritive value. The nutritional value of a protein is determined not only by the amount but also by the type of amino acids present therein. Certain food proteins such as those present in animal products contain higher amounts of essential amino acids than those present in foods of animal origin. Based on their amino acid composition, proteins are classified as follows:

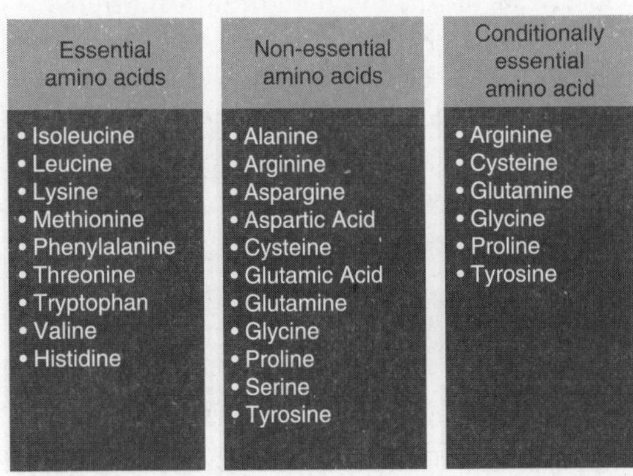

Figure 13.3 Essential, non-essential and conditionally essential amino acids

Complete Proteins: Food proteins which contain adequate amount of all essential amino acids are known as complete proteins. In general, egg, meat, poultry and dairy products have complete proteins. Such proteins can help in growth and maintenance of body.

Incomplete Proteins: If a food protein does not supply or supplies one or more essential amino acid in very small amounts, it is known as an incomplete protein. For example, proteins of wheat and pulses are incomplete, because they contain very small amount of essential amino acids lysine and methionine, respectively. Such proteins would not be able to support growth and maintenance functions adequately. The amino acid which is lacking or is present in limited amounts in a food protein is known as *limiting amino acid.*

13.3.2 Functions of Proteins

Protein is a versatile nutrient and performs a wide spectrum of functions. Amino acids supplied by diets are used for synthesizing more than 1000 types of body proteins each of which have their own specific functions. The salient functions are discussed in the following paragraphs.

Promote Growth and Provide Structure: Proteins are aptly called the building blocks of our body because they help in creating new cells. An infant weighs about 2.5–3.0 kg at birth but upon reaching adulthood the weight is found to have increased to about 50–60 kg. The increase in body size and structure is the result of cell multiplication. Proteins-form the structure of almost all cells present in our body. Hydroxyapatite in bones, collagen in skin/teeth/ligaments and tendons, and keratin in hair/fingernails are some examples of proteins providing structure to our body. This function is particularly important during periods of rapid growth such as pregnancy, infancy and adolescence.

Help in Repair: While chopping vegetables our finger often gets hurt but within a few days the cut is no more visible. Same is true after an accident or a surgery. Proteins help to heal wounds by promoting anabolic process at the site of injury. Vitamin A and C help proteins to perform this function more efficiently.

Facilitate Muscle Formation and Maintenance: Proteins are the most vital nutrients required for the formation of muscles in our body. Muscles receive oxygen during activity with the help of myoglobin, a protein. In fact, proteins are needed for movement such as those which result

GOOD TO KNOW

All pulses are good source of protein. However, pulses with husk (whole pulses) are more nutritious than their de-husked counter parts.

THINK TANK

Do athletes need more protein?

Protein supplements are increasingly being used with the hope of creating more and more muscle mass and even the popular *six-pack abs*. They are also being used by athletes to improve performance during sports activities. But are they really useful?

1. Protein supplements when consumed along with a diet supplying protein below the recommended dietary allowance (RDA) can be helpful in increasing muscle mass and activity.
2. When proteins are consumed above the RDA, they get converted to fat and increase body weight.
3. When protein supplements are consumed along with an otherwise balanced diet, they do not any further improve athletic performance as would a balanced diet do.
4. When muscle mass reaches desirable levels (normal body composition), any further increase in dietary proteins would increase body fat and not muscle because the body's feedback mechanism stops any further gain. At this stage, more muscle mass can be gained only if the body processes are stimulated to gain more muscle mass. This is not a natural process. Certain health supplements contain specific substances such as steroids which can do this. Steroid consumption can be harmful for the body.
5. Combining endurance exercises and strength training along with a balanced diet is most useful.

from contraction and relaxation of muscles; examples, movement of the muscle of the heart, kidneys and legs. Actin and myosin are two most important proteins which maintain normal muscle movements.

Facilitate Catalysis of Body Processes: Some proteins act as enzymes. Enzymes (also known as biological catalysts) facilitate chemical reactions without being changed or modified in the process. Enzymes help to break and build substances in the body. For example, enzymes help break down carbohydrates, proteins and fat into monosaccharides, amino acids and fatty acids, respectively, so that the body can absorb them. Lipase, amylase and pepsin are examples of enzymes involved in the digestion of fats, carbohydrates and protein, respectively.

Facilitate Communication: Tissues and organs communicate to each other with the help of hormones and certain cell signalling proteins. Hormones are messenger molecules which are made in one part of the body but they act in another part of the body. Not all hormones are made of protein. Hormones such as insulin and glucagon are protein based and they help to regulate blood glucose levels. Cell signalling proteins help to initiate synthesis of body proteins.

Regulate Fluid Balance: An important function of protein is to facilitate normal distribution of fluids in various compartments of the body viz. fluids outside (extracellular) and inside the cells (intracellular). Blood proteins, such as albumin, help to maintain fluid balance between the intravascular (fluids present in the blood and lymph vessels) and interstitial fluids (fluids present between the cells). During protein deficiency, the concentration of albumin is below normal which results in accumulation of interstitial fluids which can be seen as oedema (swelling in hands, feet and abdominal cavity).

Regulate Acid-base Balance: Proteins help to maintain normal pH, that is, acid-base balance in the body. In other words, proteins help to maintain the pH of blood near normal, that is, 7.4. Haemoglobin is an important regulator of blood pH.

Transporters: This is a selective function of protein, wherein it specifically binds a substance and carries it from one organ to another, generally through the blood. For example, lipoproteins transport lipids from liver to various organs and back, and haemoglobin carries oxygen from heart to various organs and carbon dioxide from organs to the heart. The protein transferrin carries iron, whereas rhodopsin carries vitamin A.

Strengthen Immune System: Proteins (antibodies) help to provide protection against infections such as common cough and cold or microbial attack. Antibodies are specialized infection-fighting protein molecules circulating in blood and other fluids of the body. They bind with or destroy pathogenic microorganisms thereby preventing the occurrence of infections. Thus, proteins help to strengthen the immune response.

GOOD TO KNOW

What Is pH?

pH refers to the concentration of hydrogen ions in a substance and is measured on a scale of 0–14. The higher is the concentration of hydrogen ions, the lower is the pH. Acids have a low pH (<7.0). Alkalis (base) have lower concentration of hydrogen ions and hence have a pH value above 7.0. Protein-rich foods upon metabolism yield hydrogen ions and are acidic. Carbohydrates and fat are metabolically neutral (pH = 7.0).

Provide Energy: Although carbohydrates and fats are the preferred source of energy, during negative energy balance the body use food and body proteins to release energy required for sustaining life. However, when proteins are used for this function, they are not able to perform other specific functions which no other nutrient can do. Excess proteins are converted to fats for long-term energy storage.

These are just some of the roles that proteins perform in our body. An adequate intake of good quality protein (refer to subsection 13.3.4) is important to maintain good health.

13.3.3 Food Sources

In terms of **quantity**, eggs, milk, certain milk products (cheese, curd), meat, fish and poultry are good sources of protein followed by pulses, legumes and nuts (Figure 13.6c). Sugars and fats/oils do not contain protein and if present, it is in traces. However, **quality** of proteins is equally important because even if a food is loaded with protein but the protein lacks essential amino acids, it alone will not be able to sustain life. Foods of animal origin such as eggs, meat, marine animals, poultry, milk, cheese, curd, yogurt and certain insects contain complete proteins. Egg protein is considered to be 'reference protein' as it contains highest amounts of essential amino acids and in the proportions/form close to the needs of the body. Pulses, legumes, nuts and cereals provide incomplete proteins. The protein 'zein' present in corn is of very low quality as it lacks most essential amino acids.

13.3.4 Quality of Protein

It is evident from the past discussions that we should incorporate adequate amount of good quality proteins in our diet. A good quality protein has the following characteristics:

1. It provides all the essential amino acids as per the need of the body.
2. It provides enough amount of non-essential amino acids which would provide base (carbon, hydrogen, oxygen, nitrogen) for the synthesis of more non-essential amino acids.
3. It should be easy to digest; that is, have a high biological value.

Proteins obtained from animal origin meet, whereas plant proteins do not meet the above specifications. However, it is not possible to sustain on animal foods alone. Further, Indian diets are largely vegetarian or lacto-vegetarian, and vegetarians are able to meet their amino acid requirements through judicious selection of foods and protein complementation. Complementation of proteins is an easy, healthy way of improving protein quality in our diets (refer to Box 13.4). It can be achieved by:

1. combining an incomplete protein food source with a complete protein source and
2. combining two food sources containing incomplete protein.

In addition to mutual supplementation, fermentation and sprouting also help in improving the quality of proteins.

13.3.5 Deficiency and Excess

Proteins should constitute 10–35 per cent of the total energy intake; 15–20 per cent being an ideal intake. As in case of carbohydrates and fats, deficient and excessive intakes can have serious consequences on health.

Deficiency of protein is common in children and adults living in underdeveloped and developing nations. Inadequate food intake, poverty, ignorance, infections and unhygienic surroundings are the major causes of protein-energy malnutrition (PEM) in children and at times in adults. PEM is also observed in patients

Box 13.4 Complementation or Mutual Supplementation of Protein

This process refers to combining two or more proteins (foods) so that the excess of one or more essential amino acids present in a particular protein can make up for the deficiency of these amino acid(s) in another protein (food). The amino acid profile of some commonly consumed foods is given in Table 13.4.

Table 13.4 Protein content and amino acid profile of some commonly consumed foods

EAA (essential amino acid) (mg/g N)	Foodstuffs					
	Wheat (whole)	Rice (milled)	Soybean	Milk (cow)	Egg (hen)	Spinach
Protein (g/100 g)	11.8	6.8	43.2	3.2	13.3	2.0
Total Nitrogen (g/100 g)	1.89	1.02	6.91	0.51	2.13	0.32
Lysine	170	230	400	500	440	400
Tryptophan	070	080	080	090	090	100
Phenylalanine	280	280	300	320	360	330
Methionine	090	150	080	160	210	110

Protein supplementation is particularly important for improving the protein quality in the diets of populations with limited economic resources or those having limited access to good quality or high biological value proteins such as animal proteins as is observed in Indian diets. Cereals lack or have low amount of lysine, which is an essential amino acid for cell multiplication (growth, convalescence). Lysine is the most limiting amino acid in wheat. *A limiting amino acid is an indispensable amino acid which is present in the lowest quantity in the food, in comparison with the same amino acid in a reference protein such as egg or milk.* Different foods have different limiting amino acids in varying amount. For example, pulse lack methionine which is present in comparatively adequate amount in cereals. Cereals lack lysine which is present in adequate amount in pulses and green leafy vegetables. Thus, if a processed food, dish or meal contains a combination of cereals, pulses and/or green leafy vegetables in right proportion, the resultant protein consumed would be complete or of high biological value due to the combined presence of all essential amino acids in adequate amounts. Cereal–pulse combinations such as roti-dal, chawal-dal, dosa/uttapam/idli/rice with sambhar, khichdi, soy bread, soy bun and soy biscuits contain almost all the essential amino acids. Combining cereals and pulses in 3:1 ratio helps to obtain a protein with all essential amino acids in right proportion.

Some other examples of mutual supplementation include the following:

1. Combining cereals (low in lysine) with pulses, legumes, milk, egg, meat, fish or poultry which have ample amount of lysine.
2. Similarly, combining soybean protein which is deficient in methionine but high in lysine with sesame protein (high in methionine, low in lysine).

suffering from chronic diseases such as tuberculosis, cancer, AIDS and anorexia nervosa. Alcohol and drug addiction can also result in PEM. PEM exhibits itself in two distinct forms viz. marasmus and kwashiorkor, which exhibit the extremes of a spectrum of symptoms and most frequently it is marasmic kwashiorkor that is observed in such children. The salient feature of marasmus and kwashiorkor are given in Figure 13.5. Chronic deficiency of protein can increase the risk of developing osteoporosis in adults because protein is required for calcium absorption.

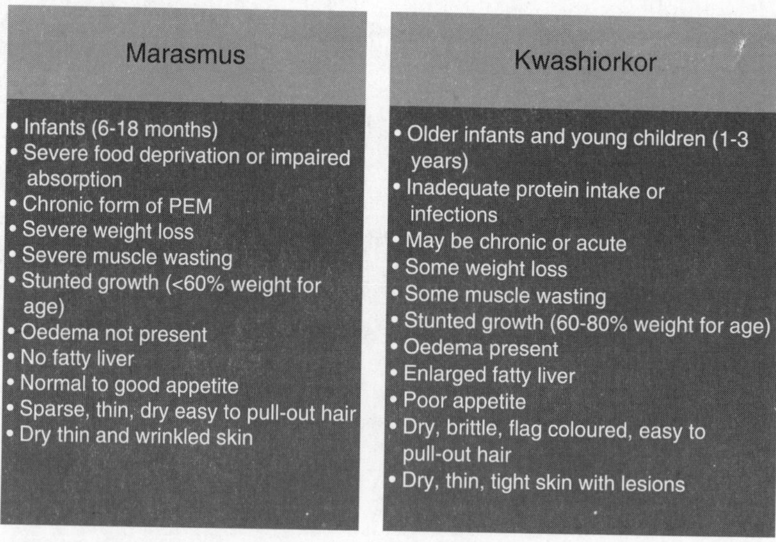

Figure 13.5 Features of marasmus and kwashiorkor

The upper limit for protein intake is 35 per cent of the total energy. When the protein intake exceeds this value, the diet is not able to provide adequate amount of other nutrients especially carbohydrates and fat. This can result in the development of nutritional deficiencies. Though a high-protein intake does not cause any kidney disease, it does increase the excretory work of kidneys and accelerates deterioration of the organ, especially in patients with chronic kidney disease. Excessive protein intake enhances calcium excretion which can result in bone loss. Increased intake of animal proteins enhances the risk of developing heart diseases and cancer of the colon. Low-carb high-protein weight-reduction diets are not nutritionally balanced and can result in the development of nutrient deficiencies.

Thus, keeping in mind the types, functions, typical food sources (Figure 13.6 (a,b and c)) and effects of deficiency/excess of each macronutrient, the following dietary guidelines have been developed to improve health and prevent disease.

1. Maintain a healthy body weight, that is, balance the energy intake with energy expenditure.
2. Consume plenty (around five servings) of fruits and vegetables.
3. Choose whole grains and high-fibre foods.
4. Limit the intake of total fat, especially saturated fat and cholesterol.
5. Prefer lean meats (fish, seafood and poultry) and egg white over red meats.
6. Minimize the intake of beverages and processed foods especially with 'added sugars'.
7. Eat out less frequently and choose foods wisely.

Figure 13.6 (a) Typical food sources of carbohydrates

Figure 13.6 (b) Typical food sources of fats

Figure 13.6 (c) Typical food sources of proteins

Let us review the three situations given in the beginning, which will help you to recapitulate this chapter.

Situation 1: No fat diets create deficiency of essential fatty acids and should never be tried. Weight reduction dietary regimes involve the consumption of a well-balanced diet containing all the three macronutrients in the right proportion. It should be coupled with an active lifestyle.

Situation 2: Increasing the intake of dietary fibre by incorporating whole grains (cereals, pulses), legumes with husk, seeds, not removing the edible peels of fruits and vegetables can prove to be helpful in alleviating constipation and managing diabetes.

Situation 3: Flaxseeds are good source of omega-3 alpha-linolenic acid. Omega-3 fatty acids are deficient in Indian diets and are also helpful in preventing heart diseases. Oat flour is a good source of soluble fibre (β-glucan). This type of fibre lowers blood cholesterol. Thus, the bun prepared by the chef was really heart-healthy!

Know Your Progress

True or False Questions

1. Oats are good sources of gums, an insoluble fibre.
2. Fructans are prebiotics.
3. Oleic acid is a type of PUFA.
4. Linoleic acid is a type of omega-3 fatty acid.
5. One teaspoon of oil would provide 20 kcals energy.
6. Methionine is an essential amino acid lacking in pulses.

Very Short Answer Questions

1. Define the term 'sugars'.
2. What are available carbohydrates?
3. Describe omega-3 fatty acids.
4. What should be the desirable lipid profile of an adult?
5. What are essential amino acids?

Short Answer Questions

1. Differentiate between soluble and insoluble fibres.
2. How can carbohydrates prevent or help in managing heart diseases and diabetes?
3. Give the food sources of various fatty acids.
4. Give five important functions of protein.
5. Describe the consequences of protein and fat deficiencies.

Long Answer Questions

1. Give the functions of carbohydrates.
2. What is the role of fatty acids in health and disease?
3. What is protein quality? How can we improve the quality of proteins in our diets?
4. How much amount of carbohydrate, fat and protein should we consume in a day? Calculate your macronutrient requirements for a day.

Beyond the Text

- Suggest dietary guidelines for macronutrient intake to the following college going students:

1. 'G' is fond of fast food and consumes burger or pizza at least thrice a week. He consumes one can of soft drink every day. Lately, he has been experiencing constipation.
2. 'U' is actively involved in sports activities. She will be taking part in a marathon next month and is undergoing training.
3. 'H' does not take breakfast before coming to college and often feels lethargic in the first class. After lunch he is active and alert.

Chapter 14

Micronutrients

Chapter Objectives

After reading this chapter, the reader should be able to

- Differentiate between macronutrients and micronutrients.
- Understand the myriad functions of various vitamins and minerals.
- Know the typical food sources and factors influencing the bioavailability of vitamins and minerals.
- Identify the effects of deficient and high intakes of each micronutrient on health.
- Appreciate the significance of optimum micronutrient intake.

Consider the following situations:

Situation 1: Preschoolers in a slum area are not able to play late evening because most of them cannot see clearly after dark. Responding to this observation, a non-government organization started feeding them food cooked in red palm oil.

Situation 2: A group of adolescents were having pizzas, burgers and pastas very frequently. Of late some are feeling anorexic and lethargic with mild gastro-intestinal disturbances.

Situation 3: A consumes spinach parantha four times a week with tea during breakfast. B consumes cauliflower leaves or fenugreek leaves parantha with same frequency along with a guava. A is anaemic while B is not.

Situation 4: During a medical check-up, the blood pressure (BP) of A was 156/92 mmHg while that of B was 110/65 mmHg. A reduced while B increased salt intake. After a month, A had normal while B still had low BP.

Situation 5: A survey reported high prevalence of mentally retarded children in a hilly area. The soil was analysed for its iodine content.

In Chapter 13, we discussed about the energy-yielding macronutrients (carbohydrates, proteins, fats) and their varied roles in maintaining health. In addition to macronutrients, micronutrients are equally important in sustaining and maintaining a healthy life. While macronutrients are the source of energy, micronutrients facilitate the availability and utilization of this energy. They are required in very small amounts (milligram/microgram) by our body (Annexure 2). Vitamins and minerals are the key micronutrients (Figure 14.1) and they must be present in optimum amounts in our daily diets. This chapter will briefly discuss some significant vitamins and minerals.

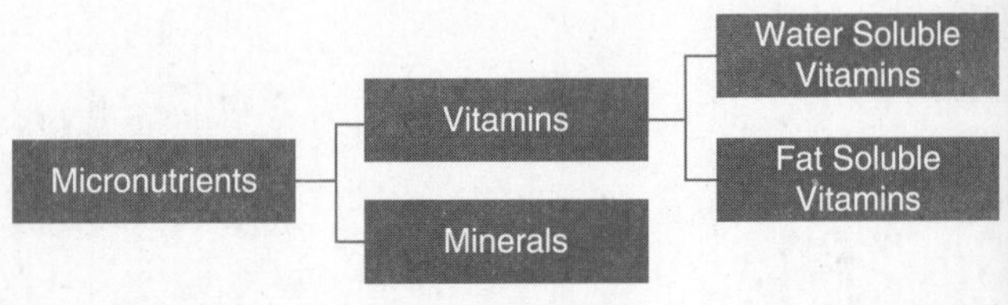

Figure 14.1 Micronutrients present in our food and body

14.1 Vitamins: Introduction and Classification

The word vitamin literally means 'vital to life'. *Vitamins are organic essential nutrients required in small amounts to perform specific functions that result in survival, maintenance, growth and reproduction of the body.* Vitamins at times are present in their inactive forms in foods. Such inactive forms of vitamins are known as *provitamins or vitamin precursors.* They are converted to active forms in the body. Contrary to common belief, vitamins are present in all food groups and not just fruits and vegetables. Box 14.1 gives the general characteristics of vitamins.

Box 14.1 General Characteristics of Vitamins

- Organic compounds with structures different from macronutrients.
- Natural components of food.
- Present in very small amounts (mg or µg).
- Not synthesized in the body in sufficient amounts to meet daily requirements.
- Essential for survival and normal physiological functions.
- Low intake results in states of deficiency.
- Excess intake may cause toxicity.

Based on their solubility in water, they are classified into two categories:

1. Fat-soluble vitamins
2. Water-soluble vitamins

Both categories of vitamins are equally important for health. However, their digestion, absorption, transportation, utilization, functions, storage, rich food sources and states of deficiency or toxic intakes are different. Figure 14.2 gives a comparative overview of fat-soluble and water-soluble vitamins.

Fat soluble vitamins	Water soluble vitamins
• Soluble in fat, insoluble in water. • Vitamins A, D, E, K. • Absorbed directly into the blood stream. • Being hydrophobic, they travel bound to transport proteins. • Excess amount is stored in liver and fatty tissues. • Chances of toxicity are high when consumed in excess (dietary supplements). • States of deficiency arise after a few weeks or months of low intake. • Should be present in daily diets.	• Soluble in water, insoluble in fat. • Vitamins B, C. • Absorbed with the help of lymphatic system into the blood stream. • Being hyrdrophillic, they travel freely in blood. • Much of the excess amount is excreted through urine. • Chances of toxicity are low when consumed in excess (dietary supplements). • States of deficiency arise within a few days of low intake. • Must be present in adequate amounts in daily diets.

Figure 14.2 Fat and water soluble vitamins: comparative overview

14.2 Fat-Soluble Vitamins

There are four fat-soluble vitamins, *viz.* A, D, E and K. As mentioned in Figure 14.2, they are hydrophobic, that is, insoluble in gastric juice and require bile juice for their digestion and absorption. They can be stored in considerable amounts in our body. The stores are utilized during periods of low intake. Thus, even after having a low intake for several weeks or months, the state of deficiency does not occur. It must however be remembered that over a period of time, the average daily intakes must approximate the recommendations. Refer to Annexures 1 and 2 for recommendations made by Indian Council of Medical Research regarding the requirements of various vitamins for males/females, various age groups and physiological conditions (pregnancy/lactation).

14.2.1 Vitamin A

Vitamin A, a fat-soluble vitamin, was discovered in 1916 by Elmer V. McCullum and Marguerite Davis. It refers to a series of biologically active compounds called retinoids. **Retinoids** include **retinol, retinal** and **retinoic acid** as well as precursor of vitamin A, viz. **carotenoids** (Figure 14.3). Several types of carotenoids exist in nature bull all cannot be converted to a biologically active form of vitamin A. Alpha-carotene, beta-carotene and beta-cryptoxanthin have vitamin A activity, that is, they can be converted in human body into biologically active forms. They are therefore called **provitamin A carotenoids**. Certain carotenes such as lycopene, leutin and zeaxanthin do not have vitamin A activity and are called **nonprovitamin A carotenoids**.

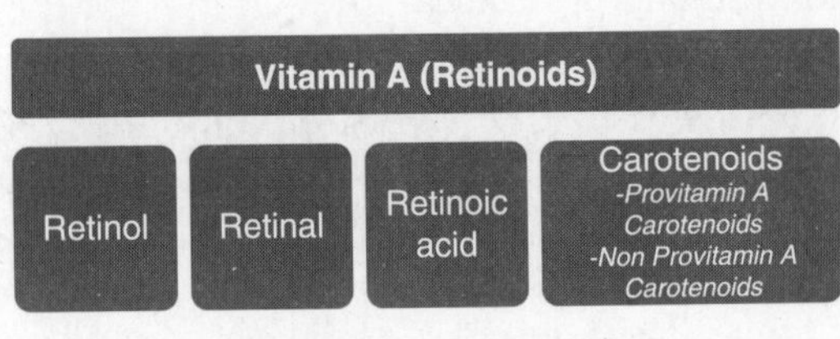

Figure 14.3 Forms of vitamin A

Functions: Vitamin A performs several functions; the salient ones are explained below (Figure 14.4):

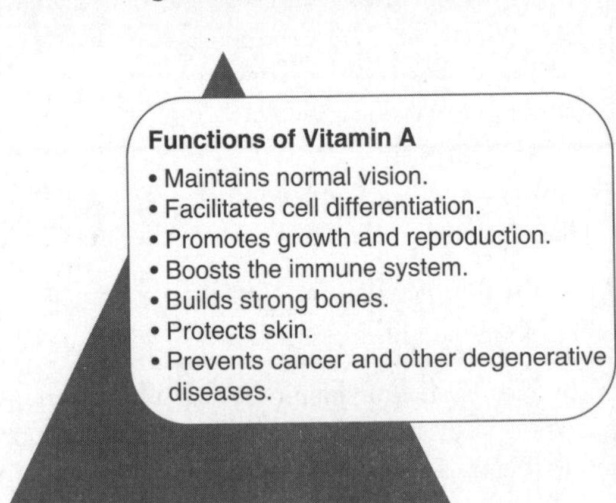

Figure 14.4 Functions of Vitamin A

Maintain normal vision: Our eyes contain about 0.1 per cent of the total vitamin A present in the body, yet this small amount plays a very crucial function – the function of maintaining normal vision. Vitamin A is present in rods and cones–special cells located in the retina of the eye. Both rods and cones are required for vision. Cones help us to differentiate between colours while rods are associated with black and white as well as vision in dim light or at dark. This is possible due to the presence of rhodopsin, a compound composed of a protein (opsin) and vitamin A. It is present in the rods. When we enter a dark room, our eyes get dilated and they also produce rhodopsin. Rhodopsin helps to see in dark. The time required to form rhodopsin is known as **dark adaptation**. On switching on the light or going into a brightly illuminated area, rhodopsin breaks down to vitamin A and protein. This process of formation and breakdown of rhodopsin is known as the **'visual cycle'** (Figure 14.5). Further, vitamin A is also required for maintaining the health of cornea–the outermost tissue layer of the eyes.

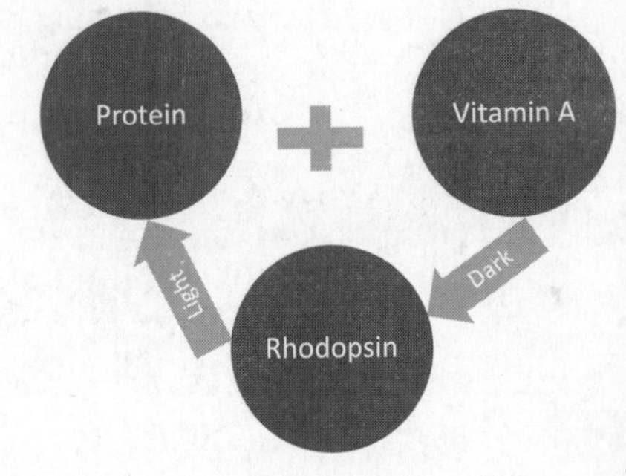

Figure 14.5 The visual cycle (dark and light adaptation)

Facilitates Cell Differentiation: Why do the worn out cells of the heart get replaced by the same type of cells and not by that of other organ (such as kidney)? This is so because vitamin A helps in cell differentiation and regulation of gene expression–a process by which each type of cell develops to perform a specific function.

Promote Growth and Reproduction: Vitamin A is required for growth, repair and reproduction. It promotes bone growth. Vitamin A supports sperm production in males and fertility among females. Fetal development during pregnancy is largely dependent on the vitamin A status of pregnant women.

Boosts the Immune System: Vitamin A promotes our immune system, that is, resistance to infections, through the production of lymphocytes and antibodies (special cells that fight against disease-causing pathogens) as well as by maintaining the protective barriers of the body such as the skin, lining of respiratory and gastro-intestinal tract (hinder the entry of pathogens in the body). During vitamin A deficiency, these protective barriers become weak resulting in the increased incidence of infection-related pneumonia and diarrhoea.

Build Strong Bones: When consumed in adequate amounts, vitamin A facilitates growth and remodelling of bones. It helps to make bones longer and stronger.

Protect the Skin: In view of its antioxidant role, vitamin A protects the skin from oxidative damage caused by free oxygen radicals especially in the presence of ultraviolet (UV) light; a reason why this vitamin is often added to beauty products such as creams and lotions.

Prevent Cancer and Other Degenerative Diseases: High intake of carotenoids through the consumption of vitamin A-rich fruits and vegetables (not supplements) has been associated with lower risk of developing prostrate cancer. Non-provitamin A such as lycopene lowers the risk of developing cancer, heart diseases and osteoporosis.

Food Sources: Retinoids are present in various forms in several foods. The sources of vitamin A can be categorized as:

Animal Foods: They are particularly rich in preformed vitamin A, viz. retinol, retinal and retinoic acid, though some amount of carotenoids may also be present. Liver (especially beef, fish and chicken), organ meats, fatty fish, fish liver oil and dairy products are rich sources of vitamin A. Whole fat dairy products (curd, yogurt, cheeses, butter, cream, ghee, khoa etc.) are richer as compared to those prepared from reduced fat milk. Reduced fat or toned milk fortified with vitamin A should be preferred. Fortified margarines and table spreads are also available in the market.

Plant Foods: Orange-red coloured fruits and vegetables as well as green leaves are good sources of carotenoids. Green leafy vegetables such as amaranth, colocasia, fetid cassia, turnip greens and curry leaves are exceptionally good sources of carotenoids. Mango, apricot, persimmon, passion fruit juice, papaya, orange, raspberries, carrots, tomato juice and broccoli are excellent sources. Red palm oil and spirulina (a blue-green algae) are particularly good sources of vitamin A.

Deficiency: Vitamin A deficiency is either primary or secondary. Primary deficiency is caused by under nutrition while secondary deficiency is caused by underlying diseases (mal absorption, diarrhoea, liver disorders, infections, food allergies etc.). Vitamin A deficiency disorders (VADD) include:

Impaired Vision and Blindness: Mild vitamin A deficiency causes impaired ability to see during dim light or at dark, which is known as **night blindness**. It gradually progresses to cause damage to cornea, dry eyes, scarring and even blindness. The World Health Organization has proposed an ocular (eye) symptom classification of vitamin A deficiency (Figure 14.6). Due to wide prevalence of night blindness/xerophthalmia and its advanced stages, the Government of India has implemented ***National Prophylaxis Programme*** for prevention of nutritional blindness due to Vitamin A deficiency, which aims at providing

1. supplementary doses of vitamin A to infants and preschoolers,
2. easy access to vitamin A-rich foods,
3. improved agricultural practices and
4. nutrition education to the masses at large.

Keratinization or Hyperkeratosis: Due to vitamin A deficiency, the skin and nails produce large amount of keratin (a protein) due to which the skin and nail surface becomes hard, rough, scaly and bumpy. The changes on the skin surface are also described as 'goose flesh'.

Compromised Immune Function: Lack of vitamin A increases the vulnerability especially of children to infections. Respiratory tract infections (pneumonia), diarrhoea, measles and malaria are associated with low serum levels of vitamin A.

Other Effects: Deficiency of this vitamin results in blocking of perspiration, reduced secretions of gastrointestinal tract, respiratory tract and genital organs and increased risk of developing kidney stones. Children are at a higher risk of growth retardation, bone deformities and defective teeth formation.

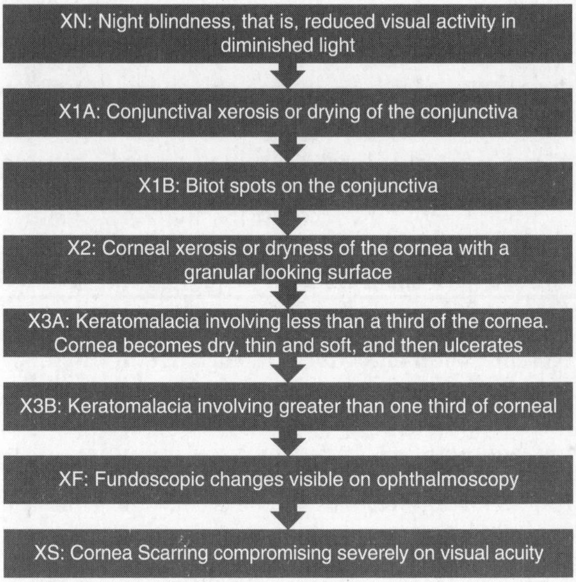

XN: Night blindness, that is, reduced visual activity in diminished light

↓

X1A: Conjunctival xerosis or drying of the conjunctiva

↓

X1B: Bitot spots on the conjunctiva

↓

X2: Corneal xerosis or dryness of the cornea with a granular looking surface

↓

X3A: Keratomalacia involving less than a third of the cornea. Cornea becomes dry, thin and soft, and then ulcerates

↓

X3B: Keratomalacia involving greater than one third of corneal

↓

XF: Fundoscopic changes visible on ophthalmoscopy

↓

XS: Cornea Scarring compromising severely on visual acuity

Figure 14.6 Progressive eye changes affecting vision due to vitamin A deficiency

Toxicity: Consumption of vitamin A through foods cannot cause toxicity. Overindulgence in vitamin A supplements can result in **hypervitaminosis A**, which is characterized by birth defects, skin disorders, bone pain, increased susceptibility to bone fractures, liver damage and in extreme cases death.

14.2.2 Vitamin D

Vitamin D is a fat-soluble vitamin that is often referred to as 'sunshine vitamin' or a 'prohormone'. It is called sunshine vitamin because the body can synthesize it with the help of sunlight and a prohormone because it can be produced in the skin, activated in the liver/kidneys and transported/utilized in various other tissues of the body–the characteristics typical to a hormone.

Vitamin D is a group of 10 related compounds. However, only two forms are widely present in food, *viz.* ergocalciferol (vitamin D_2) and cholecalciferol (vitamin D_3). While ergocalciferol is found in plant, cholecalciferol is present in animal foods. Cholecalciferol is also the form in which our body synthesizes vitamin D by the help of UV rays of the sun (Figure 14.7).

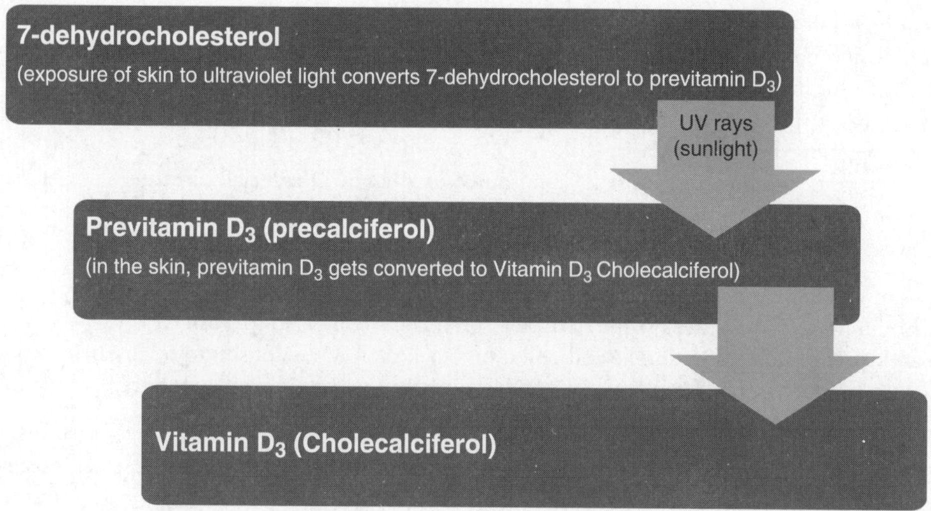

Figure 14.7 Synthesis of cholecalciferol in the skin

Functions: Vitamin D helps in the regulation of blood calcium concentrations, expression of genes, differentiation of cells and performs several other functions. The salient functions (Figure 14.8) include:

Maintain Normal Blood Calcium Levels: Vitamin D helps in maintaining normal calcium levels in blood even during states of deficient calcium intake. Vitamin D is able to do this through three ways:

1. increase calcium absorption in the small intestine,
2. decrease calcium excretion in the urine and
3. facilitate release of calcium from bones.

Thus, vitamin D helps to develop strong, dense bones. Children having normal vitamin D status, tend to be taller and achieve a higher growth potential as compared to those with vitamin D deficiency.

Gene Expression and Cell Differentiation: This vitamin has been found to be associated with enhancing and suppressing the activity of genes that regulate cell growth. It stimulates immature cells to become mature and functional. This function of vitamin D can help in preventing certain types of cancers such as those of the colon, breast, skin and prostrate.

Other Roles: Vitamin D has been found to be helpful in maintaining the structure and functional capacity of cells of the immune system, brain and nervous system, pancreas, skin, muscles/cartilage as well as reproductive organs. According to recent research, vitamin D provides protection against tuberculosis, inflammation, multiple sclerosis and hypertension.

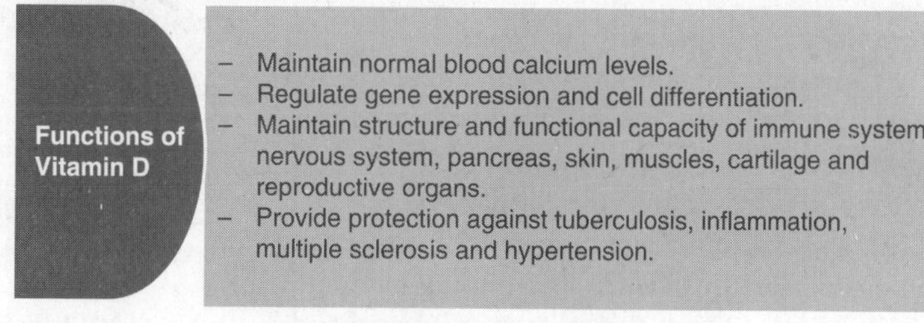

Figure 14.8 Functions of vitamin D

Sources: Foods of animal origin contain vitamin D. Cod liver oil, fatty fish (cat fish, tuna, halibut), oysters, salmon, shitake mushrooms, milk, butter, egg yolk, beef liver, margarine and cheese are good sources of this vitamin. Reduced fat milk and products made from it are not good sources of vitamin D. Reduced fat milk should therefore ideally be fortified with vitamin D to ensure that calcium present therein is absorbed sufficiently. Several dairy and bakery products as well as breakfast cereals are fortified with vitamin D and the information regarding this is mentioned on the labels.

Sunshine is however the best and most reliable source of vitamin D especially in case of vegetarians. Sun exposure of 5–30 minutes between the hours of 10.00 am to 3.00 pm twice a week is usually adequate. The exact amount of sun exposure required depends upon the following factors:

Age: Synthesis of vitamin D in skin decreases with advancing age.
Season: Lesser duration of exposure is required during summers as compared to winters.
Environment: Heavy clouds, smog, smoke and pollution block the UV rays of sun to reach earth.
Pigmentation: Dark skin contains more melanin (a skin pigment), which hinders synthesis of precalciferol.
Sunscreen: Use of creams with sun protection factor above eight, reduce vitamin D synthesis.

Deficiency: Vitamin insufficiency is increasingly becoming common. Under nutrition, repeated pregnancies, inadequate breast feeding, limited or no sun exposure (lifestyle), vegetarian dietary habits and dark skin lead to states of vitamin D deficiency. Severe vitamin D deficiency causes rickets, osteomalacia and osteoporosis.

Rickets: Infants and children who are born and breastfed on under nourished mothers are at risk of developing rickets. Vitamin D insufficiency during early years results in inadequate mineralization of bones that are weak, porous and do not grow to their full potential. The long bones of the legs are not able to handle weight of the body and tend to curve. This symptom is known as *bowed legs*. Such legs have knees protruding outward, which are termed *knock-knees*. Inadequate calcification causes bowing of the ribs, which resemble beads of a rosary. This symptom is known as rachitic rosary. Other symptoms include delayed closing of fontanel, protrusion of abdomen and muscle spasms.

Osteomalacia and Osteoporosis: Poor mineralization and increased demineralization causes bones to become soft, flexible, brittle and deformed. There is diffuse bone pain and muscle weakness. This condition is known as osteomalacia and is seen among adults.

Osteoporosis is characterized by increased bone demineralization, loss of bone mass and low bone mineral density (BMD). It is a chronic stage of vitamin D insufficiency.

Vitamin D deficiency has also been found to be associated with Alzheimer's disease – a state of poor mental health, declining cognitive performance and personality changes.

Toxicity: Sun exposure and vitamin D intake from foods does not cause toxicity. Mega doses of vitamin D supplements can however be toxic. The tolerable Upper Limit (UL) for adults is 50 μg (2000IU) per day. Consuming vitamin D above these levels causes toxicity, that is, hypercalcemia (blood calcium levels above normal) and hypercalciuria (high amount of calcium in urine). This causes increased urination, thirst and calcium deposition in the tissues of various organs (liver, heart etc.) causing pain and organ damage. Depression, nausea, vomiting and loss of appetite are also observed.

GOOD TO KNOW

Are Vitamin D Supplements Good for Health?

Vitamin D supplements should be taken by those who do not receive adequate exposure to UV rays of the sun and if the diets are also lacking in milk, eggs and meat products. Long-term vitamin D supplementation can have serious consequences on health and should be avoided.

14.2.3 Vitamin E

Vitamin E, a fat-soluble vitamin, was discovered by Evans and Bishop in 1922 while studying the association between fertility and nutrition. They found that addition of even small amounts of wheat germ, fresh lettuce or dried alfalfa could prevent death of foetus during pregnancy. This **anti-sterility factor** was termed tocopherol meaning 'to bring forth child birth'. Vitamin E is of two types, viz. **tocopherol and tocotrienol**; both existing in four forms each, that is, alpha, beta, gamma and delta. Alpha tocopherol is the only form that has vitamin E activity in the human body. Vitamin E is present in high concentrations in organs with high fatty acid content (liver, brain), adipose tissues and some are present in cell membranes. Normal plasma concentration of vitamin E is 12–30 μmol/l (5–13 mg/l).

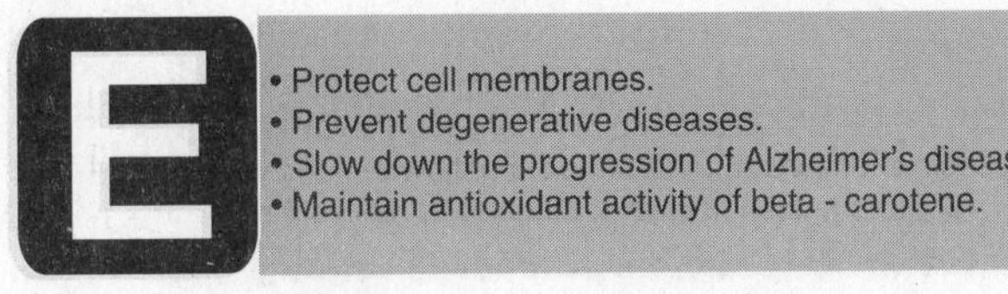

Figure 14.9 Functions of vitamin E

Functions: Vitamin E is recognized as a potent antioxidant and hence at times also called 'anti-ageing vitamin'. It is essential for normal reproduction, maintenance of cell membranes and prevention of degenerative diseases (Figure 14.9).

Protect cell membranes: Polyunsaturated fatty acids are abundantly present in cell membranes and are prone to oxidative damage. If adequate amount of vitamin E is present in cell membranes, it prevents oxygen-derived free radicals (Box 14.2) to react and cause oxidative damage to the membranes. This type of protection is particularly important for cells which come in contact with high amount of oxygen such as those of the lungs and RBCs. This function of vitamin E helps in prolonging cell life or in other words delaying the ageing process.

Box 14.2 What Are 'Oxygen-Derived Free Radicals'?

Every cell of our body utilizes oxygen to perform various functions. While doing so, oxygen sometimes reacts with body compounds and produce highly unstable molecules, which are known as 'oxygen-derived free radicals'. Smoke, pollution and UV radiations can also produce such free radicals. Oxygen-derived free radicals hinder with cell functioning or destroy cell components. They can thus cause heart diseases, cancer, diabetes, arthritis, cataracts etc. **Antioxidants** such as vitamin A, C, E, beta-carotene, cryptoxanthin and lycopene help in scavenging free radicals thereby protecting against such diseases.

Prevent Degenerative Diseases: Vitamin E has been found to be associated with reduced risk of developing cancer. Vitamin E-rich foods rather than supplements have been found to decrease the risk of developing prostrate cancer, no effect has been observed in cancer of the breast, lung and gastrointestinal tract. Vitamin E-rich diets have been found to be associated with reduced risk of developing cataracts among elderly, especially among smokers. Vitamin E reduces the progression of atherosclerosis–an early stage of cardiovascular disease. It, thus, helps in maintaining heart health.

Other Functions: Research is indicating that vitamin E may slow down the progression of Alzheimer's disease. It is helpful in maintaining the antioxidant activity of beta-carotene–a precursor of vitamin A.

Food Sources: Wheat germ oil (137 mg/100 ml) is an exceptionally good source of vitamin E. Oils, margarines and salad dressings rich in poly unsaturated fatty acids (sunflower, cottonseed, safflower, corn), green leafy vegetables (spinach, turnip greens, collard greens), broccoli, tomato paste, whole grains, organ meats (liver), egg yolks, nuts and oilseeds (almonds, hazelnuts, peanuts) are good sources of vitamin E.

Deficiency: Under nourished individuals or those who are suffering from fat mal absorption or genetic abnormalities are at risk of developing vitamin E deficiency. During the deficiency of this vitamin, the RBCs weaken and rupture, that is, their lifespan becomes shorter (<120 days). This results in low RBC count. This condition is known as haemolytic anaemia. Premature infants are particularly at risk of developing haemolytic anaemia because vitamin E stores are transferred from the mother to the foetus during the last phase of pregnancy. Severe vitamin E deficiency can result in failure to undertake a successful pregnancy, neuro-muscular dysfunction, cramps in legs and certain types of breast diseases.

Toxicity: Vitamin E toxicity is uncommon because the liver helps to regulate blood concentrations of this vitamin. Intake of vitamin E rich foods do not cause toxicity but prolonged intake of supplements can lead to toxicity. The UL for adults is 1000 mg per day, which is about 65 times above the recommendations for adults. Extremely high doses interfere with the blood clotting action of vitamin K, which in certain cases may lead to haemorrhage.

14.2.4 Vitamin K

Vitamin K was discovered by Henrik Dam in 1935 when chicken fed on fat-free diets developed haemorrhages and their blood clotted very slowly. It was found that feeding alfalfa or decayed fish meal corrected this condition. The dietary substance found to be effective in the treatment of this condition was named vitamin K as it was the next letter of the alphabet not already used for a vitamin at that time. Vitamin K is also known as 'coagulation' or 'koagulation' vitamin. Vitamin K refers to three compounds that have similar structures and functions

1. **Phylloquinone (K_1):** present in plant foods,
2. **Menaquinone (K_2):** produced by bacteria present in large amounts and
3. **Menadione (K_3):** produced commercially.

While considerable reserve amounts of other fat-soluble vitamins are present in the body, the same is not true in the case of vitamin K. Though the liver holds vitamin K consumed in excess of the requirement, it rapidly breaks down and eliminates it, and thus the reserves get depleted quickly.

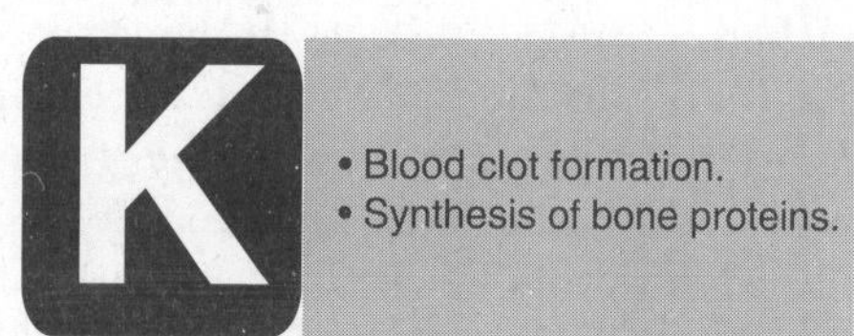

Figure 14.10 Functions of vitamin K

Functions: The following functions of vitamin K (Figure 14.10) have been identified:

Blood Clot Formation: This is the most significant function of vitamin K because it can differentiate between life and death. It is vital to the formation of blood clots, which are necessary to stop bleeding. Vitamin K in the liver promotes the synthesis of γ-carboxy-glutamic acid–an amino acid. This amino acid is required for the synthesis of blood coagulation factors II, VII, IX and X. Figure 14.11 gives the series of events required for blood clot formation and their dependence on vitamin K.

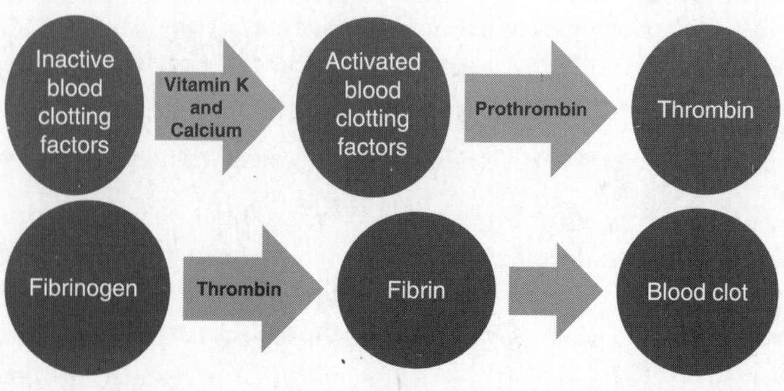

Figure 14.11 The process of blood clot formation

Synthesis of Bone Proteins: Vitamin K is associated with bone formation and improved bone mineral density (BMD). It is required for the synthesis of osteocalcin–a protein present in bones. Osteocalcin is not able to bind with minerals required for bone formation (calcium, phosphorus) during vitamin K deficiency, resulting in low bone density. Vitamin K has particularly been associated with decreased risk of hip fractures.

Food Sources: Vitamin K can be obtained from food and it is also synthesized by bacteria present in our intestine. Dark green vegetables, legumes, oils and fish are good dietary sources. Spinach, brussels sprouts, broccoli, asparagus, cabbage, cucumber, peas and certain oils (soybean/canola), margarines, spreads and salad dressings contain good amount of vitamin K.

Deficiency: The daily requirement of vitamin K is very small and therefore primary deficiency due to low dietary intake is very rare. Secondary deficiency is however observed during states of fat mal absorption or drug-nutrient interactions. Antibiotics kill vitamin K-producing bacteria present in the intestine while anticoagulants interfere with vitamin K activity. During the deficiency of this vitamin there is severe or uncontrolled bleeding at the site of wound. If prompt medical treatment is not received by such patients, it results in death due to excessive blood loss. Newly born infants have very low stores of vitamin K; their large intestines lack vitamin K-producing bacteria and human milk is also not a good source of this vitamin as a result they are at risk of developing vitamin K deficiency. This can result in haemorrhages or uncontrolled internal bleeding. Certain organizations therefore recommend a dose of vitamin K to the infant at the time of birth.

Table 14.1 Fat-soluble vitamins: Typical food sources deficiency and toxicity

Vitamin	Typical food sources	Deficiency	Toxicity
Vitamin A	Liver (especially beef, fish and chicken), organ meats, fatty fish, fish liver oil, dairy products, orange-red and green coloured fruits/vegetables, fortified margarines and table spreads	Impaired vision, blindness, keratinization or hyperkeratosis, lowered immunity, growth retardation	Hypervitaminosis A
Vitamin D	Cod liver oil, fatty fish (cat fish, tuna, halibut), oysters, salmon, shitake mushrooms, milk, butter, egg yolk, beef liver, margarine and cheese	Rickets, osteomalacia and osteoporosis	Depression, nausea, vomiting and loss of appetite due to hypercalcemia and hypercalciuria resulting in organ damage
Vitamin E	Wheat germ oil, sunflower, cottonseed, safflower, corn oil, green leafy vegetables, broccoli, tomato paste, whole grains, organ meats (liver), egg yolks, nuts and oilseeds (almonds, hazelnuts, peanuts)	Haemolytic anaemia, unsuccessful pregnancy, neuro-muscular dysfunction, cramps in legs and certain types of breast diseases	Rare. Interference with the blood clotting action of vitamin K
Vitamin K	Dark green vegetables, legumes, oils, fish, spinach, Brussels sprouts, broccoli, asparagus, cabbage, cucumber, peas and certain oils (soybean/canola)/margarine	Haemorrhages and uncontrolled bleeding at the site of wound	None

Toxicity: No adverse effects have been associated with high intakes of vitamin K. No recommended dietary allowance (RDA) and UL has been set for this vitamin. A total of 90 μg and 120 μg per day is generally considered to be an acceptable intake (AI) for adult women and men, respectively.

Thus, all the four fat-soluble vitamins influence the health of almost all organs (eyes, skin, brain, gastrointestinal tract, blood, etc.). Some of their functions are interrelated and certain food sources are also common. The typical food sources and the effects of deficiency as well as excess intake of fat-soluble vitamins have been summarized in Table 14.1.

14.3 Water-Soluble Vitamins

The B-group vitamins and vitamin C constitute the water-soluble vitamins. They are distinctly different from fat-soluble vitamins as mentioned in Figure 14.1. There are nine essential water-soluble vitamins: eight B group vitamins and vitamin C. The eight B group vitamins include thiamin, riboflavin, niacin, pantothenic acid, vitamin B_6, biotin, folate and cyanocobalamin. Water-soluble vitamins are often falsely claimed to be the source of energy. On the contrary, they help the body to utilize the energy provided by the macronutrients (carbohydrate, proteins and fats) and they themselves do not provide energy. Some common characteristics of water-soluble vitamins are listed below:

1. They are soluble in water but not in fats.
2. Many a times these vitamins are found bound to protein in food. This bond needs to be broken down to facilitate their digestion.
3. A major proportion of the water-soluble vitamins are absorbed in the small intestine.
4. Existing nutritional status, presence/deficiency of other nutrients, drugs, physiological status, age, gender, illness etc. influence their bioavailability in the body.
5. They cannot be stored in appreciable amounts in he body and therefore adequate amount of each of these vitamins should be consumed on a daily basis.

We will first begin our discussion with B-group vitamins. Apart from having the properties of water-soluble vitamins, a characteristic common to all B-group vitamins is that they act primarily as coenzymes. Coenzymes are biologically active molecules that help enzymes in enhancing the speed of a reaction in the body. Certain enzymes cannot function in the absence of a coenzyme (refer to Box 14.3).

GOOD TO KNOW

The Story – Discovery and Nomenclature of B Group Vitamins

During the discovery of vitamins, vitamin B was the second on the list, the first being vitamin A. It thus received the name 'vitamin B'. Research in the forthcoming years led to the identification of several other vitamins, which were named vitamin C, D, E and K. Sometime later it was however found that vitamin B was not a single but a group of compounds with similar characteristics. They were therefore named as B_1, B_2, B_3, B_5, B_6, B_{10} and B_{12}. However, today they are identified by their scientific names, viz. thiamin, riboflavin, niacin, pantothenic acid, Vitamin B_6, biotin, folic acid and cyanocobalamin.

Box 14.3 What Are Enzymes and Coenzymes?

Enzymes are proteins present in our body (and also food) that speed up the rate of chemical reactions but they themselves do not get altered in the process.

Coenzymes are organic compounds, often derived from B-group vitamins. They combine with inactive enzymes to form active enzymes. There are several enzymes in our body that do not function in the absence of coenzymes.

14.3.1 Thiamin

Thiamin was the first B-group vitamin discovered as early as in 1897 by a Dutch medical officer Eijkmann while making efforts to cure 'polyneuritis'–a symptom resembling beriberi by feeding rice bran and rice polishings. It is also known as vitamin B_1 or anti-beriberi factor. Thiamin is a white crystalline water-soluble vitamin that has a faint yeast-like odour and salty nut-like taste. It is absorbed in the upper part of the small intestine from where it enters the blood stream and is utilized to perform several functions. When vitamin B_1 is consumed in amounts above the requirements, excess is excreted through urine. Our body contains only 25–30 mg thiamin; mostly in the brain, heart, liver, kidneys and skeletal muscles. Vitamin B_1 is present in our body in three forms, viz. free thiamin, thiamin monophosphate and thiamin triphosphate. Since there is no significant store of thiamin, it necessitates its adequate daily dietary intake.

Functions

1. Thiamin constitutes a coenzyme–thiamin pyrophosphate (TPP). This coenzyme helps in yielding energy molecules from carbohydrates. These energy molecules are known as ATP (adenine triphosphate). They help to provide energy for all voluntary and involuntary activities of our body.
2. Thiamin in the form of TPP plays an important role in the synthesis of DNA (de-oxy ribonucleic acid) and RNA (ribonucleic acid). DNA and RNA in turn are required for the synthesis of proteins in our body.
3. It plays an important role in the synthesis of a coenzyme–NADP (nicotinamide adenine dinucleotide phosphate). This coenzyme is needed for synthesis of triglycerides.
4. Thiamin triphosphate is required for maintaining normal nerve function. It facilitates generation of nerve impulses and neuro-transmission.
5. It is also associated with muscle activity.

Food Sources: It is present in the food mostly as thiamin hydrochloride and is measured in milligrams. Thiamin is present in all natural whole foods. Thiamin is particularly concentrated in the husk/bran and germ of whole grains. Whole cereal and their products therefore contribute significant amount of thiamin to cereal-based Indian diets. Yeast, pork, ham, tuna, beans, legumes, nuts and seeds (particularly sunflower seeds) are good sources of thiamin. Vegetables, fruits, milk and milk products are fair sources. Enriched or fortified processed cereal products such as breakfast cereals, biscuits etc. contain good amount of vitamin B_1.

Bioavailability of thiamin is influenced by several factors. For instance, its absorption gets hindered by 'antithiamin factors' such as compounds present in raw fish, coffee, tea, berries, cabbage and brussels sprouts. Alcohol also interferes with its absorption. Vitamin C prevents destruction of thiamin. When the dietary intake is low, the absorption of thiamin proportionally increases and vice versa.

Deficiency: Thiamin requirements are based on energy needs; higher the energy intake, greater is the need for thiamin. Inadequate food intake or preference for processed cereals can impose the risk of deficiency. Thiamin deficiency is known as 'beriberi', which literally means 'I can't I can't'. Moderate state of beriberi result in degeneration of nerve tissues, muscle weakness, shortness of breath, headache, irritability, depression along with loss of appetite and gastrointestinal disturbances. Severe state of beriberi can be categorized as under:

DO YOU KNOW

What Contains More Thiamin: Whole Grains or Processed Cereals?

Whole grains contain more thiamin as compared to their processed counterparts. Thiamin is concentrated on the inner surface of bran. During processing such as refining of wheat flour to produce refined wheat flour, the bran is increasingly removed. Thiamin is sensitive to heat and alkali. Thus, processing results in increased losses of this vitamin.

Polished rice

Pasta

White bread

Polished rice and other processed cereal products are low in B-group vitamins, particularly thiamin

1. **Dry Beriberi:** It is characterized by damage to gastrointestinal system, which results in diarrhoea and vomiting. Muscle and nerve damage results in muscle wasting, leg cramps, tenderness and pain in legs and feet. Impaired neuro-muscular coordination causes pins and needle sensation in hands and feet.
2. **Wet Beriberi:** It is characterized by oedema (fluid retention) in arms and legs, enlargement of the heart and respiratory problems.
3. **Infantile Beriberi:** Babies breast fed by mothers having thiamin deficient diets develop this type of beriberi. They are at increased risk of cardiac arrest.
4. **Cerebral Beriberi:** It is also known as Wernicke–Korsakoff syndrome and is seen among chronic alcoholics. Typical symptoms include poor muscle control, paralysis of the eye muscles, confusion and short-term memory loss.

Toxicity: Since thiamin is water soluble, excess intake gets excreted through urine. Toxicity has therefore not been observed. No UL for thiamin has been recommended by nutrition authorities.

14.3.2 Riboflavin

Riboflavin is a water-soluble vitamin, which is also known as vitamin B_2. It is a fluorescent bluish-green pigment that is sensitive to light. Similar to thiamin, riboflavin is absorbed in the upper part of the small intestine from where it enters the blood stream and is taken to various body tissues to perform specific functions. Excess riboflavin is excreted in the urine. Our body contains only 1 g (approx.) riboflavin, the majority being in the muscle while only a small amount is in the liver and kidneys. It must therefore be supplied in the diet regularly to avoid states of deficiency.

Functions

1. Riboflavin forms part of two coenzymes: FAD (flavin adenine dinucleotide) and FMN (flavin mono nucleotide). These coenzymes perform several functions such as
   (a) FAD and FMN are required for the metabolism of carbohydrates, fats and proteins.
   (b) FAD is required for the biosynthesis of niacin (a b-group vitamin) from tryptophan (an essential amino acid).
   (c) FMN helps in the activation of vitamin A, folic acid and pyridoxine.
2. Studies have shown a positive association between riboflavin supplementation and treatment of cataracts.
3. It is required for the functioning of certain neuro-transmitters such as dopamine.

> **GOOD TO KNOW**
>
> **Milk and Riboflavin**
>
> Earlier, milk was sold in glass bottles that were left at the front door step by the milkman. The front door of the house usually receives sun rays early in the morning. Exposure of milk contained in transparent packs or bottles to sunlight led to considerable loss of riboflavin. Now-a-days, milk is sold in translucent packets that prevents the loss of riboflavin.

Food Sources: Riboflavin is measured in milligrams (mg) in foods. It is present in most foods. Riboflavin present in animal foods has higher bioavailability than plant foods. Alcohol, sedatives and barbiturates inhibit absorption of riboflavin. Liver and dried yeast are the most concentrated source of riboflavin. Milk, cottage cheese, curd and yogurt are excellent sources. As compared to milk, liver contains 16 times more riboflavin. Egg, fish, turkey, chicken, beef, ham and almonds are good sources. Cereals, fruits, vegetables, roots and tubers are poor sources. Enriched cereals contain good amount of this vitamin. Fats and oils do not contain vitamin B_2.

Thiamin	• Dry beriberi • Wet beriberi • Infantile beriberi • Cerebral beriberi
Riboflavin	• Ariboflavinosis (Glossitis, Angular stomatatis, Cheilosis, Seborrheic dermatitis, Nasolabial dyssbacea)
Niacin	• Pellagra Disease of 4D's: Diarrhea, Dermatitis, Dementia, Death)
Folic Acid	• Megaloblastic macrocytic anemia • Neural tube defect (spina bifida)
Cyanocobalamin	• Megaloblastic microcytic anemia (pernicious anemia)

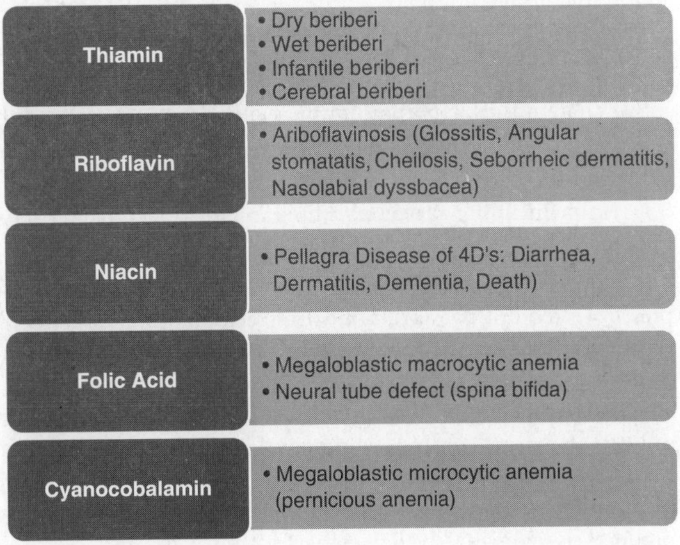

Figure 14.12 Common deficiency diseases of B-group vitamins

Deficiency: Ariboflavinosis occurs due to the deficiency of riboflavin. It usually occurs in combination with other nutrient deficiencies especially that of B-group vitamins and manifests after several months of deprivation of this vitamin. Inadequate dietary intake, chronic alcoholism, long term use of sedatives and barbiturates are associated with deficiency of riboflavin. The initial symptoms include photophobia, burning, itching and watering of eyes, reduced vision, soreness and burning of lips mouth and tongue. Important clinical symptoms include:

Glossitis: Shiny, smooth, inflamed, magenta coloured tongue.

Angular stomatitis: Painful and sore cracks in the skin at the corners of the mouth.

Cheilosis: Cracked and inflamed lips.
Seborrheic dermatitis: Clogging of the oil-producing glands of the skin.
Nasolabial dyssbacea: Cracked tough skin around the folds of the nose

Since riboflavin influences the function of other vitamins, its deficiency enhances the risk of developing anaemia.

Toxicity: Being water soluble, excess riboflavin is readily excreted. No toxicity has been associated with high doses of riboflavin.

14.3.3 Niacin

Niacin or vitamin B_3 is also a water-soluble vitamin belonging to the B-group family. It is a white crystalline needle-like compound with a bitter taste. Niacin exists in two forms, viz. nicotinic acid and nicotinamide (also known as niacinamide). Our body can easily convert nicotinic acid to nicotinamide. Nicotinic acid is the principle form of niacin present in our blood. Vitamin B_3 is absorbed in the stomach and the small intestine. It is then transported by the blood to the liver where it gets bound to a protein that facilitates its transfer to various cells of the body.

Functions

1. Nicotinamide is a component of two coenzymes nicotinamide adenine dinucleotide (NAD) and nicotinamide adenine dinucleotide phosphate (NADP). Both these coenzymes perform about 200 functions in our body.
2. Niacin is required for the metabolism of macronutrients to yield energy in usable form.
3. It plays a role in the synthesis of fatty acids.
4. Vitamin B_3 is important for the synthesis of DNA (deoxyribonucleic acid).
5. Mega doses of niacin have been found to lower the risk of developing coronary heart diseases. It lowers LDLc (bad cholesterol) and increases HDLc (good cholesterol) in blood thereby helping in maintaining a favourable lipid profile (refer Chapter 13).

Food Sources: Niacin is measured in food as niacin equivalents or NE. (refer Box 14.4). Niacin is present in food in two forms–nicotinic acid and nicotinamide. It can also be made in our body from tryptophan (an essential amino acid). As such, both niacin and tryptophan are considered 'dietary sources' of niacin.

Meat particularly beef liver, chicken (breast, liver), tuna, halibut, turkey and salmon are excellent sources of niacin. Sardines, shrimp, cod, mushrooms, legumes, nuts and sunflower seeds are good sources. Milk, curd, yogurt, cheese and eggs are fair sources. Overall, animal foods are richer sources of niacin than plant foods. Niacin present in animal products is more bio-available than plant foods. In plant foods, niacin is bound to protein that makes it unavailable for absorption. Fermentation, sprouting or soaking the grains in alkali such as lime water helps to improve bioavailability of niacin. However, it must be remembered that alkaline pH destroys other B-vitamins and should therefore not be practiced regularly (refer to Table 18.1 in Chapter 18).

Box 14.4 What Are Niacin Equivalents?

When inadequate amount of niacin is consumed, it is synthesized by our body from tryptophan (an essential amino acid). Approximately 60 mg of dietary tryptophan is needed to make 1 mg niacin. Therefore, niacin is measured in food as NE. Thus, if a food contains 1 mg niacin and 60 mg tryptophan, it provides two NE.

Deficiency: Niacin deficiency causes **pellagra** and is not seen frequently. It occurs among populations consuming corn as the only or major food in their protein deficient diets. Pellagra is therefore associated with poverty, under-nutrition and alcoholism. It is characterized by three major clinical features, viz. dermatitis, diarrhoea, dementia and death; a reason why pellagra is often called the disease of 4D's.
Dermatitis: It is dry, rough, thick and dark skin, which gets inflamed upon exposure to sun. Since the inflammation worsens on areas exposed to sun, niacin deficiency was earlier named 'mal del sol', which means illness of the sun in Italian.
Diarrhoea: Gastrointestinal disturbances result in red and swollen tongue, loss of appetite and diarrhoea. All these symptoms worsen nutrient deficiencies existing in the patient.
Dementia: Neurological problems are observed during advanced state of deficiency and are characterized by irritability, inability to concentrate, anxiety and depression. In very severe cases, it can result in complete mental imbalance and resultant madness.
Death: This occurs when the above symptoms are not cured.

Toxicity: The UL of niacin is 35 mg/day. 'Niacin flushes' characterized by painful flushes, hives and rashes are observed when the vitamin is taken even two to three times above the recommended levels. Nausea, vomiting, liver damage and impaired glucose tolerance are observed when higher doses are consumed.

14.3.4 Pantothenic Acid

Pantothenic acid or vitamin B_5 received its name from a Greek word 'pantos', which means 'everywhere'. It was given this name because this vitamin is present in almost all plant and animal tissues though in varying amounts. Its discovery dates back to World War II, when soldiers complained about 'burning feet' – a symptom that was relieved upon giving large doses of calcium pantothenate. It is a white crystalline compound that gets destroyed easily upon exposure to heat. Pantothenic acid is concentrated in the liver, adrenal gland, kidney, brain, heart and testes.

Functions

1. Pantothenic acid is a component of coenzyme A. This coenzyme is involved with the extraction of energy from macronutrients.
2. It is required for making non-essential fatty acids in the body.
3. Synthesis of neurotransmitters, steroid hormones and haemoglobin depends upon this vitamin.

Food Sources: Pantothenic acid is measured in milligrams in food. The mean bioavailability is 50 per cent on a mixed diet. It is present in almost all foods. Beef liver, shitake mushrooms and sunflower seeds are very good sources. Dairy products (milk, yogurt), turkey, fish, eggs, whole grains, potatoes, tomatoes, broccoli and coffee are good sources.

Deficiency: It is rare and is often associated with other nutrient deficiencies. Pantothenic acid deficiency is generally observed among people who depend largely upon processed non-fortified foods. Symptoms include vomiting, nausea, stomach cramps, sleeplessness, fatigue, depression, irritability and restlessness in mild cases. Severe deficiency results in hypoglycaemia and increased sensitivity to insulin. Numbness and muscle cramps, especially in the feet, makes it difficult to walk.

Toxicity: The daily requirement of vitamin B_5 is only 35 μg per day. The acceptable intake levels have been set at 5 mg per day. No UL has been given. No evidence exists related to toxicity of this nutrient.

14.3.5 Vitamin B_6

It is a water-soluble B-group vitamin that occurs in nature in three forms: pyridoxine, pyridoxal and pyridoxamine. All the three forms are biologically active and inter-convertible in body. The total vitamin B_6 content present in an adult human body is 50–150 mg. It is stored in the liver, brain, kidney, spleen and muscles. Similar to thiamin and riboflavin, it is absorbed in the small intestine and passed to the liver through blood. In the liver, vitamin B_6 is used to form pyridoxal phosphate (PLP)–the coenzyme form of this vitamin.

Functions

1. PLP helps to synthesize non-essential amino acids. In the absence of PLP, all non-essential amino acids become essential as they cannot be made in our body.
2. During states of negative energy balance (fasting/starvation etc.), the body breaks glycogen (present in liver) and muscle proteins to release glucose–a source of energy to the body (refer Chapters 12 and 13). These processes are facilitated by the help of vitamin B_6.
3. It is required for the synthesis of white blood cells and the maintenance of a healthy immune system.
4. Vitamin B_6 is essential for the synthesis of hem part of haemoglobin and hence also RBCs because haemoglobin is part of RBCs.
5. It participates in the formation of neurotransmitters (such as adrenalin) and taurine. They positively influence the function of brain and eyes.
6. Mega doses of this vitamin help to lower homocysteine levels. Elevated homocysteine levels are a marker for heart disease.

Food sources: Vitamin B_6 is measured in milligrams in food. Liver of beef and chicken, tuna, turkey, beef, pork and chick peas are very good sources. Halibut, herring, turkey, chicken meat, tomato, potato, sweet potato, bananas, sesame and sunflower seeds are fair sources. On a mixed diet, about 75 per cent of vitamin B_6 is absorbed by the body. Isoniazid, a drug used to treat tuberculosis, and alcohol interfere with the absorption of this vitamin.

Deficiency: It is rare and is usually observed along with the deficiencies of other vitamins or protein. Mild deficiency results in general weakness, sleeplessness, changes in skin and brain function. Chronic or severe deficiency of vitamin B_6 results in inadequate hem production and hence lower levels of haemoglobin. This condition is known as **microcytic hypochromic anaemia**. Deficiency of vitamin B_6 also damages the nervous system causing depression, headaches, confusion, convulsions and at times seizures. It also increases the risk of developing heart diseases.

Toxicity: The daily requirement of pyridoxine is 2.0 mg per day for adults. The UL for vitamin B_6 is set at 100 mg per day. Mega doses cause tingling/prickling or burning sensation, numbness in feet and hands, difficulty in walking, headache, stomach upset and neurological damage.

14.3.6 Biotin

Another water-soluble vitamin of the B-group family is biotin or vitamin B_7, which was discovered in 1930s and was studied with relevance to its presence in egg white. Biotin can be obtained from diet and is also produced by bacteria present in the large intestine. The amount of biotin produced in the large intestine is not sufficient enough to meet our daily needs. Biotin is carried from the intestines to the liver from where it is distributed to various parts of the body. Very small amount of biotin is present in our body. It is stored in muscles, liver and brain tissues.

Functions

1. Biotin acts as a coenzyme in the breakdown process of amino acids and fatty acids to release energy during states of negative energy balance.
2. It is required for the synthesis of DNA and hence influences growth and development.

Food Sources: Cooked eggs are a good sources of biotin but not raw. Liver of all animals, fish, egg yolk, cheese, soybean, peanut, cauliflower and whole grains are good sources. Fruits and meat have low biotin content.

Deficiency: Signs and symptoms of biotin deficiency are poorly understood. Inadequate intake of biotin results in lethargy, depression, hallucinations, skin irritations, red scaly rashes around eyes nose and mouth, infections, hair loss (alopecia), poor muscle control, seizures and developmental delays in infants.

Toxicity: Insufficient information is available regarding daily requirements. An AI level suggested by nutrition authorities is 30 μg per day. Cases of toxicity have not been reported.

> **GOOD TO KNOW**
>
> **Biotin and 'Egg White Injury'**
>
> In raw egg white, a protein called avidin binds biotin and prevents its absorption. Cooking denatures avidin, preventing it from binding to biotin. Deficiency of biotin due to consumption of raw egg whites is known as 'egg white injury'. It is characterized by hair loss, dermatitis and paralysis. Theoretically, consumption of about 12 raw egg whites for a prolonged duration of time can precipitate biotin deficiency and 'egg white injury'.

4.3.7 Folate

The word folic acid was derived from the Latin word 'folia' (leaf) meaning foliage in 1941 when it was extracted from spinach leaves. The word folic acid or folate or folacin refers to a group of compounds having structure and functions similar to pteroglutamic acid. It is a bright yellow crystalline B group vitamin that is slightly soluble in water. Gastric and pancreatic juice facilitate its absorption, which is primarily in the small intestine. Our body contains about 6–10 mg folate. The reserves are sufficient to last for 2–4 months without any additional intake.

Functions

1. Folate plays a crucial role in the synthesis of DNA and RNA and hence cell division. It is therefore essential for growth, repair and maintenance of all body tissues.
2. Along with vitamin B_6 and vitamin B_{12}, folate promotes synthesis of RBCs. As such it is required for maturation of RBCs.
3. Folate is required for the normal growth and development of nerve tissues in the growing foetus. It is required for the development of neural tube (a flat sheet of nerve tissues) – a structure that eventually forms the brain and spinal cord.

4. It helps to maintain integrity of gastrointestinal tract tissues and promotes their efficient repair.
5. Formation of methionine from homocysteine requires folic acid.
6. Folic acid supplements help to control homocysteine levels thereby reducing the risk of heart disease
7. Mega doses have been found to lower the risk of developing cancer of the oesophagus, pancreas, colon and rectum.

Food Sources: Dietary folate is expressed in micrograms (mcg) and so is the folic acid present in supple ments or fortified food. However, when a person is consuming folate from both diet and supplements/fort fied food, it is referred to as 'dietary folate equivalents' because of the difference in their bioavailability (refe Box 14.5). Alcohol, smoking, oral contraceptives, anti-cancer drugs, aspirin and antacids interfere with th absorption or utilization of folate.

Liver of goat, sheep and beef are exceptionally good sources followed by Bengal gram whole, cluster bean pinto beans, lentils, black gram dal, cowpea, guava, asparagus, turnip greens and broccoli. All dark gree leafy vegetables are rich in folate. Wheat germ, kidney beans, sunflower seeds, beets, tomato and orange juice are fair source.

Box 14.5 What Are Dietary Folate Equivalents?

Synthetic folate from fortified foods and supplements is 1.7 times more bioavailable than that present in natural foods. Thus, to know how much folate would be available to the body when on a diet and supplement regime, we use the 'dietary folate equivalents' (DFE). it can be calculated by the following equation:

$$DFE = \mu \text{ food folate} + (1.7 \times \mu \text{ synthetic folate})$$

Deficiency: An adult requires around 200 μg folate per day. Inadequacy of folate impairs cell division and protein synthesis. This affects the formation of RBCs and replacement of broken tissues in gastrointestinal tract. Thus, the two most important symptoms of folate deficiency are anaemia and stomach upset. The RBCs are larger than normal in size ar therefore the condition is known as **megaloblastic** or **macrocytic anaemia**. It results in weakness, irritab ity, shortness of breath and mental confusion. Other symptoms include smooth red tongue and elevate homocysteine levels. Folate deficiency during pregnancy causes **neural tube defect** in some new born, t most common form of defect is **spina bifida**. Such children have mental impairment, partial paralysis ar poor bladder/bowel control. It is therefore recommended that all women must take folic acid supplemer prior to conceiving and during pregnancy.

Toxicity: Dietary folate does not cause toxicity. Supplements may however hinder the diagnosis for de ciency of cyanocobalamin–another B-group vitamin. The UL for folate is 1000 μg per day.

14.3.8 Cyanocobalamin

Cyanocobalamin or vitamin B_{12} is a deep-red needle like crystalline compound, which is slightly soluble water. It was identified as an extrinsic factor responsible for the treatment of pernicious anaemia in 194 Cyanocobalamin is a unique vitamin because it cannot be made by plants or higher animals (includi humans). It can however be made by micro-organisms. The presence of vitamin B_{12} in food is the result micro-organisms present in that food or its being synthesized by the microbes in the gastrointestinal tr of an animal. It is absorbed in the ileum with the help of intrinsic factor. Intrinsic factor is a muco-prote enzyme produced by the gastric mucosa. Cyanocobalamin is stored in the liver. The total body store is or around 3–5 mg, but this small amount can meet our requirement for several years. Commercially, t vitamin is obtained as a by-product during the fermentation-based production of certain antibiotics such aureomycin and streptomycin.

Functions

1. Cyanocobalamin helps the body to use some amino acids and fatty acids for energy (ATP) production.
2. It helps to convert homocysteine to methionine. Methionine is an essential amino acid (refer to Chapter 13).
3. Vitamin B_{12} is essential for the maturation of RBCs in the bone marrow.
4. It facilitates synthesis of nucleic acids and nucleoproteins.
5. The presence of this vitamin is essential for the metabolism of nervous tissues and maintenance of myelin sheath on the nerves.
6. Along with folate and vitamin B_6, it is useful in the treatment of elevated homocysteine levels (a risk factor for development of cardiac disease).

Food sources: Shell fish (particularly clams and oysters), beef liver, chicken liver, herring, crab and fortified cereals are excellent sources. Salmon, sardines, lobster, beef, tuna, shrimp, halibut, yogurt and fortified soymilk are good sources. Squid, cod, milk, cottage cheese and pork loin chops are fair sources.

Deficiency: Inadequate dietary intake (a strictly vegetarian diet), increased requirement (pregnancy, lactation) or poor absorption (lack of intrinsic factor, old age, mal-absorption, gastric surgery) can result in vitamin B_{12} deficiency over a period of time. Deficiency of cyanocobalamin causes **megaloblastic microcytic anaemia**. If the anaemia occurs due to an inappropriate immune response, which destroys certain cells lining the stomach that are responsible for producing intrinsic factor, it is known as *pernicious anaemia*. In both the cases, the RBCs are large, nucleated and fragile. This symptom is similar to that of folic acid deficiency. It is important to note that cyanocobalamin deficiency very often triggers that of folic acid and therefore the treatment of megaloblastic microcytic anaemia needs to be done carefully. Irreversible degeneration of nerve cells is an important characteristic symptom of vitamin B_{12} deficiency. It is characterized by degeneration of spinal cord, spastic weakness, loss of position sense in the lower limbs and neuro-psychiatric disorders. Other symptoms include sore tongue, loss of appetite, constipation and reduced cognitive capacity.

Toxicity: No toxicity has been observed even upon consuming 1000 mcg cyanocobalamin per day. No UL has been recommended by nutrition authorities.

GOOD TO KNOW

Fortified and Enriched Foods

Fortified and enriched foods are increasingly becoming popular. These are methods of improving nutritive value of food.

Fortification refers to addition of nutrients during processing. For example, polishing of rice during milling results in considerable loss of thiamin. If this vitamin is added during the production of milled polished rice, it is known as fortification. It also refers to addition of a nutrient not otherwise lost during processing.

Enrichment is a type of fortification wherein nutrients are added to food to make it more nutrient dense (than its unprocessed counterpart). Thiamin, riboflavin, niacin, folate and iron are often added to cereals to make them 'enriched'.

14.3.9 Ascorbic Acid

Vitamin C or ascorbic acid is the most popular water-soluble vitamin. It is also known as 'fresh fruit vitamin' or 'anti-scorbutic factor'. Vitamin C is a white crystalline compound that can be made from glucose by all plants and most animals, but not humans. Ascorbic acid is rapidly absorbed from the intestines.

The degree of absorption decreases as the intake increases and may vary from 15 per cent to 98 per cent (average absorption is around 80–90 per cent). Unabsorbed vitamin C is metabolized by intestinal flora. When it is consumed above the requirement, excess is excreted in the urine. Vitamin C has significant interactions with vitamin A, thiamin, riboflavin, pyridoxine, folic acid and vitamin E. It performs a wide array of functions in our body (Figure 14.13).

Functions

1. **Formation and Maintenance of Collagen:** Ascorbic acid is essential for the formation and maintenance of collagen. Collagen is a fibrous protein that acts as a matrix or cementing material that holds the cells of the body together. Thus, it binds together cells of the skin, bones, teeth, tendons, cartilage, blood vessels etc.
2. **Healing of Wounds:** This vitamin facilitates early healing of cuts and wounds such as those formed during injury and surgery.
3. **Absorption of Minerals:** Vitamin C enhances absorption of iron, copper and chromium. The most significant is the intestinal absorption of iron, which is carried out by converting ferric form of iron into ferrous.
4. **Acts as an Antioxidant:** Similar to vitamin E and unlike B-group vitamins, vitamin C acts as an antioxidant. It protects several water-soluble substances and tissues from oxidative damage. This property helps in reducing the risk of several degenerative diseases such as cataract, heart diseases and cancer.
5. **Production of Hormones:** Ascorbic acid is required for the production of several hormones such as thyroxin and steroid hormones. It thus helps to regulate the metabolic process.
6. **Formation of Carnitine:** Vitamin C assists in the hydroxylation of carnitine. Carnitine is a compound that transports fatty acids within the cells.
7. **Formation of Neurotransmitters:** It is necessary for the conversion of tryptophan and tyrosine to neurotransmitters, namely serotonin and nor epinephrine.

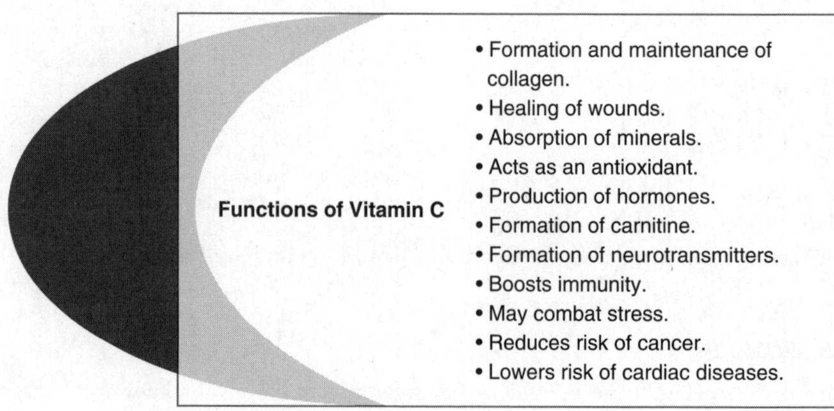

Figure 14.13 Functions of vitamin C

8. **Boosts Immunity:** The role of vitamin C in enhancing immunity is related to prevention of common infections particularly common cough and cold. A daily dose of at least 200 mg reduces the frequency of developing common cold by 8 per cent. It helps to provide relief from nasal congestion by deactivating histamine. Histamine is a substance produced by cells of the immune system in response to viral infection and is related to nasal congestion.
9. **May Combat Stress:** The requirement of vitamin C increases during physical stresses such as infections, burns, extremely high or low temperatures, cigarette smoking and consumption of heavy metals (lead, mercury, cadmium) as well as certain drugs (aspirin, oral contraceptives), thereby indicating its role in dealing positively with stress.
10. **Reduces Risk of Cancer:** Higher intake of vitamin C has been found to be associated with lowered risk for cancer especially of the lungs, mouth, oesophagus, stomach and colon. For instance, a high concentration of this vitamin in gastric juice prevents formation of N-nitroso compounds (refer Chapter 3), which are mutagenic.
11. **Lowers Risk of Cardiac Diseases:** It reduces the risk of developing heart diseases by preventing oxidation of low-density lipoprotein cholesterol.

Food Sources: Vitamin C content in food is expressed in milligrams. Pectin and zinc have been found to hinder with the absorption of ascorbic acid. Amla is an exceptionally good source as it contains 600 mg of ascorbic acid per 100 g edible portion, which is 1.5 times the day's requirement for an adult. Fruits such as gooseberry, guava, cashew fruit, zizyphus, lime, orange, sweet lime, strawberry, raspberry, papaya, pineapple and mahua are good sources while pomegranate, tomato, phalsa, passion fruit, melon and grapes are moderate sources. Vegetables such as drumstick leaves, parsley, amaranth, turnip greens, agathi, cabbage, broccoli, radish leaves, capsicum, chillies, knol-knol, bitter gourd and fetid cassia are very good source while carrot leaves, beet greens, colocasia leaves, fenugreek leaves and cluster beans are moderate source. Milk and milk products, meat, fish, poultry, cereals, pulses and nuts are poor sources of ascorbic acid. Sprouting and fermentation increases the vitamin C content of cereals and pulses by 5–15 per cent.

GOOD TO KNOW

Vitamin C Facts

- Fresh amla juice contains 20 times as much vitamin C as orange juice.
- Vitamin C content of apple is only 1.0 mg/100 g while that of guava is 212 mg/100 g.
- All sour foods are not rich in vitamin C; tamarind pulp contains only 3.0 mg ascorbic acid per 100 g while lemon contains 40 mg/100 g.

Deficiency: Inadequate intake or impaired absorption for about a month results in the deficiency of vitamin C, which is known as scurvy. Adult scurvy is characterized by the following symptoms:

1. The body begins to lose its ability to synthesize collagen resulting in bleeding of gums and tissues around the joints. Small blood vessels break and tiny pinpoint haemorrhages appear just under the skin. These can be seen particularly in calf, buttocks and forearm. There is swelling and pain around the joints due to haemorrhages.
2. The gums become swollen, blue-red, spongy and friable. There is loosening of teeth.
3. Muscles, including that of the heart degenerate.
4. The skin becomes rough, brown, scaly and dry. This results in impaired healing and at times opening of wounds in the process of healing.

5. Bone rebuilding falters; the ends of long bones become softened, malformed and painful.
6. Anaemia develops due to impaired absorption of dietary iron.
7. The frequency and severity of infections increases.
8. Hysteria and depression have been observed as the characteristic psychological symptoms.
9. Sudden death may occur due to massive internal bleeding.

Infantile scurvy is characterized by

1. Swelling and haemorrhages of the gums surrounding the erupting teeth.
2. Extreme pain on movement and development of haemorrhages particularly around the joints. This results in paralysis of limbs.

Scurvy is seen among infants who are born to mothers having low vitamin C status, inadequate vitamin C intake/absorption during lactation or if the weaning food lacks fresh sources of this vitamin.

GOOD TO KNOW?

How Was the Treatment of Scurvy Discovered?

James Lind, a British medical doctor, discovered the treatment of scurvy in the mid-1700s. He provided treatment to 12 sailors (6 groups of 2 sailors each) suffering from scurvy by giving them lemons, limes, cider, nutmeg, sea water or vinegar. Sailors given lemons and lime got cured but others did not. Though this led to identification of treatment, the factor responsible for curing scurvy was however isolated after almost 200 years and was then named 'ascorbic acid'.

Toxicity: The safe UL for ascorbic acid is 2 g (2000 mg) per day. The recommended daily dietary intake fc men and women is 40 mg per day. It is recommended that smokers should increases their vitamin C intak by an additional 35 mg per day. Mega doses, however, increase the risk of developing nausea, abdomin cramps and diarrhoea. The risk of developing kidney stones, gout and iron overload (hemochromatosis) als increases. Vitamin C acts as a pro-oxidant when quantities exceed the needs of the body.

The functions, food sources, deficiency and effects of toxicity for various water-soluble vitamins discusse so far have been summarized in Tables 14.2(a) and 14.2(b).

Table 14.2 (a) Water soluble vitamins: Salient functions, typical food sources, deficiency and toxicity

Vitamin	Salient Functions	Typical Food Source	Deficiency	Toxicity
Thiamin	Constitutes a co-enzyme – thiamin pyrophosphate (TPP), synthesis of DNA/ RNA/ NADP, maintain nerve and muscle function.	Husk/bran and germ of whole grains, Yeast, pork, ham, tuna, beans, legumes, nuts and seeds (particularly sunflower seeds).	Dry beri beri, wet beri beri, infantile beri beri cerebral beri beri	None reported
Riboflavin	Forms part of two coenzymes FAD and FMN, conversion of tryptophan to niacin, treatment of cataracts, function of dopamine, activation of vitamin A, folic acid and pyridoxine.	Liver, dried yeast, milk, cottage cheese, curd, yogurt, egg, fish, turkey, chicken, beef, ham and almonds.	Ariboflavinosis (glossitis, Angular stomatitis, cheilosis, Seborrheic dermatitis, Nasolabial dyssbacea), anemia	None reported

Niacin	Component of two coenzymes–NAD & NADP, metabolism of macronutrients to yield energy, synthesis of fatty acids and DNA, protect fro heart diseases.	Beef liver, chicken (breast, liver), tuna, halibut, turkey, salmon, Sardines, shrimp, cod, mushrooms, legumes, nuts and sunflower seeds.	Pellagra or Disease of 4D's: Diarrhea, Dermatitis, Dementia, Death	Niacin flushes, Nausea, vomiting, liver damage and impaired glucose tolerance.
Pantothenic acid	Component of coenzyme A, synthesis of non essential fatty acids, neurotransmitters, steroid hormones and hemoglobin.	Beef liver, shitake mushrooms and sunflower seeds, dairy products (milk, yogurt), turkey, fish, eggs, whole grains, potatoes, tomatoes, broccoli and coffee.	Mild: vomiting, nausea, stomach cramps, sleeplessness, fatigue, depression, irritability & restlessness. Severe: hypoglycemia, increased sensitivity to insulin.	None reported
Vitamin B_6	Synthesis of non-essential amino acids, white blood cells, hemoglobin, neurotransmitters, lower homocysteine levels, facilitate release of energy from adipose tissues and lean tissues.	Liver of beef and chicken, tuna, turkey, beef, pork and chick peas	Microcytic hypochromic anemia, damage to nervous system (depression, headaches, confusion, convulsions, seizures)	tingling/prickling sensation, numbness in feet & hands, difficulty in walking, headache, stomach upset and neurological damage.

Table 14.2 (b) Water soluble vitamins: Salient functions, typical food sources, deficiency and toxicity

Vitamin	Salient Functions	Typical Food Sources	Deficiency	Toxicity
Biotin	Acts as a co-enzyme in the breakdown process of amino acids and fatty acids to release energy, Synthesis of DNA.	Cooked eggs – egg yolk (not raw), liver of all animals, fish, cheese, soybean, peanut, cauliflower and whole grains	lethargy, depression, hallucinations, skin irritations, red scaly rashes around eyes nose and mouth, infections, hair loss, poor muscle control, seizures & developmental delays in infants	None reported
Folate	Synthesis of DNA, RNA, methionine, maturation of red blood cells, development of neural tube, maintain digestive system health, control homocysteine levels, lower risk of cancer.	Liver of goat, sheep and beef, bengal gram whole, cluster beans, pinto beans, lentils, black gram dal, cowpea, guava, asparagus, turnip greens, broccoli, all dark green leafy vegetables.	Megaloblastic or macrocytic anemia, neural tube defect.	None reported

Cyanocobalamin	maturation of red blood cells, release energy from amino acids and fatty acids, synthesis of nucleic acids and nucleoproteins, maintenance of myelin sheath, lower elevated homocysteine levels	Shell fish (particularly clams and oysters), beef liver, chicken liver, herring, fortified cereals, salmon, sardines, lobster, beef, tuna, shrimp, halibut, yogurt and fortified soymilk	Megaloblastic microcytic anemia, pernicious anemia, degeneration of spinal cord, spastic weakness, loss of position sense in the lower limbs and neuro-psychiatric disorders	None reported
Ascorbic Acid	Formation and maintenance of collagen, healing of wounds, absorption of minerals, acts as antioxidant, Production of hormones, formation of carnitine & neurotransmitters, boosts immunity, combat stress, reduces risk of cancer and cardiac diseases.	Fruits: Amla, gooseberry, guava, cashewfruit, zizyphus, lime, orange, sweet lime, strawberry, raspberry, papaya, pineapple & mahua Vegetables: drumstick leaves, parsley, amaranth, turnip greens, agathi, cabbage, broccoli, raddish leaves, capsicum, chillies, knol-knol, bitter gourd and fetid cassia.	Scurvy (bleeding gums, pin point hemorrhages, abnormal bone growth and joint pain).	Increased risk of nausea, abdominal cramps, diarrhea, kidney stones, gout and iron overload.

14.4 Minerals: Introduction and Categorization

Minerals are inorganic elements that remain largely as ash when plant or animal tissues are burned. They constitute about 4–5 per cent of the total body weight or 2.7 kg and 3.6 kg in a healthy adult woman and man, respectively. Approximately 50 per cent of this weight is constituted by calcium, 25 per cent by phosphorus and the remaining 25 per cent comprise magnesium, sodium, potassium, chloride, sulphur, iron, zinc, iodine, selenium, manganese, fluoride, molybdenum, copper, chromium, cobalt and boron. Similar to vitamins, minerals are classified as micronutrients because they are required in small amounts (lower than that of carbohydrates, proteins and fats) by our body. Minerals share several common characteristics and yet they are quite different from vitamins. Some of the typical similarities and differences between minerals and vitamins are highlighted in Figure 14.14.

How do minerals differ from vitamins?	What do minerals and vitamins share in common?
• Minerals are inorganic while vitamins are organic compounds. • Vitamins get destroyed easily but the same is not true in case of minerals. Minerals do not get destroyed by heat, acid or air. • Unlike vitamins, comparatively little care is required to conserve minerals in food. • The daily requirement of each mineral is different from that of each vitamin. • Good food sources of each mineral varies from vitamins and so do their deficiencies and toxicities.	• They are micronutrients, that is, required in amounts lesser than macronutrients. • While certain minerals such as potassium are absorbed, transported and excreted much like water-soluble vitamins, others such as calcium show similarities with fat-soluble vitamins in these processes. • They interact with other nutrients to perform vital physiological functions. • Inadequate intake, malabsorption and reduced bioavailability result in states of their respective deficiencies.

Figure 14.14 Differences and similarities between minerals and vitamins

Based on their requirements by our body, minerals are classified as major minerals and trace elements.

Major Minerals: Minerals that are required by humans in amounts more than or equal to 100 g/day for maintaining good health are known as major minerals or macro minerals. Minerals under this category include calcium, phosphorus, magnesium, sulphur, sodium, chloride and potassium.

Trace Elements: Minerals that are required by humans in amounts less than or equal to 100 mg/day for maintaining good health are known as micro minerals or trace elements. Minerals under this category include iron, zinc, selenium, copper, chromium, fluoride, iodine, manganese and molybdenum.

Ultra Trace Elements: Minerals that are required in small quantities, typically in micrograms or lesser amount, are referred to as ultra trace elements. These include arsenic, aluminium, tin, nickel, vanadium and silicon.

Although each mineral performs a characteristic function, some of the typical or general functions of minerals have been highlighted in Box 14.6. The daily recommended dietary intake for various minerals across all age groups are given in annexures.

Whether a mineral is classified as macro mineral, trace element or ultra trace element, they all are equally important to maintain health. Since it is beyond the scope of this book to discuss all minerals, a few important ones are discussed briefly below.

Box 14.6 General Functions of Minerals

- maintain acid–base balance,
- regulate the osmotic pressure within and outside the cells,
- facilitate the transfer of essential nutrients and other molecules accross the cell membranes,
- contribute towards the maintenance of normal blood composition,
- maintain nerve and muscular irritability,
- involved in growth and development,
- promote immune function,
- regulate activities of several enzymes and
- interact with DNA to control transcription of certain proteins.

14.4.1 Calcium

Calcium constitutes only about 1.5–2.0 per cent of the total body weight, but it is the most abundant mineral present in our body. Approximately 99 per cent of this amount is present in the skeleton while the remaining is in blood and other tissues. Calcium absorption occurs most efficiently in the duodenum but because the food remains in contact with ileum and jejunum for a longer time, maximum absorption occurs in these parts of the gut. Some absorption also occurs through the colon. Calcium absorption and excretion is influenced by several factors such as diet, activity, age, gender and physiological stage (Figure 14.15). These factors influence calcium homeostasis, which is vital to maintaining health. When blood calcium levels begin to fall, the body diminishes the excretion of this nutrient through urine (*reabsorption*), enhances its absorption in the gut (*absorption*) and stimulates the breakdown of bones to release calcium into the blood (*resorption*). On the other hand, if calcium begins to rise, there is enhanced excretion, lowered absorption and deposition of calcium in bones (*mineralization*). An adequate intake is important for developing strong bones and teeth, performing metabolic functions, preventing kidney stones, obesity, hypertension, diabetes and certain types of cancer.

Functions

1. **Form and Maintain Bones and Teeth:** Calcium along with phosphorus and vitamin D are vital for bone and teeth health. Calcium gets deposited in the bones in the form of hydroxyapatite, which are crystals of calcium and phosphorus. These crystals gradually become denser and they

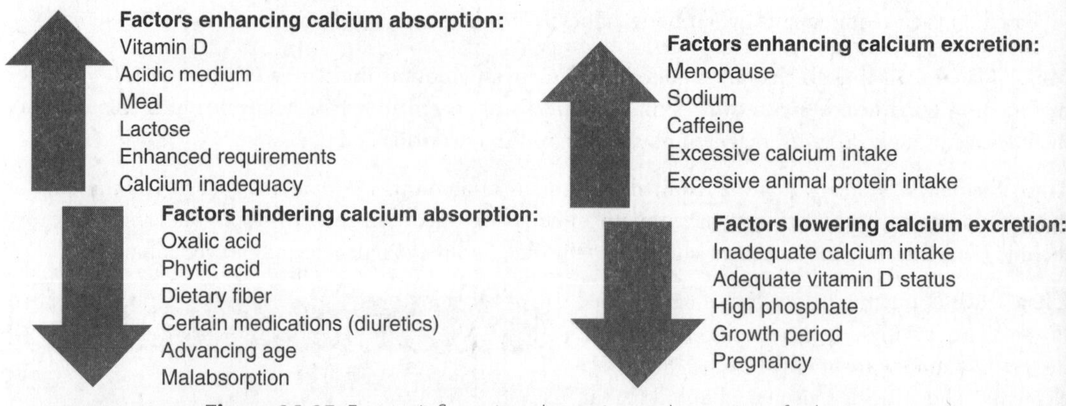

Figure 14.15 Factors influencing absorption and excretion of calcium

provide strength and rigidity to the bones through the process of mineralization. Bone is a living tissue that gains and loses calcium throughout life and this process is known as remodelling. Bones are therefore considered calcium banks from where the body takes out calcium whenever required and deposits this mineral whenever feasible.

2. **Maintain Muscle Contraction:** Calcium ions present inside the cells of the muscles promote their contraction and relaxation. Low calcium levels can result in muscle fatigue during exercise or any other physical activity.
3. **Promote Blood Clot Formation:** As discussed earlier (refer functions of vitamin K), along with vitamin K and protein, calcium is also required for blood clot formation. Calcium participates in several steps associated with the formation of fibrin – a protein necessary for blood clot formation.
4. **Facilitate Nerve Function:** Calcium is required for transmission of nerve signals. An adequate movement of calcium ions across the nerve cell membrane ensures the efficiency of this function.
5. **Promote Cell Metabolism:** Calcium plays an important role in cell division, cell multiplication, their secretions and specific movements.
6. **Prevent Diseases:** Diets rich in calcium, magnesium and potassium have been found to prevent hypertension. Adequate calcium intake also lowers the risk of elevated blood cholesterol levels, diabetes and colon cancer. It is also associated with maintenance of a healthy body composition and prevention of obesity.

GOOD TO KNOW

Bone Facts

Fact 1. Bone mineralization is greatest during childhood.

Fact 2. Peak bone mass is reached between the age of 20 and 30.

Fact 3. Bone mass begins to decline between the age of 30 and 40 and continues throughout the remainder of life. Women after menopause lose bone calcium much faster.

Food sources: Only approximately 30 per cent dietary calcium gets absorbed. Several dietary factors hinder or promote calcium absorption (Figure 14.14). Milk and certain milk products such as cheese (cottage and processed), yogurt, curd, lassi, khoa, condensed milk and milk powder are the best sources of calcium. Sardines (with bones) and Bok Choy (a Chinese cabbage) contain more calcium than milk, but are not commonly consumed in India. Green leafy vegetables particularly fetid cassia, colocasia leaves, nerringi and ape leaves agathi are excellent source of calcium followed by amaranth, cauliflower greens, curry leaves, knol-knol greens and turnip greens. Bioavailability of calcium from milk is approximately 30 per cent while hat from green leafy vegetables varies between 5 per cent and 50 per cent. Among nuts and oilseeds, gingelly and coconut (dry) are excellent source of this mineral. While meat and poultry are poor source, fish and other sea food such as dried bhagon/chela/chingri/mutijella, shrimp (dried) and crab are excellent source.

GOOD TO KNOW

Spinach Is a Good Source of Calcium: Fact or Myth!

Spinach Is not a good source of calcium.

1. 100 g of spinach contains 73 mg while a similar amount of spinach stalks contain 90 mg calcium.
2. To name a few, the following leafy vegetables contain more calcium than spinach and its stalks: agathi, amaranth, bathua, beet greens, Bengal gram leaves, carrot leaves, cauliflower greens, celery, lettuce, mint, coriander, fenugreek leaves, colocasia leaves, fetid cassia, garden cress, rape leaves, nerringi, turnip greens, radish leaves and so on.
3. Bioavailability of calcium from spinach is <5 per cent while that from certain other leafy vegetables such as turnip greens is >50 per cent.

Deficiency: Hypocalcaemia or calcium deficiency initially causes muscle pain, spasms and tingling sensation in hands and feet. Chronic deficiency in children results in stunted growth because the bones do not grow to their full potential. Among adults there is gradual bone loss resulting in osteoporosis. Osteoporosis is characterized by porous and fragile bones resulting in increased risk for fractures. It is of two types:

Osteoporosis I or post menopausal osteoporosis: Rapid loss of oestrogen among women or gradual loss of testosterone in men results in bone loss especially of the wrist and spine. It is most commonly observed in the age of 50–70 years.

Osteoporosis II or senile osteoporosis: Reduced calcium absorption results in increased bone mineral loss particularly of the hip. It is usually observed among elderly above 70 years of age.

GOOD TO KNOW

Are Calcium Supplements Good for Health?

A well-balanced diet that contains good amount of calcium, vitamin D, phosphorus and magnesium fulfils the need for calcium. An adult woman and man need calcium equivalent to that present in 3 1/3 and 4 cups of milk, respectively. Thus, calcium intake need not be fulfilled only by taking milk but other foods rich in this nutrient as well.

Calcium supplements should be taken only if the diet is not nutritionally adequate or if there is an underlying health problem such as fat mal absorption (food allergy, cancer, surgery of gastrointestinal tract). Excessive intake of supplements can cause toxicity (see below).

Toxicity: The tolerable UL for calcium is 2500 mg/day for adult ≤51 years of age and 2000 mg for those above 51 years. Excessive calcium interferes with absorption of other nutrients and tends to get deposited in the soft tissues particularly of the muscles, heart and kidney. Hypercalcemia increases the risk of kidney dysfunction and renal stone formation.

14.4.2 Phosphorus

Phosphorus is the second most abundant mineral in the body. About 85 per cent is found combined with calcium in the form of hydroxyapatite crystals in the bones and teeth. Remaining is found in the body fluids and soft tissues. Phosphorus is readily absorbed in the small intestine. Similar to calcium, when the blood phosphorus levels fall, the absorption from small intestine increases and there is also bone resorption and vice versa.

Functions

1. **Forms Structural Component of Body:** Phosphorus is part of the structure of biological membranes (membranes surrounding various cells), DNA, ATP (the energy molecule) and certain lipoproteins.
2. **Performs Functional Roles:** Phosphorus performs several functions such as
   (a) It is essential for protein synthesis.
   (b) Phosphorus is required for release of energy from carbohydrates, proteins and fats.
   (c) This mineral helps to maintain normal blood pH.
   (d) It is essential for growth, development and maintenance of skeleton.
   (e) It facilitates transport of fat in the circulation (blood).

Food Sources: Phosphorus is present in almost all foods. Protein-rich foods are particularly rich in this nutrient. Dairy products, meat, poultry, seafood, nuts and seeds are good source. Bioavailability is lower in seeds and grains because they contain phytates that hinder with the absorption of phosphorus. It is also present in processed foods and carbonated beverages. A can of cola contains about 50 mg phosphorus (mentioned as phosphoric acid on the label).

Deficiency: It is rare. Characteristic symptoms include loss of appetite, anaemia, muscle weakness, poor bone development, bone pain and in extreme cases death.

Toxicity: The UL for adults (19–70 years) is 4000 mg per day. It is characterized by muscle spasms, convulsions and mineralization of soft tissues especially of the kidneys. Excessive intake of phosphorus (heavy consumption of cola drinks) can result in demineralization of bones because of disturbance in the calcium/phosphorus ratio.

14.4.3 Sodium

Sodium is the mineral that along with chloride provides savoury taste to our food and sweat. Sodium exists along with chloride and potassium and they constitute 2 per cent, 3 per cent and 5 per cent (respectively) of the total mineral content in our body. Sodium constitutes 325 g of an adult body weight. Nearly 50 per cent of this amount is dissolved in the extracellular (outside the cell) fluid, 5 per cent in intracellular (inside the cell) fluid and 45 per cent in skeleton. Sodium is consumed through diet: water, natural and processed foods being the primary sources. This macro mineral is readily absorbed from the intestine and carried to the kidneys, where it is filtered and returned to the blood to maintain normal blood sodium levels. Excess sodium is excreted by the kidneys through urine (1–500 mmol/day). Some sodium is also lost daily through faeces (5–10 mmol) and sweat (20–80 mmol).

GOOD TO KNOW

Sodium and Hypertension (High Blood Pressure)

Sodium increases blood volume by retaining water in the body. Increased blood volume results in elevation of BP. BP above 120/80 mmHg is described as hypertension. Since salt is the major contributor of sodium in the diet, salt-restricted diets lower BP and are beneficial in the prevention as well as treatment of hypertension. 'It is recommended to limit daily salt intake to ~ one teaspoon'. Dietary Approaches to Stop Hypertension (DASH) is an authentic diet plan that suggests dietary habits related to maintenance of normal BP

- Consume fruits and vegetables liberally.
- Prefer low fat milk and its products.
- Include whole grains, nuts, poultry and fish.
- Reduce the intake of red meat, butter and other high fat/high salt-processed products.
- Avoid alcohol consumption and smoking.

Sodium and Hypotension (Low Blood Pressure)

Hypotension is a condition characterized by BP below 120/80 mmHg. Sodium per se can increase BP by increasing blood volume. This is effective for individuals who have suffered from sodium deficiency and hence a drop in BP due to excessive loss through diarrhoea, vomiting or sweating. Individuals with a low blood count such as low levels of haemoglobin, red and white blood cells would not benefit by increased sodium intake. They need to consume a nutrient dense diet to maintain a normal BP.

Functions

1. **Regulate Body Fluids, BP and pH:** Sodium performs these functions with the help of chloride and potassium. Sodium is the major cation while chloride an anion, present in extracellular fluid. Potassium is a predominant cation in intracellular fluids. Thus, these three electrolytes maintain hydrogen ion concentration or normal pH of the blood. They also regulate the water balance and hence the BP. Chapter 4 discusses this concept in detail.
2. **Promote Muscle and Nerve Function:** Sodium stimulates muscle contraction, an action essential to perform any type of work. It helps in the efficient transmission of nerve impulses. This is one of the reasons for lethargy observed after excessive sweating during summer.

DO YOU KNOW?

Are Calcium Supplements Good for Health?

Low sodium salt contains 10 per cent less sodium than common salt. It can be consumed in daily diets by all, except those suffering from elevated blood potassium levels (kidney disorders, congestive heart failure).

Food Sources: Salt is the most concentrated and principal dietary source of sodium. Common salt or table salt is actually sodium chloride (NaCl). One teaspoon of salt or NaCl contains about 2300 mg sodium or, in other words, table salt contains 40 per cent sodium and 60 per cent chloride.

Processed foods contain a higher amount of sodium as compared with natural foods. A 100 g edible portion of natural foods on an average provides only 0.1–3.0 mmol but a similar amount of processed food would provide 10–80 mmol sodium. This is so because processed foods contain additives such as sodium bicarbonate, sodium benzoate, sodium citrate and mono sodium glutamate. Thus, savoury biscuits, crackers, chips, salted nuts/popcorn/snacks, most bakery products, salted butter, processed cheese, ketchup, sauces, chutneys, mayonnaise, salad dressings and ready-to-eat breakfast cereals/pastas have high sodium content. Labels should be read and interpreted carefully. Commonly used terms on label include

1. Sodium free and Salt free: <5 mg sodium per serving.
2. Low sodium: ≤140 mg sodium per serving.
3. Very low sodium: ≤35 mg sodium per serving.

Natural foods such as sea food, milk, meat, most green leafy vegetables, lotus stem, knol-knol, litchi and musk melon are good source of sodium.

Deficiency: It occurs as a result of diarrhoea, vomiting or excessive sweating. Deficiency of sodium (hyponatremia) results in dry mouth and tongue, shrunken skin that lacks turgor, sunken eyes, weak but rapid pulse, low BP (hypotension), muscle cramps, loss of appetite and mental apathy. Severe deficiency of sodium such as observed in acute diarrhoea (such as food poisoning) can result in circulatory failure.

Excess: The tolerable UL for adult intake is 2300 mg. Excess of sodium in the body is known as hypernatremia. Initial symptoms are irritability, restlessness and elevated BP (hypertension). The severity of hypertension is directly proportional to sodium overload. In extreme situations seizures and death may occur.

14.4.4 Iron

Iron is a trace element. It is most commonly associated with anaemia, which arises due to its deficiency. In a healthy adult, total iron content is approximately 50 mg/kg body weight. Nearly 60 per cent is present in haemoglobin, 8 per cent in myoglobin, 5 per cent in enzymes and 2 per cent in the form of transferrin. The remaining 25 per cent is stored – primarily in the liver and some in the spleen, kidney and bone marrow. Iron is stored in the form of ferritin, hemosiderin and siderophillin. During periods of low intake or high requirements, iron absorption increases and vice versa. The bioavailability of iron is quite low. Overall 18 per cent of dietary iron is absorbed from mixed diets and only about 10 per cent from typical Indian vegetarian diets. During disease of the digestive system, the absorption may fall as low as 2 per cent only while it may increase to 35 per cent during periods of growth.

Functions

1. **Transport Oxygen and Carbon-Dioxide Through Blood:** Haemoglobin is composed of 'hem' and 'globin'. 'Hem' comprises iron while 'globin' is a protein. The synthesis of haemoglobin requires iron and protein. The function of haemoglobin is to carry oxygen from the lungs to the cells of our body and bring carbon-dioxide from various cells to lungs. This function makes iron essential for survival.
2. **Transport Oxygen and Carbon-Dioxide Through Muscles:** Iron is a constituent of myoglobin. Myoglobin is present in cells of the muscles. Myoglobin supplies oxygen to muscle cells during physical activity and takes back carbon-dioxide. Lack of oxygen reduces the capacity to perform work (such as writing, walking, running) and causes muscle pain.
3. **Promote Enzyme Function:** It acts as a cofactor for several important enzymes, which are collectively known as cytochrome P450. These enzymes perform several functions. They help to metabolize drugs and remove toxins. Some of these enzymes reduce the risk of developing cancer and coronary heart diseases.
4. **Maintain Brain Function:** Iron is essential to normal brain function. It is necessary for the synthesis of myelin sheath – a protective covering that surrounds the nerve cells. It is also required for the

production of neurotransmitters – substances that carry messages between nerve cells. Thus, our capacity to learn, understand and remember is influenced to some extent by iron.

5. **Other Functions:** It protects DNA, cell membranes and proteins from oxidative damage. It is also involved in DNA synthesis. Thus, iron is important for normal growth and development. It is also necessary for maintaining optimal immune function.

Food Sources: Food contains two types of iron: hem iron and non-hem iron.

Hem Iron: It is present only in animal foods because hem iron is part of haemoglobin and myoglobin. It is not present in plant foods. Hem iron is more bioavailable than non-hem iron. Meat, fish, sea food and poultry are good source of hem iron. They contain about 40 per cent hem and 60 per cent non-hem iron. Liver of all animals are concentrated source especially beef liver. Among sea food, dried – Chingri, bhekti, parsey, tapra, fresh – Bombay duck, crab (muscle) and ribbon fish are good source of non-hem iron.

Non-Hem Iron: It is present only in plant foods. Non-hem iron has a lower bioavailability than hem iron. Garden cress seeds and Niger seeds are excellent source. Green leafy vegetables such as amaranth, garden cress, turnip greens, chekur manis, Bengal gram leaves, cow pea leaves, mustard leaves, parsley, radish leaves and shepu are rich in iron. Cauliflower greens are an excellent source of iron. Pulses/legumes (such as soybean, moth beans, Bengal gram, cowpea, lentil and horse gram) and fruits (such as currants, dates, apricots, water melon, raisins and sethaphal) are moderate source of iron. Milk, milk products and most cereals (except rice bran and rice flakes) are poor source.

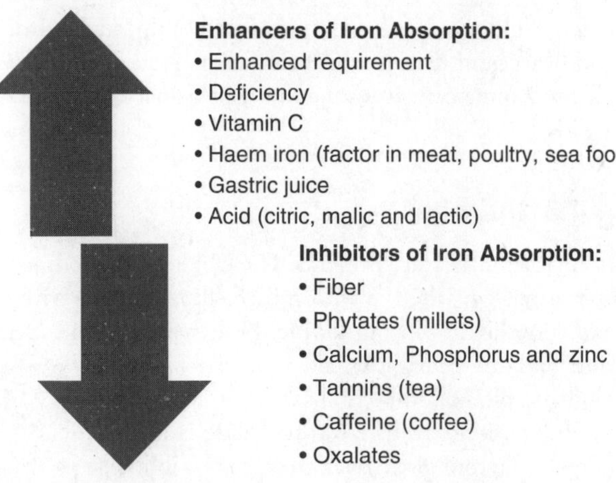

Figure 14.16 Factors influencing absorption of iron

GOOD TO KNOW

Spinach and Apples are Good Sources of Iron: Myth or Fact!

It is a myth. Spinach, apple and pomegranates are not good sources of iron. Spinach contains on an average 1.17 mg iron per 100 g. To name a few, amaranth, garden cress, turnip greens, chekur manis, Bengal gram leaves, cow pea leaves, mustard leaves, parsley, radish leaves, shepu, lettuce, mint and drumstick leaves contain more iron. Further, the bioavailability of iron present in spinach is <5 per cent.

On an average 100 g apple contains only 0.67 mg iron. To name a few amla, apricot, pear, currants, dates, figs, guava, loquat, water melon, mulberry, peaches, pineapple, raisins, raspberry, sapota, seethaphal, tomatoes, etc. contain more iron than apple. Besides being richer in iron, many of these fruits are good source of vitamin C (enhancer) and poor source of fibre (inhibitor), indicating greater bioavailability of iron present in them.

It is important to remember that, it is not only the iron content of a food that is important. Equally important is the composition of the meal and other environmental factors that influences its absorption (Figure 14.16). Cooking in cast iron utensils especially in an acidic (sour) medium increases the iron content of the dish considerably. However, the absorption of this iron is very low (1–2 per cent).

Deficiency: Inadequate iron status is the most common nutritional deficiency across the world. ***Mild iron deficiency*** is characterized by fatigue, listlessness, irritability, reduced work capacity, lowered intellectual performance, cold hands and feet as well as poor immunity. It is characterized by serum ferritin levels below 12 μg/l; the haemoglobin levels may be normal. In ***severe iron deficiency*** there is anaemia. Anaemia is a condition where in there is decreased capacity of blood to carry oxygen. The haemoglobin levels fall below 12 and 13 g per 100 ml in women and men, respectively. In addition to the above symptoms, there is breathlessness, giddiness, dimness of vision, headache, insomnia, palpitation, loss of appetite, indigestion, poor hair growth, premature greying of hair and atrophy of the tongue papillae. The nails become brittle, longitudinally ridged with fissures and spoon shaped. This condition is known as koilonychia. Iron deficiency in pregnant women often leads to pica (a behaviour characterized by eating of non-edible items such as chalk, sand, mud, clay etc.), increased risk of miscarriages, low birth weight and premature delivery. Severe iron deficiency can be fatal.

Toxicity: The safe UL is 45 mg/day for healthy adults. Iron toxicity results in nausea, vomiting, diarrhoea, rapid heart beat, dizziness and confusion. Iron overload, which arises due to a genetic disorder, is known as hemochromatosis. It results in organ damage, increased risk to heart disease, liver cancer, diabetes and arthritis.

14.4.5 Iodine

Iodine is an ultra trace mineral. The deficiency of iodine was identified as early as in 2700BC by Chinese civilizations who used to treat it by feeding patients with sea weed and marine animals. An adult human body contains 15–45 mg iodine. Nearly 75 per cent of the total body iodine is present in the thyroid gland and the remainder is in the circulating blood and fluids. Iodine is absorbed in the form of iodide. On an average 60 μg iodide is taken up by the thyroid gland every day. Excess iodine is readily excreted in the urine (90 per cent of the ingested iodine). Small amount is also lost from faeces and sweat.

DO YOU KNOW?

Only 150–200 μg iodine is required per day, which amounts to a pinhead to meet the requirement for a month. An average lifetime requirement would add up to less than a teaspoonful of iodine. Yet, millions of people suffer from iodine deficiency across the world.

Functions: The only but a major function of iodine is associated with the thyroid gland. Iodine is stored in the thyroid gland where it is used for the synthesis of thyroid hormones, viz. triiodothyronine (T_3) and thyroxin (T_4). Thyroid hormones help to regulate body temperature, basal metabolic rate, reproduction, growth and development.

Food Sources: The most significant source of iodine in our diets is iodized salt. It contains 30 ppm iodine. Salt is iodized with potassium iodate, which is easily absorbed by the human body. Fish, sea food, sea weeds and mother's milk are particularly rich in iodine. Fresh water animals have lower iodine content as compared to those obtained from sea or ocean. The iodine content of plant foods depends upon the iodine content of soil. Thus, crops grown in hilly areas are deficient while those grown in plateau areas are rich in iodine. Milk is also a good source of iodine but heat processing results in considerable loss of this mineral. Certain foods contain goitrogens, which hinder with the absorption of iodine (Box 14.7).

Box 14.7 What Are Goitrogens?

Compounds found in some vegetables decrease iodine utilization by the thyroid gland. They are present naturally in certain foods such as cabbage, kale, Brussels sprouts, cauliflower, broccoli and kohlrabi. Cooking inactivates goitrogens.

Proper selection, use and care of iodized salt is equally important because iodine gets lost easily in hot humid environment. Some important tips that should be kept in mind include

1. Do not buy loose salts. Always buy in sealed packets. The packet must mention 'iodized salt' or 'fortified with iodine' or have the symbol of 'smiling sun'.
2. Always store salt in air tight containers. Ensure to close the lid carefully each time. Do not store in clay, wire or bamboo pots as they cannot be closed tightly.
3. Always use dry spoon or hands while taking out salt from the container.
4. Add salt towards the finishing steps of cooking a dish.
5. Never store salt near a source of heat such as gas stove, heater and sunlight.

Deficiency: Iodine deficiency is prevalent all across our country though its incidence is higher in hilly areas. The sub-Himalayan region has been identified as the 'goitre belt' in India. Two major clinical conditions that arise due to iodine deficiency include goitre and cretinism. Goitre develops when the iodine intake is <50 μg/day and cretinism occurs when the mother's diet contains <30 μg/day.

Causes of iodine deficiency include

1. Lack of iodine in the diet due to poor content of iodine in cultivated soil. If the iodine content of water is <2 μg/l, the area is considered to be 'iodine deficient'. The crops grown in such area would be low in iodine.
2. Inadequate iodine intake due to low consumption of sea food, high cost or low availability of sea food.
3. Inadequate utilization of iodine due to the presence of goitrogens (such as phenols, lithium, thiocyanates) in natural foods, which interfere with the absorption of iodine.

Consequences of iodine deficiency include

Iodine deficiency disorders: Iodine deficiency results in a spectrum of ill effects on health, growth and development across all age groups:

Foetus: abortions, still births, congenital anomalies, cretinism and increased infant mortality.

Newborn: neonate hypothyroidism.

Children and Adolescents: retarded mental and physical development.

Adults: goitre and its complications and hyperthyroidism (iodine induced).

All Ages: goitre, hypothyroidism, impaired mental function and increased susceptibility to nuclear radiations. Mild to moderate iodine deficiency decreases IQ (intelligence quotient) by 10–15 points.

Goitre and cretinism are the two most important disorders that arise due to iodine deficiency.

Goitre: It is described as non-neoplastic, non-inflammatory and non-toxic enlargement of the thyroid gland. The thyroid enlarges as a result of the efforts of the gland to trap more iodine. The different stages of goitre are:

Stage 0: No goitre (thyroid neither palpable nor visible).

Stage 1: Thyroid palpable but not visible in the normal neck position.

Stage 2: Thyroid is visible in normal neck position.

Cretinism: It is characterized by mental deficiency/retardation (impaired learning and lowered IQ-intelligence quotient), delayed motor development, growth failure (dwarfism), muscular disorders, spastic diplegia (paralysis of lower limbs), speech and hearing defects (deaf mutism), squint and hypothyroidism.

Iodine deficiency disorders are preventable. Prevention raises the learning capacity and improves the school performance of children in school. To improve iodine nutrition of its population, the Government of India is running National Iodine Deficiency Disorders Control Program – NIDDCP (formerly known as National Goitre Control Program, 1962). It is also facilitating universal access to iodized salt since 1995 and has banned sale of non-iodized salt for human consumption since 1997.

Toxicity: The tolerable UL for iodine is 1100 μg/day. Large amounts of iodine inhibit synthesis of thyroxin hormone and cause goitre. Thus, too much and too little have similar effects on the thyroid gland.

14.4.6 Fluoride

Fluoride is a trace element that plays an indispensable role in maintaining bone and teeth health. It is different from fluorine – which is a poisonous gas. Traces of fluoride are present in bones, teeth, thyroid gland and skin. Water is the most important source of fluoride. It is rapidly and completely absorbed by the body. Soon after consumption of fluoride, there is a transitory rise in the level of this mineral, which soon returns to 0.1 ppm. Some fluoride is retained by the bones. On an average, an adult skeleton contains 2.5 mg fluoride.

Functions: The only major role of fluoride identified so far is the strengthening of bones and teeth as well as prevention of carries. Fluoride combines with calcium phosphate to form fluoroapatite. As compared to hydroxyapatite, fluoroapatite is harder and more resistant to resorption by acid produced by bacteria in the mouth. It is an anti-bacterial agent and hence prevents the growth of acid-producing bacteria in the mouth. Thus, fluoride helps to prevent dental caries.

GOOD TO KNOW

Water having fluoride in the range of 0.7–1.0 ppm is considered safe for human consumption.

Food Sources: Drinking water contains fluoride. Plant foods (1 ppm) generally contain more fluoride than animal foods (0.1 ppm) except marine fish/sea food (2–5 ppm). Potatoes, legumes and tea are considered to be good source of fluoride. Tea leaves and their infusions contain about 1–2 ppm fluoride. If fluoridated water has been used for preparing a processed food, then it is a good source of this element.

Toxicity: The adequate intake of fluoride is 3–4 mg/day for women and men. The UL has been advocated to be 10 mg/day. Liberal use of fluoride tablets, swallowing of fluoride containing toothpaste (100 ppm) in liberal amounts or long-term intake of drinking water with fluoride content >1 ppm are the primary causes of toxicity. It results in loss of appetite, gastrointestinal upset, excessive production of saliva, watering of eyes and heart problems. The enamel on the teeth loses its lustre, becomes patchy, chalky white and pits appear on its surface. This condition is known as *dental fluorosis*. Skeleton is also affected by chronic high intake levels, that is, 20 ppm in drinking water. There is weakening of the skeleton particularly of the spine, the pelvis and the limbs. This condition is known as *skeletal fluorosis*.

Salient functions, typical food sources and effects of deficient/excessive intake of calcium, phosphorus, sodium, iron, iodine, fluoride, copper and iodine have been summarized in Table 14.3. In addition to the above minerals discussed, there are several other minerals that have been found to be essential for our health. These include chloride, potassium, zinc, selenium, copper, manganese, chromium, molybdenum. Their salient functions, typical food sources, states of deficiency and excess have been summarized in Table 14.3(a) and 14.3(b). Other trace minerals that have been identified to be beneficial for health include nickel, vanadium and boron. These elements have been found to play a role in the maintenance of bone health.

Let us review the situations given in the beginning to recapitulate the chapter.

Situation 1: Vitamin A deficiency is a leading cause of xerophthalmia (night blindness) seen during preschool age. Red palm oil is an excellent and economical source of vitamin A. It is therefore a good nutrition intervention to improve vitamin A nutriture.

Table 14.3 (a) Minerals: Salient functions, typical food sources, deficiency and toxicity

Mineral	Salient functions	Typical food sources	Deficiency	Toxicity
Calcium	Mineralization of bones and teeth, contraction and relaxation of muscles, blood clotting, nerve function, cellular metabolism, prevent hypertension, diabetes, cancer.	Milk, cheese (cottage and processed), yogurt, curd, lassi, khoa, condensed milk, milk powder, small fishes (with bones), gingelly, certain vegetables (bok choy, broccoli, chard, fetid cassia, colocasia leaves, nerringi, rape leaves agathi).	Stunted growth in children, Osteoporosis in adults.	Constipation, kidney stones, kidney dysfunction, interference with absorption of other minerals.
Phosphorus	Mineralization of bones and teeth, maintain normal blood pH, facilitate transport of fat in blood, protein synthesis, release of energy from macronutrients.	Dairy products, meat, poultry, seafood, nuts and seeds.	Muscle weakness, poor bone development, bone pain	Muscle spasms, convulsions and mineralization of soft tissues
Sodium	Regulate body fluids, blood pressure and pH, promote muscle and nerve function.	Salt, processed food, preserves etc., sea food, milk, meat, most green leafy vegetables, lotus stem, knol knol, litchi, musk melon.	Loss of appetite, muscle cramps, mental apathy.	Edema, acute hypertension.
Chloride	Regulate body fluids, forms a component of gastric juice.	Salt, processed food, preserves, meat, milk, eggs.	Rare – low blood pressure	Vomiting
Potassium	Regulate body fluids, nerve and muscle function, supports cell integrity.	Unprocessed fruits, vegetables, milk, certain grains.	Muscle weakness, irregular heart beat, glucose intolerance.	Vomiting, muscle weakness, cardiac arrest.
Iron	Transport oxygen and carbon dioxide in blood and muscles, promote DNA synthesis, enzyme and brain function and maintain immunity.	Hem iron: Meat, fish, sea food and poultry (especially liver) Non-hem iron: Garden cress seeds, niger seeds, Green leafy vegetables (amaranth, garden cress, turnip greens, chekur manis, bengal gram leaves, cow pea leaves, mustard leaves, parsley, radish leaves and shepu).	Anemia (breathlessness, giddiness, dimness of vision, headache, insomnia, palpitation, loss of appetite, indigestion, poor hair growth, premature greying of hair and atrophy of the tongue papillae, koilonychias, pica).	Stomach upset, fatigue, joint pain, skin pigmentation, organ damage, death.

Table 14.3 (b) Minerals: Salient functions, typical food sources, deficiency and toxicity

Mineral	Salient functions	Typical food sources	Deficiency	Toxicity
Zinc	Component of insulin, certain enzymes, proteins and genetic material, facilitate immune reactions, taste perception, wound healing, fetal development, transport of vitamins.	All foods especially protein rich foods, whole grains.	Growth retardation, late maturation, hair loss, eye and skin problems, impaired immune function.	Loss of appetite, lowered immunity, copper and iron deficiencies.
Selenium	Support antioxidant function of vitamin A, E and C. Regulate thyroid hormone.	Sea food, organ meats, depending upon composition of soil – fruits and vegetables.	Predisposes to Keshan's disease (heart disease).	Brittleness and loss of hair and nails, skin rashes, nervous system problems and garlic-breath odor.
Copper	Interact with iron and selenium to form hemoglobin, collagen synthesis, inactivate histamine.	Seafood, whole grains (cereals, legumes), nuts and seeds.	Rare. Anemia, damage to blood vessels.	Rare. Liver damage.
Iodine	Essential for synthesis of thyroid hormones which regulate growth, development and basal metabolic rate.	Iodized salt, seafood and plant foods depending upon iodine content of soil.	Iodine Deficiency Disorders: Goiter, cretinism.	Hypothyroidism, goiter.
Fluoride	Promote bone and teeth health, prevents dental carries.	Drinking water in certain areas, fluoridated water, seafood, tea, potatoes.	Rare. Increased susceptibility to tooth decay.	Dental and skeletal fluorosis.
Manganese	Help in the metabolism of carbohydrates proteins and fats, bone formation	Whole grains, nuts, leafy vegetables.	Rare	Disorders of the nervous system.
Chromium	Help to maintain normal glucose levels.	Brewer's yeast, whole grains, liver.	Condition similar to diabetes.	None reported
Molybdenum	Facilitate function of several enzymes.	Legumes, pulses, cereals.	None reported	None reported in humans. Kidney damage in animals.

Situation 2: Pizzas, burgers and pastas have low vitamin and mineral content especially that of B-group vitamins, ascorbic acid and iron. Nutrient deficiencies are resulting in loss of appetite, lethargy and mild gastrointestinal disturbances.

Situation 3: A is anaemic while B is not because spinach is not a good source of iron, bioavailability of iron from spinach is low and tea hinders absorption of iron. Cauliflower leaves and fenugreek leaves are richer source of greater bioavailable iron. Guava is a food source of Vitamin C which favours iron absorption.

Situation 4: A had normal while B still had low BP because of poor nutritional status. Sodium deficiency is not always the cause of hypotension.
Situation 5: Crops grown on iodine deficient soil have low content of this element. Top layer of soil is rich in iodine but in a hilly area, during rainfall, the top layer of soil gets drained off. This is a major reason of iodine deficiency in local population.

Know Your Progress

True or False Questions

1. Cholecalciferol (vitamin D_3) is present in food.
2. Sunflower seeds are a good source of pantothenic acid.
3. Egg white injury is associated with biotin.
4. Riboflavin deficiency causes neural tube defect.
5. Vitamin C hinders with the absorption of iron.

Very Short Answer Questions

1. Describe the following:
   (a) Goitrogens (b) Anti-thiamin factors
   (c) Macromineral (d) Micronutrient

Short Answer Questions

1. Differentiate between water-soluble and fat-soluble vitamins.
2. Enumerate, giving suitable examples, the similarities and differences between vitamins and minerals.
3. List the factors that influence absorption of calcium.
4. Explain briefly the association between sodium and hypertension.
5. What happens due to excessive intake of vitamin E?

Long Answer Questions

1. Give the functions of following nutrients:
   (a) Vitamin D (b) Cyanocobalamin (c) Iron
2. What would happen due to deficient intake of the following nutrients:
   (a) Folate (b) Vitamin A (c) Iodine
3. Give typical food sources of the following nutrients:
   (a) Thiamin (b) Sodium (c) Calcium
4. What happens due to excessive intake of following nutrients:
   (a) Vitamin K (b) Riboflavin (c) Fluoride

Beyond the Text

- Develop two nutritious snacks for preschoolers belonging to low-income families. The snacks should be rich in calcium and vitamin A. Employ fermentation and germination while developing your recipes.

Chapter 15

Water

Chapter Objectives

After reading this chapter, the reader should be able to:

- appreciate the presence of water in various body compartments.
- know the various functions of water.
- choose foods and develop dishes keeping in mind the water requirements.
- describe the concept of water balance.
- identify symptoms associated with dehydration and water intoxication.
- ensure optimal intake of water.

Consider the following situations:

Situation 1: A and D have similar food consumption. A consumes 6 to 8 glasses of fluid a day, whereas D manages to take only 1 to 2 glasses of fluid a day. D complains of chronic constipation.

Situation 2: Y and X are office colleagues. Y drinks water only when thirsty. X drinks a glass of water every 15 minutes. After about eight months, X suffered from mild convulsions.

Situation 3: R and Q went for picnic with their friends. R drank water, whereas Q took one serving each of beer and wine. At the end of the day, R was active and energetic, whereas Q felt lethargic. The hands and feet of Q were cold.

Situation 4: Sports coach of Team A advised the mess cook to incorporate lemonade, buttermilk, soup and dishes with thin gravy daily in the menu for his students. Sports coach of Team B advised the students to drink lot of water but did not alter the menu. Team A performed better. The participants of Team B experienced loss of appetite and lack of alertness.

Water has ubiquitous presence on earth and our body. It is essential to our survival and we cannot live without water for more than a few days. Water not only sustains life but also performs several functions in our body. As can be observed in the given situations 1–4, there are several opinions related to water consumption but do they have a scientific background? This chapter will help us to understand the role of water in both health and disease.

15.1 Body Water and Its Distribution

Water is the single most abundant molecule in human body. It constitutes nearly 60 per cent (ranging from 45 per cent to 70 per cent) of the total body weight in adults. The amount of water present as percentage of total body weight varies from one individual to another. This variation can be attributed to body composition and age. Metabolically active tissues (also called lean tissues) such as tissues of the muscle and viscera have a higher concentration of water as compared to other tissues (adipose and calcified tissues). Therefore, individuals with higher lean tissue mass would have higher content of water in their body. Athletes, men and physically active individuals have higher water content than non-athletes, woman and sedentary workers, as the former have higher lean tissue mass. Body water as percentage of total body weight also varies with age. Infants and children have higher water content as compared to adults (Table 15.1).

Table 15.1 Distribution of body water

Age/ Gender	Body Water (% of Total Body Weight)	
	Intracellular	Extracellular
Infant	27.0	42.0
Child (one-year old)	28.0	32.0
Female (60 kg)	25.9	22.0
Male (70 kg)	30.9	23.4

Pure water does not exist in body; it is present along with minerals or electrolytes and other substances and is therefore referred to as body fluid. As can be observed in Table 15.1, nearly two-thirds of body water exists as intracellular fluid (within the cells) whereas about one-third is present as extracellular fluid (outside the cells). Extracellular fluid includes interstitial or intercellular fluid and intravascular fluid. Interstitial fluid refers to the water present between and around cells. Intravascular fluid refers to the water present in plasma, lymph and spinal fluids. Interstitial fluids comprise three to four times greater amount of water as compared to water present in intravascular fluids.

15.2 Functions

Water is a universal molecule present in our body. It is vital to survival, we can live longer without food than without water. Water performs several functions by taking part in more than a hundred metabolic processes in our body. The salient functions are discussed in the following paragraphs (Figure 15.1).

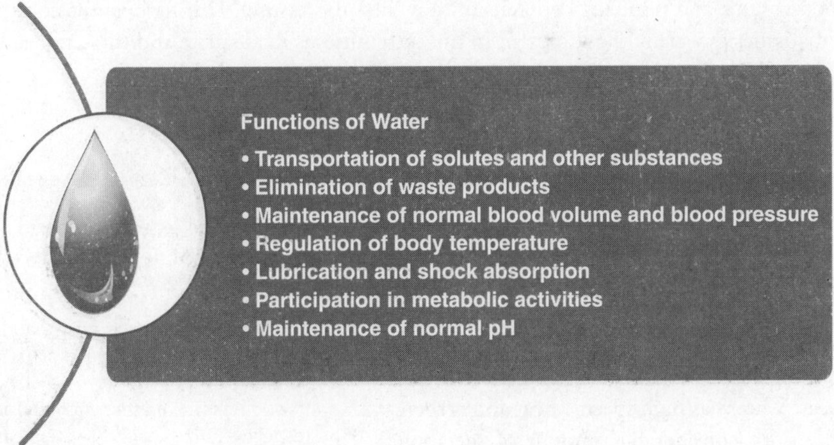

Figure 15.1 Functions of water

Transportation of Solutes and Other Substances: Water is often called a universal solvent because a host of substances in our body can get dissolved in various body fluids containing water such as blood, saliva and secretions of the gastrointestinal tract. For example, several water soluble vitamins, minerals (such as sodium), amino acids and glucose get transported from one organ to another through blood in dissolved state.

Substances which are insoluble in water such as lipids and fat soluble vitamins are either metabolically modified or bound to another substance (lipids get bound to proteins to form lipoproteins) to facilitate their easy transport in the body fluids such as blood.

Elimination of Waste Products: Urine and faecal matter contain water; both of them help in maintaining normal composition of blood and to some extent other secretions. Several waste products are produced during the process of digestion and metabolism. For example, urea is a by-product of protein metabolism, whereas ketones are formed after the breakdown of fat. Such waste products get accumulated in the blood and are subsequently excreted by the kidneys through urine. Urine also helps to excrete certain nutrients such as water soluble vitamins and sodium/potassium when they are consumed in amounts above the body's requirement/storage capacity.

THINK TANK

What is the association between acne and water?

Acne develops to a great extent due to the presence of toxic waste products in blood and clogged pores of the skin. Water helps to excrete toxic waste products from the body through urine and faecal matter. It helps to cleanse the skin through the process of sweating. Thus, water can help in preventing acne to some extent.

Along with fibre, water helps to make the faecal matter soft, bulky and hence easy to evacuate. Therefore, water is useful in preventing constipation and maintaining normal gut health. Thus, water is aptly called as 'blood purifier'.

Maintenance of Normal Blood Volume and Blood Pressure: Water is a normal constituent of blood. Adequate water consumption helps in maintaining normal blood volume. This ensures that the pressure exerted by blood on the walls of blood vessels (arteries, veins and capillaries) while flowing through them would be within the normal range. A normal blood pressure ensures adequate supply of blood to all body organs and parts, even those present in the extremities. Water consumption below requirements can lead to lower blood volume and resultant hypotension (low blood pressure). During hypotension, the supply of blood to body tissues especially those present in the extremities is inadequate and this can result in necrosis or death of the cells.

Regulation of Body Temperature: There are two sources of heat available to our body; one is the heat produced during the metabolism of carbohydrates, fats and proteins and another is the heat present in the atmosphere surrounding us (such as during summers). The heat available through both these sources contributes to the temperature of body. Water helps to regulate body temperature near 98.6°F by retaining or eliminating body heat, as the need be. This is based on the principle of specific heat. Water has high specific heat and is therefore able to absorb considerable amount of metabolic energy without raising the body temperature. This property of water is very useful in retaining the body heat during winters. Excess body heat gets eliminated by perspiration and sweating. Heat energy contained in sweat is eliminated by a process often referred to as 'evaporative cooling'.

THINK TANK

Why are the extremities cold after an episode of diarrhoea, vomiting or fever?

Our extremities such as nose, hands and feet feel cold after an episode of acute diarrhoea, vomiting or fever due to loss of body water through faecal matter, vomitus and sweat, respectively. Loss of body water results in

1. reduced retention of metabolic heat.
2. lower blood volume and resultant hypotension (low blood pressure).

To overcome this symptom, the water consumption through diet should be immediately increased.

Lubrication and Shock Absorption: Several secretions of the body contain water and these secretions act as lubricants/shock absorbers. Some of the salient secretions along with their function are listed below:

1. Saliva, present in the mouth, facilitates swallowing.
2. Mucus keeps the lungs moist and protects them from physical irritants.
3. Mucin, gastric and bile juice are secretions of the gastrointestinal tract. Mucin protects the stomach lining from developing ulcers, whereas gastric and bile juice promote digestion.
4. Sweat, released by sweat glands under the skin, helps to regulate the body temperature.
5. Synovial fluid allows the joints to move smoothly.
6. Reproductive fluids, such as the amniotic fluid, provide cushion and protect the foetus during pregnancy.

Participation in Metabolic Activities: Water takes part in several metabolic activities such as hydrolysis and condensation reactions. Metabolic activities wherein chemical bonds are broken with the addition of water molecules are known as hydrolysis reactions. Such reactions occur during the digestion of carbohydrates and fats. Condensation reactions are anabolic in nature wherein molecules are joined together by the formation of chemical bonds and water is produced during this process.

Maintenance of Normal pH: Water helps proteins and electrolytes in maintaining normal pH of blood. Water along with several cations (sodium, potassium, calcium and magnesium) and anions (chloride, bicarbonate, phosphate, sulphates, organic acids and proteins) helps to regulate each other's concentration in blood. It thus helps to maintain normal composition (pH) of blood and other secretions of the body.

15.3 Food Sources

Daily dietaries must provide adequate amount of water as per our requirements to maintain good health. Water is present in the diet in visible and invisible form.

Visible water: Drinking water is the largest contributor of visible water itself. Box 15.1 describes the different types of water available in the market. In addition to this, we also consume appreciable amount of water through beverages such as tea, coffee, soft drinks, sports drinks (Figure 15.2), fruit/vegetable juices, soups and lemonade to name a few. In a normal diet, water should be consumed as such and in the form of beverages.

Box 15.1 Different Types of Water

Artesian water: It is obtained by tapping a confined underground aquifer which is below the natural water table. It contains minerals. It is sold at a price higher than mineral water.
Mineral water: It contains at least 250 parts per million (ppm) of dissolved minerals. It is obtained from a geologically and physically protected underground source of water.
Purified water: It is tap or ground water which has been treated by distillation or deionization or reverse osmosis. Distilled water is a type of purified water produced by steam distillation process. Distilled water does not contain minerals.
Sparkling water: Water after treatment and usually replacement of carbon dioxide, contains the same amount of carbon dioxide that it had at emergence from source.
Spring water: It is underground water which flows naturally above the surface of ground to form a spring. It contains minerals.

Is bottled water more nutritious or healthier than non-bottled water?

No. Potable drinking water is equally healthy, safe and nutritious as is bottled water.

Undernourished individuals benefit more by consuming water-rich beverages which also provide them with other essential micronutrients such as fruit juices, fruit drinks, soups and sports drinks (Figure 15.2). Overweight and obese individuals should preferably incorporate calorie-free plain drinking water.

Invisible Water: This is also termed as 'moisture content of food'; it is the water present inherently in food (raw and cooked state). Generally, fruits and vegetables are good sources of water. Their water content varies between 70 per cent and 98 per cent. Pulses, legumes, cereals, butter and cream have low moisture content (below 25 per cent of total weight). Sugar, jaggery, oil and margarines do not contain any appreciable amount of moisture (less than 10 per cent). The rich, moderate and poor food sources of water are given in Figure 15.3.

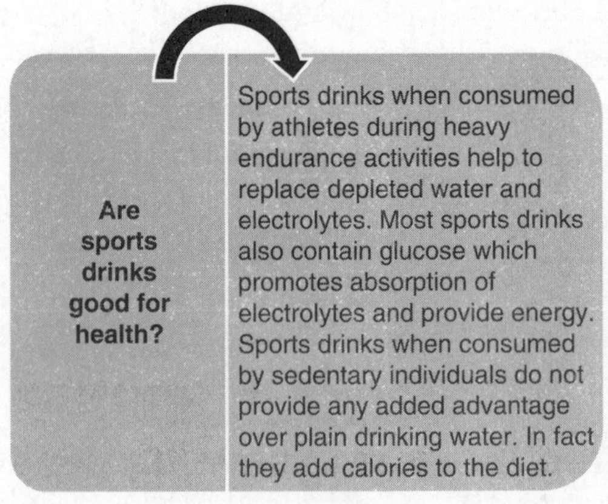

Figure 15.2 Sports drinks

Water absorption gets facilitated by the presence of electrolytes such as sodium, potassium, calcium, magnesium and chloride. Alcohol, caffeine and diuretic drugs promote excretion of water through increased urination.

Water content (per cent)	Examples of typical foodstuffs
01–10 per cent	Wheat germ, soybean, bengal gram, lotus stem (dry), seeds, almonds, walnuts, cashewnuts, beef meal, sugarcane and jaggery.
>10–20 per cent	Cereals, pulses, asafoetida, pepper (dry), apricot (dry), black currants, yeast (dried) and butter.
>20–30 per cent	Arrowroot flour, caradamom, cloves (dry), khoa and honey.
>30–40 per cent	Bread, madapu ginja, arecanut and certain dried fishes (surmai).
>40–50 per cent	Hilsa, cheese, water lily and papa.
>50–60 per cent	Beans, dates (fresh) and mango seed kernel.
>60–70 per cent	Maize (tender), red gram (tender), curry leaves and potato.
>70–80 per cent	Peas (green), eggs, meat, most fishes and sea food.
>80–90 per cent	Amaranthus, bathua, beetgreens, brussel sprouts, onions, apple, cherries, lemon and orange.
>90–100 per cent	Lettuce, cucumber, tomato, coconut water and all beverages.

Figure 15.3 Water content of various foods

15.4 Water Balance in Health and Disease

Each cell of our body contains a specific amount of water which helps to maintain its optimal composition. The amount of water required by the body is achieved by maintaining a careful balance between water lost and that received by our body.

Water received by body (average; adult):	Total = 1450–2800 ml	**Water lost by body** (average; adult):	Total = 1450–2800 ml
Liquids in diet:	550–1500 ml	**Kidneys (urine):**	500–1400 ml
Foods in diet:	700–1000 ml	**Skin (sweat):**	450–900 ml
Metabolic water:	200–300 ml	**Lungs (respiration):**	350 ml
		Digestive tract (feces)	150 ml

As can be understood by the above representation, the kidneys, skin, lungs and gastrointestinal tract contribute towards maintaining a balance between water obtained from metabolic activities/diet and that lost through urine, sweat, faeces and breath.

Water Intake: Water is obtained by our body through metabolic reactions and diet. The water produced by our body is an involuntary activity and cannot be modified or controlled. Moreover, the amount produced is also quite insignificant. Since our body does not store water, it needs to be supplied on a regular/daily basis. The requirement of water varies from one individual to another, and is influenced to a great extent by their metabolic demands, that is,

1. sensible and insensible water losses
2. solute content to be eliminated through urine, that is, dietary intake/composition and metabolic activities
3. body composition
4. age
5. gender

According to National Institute of Health, on an average, an adult man (70 kg) and woman (60 kg) must consume 3.7 and 2.7 litres of water through beverages and food every day. Usually, plain drinking water and beverages suffice for almost 80 per cent of our water requirement. Food is able to provide about the remaining 20 per cent of our daily requirement. It is not necessary to drink fruit juices or other expensive beverages to meet the water requirements; if the diet is nutritionally balanced, plain drinking water is the best option to remain healthy. However, during hot summer months, the need for water, containing electrolytes (fluids) increases. Juice and soup of fruits/vegetables, coconut water, barley water, mint water, buttermilk, lassi, tea, coffee and non-carbonated non-alcoholic drinks are good options for increasing water consumption. The water content of dishes can also be increased by making small modifications in recipes. Choice of dishes also influences water consumption. For example, cooked rice contains greater amount of water as compared to chapati. Thus, judicious selection of menu can ensure greater water consumption during the hot summer months.

Excretion of Water: Water gets excreted from our body continuously and is vital to our survival. Water lost through skin, lungs and gastrointestinal tract is often referred to as basal loss. A minimum of 500 ml of water is lost through these three routes. However, these losses may exceed 1500 ml during warm weather, excessive physical activity, illness (diarrhoea, vomiting) and certain physiological states (post-menopausal phase).

Useful Activity

After a long walk on a hot humid day around the beach, five customers have selected a beach side resort cum restaurant to have lunch. They are extremely thirsty and dehydrated. The restaurant offers the following menu:

Salad with sprouts	Grilled fish and prawns	Stuffed tomatoes
Vegetable macaroni	Apple and orange pie	Beer

What changes would you suggest to the chef to make the menu more suitable for such customers?

Water balance: The balance between water consumption and excretion is achieved through several feedback mechanisms of our body. Water balance is achieved through the following two salient ways:

1. **Regulation of Thirst:** Water insufficiency results in low blood volume and a subsequent lowering of blood pressure. Low blood pressure stimulates the thirst sensors within the brain and the person feels thirsty resulting in increased water consumption. Subsequent to water consumption, the blood volume and hence blood pressure levels improve; the thirst stimulus subsides. Thirst helps to replenish lost water but may not prevent dehydration (state of water insufficiency) because after drinking water, the body requires about 30–60 minutes to absorb and distribute it. During this time, greater water losses may already have occurred.
2. **Maintenance of Water Electrolyte Balance:** When water gets depleted from the body, feedback mechanisms are initiated by the brain and the kidneys to conserve water and sodium.

   Brain: When the water level drops in the body, there is a simultaneous rise in the concentration of sodium in the blood. This triggers the brain cells to stimulate the pituitary gland. The pituitary gland release ***antidiuretic hormone (ADH)*** which signals the kidneys to conserve or retain water, thereby diluting sodium levels in the blood. Therefore, the urine output would reduce to replenish (conserve) the water. When sodium levels reach normal values in blood, the ADH is no more released.

   Kidneys: These vital organs help in maintaining water balance by increasing the urine excretion during states of water intoxication and reducing urine output during states of water deficiency. This process is regulated by the ***renin-angiotensin-aldosterone mechanism***. Consequent to low blood volume and low blood pressure, the kidneys respond to reduced blood flow through the nephrons by releasing an enzyme renin. Renin, in turn, converts a protein angiotensin released from liver to angiotensin I. The lungs, kidney and brain convert angiotensin I to angiotensin II. Angiotensin II stimulates the release of aldosterone by adrenal glands. Aldosterone signals the kidneys to retain sodium. When sodium concentration rises in blood, water also

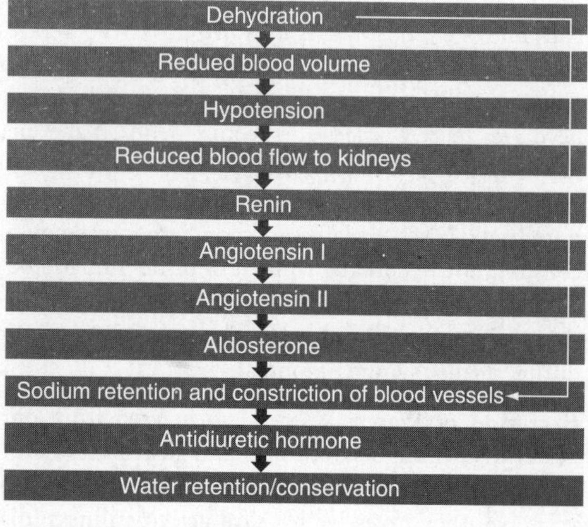

Figure 15.4 Water conserving mechanisms of our body

Water loss (per cent) of total body weight	Symptoms
0-1	Thirst
1-2	Greater thirst, fatigue, weakness, mild discomfort, loss of appetite
3-4	Dry mouth, compromised physical performance, reduced urine output, impatience, apathy
5-6	Compromised mental and physical performance, nausea, failure to regulate body temperature
7-8	Headache, irritability, sleepiness, increased respiratory rate
9-10	Laziness, difficult breathing, dizziness, muscle spasms, delirium
> 10	Losss of balance, kidney failure, collapse

Figure 15.5 Adverse effects on health of progressive dehydration

moves in. Aldosterone also promotes the release of ADH which helps to retain water in the body. If water intake is in excess of the requirements, these mechanisms do not come into play and urine output is increased (Figure 15.4).

Dehydration: It is a state of water insufficiency in the body. It develops due to either water deprivation or increased water losses. Elderly, infants and young children are at risk of delayed thirst perception and hence lesser water consumption as compared to the demands of the body. Diarrhoea (acute/chronic), vomiting, fever, burns, surgery, delivery and excessive sweating are the common reasons of dehydration. Diarrhoea due to infections, food poisoning, food allergies/intolerance and drug reactions is the leading cause of dehydration in India. Children suffering from undernutrition such as marasmus and kwashiorkor are at a higher risk of developing dehydration due to infection-related diarrhoea. A water insufficiency of less than 1 per cent may cause increased thirst and lethargy. These symptoms may progress to mental confusion, coma and finally death if the loss reaches 20 per cent or more of the total body water (Figure 15.5).

While mild dehydration can be overcome by drinking plain water, moderate dehydration requires oral rehydration therapy (ORT). This involves the consumption of fluids (not plain water) which can replace water and important electrolytes such as sodium, potassium and chloride (Box 15.2). The treatment of severe dehydration requires replenishing the lost water and electrolytes through parenteral route such as intravenous feeding.

Water Intoxication: Excessive intake of water, that is, above the capacity of kidneys to excrete, can result in water intoxication. It is a rare problem and often observed in athletes or heavy workers who try to quench their thirst with plain water rather than fluids containing important electrolytes. Patients suffering from nephritis, chronic renal failure and certain mental illness or those who are over obsessed with water consumption are at risk of developing water intoxication. Certain fad diets recommend consumption of large quantities of water to facilitate weight loss. Such diets have also been found to cause water intoxication and should therefore not be followed. Excessive intake of water causes blood dilution and hence lowered serum sodium levels. This condition is known as **hyponatremia**. The symptoms include headache, confusion, convulsions/seizures and even death in extreme cases. Death has been observed when the water consumption exceeds 10 litres in 24 hours.

Box 15.2 What Is Oral Rehydration Therapy?

Oral rehydration therapy (ORT) refers to providing fluids and/or oral rehydration salt solutions to patients suffering from dehydration, by mouth. It is given during episodes of severe diarrhoea and vomiting as observed during infections (cholera, typhoid, hepatitis), food poisoning episodes (botulism) and chemotherapy (given during cancer). It is useful during excessive sweating such as observed among sports persons/heavy workers (carpenters, road workers). It is not given during renal distress and acute fevers. ORT has the following components:

1. Clear liquid diets: These include beverages which are transparent at room temperature such as clear fat-free soups/broths, light tea/coffee (without cream/milk), strained fruit juices, green coconut water, whey water, barley water, gelatin, fruit ice, lemonade, sugar and salt-based water.
2. Full liquid diets: These include foods and fluids translucent at room temperature such as unstrained soups/juice/broths, cereal-based porridges (semolina/rice), milk and milk-based beverages, yogurt, plain puddings, custards, ice cream jelly and soufflé.
3. Oral rehydration salt (ORS) solution: It is a form of solution prepared by dissolving the contents (powder) of one sachet of ORS in one litre of potable water. The standard composition of the ORS powder and solution is given in the following table.

ORS Powder		ORS Solution	
1. Sodium chloride (common salt):	3.5 g	1. Glucose:	2.0 g/litre
2. Glucose:	20 g	2. Sodium:	90 mEq/litre
3. Tri-sodium citrate:	2.9 g	3. Potassium:	20 mEq/litre
4. Potassium Chloride:	1.5 g	4. Chloride:	80 mEq/litre
		5. Bicarbonate:	30 mEq/litre
To be dissolved in one litre of potable water.		Osmolality is 330 mOsm/litre (favours rehydration).	

Hard and Soft Water: Hard water has a high concentration of calcium and magnesium, whereas soft water has a higher concentration of sodium and potassium. The amount of these minerals in water is very small and does not contribute significantly towards the daily consumption. Calcium obtained through hard water consumption has been found to have small beneficial impact in lowering blood cholesterol and maintaining normal blood pressure. It can, to a very small extent, increase risk of developing stones in kidney/gall bladder among susceptible individuals. Soft water which, on the other hand, contains sodium and potassium is not advisable for patients suffering from congestive heart failure or acute renal failure/end-stage renal disease. The amount of these minerals present does not significantly influence good health in normal individuals.

Thus, an adequate consumption of water is essential to remain healthy. Under ordinary conditions, we need about 8–10 glasses (250 ml each) of plain water a day. Drinking plain potable water is the simplest, healthiest and economical method of maintaining water balance. However, during adverse conditions such as illness, drug reactions and excessive sweating, there would be a need to prefer fluid over plain water. Fluids refer to juices/extracts/soups of natural foods and beverages made from them. During extreme condition, ORS solution can also be consumed.

Let us review the three situations given in the beginning which will help you to recapitulate this chapter.

Situation 1: D complained of chronic constipation due to low water consumption. Water along with fibre helps to make faeces soft, bulky and easy to evacuate.

Situation 2: X suffered from mild convulsions due to hyponatremia arising as a result of water intoxication. We should not be obsessed about water consumption; it is best to drink a glass of water whenever there is a feeling of mild thirst.

Situation 3: Q felt lethargic and his hands and feet were cold due to dehydration. Q consumed alcoholic beverages which act as diuretics, that is, they increase the excretion of water through urine. Excessive urine excretion, if not accompanied by adequate water consumption, can cause water insufficiency. This results in low blood volume and blood pressure. Q exhibited symptoms of water deficiency. Alcoholic beverages must be avoided during dry and or warm weather, during excursions and air travel.
Situation 4: Team B experienced water imbalance, whereas Team A had adequate consumption of water and electrolytes. During excessive sweating, there is loss of both water and electrolytes. Thus, both need to be replaced. Plain water cannot replenish the amount of electrolytes lost during heavy sweating. Consumption of clear fluids, full fluids and dishes with high water content can be very helpful.

Know Your Progress

True or False Questions

1. As percentage of total body weight, an infant's body contains higher amount of extracellular water as compared to that of an adult.
2. Mucin is present in lungs.
3. Intravascular fluid refers to water present in plasma, lymph and spinal fluids.
4. Body absorbs and distributes water within 10 minutes of ingestion.
5. According to 'WHO formula', each sachet of ORS contains 2.5 gm sodium chloride.

Very Short Answer Questions

Describe the following terms:

1. Hyponatremia
2. ORS solution
3. Condensation reactions
4. Mineral water
5. Sports drinks

Short Answer Questions

1. Enlist any five secretions of human body which contain water.
2. What is the role of renin in maintaining normal blood volume?
3. What is water intoxication?
4. Identify 10 foods which are moderate sources of water.
5. What is a clear liquid diet?

Long Answer Questions

1. Describe the various functions of water.
2. What is water balance? How is it maintained in our body?
3. What is water insufficiency? Discuss its consequences. How can it be treated?

Beyond the Text

Keeping the water requirements in mind plan a breakfast and lunch menu for the following:

1. Athletes going for a marathon.
2. College students involved in excursion activities in a dessert area for seven days.

Chapter 16

Balanced Diet

Chapter Objectives

After reading this chapter, the reader should be able to:

- understand the basic concepts of a balanced diet.
- describe the various dietary reference intake standards.
- differentiate between nutrient requirement, recommendation, adequate intake and tolerable upper limits.
- critically evaluate and use of various dietary educational tools.
- apply the knowledge of recommendations and guidelines for the preparation or consumption of a balanced meal or diet.

Consider the following situations:

Situation 1: A health conscious customer walks into a bakery store for his regular need for bread. He browses through white, brown, garlic and French breads on the shelves, walks out without purchasing.

Situation 2: A chef has been requested to prepare a nutritionally balanced menu for a group of delegates. The menu prepared includes minestrone soup, Russian salad, grilled fish, baked vegetables, soy bread, milk bread, dips (tomato, mint) and orange soufflé.

Situation 3: A and B read about the role of salads in maintaining good health and decided to have a large serving instead of a regular lunch in their office. The salad prepared by A had lettuce, carrot, beetroot and tomato. B added some macaroni, chicken pieces, cheese powder (prepared from whole milk), lettuce, tomato and sprouted flaxseeds to the salad.

Situation 4: A pregnant woman continues to keep her food intake same as was prior to conception except for increasing the intake of chapati and ghee. At the end of sixth month, she was diagnosed to be suffering from anaemia and obesity. She had an optimum health status prior to pregnancy.

Most of us aspire to consume a balanced diet, but what constitutes such a diet? Even if we have knowledge about nutrients, their functions, requirements and food sources, it is often a challenging task to make appropriate food choices and that too in the right amount. To know what makes a meal or diet nutritionally balanced, it is important to practically apply the knowledge of nutrient intake or recommendations and dietary guidelines for purchasing groceries, cooking, serving and consuming meals.

16.1 Concept of Balanced Diet

A balanced diet can be described as 'one which contains different types of foods in such quantities and proportions that the need for calories, proteins, vitamins, minerals and other nutrients is adequately met and a small provision is made for extra nutrients to withstand short duration of lean intake'. In other words, a diet which helps to maintain an optimum nutritional status during the various phases of life cycle is known as a nutritionally balanced diet'. Such a diet should be nutritionally adequate and have variety of foods but each in moderate amounts.

Characteristics of a Balanced Diet: A balanced diet has the following features (Figure 16.1):

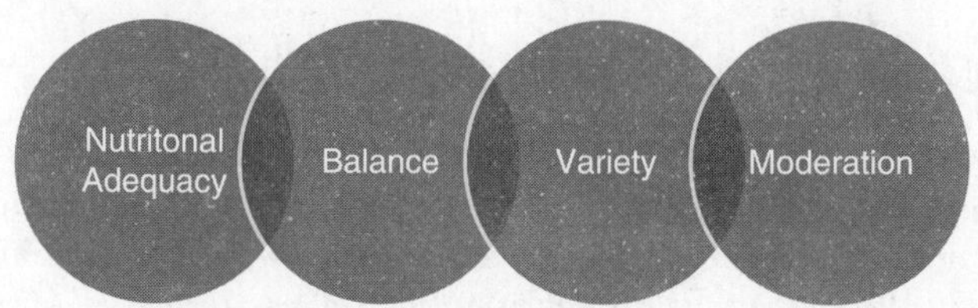

Figure 16.1 Characteristics of a balanced diet

Nutritional adequacy: A balanced diet is nutritionally adequate, that is, it provides energy and all the nutrients required by our body to meet the day-to-day needs and suffice for periods of lean intake.

Day-to-day needs refer to the requirements of body for survival, to carry out various metabolic processes and to promote repair/growth (if any). Lean intake refers to periods of low or insufficient intake. This is an important characteristic of balanced diets because neither do we consume exactly the same amount nor are our requirements exactly the same every day. Yet, we do not suffer from any symptoms of nutrient deficiency on one day and that of excess on another day. This can be understood through the following example. If iron is consumed in amount above our requirement, it is stored in the form of serum ferritin in our body. On certain days, if insufficient amount of iron is consumed, the stored iron is utilized and we do not suffer from signs of iron-deficiency anaemia for some time. The same is true for several other nutrients which can be stored in our body during times of ample intake and utilized during episodes of deficient intake.

Balance: All nutrients are equally important for our body. Laying too much stress on foods rich in a particular nutrient while avoiding others results in dietary excess and deficiencies. For instance, excessive consumption of protein-rich foods can result in reduced intake of carbohydrates and hence ketosis. Excessive consumption of calcium-rich foods and supplements can hinder with the absorption of iron. Consuming vegetable-based salad only while avoiding cereals and legumes in a meal tends to make the meal nutritionally unbalanced because such meals would provide ample of vitamins, minerals and fibre but insignificant amount of carbohydrates, fats and protein. A balanced diet provides foods in proportion to one another and in proportion to the needs of the body.

Variety: Many individuals tend to limit their food choices to a few foods, resulting in repetitive intake of certain foods. Such diets may provide all nutrients but not in sufficient amount or they may lack a particular nutrient especially micronutrients/trace elements. Narrowing the choice of consuming pulses to only those without husk can result in deficient intake of fibre and thiamin–the nutrients present in outer husk. Consuming fruits every day is a healthy habit, but narrowing the choice to only one or two fruits such as papaya or apple is not good; both are not good sources of either vitamin C or iron. Different foods within the same group/category contain different array of nutrients. Such variations need not always be on a daily basis but it can be weekly or monthly. Changing the cooking oil every month can help to ensure a more improved intake of omega-3 to omega-6 ratio which is good for the circulatory system.

Moderation: Foods which are calorie dense but are not nutrient dense should be consumed in moderation, that is, in limited amount. Foods rich in fats and sugars (monosaccharides and disaccharides) should be consumed with caution because they are poor sources of other nutrients. For example, a sugar candy is more calorie dense as compared to an orange which is more nutrient dense than the former.

However, to plan a balanced diet, it is important to determine the nutrient requirements of a person Nutrient requirements can be determined based on expensive clinical examinations which are not feasible for the general population. Thus, balanced diets are often based on dietary recommendations given by medical/dietary authorities for a particular population. Dietary recommendations include dietary reference intake standards and dietary guidelines. These are discussed below

16.2 Dietary Reference Intake Standards

Dietary reference intake standards are the tools used for developing balanced diets. They provide nutrient intake values which form the base for identifying the type and amount of foods to be incorporated in a nutritionally balanced diet. *Dietary standards give the average daily amount of essential nutrients estimated based on available scientific knowledge, and are sufficiently high to meet the physiological needs of practically all healthy persons in a group (or population) with specified characteristics.* These standards are based on the nutrient requirements (Box 16.1) of our body. During the development of such reference intakes for a nutrient the following points are kept in mind:

1. Function and metabolism of the nutrient
2. Dietary intake patterns
3. Requirements for each category of population
4. Safe level of intake
5. Tolerable upper intake and toxicity
6. Individual variations

Box 16.1 What Do You Understand by the Term 'Nutrient Requirement'?

It is the lowest intake level of a nutrient which supports basic physiological functions and promotes optimal health. Nutrient requirements are influenced by age, gender, physiological status, genetics, medications, lifestyle and environment.

Dietary reference intake standards are given by different expert authorities of different countries such as Food and Agriculture Organization (FAO), World Health organization (WHO) and National Academy of Sciences (NAS). In India, these standards are given by Indian Council of Medical Research (ICMR). Dietary standards provide four different set of values for nutrient requirements/intake. These include the following (Figure 16.2):

1. Estimated Average Requirement (EAR)
2. Recommended Dietary Allowance (RDA)
3. Acceptable Intake (AI)
4. Tolerable Upper Intake Level (UL)

These four values are applicable for all nutrients but not energy. Thus, in addition to these, we also have recommendations for energy which are as follows:

1. Estimated Energy Requirement (EER)
2. Acceptable Macronutrient Distribution Ranges (AMDR)

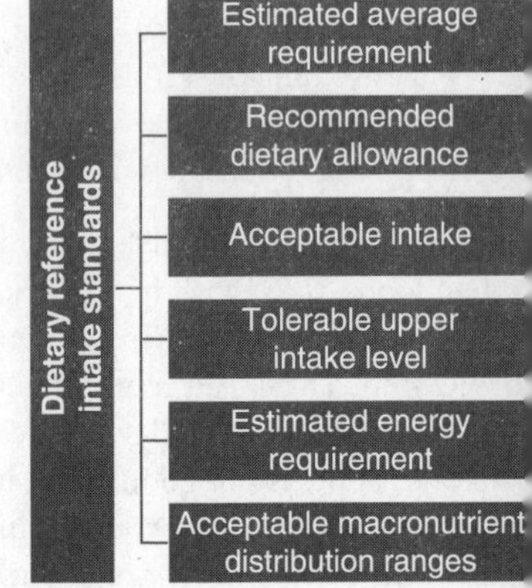

Figure 16.2 Dietary reference intake standards

16.2.1 Estimated Average Requirement (EAR)

The average daily nutrient intake level estimate to meet the requirement of half of the healthy individuals in a particular life stage and gender group. Thus, EAR values refer to intakes which are insufficient to meet the nutrient requirements of the remaining half of the population. Therefore, EARs are not used as dietary intake goals. These values are generally used for assessing the diets and developing targets for establishing RDAs or for supplementary feeding programs.

16.2.2 Recommended Dietary Allowance (RDA)

It is the average daily dietary intake level sufficient to meet the nutrient requirements of nearly all (97–98 per cent) healthy individuals in a particular life stage and gender group. The RDA for a particular nutrient includes the amount sufficient to meet the daily requirement as well as an extra margin to suffice for individual variability, periods of low or lean intake, composition of diet/food choices and losses during cooking (Figure 16.3).

1. **Individual Variability:** A population comprises of variations in height, weight, body composition, activity level, genetics and environment. An RDA for any nutrient takes into account individual variability. Therefore, RDA for a particular nutrient corresponds to the requirements which cover most of the individuals (97.5 per cent) in a given population. This corresponds to mean +2SD. This is termed as 'safe level of intake of a nutrient', that is, the chances of individuals having requirements above the RDA would be only 2.5 per cent. Thus, RDA is 25 per cent higher than the mean requirement; 12.5 per cent being considered as the extent of individual variability in the requirement of most nutrients. This concept does not apply to energy.
2. **Periods of Low or Lean Intake:** Nutrient intake does not remain consistent over a period of weeks and months; the intake is many-a-times above or below the requirements. An RDA takes into consideration periods of low intake. It provides nutrient in an amount sufficient enough to make over the periods of low intake in addition to meeting the daily requirement and at the same time not cause toxicity.

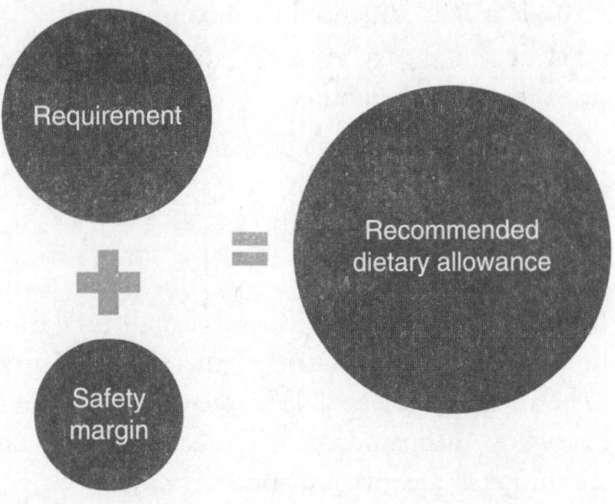

Figure 16.3 Components of recommended dietary allowance

3. **Composition of Diet/Food Choices:** The RDA for any group (gender/age/physiological state) of people varies from one country to another. This is so because dietary patterns are region specific. Food choices influence bioavailability of nutrients; for example, the bioavailability of iron from cereal-based Indian diets is low and therefore the RDA for iron is higher for Indians as compared to their counterparts in the United States or United Kingdom.
4. **Losses During Cooking:** Nutrients get lost during pre-preparation, cooking, serving and storing. Several nutrients especially protein, water soluble vitamins and certain minerals get lost or destroyed during processing. For example, nearly 50 per cent vitamin C gets lost during cooking. Therefore, the RDA for vitamin C is double the actual requirement for any category of population.

General Characteristics of Recommended Dietary Allowances

1. RDAs are high enough to meet the needs of 97.5 per cent of a population.
2. They are applicable only to healthy individuals and not for those suffering from any type of disease or physical/mental challenge.
3. RDA value of a nutrient is valid only when all other dietary nutrient intakes are satisfactory and the individual is in nutritional homeostasis.
4. It is based on age (infancy to adulthood), gender (male/female), activity level (sedentary, moderate and heavy) and physiological status (pregnancy/lactation).

5. RDA for energy is expressed in kilocalories, macronutrients in grams and micronutrients in milligrams or micrograms.
6. The recommendations for protein are based on body weight whereas those of thiamin, riboflavin and niacin are dependent on the RDA for energy.
7. RDA for energy, thiamin, riboflavin and niacin are given as additional intakes during pregnancy and lactation.
8. ICMR provides RDA values for energy, protein, visible fat, calcium, iron, zinc, magnesium, vitamin A, thiamin, riboflavin, niacin, pyridoxine, folic acid and cyanocobalamin (refer to annexures).
9. RDA is a guide for meeting nutrient needs. It is not meant for evaluating whether a person is eating as per requirement.
10. It is not influenced by income.
11. It would vary for each category (age/gender/activity/physiological status) of population from one country to another.

The recommendations given for energy and nutrients are the most frequently used tools for planning balanced diets. These values help to establish goals for adequate intake by individuals.

16.2.3 Acceptable Intake (AI)

It refers to a recommended average daily intake level which is based on observed or experimental approximations or estimates of nutrient intake by a group or groups of apparently healthy people. They are assumed to be adequate. Acceptable intake values are given (used) when an RDA cannot be established. They are also known as Adequate Intake. The acceptable intake levels as given by NAS (2005–2011) for water, vitamins and minerals are given in annexures.

16.2.4 Tolerable Upper Intake Level (UL)

It refers to the highest average daily nutrient intake level that is likely to pose no risk of adverse health effects for almost all individuals in the general population. As the intake increases above UL, the potential risk of adverse health effects increases. The tolerable upper intake levels as given by NAS (2005–2011) for various vitamins and minerals are given in annexures. Figure 16.4 illustrates the levels of various types of dietary reference intakes.

The reference intake values, given in Figure 16.4, are relevant for nutrients but not for energy and hence also the contribution by macronutrients towards total energy intake. Thus, recommendations for energy are given as EER and the acceptable distribution of macronutrients to provide energy is referred to as AMDR (Acceptable Macronutrient Distribution Ranges).

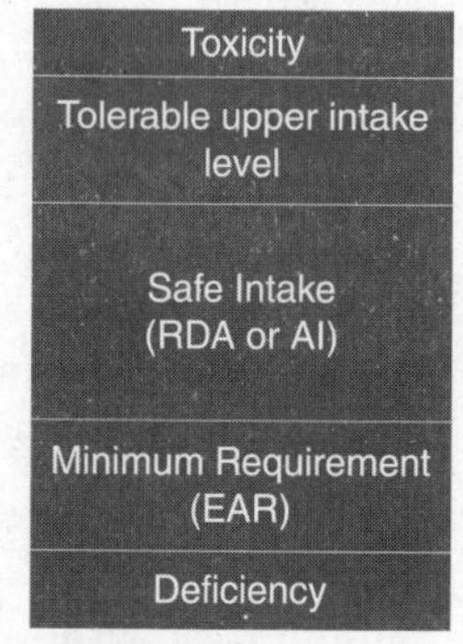

Figure 16.4 Variable levels of intake

16.2.5 Estimated Energy Requirement

Energy intake needs to be based on energy expended on various voluntary and involuntary activities. The recommendations of energy are restrictive because excess energy intake can cause obesity and its related health problems. Therefore, EER is defined as the energy intake which is estimated to maintain energy balance in healthy, normal-weight individuals of defined age, gender and level of physical activity consistent with good health.

16.2.6 Acceptable Macronutrient Distribution Ranges

The recommended energy intake should be obtained through desirable or proportionate intake of all the three macronutrients. AMDR refers to the values of macronutrients (as energy source) which when translated to food would provide adequate amount of all essential nutrients consistent with low morbidity and mortality especially due to chronic diseases. Table 16.1 gives the AMDR values for different age groups.

Table 16.1 Acceptable macronutrient distribution ranges (Institute of Medicine, 2005)

Nutrient	Total energy percent		
	Children (1–3 years)	Children (4–18 years)	(19+years Adults)
Protein	15–20	15–30	15–35
Carbohydrates	45–65	45–65	45–65
Fats	30–40	25–35	20–35
Omega-6 polyunsaturated fatty acids (linoleic acid)	5–10	5–10	5–10
Omega-3 polyunsaturated fatty acids (α-linolenic acid)	0.6–1.3	0.6–1.3	0.6–1.3

16.3 Dietary Guidelines

Computing nutrient intake data requires precision and several calculations; it needs to be done by a professional and is also time consuming. Therefore, several dietetics and medical professionals across the country have meticulously developed dietary guidelines for a common man to follow. These dietary guidelines are simple, and easy to understand/follow statements about food choices. They are made with the objective of improving and maintaining the health of the masses or are specific for a particular target group (pregnant women, infants, children, elderly, cardiac/diabetic patients to name a few). Given below are general dietary guidelines which should be followed by people who are not suffering from any disease and would help them to maintain optimal health.

1. Maintain a desirable body weight. Maintenance of body weight itself is an indication that the body is in energy balance (energy intake is equal to energy expended).
2. To avoid gradual weight gain over a period of time (especially between the age group of 40–60 years), make a very small decrease in the intake of food and an increase in physical activity.
3. At least 50 per cent of the cereals being consumed should be in the form of whole grains.
4. Choose cereal-based products which are low in sugar, fat and salt (chapati, bread, bun). Consume bakery products such as biscuits, cookies and cakes less frequently.
5. Prefer whole pulses (soybean, horse gram, bengal gram, black gram dal and cowpea) over de-husked ones (washed moong dal and washed redgram).
6. Sprouted pulses cereals and seeds are healthy. Raw pulses cereals and seeds do not contain any appreciable amount of vitamin C but sprouting for 3–5 days increases this vitamin by 5–15 per cent. Sprouting also improves the content of B-group vitamins and improves protein bioavailability.
7. Consume plenty of fruits and vegetables every day. Choose a combination of energy-dense and nutrient-dense fruits and vegetables. Two cups of fruits and 2.5 cups of vegetables are recommended for a reference 2000 kcal diet. One dark green and one orange coloured vegetable should be consumed daily.

8. Begin at least one meal with salads. Do not remove the edible peels of vegetables added to salad, provided they are clean. Salads can be made more nutritious and wholesome by adding sprouts, chees and chicken to vegetables and fruits. Use nutritious dressings rather than salad oil.
9. Prefer fruits and vegetables rather than their juice and soup.
10. Consume 3 cups of low fat milk or equivalent milk products every day.
11. Prefer lean meat (water animals and poultry) over red meat. Consume whole eggs in moderation Egg white is cholesterol free, whereas egg yolk is a highly concentrated source of this nutrient.
12. Consume moderate amount of fat. About three to four teaspoon of fat (cooking/table) is sufficien for a 2000 kcal diet. Change cooking fat frequently (e.g. monthly). Use oils obtained from plant (especially those rich in Mono Unsaturated Fatty Acids). Avoid animal and hydrogenated fat.
13. One teaspoon of oil is equivalent to one tablespoon of low-fat mayonnaise; two tablespoons of ligh salad dressing, one teaspoon of vegetable oil or one teaspoon of soft margarine.
14. Avoid consuming more than one teaspoon (2,300 mg sodium) of salt per day. Therefore, avoi using table salt, processed food, preserves and salted snacks. Low-sodium salt can be used b hypertensive patients (refer to DASH diet guidelines in Chapter 14).
15. Satisfy thirst with water. Drink plenty of plain water (8–10 glasses) every day. During states o excessive sweating, drink fluids rather than only plain water.
16. Limit the intake of foods and beverages which are high in calories, fat, sugar or salt.
17. Prefer grilling, baking, steaming, pressure cooking over sautéing and frying. Use microwave an non-stick cookware to minimize use of cooking fat.
18. Minimize the frequency of eating out. Order nutrient-dense dishes for yourself.
19. Alcoholic beverages should be strictly restricted. However, if the need be, they should be consume in as little amount as possible.
20. Read food labels to know the nutritive value of the product.
21. Develop healthy eating habits, that is, follow a regular meal and snack schedule, avoid nibbling an bingeing, eat with family and regulate your appetite.
22. Never follow a diet plan obtained from an unrecognized source. Consult a dietician to develop on for you and your family.
23. Follow an active lifestyle. Adults must exercise for at least 30 minutes most days of the week, whereas children and teenagers must exercise or get involved in active sports for at least 60 minutes every day.

In addition to dietary guidelines, other forms of educational tools are also developed by expert organizations. Pictorial representations of dietary recommendations are very successful for communicating message related to the balanced diet to the general public. Pictorial representations have been made in various shape such as pyramid (India, USA, Singapore and China) and rainbow (Canada).

Food Pyramid: The pictorial view of foods to be consumed liberally or in restricted amounts can best b depicted through a pyramid. The food pyramid picture given in Box 16.2 illustrates the amount of foods t be consumed in various proportions. The colours used in the pyramid are relevant to the food they indicat such as orange for grains, green for vegetables, red for fruits, yellow for oils, blue for milk and milk products whereas purple is for meat and beans. The wide bottom is indicative of the liberal consumption of nutrient dense foods. The narrow top is indicative of restricted foods such as sugar, fat and salt-rich processe products. The width of each coloured bar gives an idea about the proportions; for example, the orange ba s the widest which indicates that grains (carbohydrates) should be consumed the maximum. The green and ed bars indicate that vegetables can be consumed in greater amount as compared to fruits because on an average, fruits provide greater calories as compared to vegetables. The most narrow yellow bar is indicative of oil consumption because fat is energy dense and fats are also present in dairy foods, meat, fish, poultry, eggs, legumes and grains. This pyramid has been developed by the United States Department of Agriculture (USDA, 2005). To make this pyramid adaptable to Indian diet pattern, the last purple bar would contain more of pulses, legumes and a little of flesh products. Such pyramids have also been developed for pregnant women, nursing mothers, pre-schoolers and school kids.

Table 16.2 gives the approximate amounts of each category of foodstuffs (also known as food groups) for diets providing a particular amount of calories. The concept of food groups has been discussed in Chapter 17.

Keeping these nutrient recommendations and dietary guidelines in mind, balanced diets can be developed quite effortlessly. Let us review the three situations given in the beginning which will help you to recapitulate this chapter.

Table 16.2 Approximate daily amounts of each category of foods (Adapted from USDA, 2005)

Food Group	1600 Kcal	1800 Kcal	2000 Kcal	2600 Kcal	3000 Kcal
Fruits	1.5 cups	1.5 cups	2.0 cups	2.0 cups	2.5 cups
Vegetables	2.0 cups	2.5 cups	2.5 cups	3.5 cups	4.0 cups
Grains	140 g	168 g	168 g	252 g	280 g
Legumes/pulses/ meat	140 g	140 g	154 g	182 g	196 g
Milk	3 cups	3 cups	3 cups	3 cups	3 cups
Oils	5 tsp.	5 tsp.	6 tsp.	8 tsp.	10 tsp.
Calories from miscellaneous foods	132	195	267	410	512

Situation 1: The demand for nutritious or healthier foods is increasing. Breads having ingredients such as milk, soy flour, millet flour and flaxseeds would be more nutritious as compared to those made principally from refined wheat flour. The customer was looking for a more nutritious option.

Situation 2: The menu prepared by the chef included minestrone soup, Russian salad, grilled fish, baked vegetables, soy bread, milk bread, dips (tomato, mint) and orange soufflé. It was a nutritionally balanced meal.

Situation 3: The salad prepared by A had only vegetables which are good sources of vitamins, minerals and fibre. It was not a wholesome nutritionally balanced salad. B had added some macaroni, chicken pieces, cheese powder (prepared from whole milk), lettuce, tomato and sprouted flaxseeds to the salad. Macaroni is a good source of carbohydrates; chicken pieces and cheese powder (prepared from whole milk) would provide proteins; some fat and fat soluble vitamins, sprouted flaxseeds are good sources of omega-3 fatty acids, vitamin B and C; whereas vegetables are high in minerals and vitamins. The salad prepared by B was nutritionally balanced and wholesome. Such salads can occasionally be consumed as a partial substitution for lunch/dinner or as a wholesome snack.

Situation 4: During pregnancy, the demands for all nutrients increase proportionately (Box 16.2) Increasing the intake of chapati and ghee resulted in increased intake of only carbohydrates and fats. Her diet was nutritionally not balanced; therefore, she suffered from anaemia and obesity.

Box 16.2 Food Pyramid

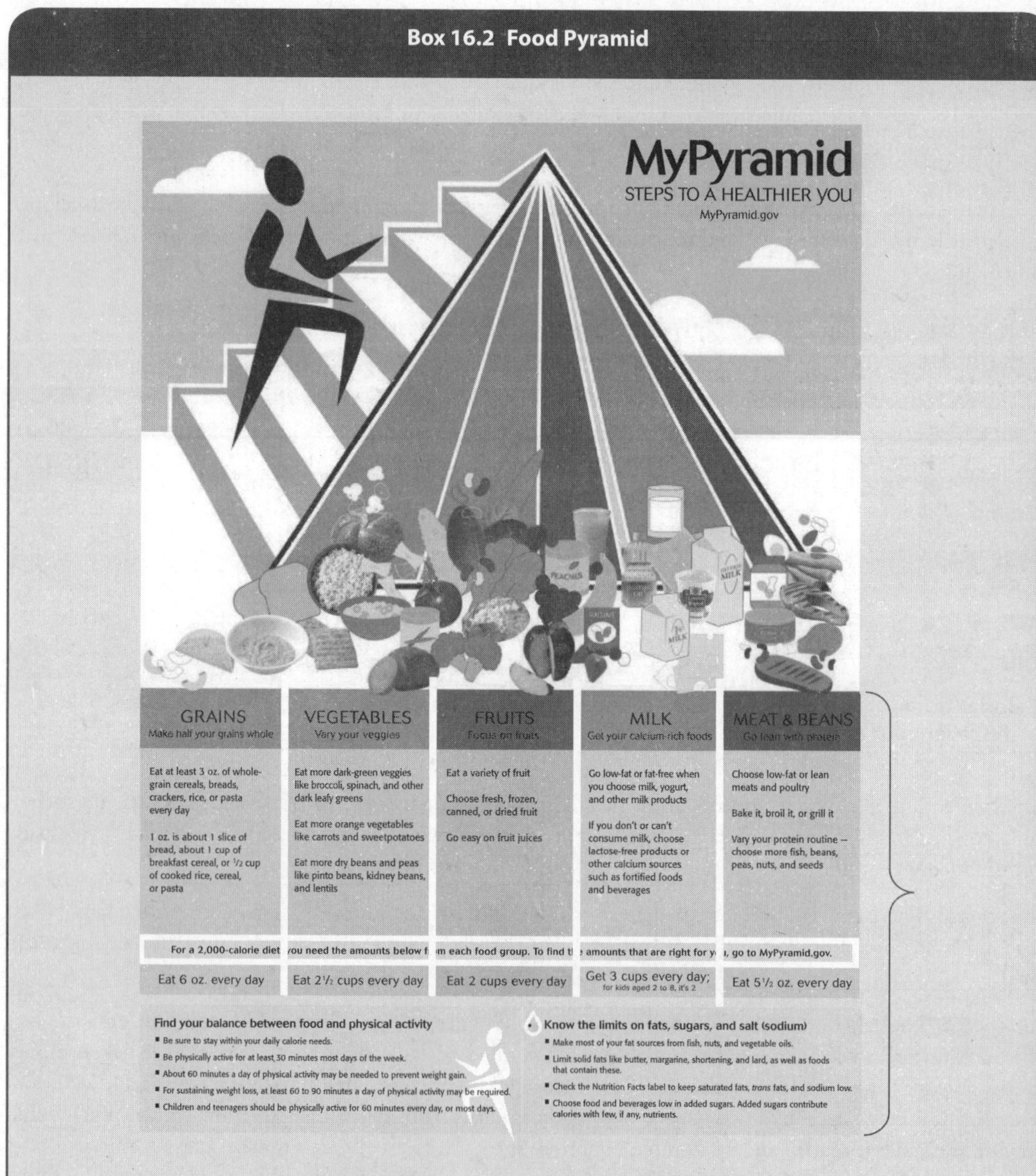

Source: U.S. Department of Agriculture. ChooseMyPlate.gov Website. Washington, DC. Food Science, Nutrition and Safety, http://www.choosemyplate.gov/print-materials-ordering/mypyramid-archive.html accessed on August 2, 2011.

Nutritionally balanced meals and diets are required for maintenance of optimal health throughout the life. To consume a balanced diet, we need not have a diet chart. Behaviour modification involving wise choice of foods is all that is required.

Know Your Progress

True or False Questions

1. A balanced diet meets the day-to-day nutrient needs.
2. RDA for a nutrient lps to meet the requirement of 97.5 per cent population.
3. RDA value of a nutrient is valid only when all other dietary nutrient intakes are satisfactory and the individual is in nutritional homeostasis.
4. One teaspoon of oil is equivalent to one teaspoon of low-fat mayonnaise.
5. Pictorial representations of dietary guidelines have been given only as a pyramid.

Very Short Answer Questions

Describe the following briefly:

1. Bioavailability
2. Requirement
3. Acceptable Macronutrient Distribution Ranges
4. Estimated Energy Requirement

Short Answer Questions

Give the criteria for developing:

1. Balanced diet
2. Recommended dietary allowance

Long Answer Questions

1. What are the salient characteristics of a balanced diet?
2. What are dietary reference intakes?
3. Enlist 10 most significant dietary guidelines for
   - College going students
   - Pregnant women
4. What is a food pyramid?

Beyond the Text

1. Search and compare the dietary guidelines of India with those given by Singapore and Canada.
2. Read food labels of the following products and identify the amount of calories, protein and fat provided by 20 gram of the product.
   (a) Low-fat bread spread
   (b) Baked potato chips

Chapter 17

Meal Planning

Chapter Objectives

After reading this chapter, the reader should be able to:

- appreciate the significance of planning meals for family and food service units.
- understand the factors which influence planning of meals.
- critically evaluate meals being served commercially.
- apply the knowledge of food groups and food exchange systems for planning meals.
- compute the nutritive value of meals as and when required.

Consider the following situations:

Situation 1: A food service outlet provides packed lunch to office goers. The meal comprising of Russian salad, grilled fish, French bread, mint dip and a small orange-flavoured waffle being wholesome is a favourite among young men and women.

Situation 2: A newly recruited cook at a senior citizen home prepared washed moong dal, stuffed bottle gourd, rice, chapati, regular salad and spinach burfi. He observed that in contrast to all other dishes, salad was not consumed.

Situation 3: A child studying in class II insists upon having jam every day in the tiffin. Growth faltering is being observed and the mother is quite worried about the food consumption pattern of her child.

Situation 4: A dinner was being hosted for 20 guests. The host started by preparing chocolate mousse, cheese soufflé and egg drop soup in the little time available. When the guests arrived, the starters and the main course were not ready.

These situations suggest that meal planning is not merely a collage of dishes; it is a careful decision-making process. It ensures judicious utilization of resources during the preparation of meals which are not only wholesome but also appealing, appetizing and attractive to the intended customers/persons.

17.1 Significance of Planning Meals

In the previous chapter, we learnt about balanced diet and the need for incorporating a variety of foods in correct proportions to ensure maintenance of optimum health. However, foods are often consumed in the form of specific meals. A group of such meals consumed throughout the day constitute a balanced diet. To incorporate adequate amount of all nutrients in the day's diet, each meal needs to be nutritionally wholesome. Therefore, meals need to be carefully planned to facilitate the availability of a variety of nutritious food. Meals are also planned for several other reasons with the ultimate objective of ensuring greater consumer satisfaction. The significance of planning meals is discussed below and is also illustrated in Figure 17.1.

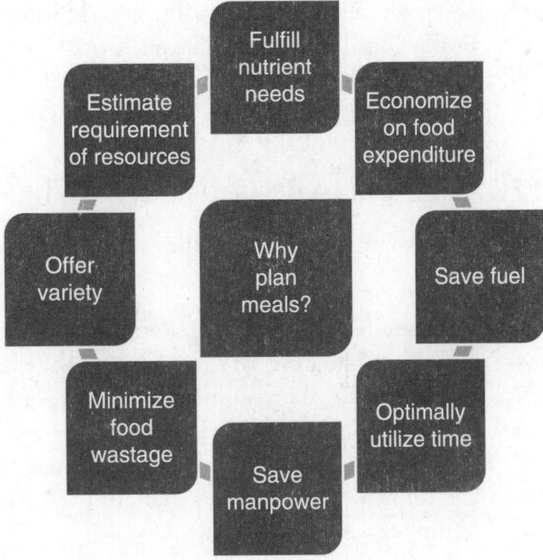

Figure 17.1 Significance of planning meals

1. Meal planning helps to ensure **fulfilment of nutritional needs** of the intended consumer. Meal platters offered by an automobile industry mess or student hostel would be rich in calories and protein, whereas those provided for office executives would be comparatively lower in calories. Vitamin A–rich meals would be required by a pre-schooler, whereas a pregnant woman would be in greater need for iron-rich foods.
2. Planning in advance helps to **estimate the requirement** of food, accessories, cooking and serving utensils. This helps to ensure availability of all ingredients and other items necessary for cooking/serving/storage at the right time.
3. Meal planning helps to **economize on the food expenditure**; if we know the food requirement in advance, it can be purchased in bulk for several days and that too at a wholesale price. For example, rice may be required every day in a hostel mess and therefore a month's requirement can be computed in advance; purchased in bulk from a wholesale store to save money. Knowing the characteristics of rice, the cook would also be more comfortable in experimenting with the recipes to ensure continued or greater customer satisfaction.
4. Meal planning helps to **save fuel,** which is an important resource for food service units. For example, soaking kidney beans a night before, helps to reduce cooking time and hence fuel considerably.
5. An intelligent meal planner creates menu for meals such that there is **optimum utilization of time**. Knowing that boiled potatoes would be required for making stuffed parathas during lunch and for preparing cream of potato soup in the evening, pressure cooking of potatoes can be carried out for both the dishes together.
6. Carefully planned meals over a period of time helps to **save manpower**. Economy of motion is often practised during the preparation of dishes which are cooked again and again in almost every meal such as breads, buns and chapati. Putting aside some warm milk for setting curds while heating milk for rabri and purchasing groceries at least one week in advance are few other examples of saving manpower.
7. Planning of meals helps to **minimize food wastage**; using leftover boiled rice in kheer or for preparing cutlets, onion raita for kadhi and drying the breadcrumbs obtained from sandwiches and using them subsequently for breading cutlets are a few examples of preventing food wastage.

THINK TANK

Vegetable poha with sprouts and beetroot raita with roasted flaxseed powder were selected over potato stuffed paratha with mint and onion chutney in a two-dish meal competition. What could be the reason(s)?

8. Above all, planned meals **offer variety in terms of colour, flavour, texture, taste** and **mouthfeel** Bottle gourd raita would have a lower visual appeal as compared to beetroot mint raita. The same would be the case with radish, cucumber and onion salad as compared to another containing lettuce, tomato, orange-bell peppers, onion sprouts and sunflower seeds.

17.2 Factors to Be Considered While Planning Meals

The following key points should be kept in mind while planning meals for home or for commercial purpose.

1. **Nutrient Needs:** The dishes and the ingredients selected should be such that they are able to provide all the necessary nutrients in the correct proportions. Table 16.2 in the previous chapter gives approximate amount of each category of foodstuffs to be incorporated in a day's diet. The amount should be divided evenly in various meals consumed throughout the day. For example, if a person is on a 2000 kcal diet and consumes four basic meals viz. breakfast, lunch, evening tea and dinner, the approximate amount of each category of foodstuffs to be consumed during lunch or dinner would be as follows:

Foodstuff	Total Intake Per Day	Breakfast	Lunch	Evening Tea	Dinner
Fruits (cups)	2 cups	1	–	1/2	1/2
Vegetables (cups)	2.5 cups	1/4	1	1/2	3/4
Cereals (g)	167 g	42	50	25	50
Legumes/pulses/meat (g)	154 g	43	45	20	45
Milk (cups)	3 cups	1	1/2	1/2	1
Oil (tsp.)	6 tsp	1.5	1.5	1.5	1.5

The amount of each foodstuff available for a meal can thus be converted into a menu. For example, the breakfast can comprise of oatmeal porridge with five almonds, poached egg-vegetable open sandwich and a bowl of mixed fruits. Lunch may comprise of a small bowl of salad with sprouts, baked beans in tomato puree, grilled chicken, breads and a sweetened yogurt. Depending upon the age, gender and physiological status of the consumer, the amount of each dish can be increased or decreased. Physically active young adults and children would require greater amount of total food as compared to a sedentary adult. A lactating mother would benefit more by consuming liberal amounts of yogurt to meet her increased needs of calcium. Thus, in a family, the food need not be cooked separately for each member, instead only the amounts of each food/dish to be served needs to be varied.

2. **Economics of Food:** While planning and preparing meals, both at household and commercial level, it is very important to keep the budget in mind. Generally, families with limited income spend a greater proportion of their monetary resource on food as compared to those with higher income. However, the absolute expenditure on food goes higher as the income increases. Thus, the choice of foods varies with the available budget. Emphasis is higher on cereals, sugar and seasonal food when the budget is small. Greater expenditure is observed on the purchase of fruits, non-seasonal foods,

non-vegetarian items and convenience foods as the budget becomes surplus. The following points can be kept in mind for optimal utilization of money:

(a) A comparative price list of commonly consumed food items at retail and wholesale level should always be kept updated. It helps to identify the source of purchase which would supply the best quality at the most competitive price.

(b) Non-perishable food items such as cereals, legumes, pulses, sugar, fats and oils should be purchased in bulk keeping in mind the storage capacity at home/food service establishment.

(c) Seasonal foods should be preferred as they are not only economical but also more nutritious and tasty as compared to their non-seasonal counterparts.

(d) Precise estimations of the edible portion such as in case of fruits and vegetables must be made before placing the market order so as to avoid over-/under-estimations as both result in loss of resources in the long run.

(e) Prepare meals according to the needs and expectations of the intended consumers. Keep a check on plate waste. Higher plate waste may be indicative of non-palatable dish or a larger than normal serving size.

(f) Use leftovers in the next meal and this would require change of menu in the subsequent meal.

(g) In house production of food accessories, maintenance of kitchen gardens and poultry farms, helps to make available cheaper and higher quality ingredients.

3. **Socio-cultural Factors and Religion:** Socio-cultural environments can influence food choices to a great extent. Meals need to be very attractive when being served to a group of invitees for a happy occasion. Sweet meats are generally not served in the first meal following death of a dear one. A new style of chopping, cooking or serving can be tried if the remaining dishes on the menu are familiar to the customer. Meals should be planned according to the cultural background of the intended consumers. Individuals belonging to southern India, China and Japan would prefer water animals over poultry in their meal. Dishes such as rice congee and deep-fried scorpion are typical to Chinese cuisine and would not be preferred by many Indians. Spicy foods are typical to Indians and Mexicans. The concept of hot and cold food and masala tea after meals are typical to the Asian culture.

   Religion also influences meal plans. Hinduism, Jainism, Buddhism, Christianity and Judaism all have distinct food laws. During the period of 'navratras', several Indians following Hindu religion prefer to consume buckwheat flour instead of wheat flour in their daily diets. Preparing vegetarian meals by incorporating buckwheat in new interesting ways would be a successful effort. Planning meals with a very high satiety value during Ramadan (period of fast from sunrise to sunset) would also be highly acceptable.

   Newer food preferences and eating behaviours are being observed due to rapid migration of populations across the world. Because of this, new meal concepts are arising and neophobia to food is decreasing. Planning of meals would thus be a challenging task in the years to come.

4. **Likes and Dislikes:** We all have food preferences and dislikes. Meals need to be planned keeping in mind the likes and dislikes of all family members so as to ensure adequate food intake. Alternative methods of incorporating the food disliked by a family member should be explored while planning meals. Soup is a good alternate to salad. Incorporating green leafy vegetables in chapatis, puris, mathris, muthias, namakparas and burfi is a good alternative to regular recipes. At the commercial level, likes and dislikes are generalized; bitter gourd would not be a vegetable of choice in a school hostel, and potato-based snacks are liked by one and all. Thus, meals should be planned keeping the inhibitions and preferences to food in mind.

5. **Availability of Food:** Meals are usually planned as per the local availability of the food and ingredients. Such meals are easy to prepare and are economical. However, commercial meals catering to high-income group can involve special foods which are typical to a particular region. This would, of course, make the meals more expensive.
6. **Availability of Manpower:** Manpower is required for purchase of groceries, pre-preparation, preparation and serving of meals. If a food service manager plans to serve 'dosa-sambhar' in a hostel mess, he or she must ensure the presence of skilled cooks. Certain dishes require elaborate preparations such as shepherd's pie, soufflé and mousse. It must be ensured before chalking out a menu that there is enough manpower available to perform the cooking/service efficiently.
7. **Variety:** Two chefs prepare cream of potato soup. One serves it with potato wedges, whereas the other one serves carrot–colocasia leaf rolls with the soup. The latter was more appealing and attractive. Family members and customers prefer meals which provide variety in terms of colour, texture, taste, flavour and mouthfeel. Some useful tips which can enhance variety in meals are given below:
   (a) Incorporate at least one food from each food group. Fruits and vegetables help to add variety of colours, flavours and textures.
   (b) Use different cooking methods. If one dish is pressure cooked, steamed or boiled, another can be fried or roasted. This introduces variety in texture.
   (c) Combine mildly flavoured foods with those which have a strong flavour such as coconut chutney with idli or mint dip with spinach pakoras. Try blending of flavours such as the flavour of roasted nuts blends well with coffee and chocolate-based desserts.
   (d) Avoid repetition of major ingredients in a meal. For example, if potato has been used for preparing pea-potato curry, then avoid using peas and potato in the accompanying dishes because it creates boredom and lowers the overall appeal of the meal.
8. **Satiety Value:** Meals should be nutritious and wholesome. They should satisfy hunger sufficient enough to last till the next meal time. Satiety value is influenced by serving size and composition of meals. Proteins and fats provide higher satiety as compared to available carbohydrates. Just like proteins and fats, non-available carbohydrates (i.e. fibre) prolong gastric emptying time and hence delay hunger. It is for this reason that salads provide greater satiety as compared to soup prepared from same amount of vegetables. Sufficient satiety provided by meals prevents nibbling and binging.

GOOD TO KNOW

1. High-moisture vegetables such as green leafy vegetables, tomatoes, bottle gourd and ridge gourd show lot of shrinkage in volume during cooking. The amount required for such vegetables per serving would be higher as compared to low moisture containing vegetables such as potatoes, colocasia, beans and peas.
2. The greater the number of dishes in a meal, the smaller is the serving size required for each dish. A meal comprising only of channa-rice with onion would require about 30 gm of channa, but if another dish such as raita is added to the meal, the amount of channa required per serving would be about 20 gm only.

17.3 Role of Food Groups in Planning Meals

Nutritional adequacy is the most significant consideration for planning meals. Including a variety of foodstuffs in optimal quantities in every meal ensure adequate intake of all nutrients. However, not all foodstuffs can be included in meals. To facilitate choice of foodstuffs, they are categorized into various 'food groups'.

Food groups or the food group system helps to convert quantitative nutrient data into food related information which can be used both by the consumer and health professionals to plan nutritionally adequate meals and diets. Several guidelines have been given from time to time by various nutrition agencies around the world (since 1894) regarding the categorization of foods into food groups. According to Indian Council of Medical Research (ICMR), foods can be grouped into five distinct groups (Figure 17.2).

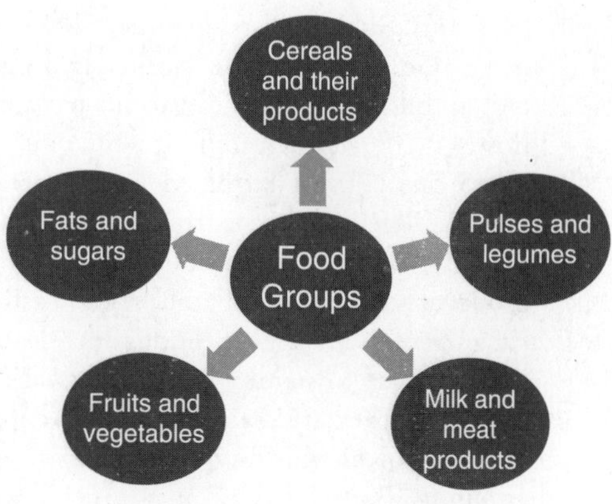

Figure 17.2 Five food groups

1. **Cereal Grains and Their Products:** These include whole wheat, cracked wheat, whole wheat flour, semolina, refined wheat flour, rice, rice flour, rice flakes, jowar, bajra, ragi, oats, maize and their processed products such as pastas and breads. They are good sources of carbohydrates and hence energy. When consumed along with bran, they provide good amount of fibre and thiamin; they do not contain vitamin C. Indian diets are primarily cereal-based diets. Cereals should constitute about 30 per cent of the total food present on the plate.
2. **Pulses and Legumes:** This group includes soybean, kidney beans, peas, beans, cowpea, bengal gram, horse gram, lentil, green gram, black gram, moth and their milled/processed products. They are good sources of protein and invisible fat in our diets. Pulses/legumes and cereals contain second-class protein but when consumed together (dal-roti, dal-rice) they provide first-class protein. When consumed with husk, they provide good amount of fibre and thiamin. Most are good sources of riboflavin, folic acid, calcium and iron. They do not contain vitamin C unless sprouted and fermented. They should comprise 10–20 per cent of the total food present on our plate; higher amount in case of vegetarians.
3. **Milk and Meat Products:** Milk and milk products here refer to milk (full cream/skimmed), condensed milk, curd, yogurt, cottage cheese, processed cheese and khoa. Meat and meat products include eggs, poultry, meat, fish, crabs, lobsters, etc., and the organ meats particularly liver; all these foods are good source of high-quality protein, egg protein being the best. Milk products are not only rich in calcium, phosphorus and vitamin D but also are good sources of fat, fat soluble vitamins and riboflavin. Meat and its products are good source of haem iron, a form of iron which has higher biological availability as compared to the form of iron present in plant origin foods (non-haem iron). Liver is the most concentrated source of the nutrients mentioned earlier in addition to certain minerals such as iron and zinc. It is recommended that irrespective of the total food intake, 2.5–3 cups of milk or equivalent amount of milk products must be consumed in a day. Milk and meat products must constitute 10–20 per cent of the total food present in our plate.
4. **Fruits and Vegetables:** This group is a reservoir of vitamins, minerals and phytochemicals; and is often referred to as the protective and regulatory group. Most fruits and vegetables are good sources of fibre also. Fruits such as papaya, mango, dates and tomato are good sources of carotenoids, whereas gooseberry, lemon, orange, lime, pear and pineapple are good sources of ascorbic acid.

   Vegetables are categorized into two groups viz. group A and group B. Group A includes all green leafy vegetables. Group B includes all other vegetables. such as ladyfinger, eggplant, cucumber, cabbage, cauliflower and onions. Vegetables must constitute 30 per cent and fruits 20 per cent of the total food present in our plate.

5. **Fats and Sugars:** Fats include butter, cream, ghee, margarine, lard, vanaspati, other hydrogenated shortenings, oils obtained from various parts of plants (sunflower, safflower, groundnut, coconut, wheat germ, olive, canola) and fat-based bread spreads and table spreads. They are concentrated source of energy, fatty acids, fat soluble vitamins especially A and D. The food group of sugars include table sugar, castor sugar, icing sugar, molasses, jaggery, honey, maple syrup, glucose and dextrose to name a few. Sugars are good sources of energy; they are considered to be empty calories as they do not provide any other nutrient in appreciable amounts. Both fats and sugars need to be taken in restricted amounts.

Source: Shutterstock.

Figure 17.3 Proportions of food groups in a plate–A pictorial presentation

The information discussed above regarding food groups has been further translated into 'My Plate' system which is a nutrition guideline for consumers. 'My Plate' system has been adopted by several nutrition advisories around the world (such as UK, USA and Canada) to give consumers a visual concept about the amount of each food group which should be present in the plate during every meal.

Figure 17.3 gives a pictorial presentation of the 'My Plate' system. It helps the consumer to understand food consumption in terms of proportions rather than quantity of food or nutrients. Such pictorial presentations are supplemented with easy to adopt messages such as, nearly half the plate should be full of fruits and vegetables; at least half of the cereals should be 'whole'; protein-rich foods should be varied in every meal and the intake of processed food and sweetened drinks should be limited.

GOOD TO KNOW

What Is 'Eat 5 a Day'?

The World Health Organization has suggested in its mandate of 'good health for all' to consume at least 400 g of vegetables and fruits in a day. This message has been adapted to 'Eat 5 a Day' which means one must take five servings/portions of fruits and vegetables in daily diets to maintain good health. 'Eat 5 a day' message has been adopted by nutrition advisories of India, Australia, New Zealand, Germany, France, UK, USA and Canada.

17.4 Nutrition Overview: Meals at Food Service Units

Consumers are frequently eating out and the demand for nutritionally adequate meals is also increasing. Meals planned by food service outlets seldom have nutritional adequacy as the foremost objective. Critical nutritional evaluation of meals served by food service outlets indicate the following:

1. Most meals do not have all the five food groups.
2. The fruit and meat/milk and pulses/legume food groups are particularly lacking.
3. Meals are rich in refined cereals, fat and sugars.
4. Even if vegetables are present, the quantity is insufficient.

Table 17.1 gives a comparative nutritional evaluation of commonly consumed meals at canteens, kiosks, roadside vendors and restaurants.

Table 17.1 Evaluation of certain commonly available snacks and meals

Menu/Meal	Energy Giving		Body Building		Protective and Regulatory	
	Cereals	Fats/Sugars	Milk/Meat	Pulses/legumes	Fruits	Vegetables
Idli-sambhar	🙂	😕	☹	🙂	☹	😕
Aloo tikki	🙂	🙂	☹	☹	☹	☹
Vegetable burger	🙂	😕	☹	☹	☹	😕
Chicken burger	🙂	😕	🙂	☹	☹	😕
Seekh kabab with onion rings and mint chutney	☹	☹	🙂	☹	☹	😕
Espania salad with bread	🙂	☹	🙂	☹	😕	🙂
Cherokee calzone	🙂	🙂	🙂	☹	☹	🙂
Chicken teriyaki subs	🙂	🙂	🙂	☹	☹	😕
Vegetable lasagne	🙂	🙂	🙂	☹	☹	🙂
Vegetable chow mein	🙂	🙂	☹	☹	☹	😕
Chicken dim sum	🙂	🙂	🙂	☹	☹	☹
Cauliflower stuffed paratha with curd	🙂	🙂	🙂	☹	☹	🙂
Dal paratha with cucumber curd	🙂	🙂	🙂	🙂	☹	🙂
Chole bhature	🙂	🙂	☹	🙂	☹	😕
Salad, dal makhni, roti	🙂	🙂	😕	🙂	☹	🙂
Pineapple sundae	☹	🙂	🙂	☹	🙂	☹

Note: 🙂 *Food group presents in adequate amount.* 😕 *Food group presents but in inadequate amount.* ☹ *Food group not present.*

However, the picture is different when meals are planned and served by professionals. Consider the following example:

Steamed fresh salmon and spinach roulade, salmon tartar and eggplant pancake
(Food groups provided: Meat, Vegetables and Cereals)

Fillet of beef pickled with coarse salt, dill juniper and coriander berries served raw thinly sliced and roasted with a tangy red wine sance
(Food groups provided: Meat and Vegetables)

Buttered baby vegetables
(Food groups provided: Vegetables and fat)

Potato cake
(Food groups provided: Cereals, fat and sugar)

Cocoa wafers layered with a white chocolate served with light custard and seasonal fruit
(Food groups provided: Milk and fruits)

Comments: The above meal provides all the food groups in relatively optimal amounts and provides variety in colour, texture, taste, flavour and mouthfeel. However, the above menu is suitable for high-income group.

Thus, carefully planned meals can provide sufficient amount of all the five food groups. Such meals are often served by railways, airlines and hostel mess run by professionals who have basic knowledge of nutrition and meal planning.

THINK TANK

Three friends went to a restaurant to celebrate the last day of their vacation. They selected the following dishes from an 'A la Carte' menu.
A selected chicken soup, chicken lasagne and chocolate sundae.
B selected cream of tomato soup, vegetable burger with extra cheese and French fries.
C selected barcelona salad, chicken burger and pineapple milk shake.
Who selected the best menu?

17.5 Calculation of Nutritive Value

The food group system helps us to develop and evaluate meals quickly but gives only an approximate idea about which nutrients would be present or absent in a meal or diet. However, if we want more specific information such as the amount of nutrients present per serving of a dish or in a meal, the nutritive value can be computed. Calculating the nutritive value helps us to:

1. provide more specific information regarding the nutrient content.
2. develop dishes or meals of desired nutritive value.
3. enhance customer satisfaction.
4. ensure survival in competitive environment.

Table 17.2 The food exchange system (Nutritive Value of Indian Foods, 2011)

S.No.	Exchange	Serving Size or Raw Weight	Carbohydrates (grams)	Proteins (grams)	Fat (grams)	Energy (kcal)
1	Vegetables:					
	Green leafy	½ cup	6.0	negligible	–	30–40
	Others	½ cup	6–10	negligible	negligible	50–60
2	Fruits	variable	10	negligible	negligible	40
3	Cereals	25	19–21	2–3	–	85
4	Legumes and pulses	25	15	6.0	–	85
5	Milk and meat	½ cup	4.0	3.5	4.0	65
		75	negligible	7.5	6.0	85
6	Fat and sugars	10	nil	nil	10.0	90
		10	10	negligible	nil	40

Nutritive value can be calculated only if we know exactly the amount of ingredients which are present in a dish or, in other words, the recipe of the dishes present in a meal. Nutritive value can be computed by two methods: (a) food exchange system and (b) food composition tables

Food Exchange System: The foods in the five food group system can be broadly categorized into six types of food exchanges as given in Table 17.2. Since the table enlists various exchanges and gives their average nutritive value, it is known as comprehensive exchange list. Each of the exchange comprises of foods of specific serving sizes and standardized energy and macronutrient content. Thus, for each of the exchange, there is a detailed exchange list of foods with similar nutritive value. Hence, in addition to comprehensive exchange list, there are six detailed exchange lists. Foods given in a particular detailed exchange list can be exchanged with one another, because they have similar (not same) nutritive value. However, foods of two different exchange lists cannot be exchanged with one another because of different (contrasting) nutritive value. For example, wheat can be exchanged with rice in the cereal exchange (similar nutritive value) but it cannot be exchanged with milk or vegetable exchange (different nutritive value).

It is a quick and easy method but gives approximate values. Another disadvantage is that this method can provide information only related to energy and macronutrient (carbohydrates, proteins and fats) content of meals. It does not provide information on micronutrients.

Food Composition Tables: Food composition tables provide accurate nutrient content of foodstuffs. The nutrient content mentioned in such tables is based on the laboratory analysis of foods. While the food exchange system gives approximate values and that too only for energy and macronutrients, these tables (National Institute of Nutrition, 2011) give information on more than 60 nutrients and related components of about 584 foodstuffs. The nutritive value calculated by this method shall, of course, be more accurate and specific. However, while the comprehensive exchange list is easy to remember, the access to food composition table for consumers may not always be possible.

Table 17.3 gives a comparative calculation of nutrients by food exchange list and food composition table.

Table 17.3 Computation of nutritive value for a sample menu–comparison between food exchange system and food composition table.

Meal: Breakfast: Poached Egg, Bread, Butter, Milk and Pear				
Ingredients	Amount (g/ml)	Nutrients		
			Food Exchange List	Food Composition Table
Egg, whole	53	Energy	489	428
Bread, white	60	Protein	17.3	21.1
Butter	06	Carbohydrate	66.0	51.2
Milk, whole	200	Fat	22.2	25.7
Pear	85	Iron	cannot be computed	2.5
		Calcium	cannot be computed	465.2
		Folic acid	cannot be computed	52.6

To recapitulate what we learnt in this chapter, let us review the situations given in the beginning of this chapter.

Situation 1: The meal (Russian salad, grilled fish, French bread, mint dip and a small orange-flavoured waffle) though fulfilling and nutritious is suitable only for a particular clientele, that is, young men and women.

Situation 2: Meals should be age specific. Senior citizens have problems related to gums and teeth. Therefore, it is difficult for them to chew hard food. Shredding, slight sautéing/short microwave cooking softens the raw vegetables, thereby making them convenient to consume. Soup is also an excellent alternative.

Situation 3: Consuming only jam with bread is not healthy. The mother was wise to try incorporating jam in pancakes, waffles, carrot-milk malpura and besan carrot leaf chapati and so on, instead of giving bread and jam.

Situation 4: Time management is very important to practically prepare what we have planned. Meals should be planned according to the resources (time, money, manpower) available. Since there was time constraint the host should have planned easy and quick to prepare dishes.

It can be concluded that meal planning is a simple yet challenging practical exercise which helps us to take decisions regarding what to prepare, how much to prepare, how much to spend, what/where/ how much to purchase, how to prepare and serve, what time to serve and so on. It requires careful understanding of fundamental aspects of nutrition and management. Thus, meal planning is an interdisciplinary multi-tasking effort.

Know Your Progress

True or False Questions

1. The exercise of planning meals should be done only for lunch and dinner.
2. Meals should be planned two days in advance.

3. It is mandatory for every meal planner to own a kitchen garden.
4. Food exchange system is based on the concept of food groups.
5. Food composition tables give more accurate information than food exchange tables.

Very Short Answer Questions

1. How is portion size related to the following:
   (a) Moisture content of vegetables.
   (b) Number of dishes in a meal.
   (c) Activity level of the consumer.

Short Answer Questions

1. Enlist the objectives of planning meals.
2. What is the difference between food exchange system and food composition tables?

Long Answer Questions

1. What factors must be kept in mind while planning meals?
2. Discuss the five food group system in detail and discuss its relevance to meal planning.
3. According to you, what is the status of meal planning in the restaurants of India?
4. Keeping the principles of meal planning in mind, develop the following:
   (a) Seven-days meals for hostel mess of school students.
   (b) Three-days meals for cricket players.
   (c) Five-meal platters for office canteen.

Beyond the Text

- Record what you ate for your breakfast and lunch for seven days. Critically evaluate your meals for nutritional adequacy as well as judicious utilization of available resources. Suggest improvements for your meals.

Chapter 18

Food Production and Nutritive Value

Chapter Objectives

After reading this chapter, the reader should be able to:

- understand the effect(s) of pre-preparation and cooking on food.
- know the influence of food production environment on nutrients.
- learn how to conserve nutrients.
- identify suitable opportunities of enhancing nutritive value of food.

Consider the following situations:

Situation 1: The incharge of a working women hostel mess sought advice from a nutritionist for planning a seven-day cyclic menu. She also gave useful tips on food cooking practices.

Situation 2: A newly recruited kitchen manager observed that all except one helper were washing vegetables after chopping. He organized a training programme in the subsequent week.

Situation 3: A chef used canned pineapple slices and processed fruit juices to prepare a dessert for a nutrition conference. The skill of the chef was highly appreciated but the same was not true for the ingredients used in the dessert.

Situation 4: Two health-conscious families (*A* and *B*) decided to consume sprouts and yogurt every day. *A* pressure cooked the sprouts and prepared savoury dishes from yogurt. *B* decided not to cook sprouts and consumed fresh yogurt. The frequency of illness was lower and gut health was better for family *B* as compared to *A*.

Situation 5: To ensure that rice grains do not stick to each other, they were cooked in excess of water which was discarded later. Habitual rice eaters developed symptoms of mild B-group vitamin deficiency.

Situations such as those just discussed emphasize the significance of cooking in our daily diets. They also signify that cooking is not just an art. The science of cooking food is equally important. The way food is cooked not only influences consumer satisfaction but also their health. Food is cooked for both household and commercial purpose. Consumption of food outside home is indicating an upward trend; use of processed foods such as tomato puree, garlic paste and canned vegetables/fruits is also increasing. Therefore, the health of consumers is being influenced to a great extent by the food they eat outside home. Hence, mass food production should be done responsibly, if we want a healthier society. This chapter shall concentrate on the effect of cooking on the nutritive value of food and the use of processed products to make the food more wholesome.

18.1 Effects of Cooking on Food

During cooking, several changes develop which result in the development of a dish or product. In addition to making the food more digestible, changes in appearance and nutritive value are the most significant ones. While changes in appearance affect the attractiveness and appeal, the other two play a significant role in maintaining health. Following changes occur in the food when it is cooked (Figure 18.1):

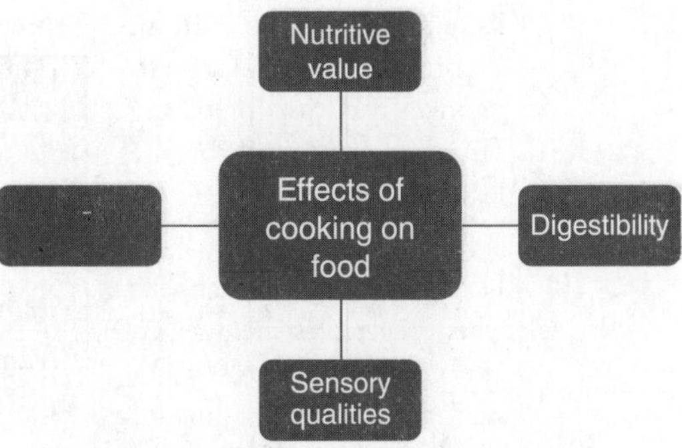

Figure 18.1 Effects of cooking on food

1. **Digestibility:** Cooking makes the food easier to digest. Application of heat softens the cellular structure of foodstuffs, making them more available to digestive juices and hence the food becomes easier to digest. The cooked food can easily be masticated which facilitates the action of salivary amylase on carbohydrates in the mouth itself. Dextrinization and gelatinization of starch, emulsification of fat as is facilitated in the presence of moist heat and softening of the muscle fibres (meat/fish/poultry) make carbohydrates, fats and proteins easier to digest. Excessive heating such as during buffet service cause protein denaturation to such an extent that it may become indigestible.
2. **Nutritive value:** While, on one hand, cooking improves bioavailability, it also results in the loss of certain nutrients and denaturation of some.

   Bioavailability: At times, some nutrients, particularly the micronutrients, are bound to certain substances (commonly known as the inhibitors) present in the food; as a result, these nutrients are not available to the human body. This is known as bioavailability of nutrients (refer to Box 18.1). However, usual cooking methods inactivate or destroy these substances (inhibitors) and, thus, release the nutrients from the unavailable complexes which can then be absorbed in the human gastrointestinal tract. For example, raw egg white contains avidin (a protein) which makes biotin (a B group vitamins) unavailable; cooking denatures avidin, thereby making biotin available to the body. Certain foods, such as cabbage, contain goitrogens (which interfere with the metabolism of iodine) which can be destroyed by heat during cooking (refer to Chapter 14). The same holds true for trypsin inhibitors present in certain pulses such as soybean and kidney beans. Other than inhibitors, dietary fibres also hinder the absorption of nutrients such as calcium and iron which are bound to fibres such as cellulose and hemicelluloses. Cooking helps to soften the fibre and, thus, makes the nutrients more available to the human body.

   > **Box 18.1 What Do We Mean by Bioavailability of Nutrients?**
   >
   > It is defined as the rate and extent to which a nutrient is digested, absorbed and utilized by our body. This is influenced by the composition of food or dish. Fibre, oxalates, phytates and phytin inhibit absorption of certain nutrients to which they are bound such as iron, calcium and protein.

   Losses and denaturation of nutrients: Nutrients are lost and at times denatured to form harmful compounds during pre-preparation and preparation of food (Table 18.1) as a result of the following:

   (a) Physical removal of a part of food such as observed during peeling, removal of bran/husk, chopping-off selected parts and trimming. De-husked cereals/their products (semolina, refined flour) and pulses without husk have lower thiamin content than whole cereals/pulses. Peels of several fruits/vegetables are richer in certain nutrients as compared to the cortex (pulp). The peel of apple is richer in riboflavin, niacin and ascorbic acid as compared to its remaining edible portion. Roasting of eggplant to remove peel for preparing *brinjal-bhurtha* results in the loss of phytochemicals in appreciable amounts.

(b) Leaching in water as happens when vegetables are washed after chopping or excess water used for soaking, boiling or cooking, brine used in canned products etc. is discarded. Water soluble vitamins and minerals such as iron, zinc, copper, calcium, magnesium and phosphorus are particularly susceptible to such losses.

Table 18.1 Effect of environmental factors on vitamins

Nutrient	Acid	Neutral	Alkali	Light	Air	Heat
Vitamin A	NS	S	S	NS	NS	NS
Vitamin D	S	S	NS	NS	NS	NS
Vitamin E	S	S	S	NS	NS	NS
Vitamin K	NS	S	NS	NS	S	S
Thiamin	S	NS	NS	S	S	NS
Riboflavin	S	S	NS	NS	S	NS
Niacin	NS	S	S	S	S	S
Folic Acid	NS	NS	S	NS	S	S
Cyanocobalamin	S	S	S	S	NS	NS

Note: NS - not stable; S - stable.

(c) Exposure to air occurs during mincing, open grinding or when cut fruits and vegetables are left uncovered. Vitamin A, D, E, ascorbic acid and cyanocobalamin are rapidly lost upon exposure to air.

(d) Exposure to light such as during concentration of milk to form condensed milk, khoa, rabri and kheer. Vitamin A/D/E, riboflavin, folate and ascorbic acid are particularly light sensitive.

(e) Evaporation, that is, nutrients are lost through water vapours such as during open pan cooking. Water soluble vitamins are susceptible to such kind of losses.

(f) Exposure to heat results in the loss of several nutrients particularly vitamins such as vitamin A, D and E, ascorbic acid, thiamin, riboflavin, niacin and pyridoxine. The amino acids tryptophan, methionine and cysteine get readily destroyed when food is exposed to 150°C and above. Prolonged and repeated heating result in greater loss of nutrients. Heating carbohydrates at very high temperatures (>100°C) such as during frying, baking, puffing and roasting can result in the formation of acrylamides; a known neurotoxicant and a weak carcinogen. Exposure of proteins to temperatures >200°C can result in the development of pyrolysis. Pyrolysis of meat proteins can result in the formation of by-products which are mutagenic.

(g) Change of pH such as observed during the addition of acid (e.g., vinegar and lemon) or alkali (e.g., sodium-bi-carbonate) can result in the loss of certain vitamins.

3. **Sensory Qualities:** Cooking helps to improve the overall appeal and taste of food. Excluding the fruits and certain vegetables, most of the other food items do not taste good in their raw state. Further, for the same food, different cooking methods can help in creating different **taste** perceptions. For instance, eggs taste very different in their raw, boiled, poached, fried and scrambled state. Cooking can enhance, suppress or create a new **flavour** in foods. In case of cabbage, cooking enhances the flavour, whereas the flavour of onions is mellowed down. Roasting of green coffee beans helps to create the characteristic coffee flavour. Judicious employment of a cooking method/technique helps to create a variety of **textures**. For example, texture of boiled potatoes is very different from that of potato chips or potato fingers. Addition of condiments/spices can also help to create **colour** in food

as is frequently done by the addition of turmeric, saffron or *Ratan Jot.* Including tomatoes, beetroot or spinach during cooking can help to add colour to dishes such as *dhokla*, rice, *chapati*, breads and *burfi*. Cooking also helps to modify the **shape/consistency** of a food item; tomatoes can be stir fried, baked or converted to juice/puree/soup. Prolonged heating as in case of mass food production may however bring deteriorative changes in texture, colour, flavour, consistency and taste such as due to repeated stirring, weight of the food itself and inadequate capacity of vessels.

4. **Food Safety:** Several harmful microorganisms get destroyed and toxins get denatured during cooking which makes the food safer to eat. Depending upon the heat treatment (temperature and duration), some, most or all of the microorganisms in vegetative form and/or their spores may be killed. Moist heat at a given temperature is always more lethal to microbes than dry heat. Temperatures in the range of 5–57°C are not sufficient for destroying pathogenic (harmful) microorganisms and is often referred to as *food danger zone*; because at this temperature, bacteria can grow rapidly. During service in large gatherings (such as marriage/birthday/other parties), the food is usually kept warm in this temperature range and hence is a common source of food infection/food poisoning.
5. **Variety:** Food can be cooked in a variety of ways and, thus, can create interest in eating, especially among the children and patients. From a single food item such as wheat, one can prepare a variety of food items such as *roti, paratha, puri*, bread, *kachori, mathri, upma, halwa*, porridge, pasta, cake, biscuits; the list can be endless. Within each of these preparations, a lot of variety can further be introduced, for example, in case of *paratha/puri*, not only its shape, size and texture can vary but a large variety can be introduced by way of different types of stuffings. Thus, it is through cooking, one can bring about a variety in colour, texture, taste, appearance and flavour of the food, creating interest in the diet; as is rightly said that 'variety is the spice of life'.

18.2 Conserving Nutrients

It is evident from the discussions in the previous section that, even if we choose nutrient-dense foodstuffs, we may end up in preparing a nutritionally poor dish or meal because of wrong cooking practices. Therefore, it is important to adopt correct practices during the purchase, storage, preparation and cooking of food. Given below are a few useful tips which can be followed to prevent the loss of nutrients present inherently in food.

1. While purchasing fruits and vegetables, preference should be given to seasonal foods. Cold storage of ripened fruits and vegetables, canning and dehydration result in the loss of nutritive value considerably.
2. Inserting nails or breaking a part of the vegetable/fruit to check its quality is a wrong practice. It destroys tissues and exposes the cortex (inner edible portion) to air, water, heat and microorganisms in the environment which results in nutrient losses and also makes the food unsafe for human consumption.
3. Storage temperature of non-perishable items influences cooking time which in turn affects nutritive value. For example, cooking time of cereals and pulses increases with storage time. Moisture content above 10 per cent may deteriorate quality of pulses when cooked.
4. Milk and milk products should not be stored in open or transparent containers especially near a source of light as this causes considerable loss of riboflavin. It is for this reason that milk is often sold in translucent poly-packs which deflect light rays.
5. Fresh fruits and vegetables should be stored whole and preferably for least possible time.
6. Vegetables and fruits should be washed before peeling and cutting to prevent loss of water soluble vitamins in particular (Figure 18.2).

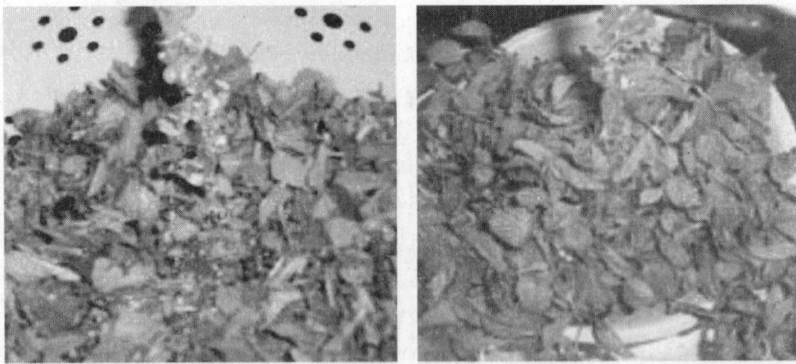

Figure 18.2 Vegetables must be washed before chopping to conserve nutrients

7. Fruits and vegetables should be chopped just before eating or cooking (Figure 18.3). Storing chopped fruits/vegetables, even at refrigeration temperature, causes loss of nutrients especially vitamins.

Figure 18.3 Exposure of cut fruits to air results in losses of vitamins

8. The finer the cuts, the greater are the loss of nutrients (macro- and micronutrients) in the juices which leaks out from fruits, vegetables or meat. Shredding, mincing and grinding cause greater loss of nutrients as compared to chopping and dicing (medium cuts).
9. Chopping should be carried out with a sharp knife to minimize damage to tissues and lessen the loss of food juices and the nutrients carried within.
10. Avoid removing edible peels of fruits and vegetables. If the dish requires removal of peels, remove thin peels (use a peeler instead of a knife) and avoid discarding them. Try to employ innovative methods of incorporating the edible peels in the dish.

USEFUL ACTIVITY

Make Your Own Nutrient Sprinkler!

Wash cauliflower/colocasia/radish/turnip leaves separately. Dry them on a muslin cloth or filter paper. Chop them in large cuts. Place them in on a microwave tray and heat till they become crisp. Upon cooling, crush them and pass through a sieve to obtain a fine powder. Your nutrient sprinkler is ready! You may add your favourite flavour or spices to it.

How to use?

Use it as a garnish. Add to savoury curd preparations, as an ingredient while preparing vegetables/pulses or while kneading dough for samosas, chapati, puri and mathri.

11. Do not discard the leaves of vegetables such as radish, cauliflower and turnip or the shells of green peas. Radish and cauliflower leaves contain more iron than radish and cauliflower itself. Such leaves can be added to any vegetable preparation, chapati, parathas, puris and mathris. The shells of green peas can be used for preparing soup, pickle and chutney.
12. Minimize cooking time. Prefer pressure cooking and closed pan cooking over open pan cooking such as sautéing, grilling, roasting, steaming and boiling
13. Do not add sodium bicarbonate; although it hastens the cooking process, there is a great loss of thiamin (also read Box 18.2).
14. Avoid prolonged cooking and repeated heating.
15. Cook or prepare just before serving or eating.
16. Add just sufficient amount of water for cooking. Any excess water should not be discarded or evaporated. It can instead be separated and used for kneading dough or preparing gravies/sauces.
17. Wash rice for minimum number of times. Use just sufficient water to cook rice by absorption method. Draining excess water after cooking rice results in the loss of starch, B-group vitamins and several minerals.
18. The addition of tamarind (acid) helps to retain certain nutrients.
19. Choice of ingredients influences the nutritive value of prepared dish or meal. Choose whole cereals (whole/cracked wheat and its flour, millets) and pulses (chick peas, horse gram, cow peas) rather than their de-husked counterparts.
20. Cook in the required quantity. Excess amount stored in the refrigerator loses its nutritive value and other quality aspects gradually.

Box 18.2 How Can We Minimize Cooking Time of Whole Pulses and Legumes?

1. Soak whole pulses overnight in cold water or in warm water for 4–5 hours. Addition of small amount of salt in the water used for soaking is also helpful.
2. If lesser time is available for soaking, boil the pulses in water for 2 minutes, soak for 1 hour and then cook as desired.
3. Putting the pulses/legumes in boiling water just before pressure cooking shortens the cooking time considerably and improves quality of cooked product.
4. Soak preferably in soft water. Hard water makes the outer coat of beans/legumes more resistant to penetration of water during cooking; this prolongs cooking time.

These are just few tips which can easily be incorporated in day-to-day cooking in any commercial food production unit and at home.

18.3 Enhancing Nutritive Value

A chef was requested to prepare an innovative Sports Drink rich in protein, phytochemicals and Q enzyme but without using milk. He used germinated green gram dal, apple, beetroot and Q enzyme powder to prepare the drink. The drink was a great success. During commercial food production, we are often confronted by situations that demand the production of dishes, meals or products which are rich sources of a particular or a group of nutrients or enhancers of nutrient absorption. Availability of certain foods, budget, time, manpower and other resources available can make such situations even more challenging. Having knowledge regarding enhancing the nutritive value of food gives a competitive edge to the food professionals. Nutritive value of food can be enhanced by combination of foods, germination, fermentation, enrichment/fortification, parboiling, parching/puffing and use of health supplements (Figure 18.4).

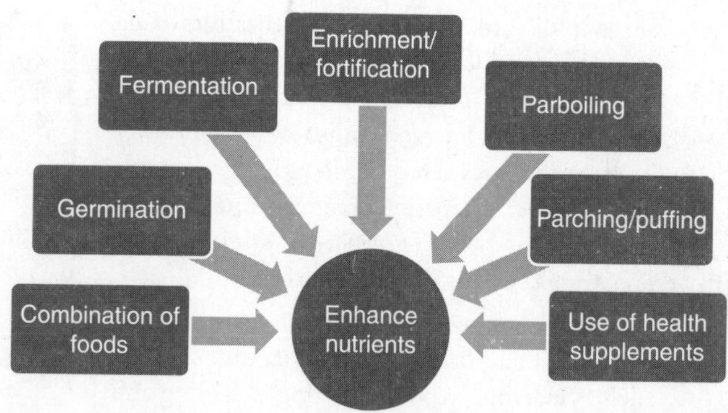

Figure 18.4 Methods of enhancing nutrients

Combination of Foods: Combination of nutrient-dense foods can help in enhancing the nutritive value of a dish or meal. Each food is a rich source of certain nutrients, whereas lacks some other. Combining foods from different food groups in a dish or meal can make it nutritionally wholesome. For example, combining energy giving foods (cereals or tubers or sugars and fats/nuts) with body building (eggs or meat or milk or pulses) and protective/regulatory foods (fruits and vegetables) would help in developing a dish which would contain almost all the nutrients. Box 13.6 in Chapter 13 discusses the concept of combining cereals and pulses to improve the quality of protein in diet. Similarly, combining vitamin-A rich fruits with those rich in ascorbic acid and iron would be healthier rather than eating any one particular fruit every day.

Germination: The process of creating a sprout in whole pulses, cereals and seeds is known as germination. It can easily be carried out in every big and small food production establishment. The three simple steps involved in germination include the following:

1. Wash and soak pulses in just sufficient water for 8–10 hours.
2. Loosely wrap the soaked pulses in muslin cloth. The water left (if any) after soaking may be used for sprinkling on the cloth to keep it moist.
3. Continue for 1–6 days, as desired. Consume raw, salted, garnished or lightly sautéed.

Sprouting the pulses has the following advantages:

(a) Sprouts are easier to digest as compared to their raw counterparts.
(b) The process of sprouting improves quality of protein, primarily due to reduction in the trypsin inhibiting factor. The ratio of essential amino acids to non-essential amino acids increases as the amount of available essential amino acids increases. For instance, the average lysine content in pulses increases from 12.7 per cent to 15.5 per cent between day 1 and day 7 of sprouting.
(c) Germinated pulses are good sources of linoleic, alpha linolenic, eicosapentaenoic and docosahexaenoic fatty acids. Thus, sprouts are helpful in preventing degenerative diseases such as hypertension, cardiac diseases, diabetes and arthritis, to name a few.
(d) Germination enhances vitamin C content of pulses and legumes by 5–15 per cent in 3–5 days of sprouting.
(e) The riboflavin, niacin, folic acid, choline and biotin content increase considerably. For example, the niacin content increases by 60–100 per cent within 48–72 hours of germination.
(f) The bioavailability of several minerals such as iron, calcium and zinc improves as they get released from the bound form.
(g) Since germination can be carried out only for whole grains, sprouts are good source of fibre. Fibre helps to alleviate constipation.

(h) Sprouts do not create flatulence (gas formation in digestive tract), a common problem associated with consumption of whole pulses/legumes. It is due to the hydrolysis of oligosaccharides-gas forming fibre.

(i) It is economical and easy to germinate grains/seeds.

(j) They require minimal cooking and therefore save time, fuel and money.

(k) They add variety to the diet.

Sprouts are a very healthy dietary food. However, care must be taken to minimize their cooking time (if necessary). Sprouts should preferably be consumed raw or slightly cooked. Cooking shall destroy vitamins which are not stable to heat. Figure 18.5 and Box 18.3 gives examples of foods which can be germinated and how we can incorporate sprouts in various dishes.

Fermentation: This is an economical and easy process which can be carried out for mass food production. It is brought about by the aid of microorganisms which hydrolyse nutrients. The breakdown of nutrients makes the food easier to digest. The by-products produced by the microbes create a characteristic taste and flavour in the dish/product and the gas (carbon dioxide) produced make the food spongy and voluminous. Fermentation has the following advantages:

1. Fermentation enhances digestibility of food.
2. It improves bioavailability of nutrients such as iron and folic acid. Iron is bound to phytin, phytates and fibre (inhibitors for iron absorption) in the raw state. It gets released from the bound form and becomes available to the body for absorption.

Figure 18.5 A variety of cereals pulses and seeds can be sprouted

Box 18.3 What Can Be Sprouted?

A wide variety of foods can be sprouted such as

Pulses/legumes: Peas, kidney beans, chickpea, cowpea, moong beans, redgram, moth and soybeans.
Cereals: Whole wheat, maize, rice, buckwheat.
Seeds: sesame, sunflower, almond, hazelnut, linseed, peanut.
Brassica (cabbage family): broccoli, cabbage, watercress, mustard, radish, daikon (*kaiware* sprouts), turnip, watercress.
Umbelliferous vegetables (parsley family): These may be used more as microgreens than sprouts, carrot, celery, fennel, parsley.
Other vegetables and herbs: Spinach, lettuce, milk thistle, lemon grass.

How can sprouts be incorporated in meals?

Salads, Chaat, filling of samosa/sandwich/patties/kachories/parathas/cheela, pulao and pancake.

3. Fermentation increases the content of methionine, vitamin B (especially folic acid) and C.
4. The slightly acidic pH of fermented foods stabilizes or prevents the loss of several vitamins such as ascorbic acid, thiamin, riboflavin, biotin and choline, to name a few.
5. Fermented foods are faster to cook and therefore loss of nutrients during cooking get minimized.
6. Several fermented dishes such as idli and dhokla have both cereals and pulses as their principal ingredient, thereby ensuring nutritional adequacy.
7. It provides variety in the diet.
8. It does not require any extra cost for most dishes (curd, yogurt, naan, bhatura, breads).

During commercial food production, at times, sodium bicarbonate is added to the dough or batter to ensure development of a spongy texture. It creates a spongy texture but results in the development of an alkaline atmosphere. Thiamin, riboflavin, pantothenic acid and ascorbic acid become unstable in the presence of alkali. Dishes such as idli, uttapam, khaman, dhokla, sauerkraut, miso and tempeh can be easily prepared by employing natural fermentation process in commercial kitchens.

Fortification/Enrichment: Enrichment refers to the addition of nutrients that were lost during processing so that the food provides nutrients as per specific standards. Fortification, on the other hand, refers to the addition of nutrients to a food which were either not originally present or present in insignificant amounts. Enriched products help in maintaining health, whereas fortification is used to prevent or correct a deficiency in a large population. In certain dishes, it is not possible to employ combination of foods or fermentation/germination such as jalebis, plain milk, boiled rice and rusk. In such dishes/products, enriched or fortified ingredients can be used instead. Enriched/fortified cereals such as rice, cereal flours, milk, butter, oils and salt are currently available in the market of India. Thus, we can use thiamin-enriched flour for preparing cakes/biscuits, and so on.

Parboiling: Parboil literally means, to boil briefly as a preliminary or incomplete cooking procedure. Although any cereal, pulse, legume and several vegetables can be parboiled, parboiling of rice and potatoes is most popular. Parboiled rice is prepared by soaking rice in water for 2–3 days. The soaked paddy is then steamed or boiled, dried and then milled to remove the bran. During this process, most of the nutrients present in the husk are absorbed by the grain. The thiamin and riboflavin content of parboiled rice is nearly double to that of polished rice. Potatoes can be parboiled by boiling for a very short period in water and keeping them in water for about half to one hour.

Parching/Puffing: It is a process of puffing-up cereals such as wheat, maize, rice or pulses, particularly bengal gram, peas and soybean. The grains are first moistened (at times soaked in water) and then suddenly subjected to high temperature. The heat converts moisture present within the grains into steam which gives the grain a crisp and porous texture. Popcorns are the most favourite parched snacks consumed in India. Salt or spices can be lightly sprayed on parched products. Such seasonings may contain the desired micronutrients or powder of green leafy vegetables such as colocasia leaves.

Use of Health Supplements: This is currently being employed when the target is to provide a specific nutrient in appreciable amounts in a dish without modifying the recipe. For example, antioxidant-rich cream of potato soup can easily be made by adding spirulina as a garnish. Similarly, vitamin A rich jalebis or semolina halwa or besan ladoos can be made by cooking/frying in red palm oil. Biscuits rich in omega-3 fatty acids can be made by using the softules of these fatty acids. Whey, milk, fish and soy protein have often been added to biscuits and breads especially those made for malnourished children, pregnant women and lactating mothers.

To recapitulate what we learnt in this chapter, let us review the situations given in the beginning of the chapter.

Situation 1: The nutritionist also suggested tips on food cooking practices to ensure conservation of nutrients in the meals.

Situation 2: The kitchen manager organized a training programme to educate his subordinates regarding correct pre-preparation and cooking practices.

Situation 3: Canned and processed products are lesser nutritious as compared to their fresh counterparts. That is why the skill and not the choice of ingredients were appreciated.

Situation 4: Sprouts and yoghurt are good sources of B-group vitamins and ascorbic acid which get destroyed upon exposure to heat. Therefore, they should preferably be consumed without being cooked.

Situation 5: Discarding excess water would result in draining away water soluble vitamins and minerals. Therefore, habitual rice eaters would develop symptoms of B-group vitamin deficiency if rice is not cooked by absorption method.

Commercial food production and service units often mention the nutritive value of food on their menu card or the label of their product. It is, thus, the responsibility of the food handlers and the chefs to ensure that the food is providing nutrients as stated by their organization. The past discussions highlight the significance and methods of conserving and enhancing nutrients with special reference to large-scale production of food. Good cooking practices not only help in maintaining the health of an individual or a family but also of the society.

Know Your Progress

True or False Questions

1. Avidin in raw egg white makes biotin unavailable.
2. Addition of sodium bicarbonate destroys thiamin.
3. Temperatures in the range of 5–67°C are referred to as 'danger zone'.
4. Fermentation increases the content of methionine.
5. Calcium is bound to phytin in raw whole pulses.

Very Short Answer Questions

1. Identify two innovative uses of bottle gourd in food production.
2. Enlist four foods which can be parched.
3. What is the correct cooking practice for preparing rice?
4. Why should we remove thin peels?
5. Is grinding of chutney in Mosel and pestle a correct method? Give justification for your answer.

Short Answer Questions

1. What factors influence bioavailability of nutrients?
2. Enlist the advantages of germination.
3. What is the difference between conserving and enhancing nutrients?
4. How can you enhance the nutritive value of burger buns?

Long Answer Questions

1. How does cooking influence nutritive value of food?
2. Suggest 10 points of conserving nutrients while cooking vegetables.
3. How can we enhance nutritive value of food? Discuss giving examples related to quantity food production.

Beyond the Text

- Select five dishes from an À la carte menu. Observe their preparation and service. Identify wrong and correct cooking practices you observed. Suggest one tip for each dish that would help to enhance its nutritive value.

Chapter 19

Nutrition and Health: Focus of the Food Industry

Chapter Objectives

After reading this chapter, the reader should be able to:

- understand the concepts of food and nutrition security.
- justify the need for health and nutrition foods.
- keep pace with the changing trends in health foods.
- evaluate the authenticity of health claims.
- use nutrition information to the best.

Consider the following situations:

Situation 1: Prior to the implementation of a national-level health programme, the government of a country undertook a health survey among some families. The data indicated prevalence of both undernutrition and overnutrition within the same families covered under the survey.

Situation 2: A leading brand of soup bar wanted to develop high-fibre tomato soup without substituting tomatoes with other vegetables. It added chitin to the soup.

Situation 3: A customer walked into a superstore to buy breakfast cereals. He spent some time reading the labels and walked out without buying one.

Situation 4: Two siblings generally consume processed foods as their lunch and evening snacks. Their haemoglobin levels are 9.2 g/100 ml and 12.5 g/100 ml. They are of the same age, gender and follow similar lifestyle.

Situation 5: A mother suffering from osteoporosis has been prescribed calcium and vitamin D supplements. The other family members have also started consuming the same supplements.

9.1 Need for Health and Nutrition Foods

7e live on a planet where around 925 million people are hungry and 1.4 billion people are verweight. A closer look indicates that undernutrition and overnutrition not only exists in a popula-on as such but also within an individual (an obese individual might be anaemic). Food security and od insecurity seem to be the origin of these problems. When all people always have physical and onomic access to sufficient, safe and nutritious food to meet their dietary needs and food preferences r an active and healthy life, it is known as **food security**. There are natural links between food security and non-communicable diseases (NCDs) such as cardiovascular disease, chronic respiratory disease, diabetes and cancer, which together cause nearly two out of three deaths in the world (80 per cent of those in developing countries). People everywhere have increasingly similar diets and increasingly similar health risks. Even very poor populations are showing alarming rates of high blood pressure, diabetes and other diet-related NCDs; while still facing hunger and traditional **food insecurity**.

Provision of sufficient quantity of food is not just sufficient to solve this problem. It is equally important to provide nutritionally balanced food to the population at large. Thus, **nutrition security** must be coupled with food security to promote optimum health (Figure 19.1 gives the causes of nutrition insecurity in a nation). Therefore, nutrition-based food products and supplements have flooded the market. To some extent, such products enable the consumers to supplement their nutrient deficient diets. However, consumers are not always sufficiently educated to make judicious food choices. Therefore, it is important to keep oneself aware of the advancements in the food and nutraceutical industries.

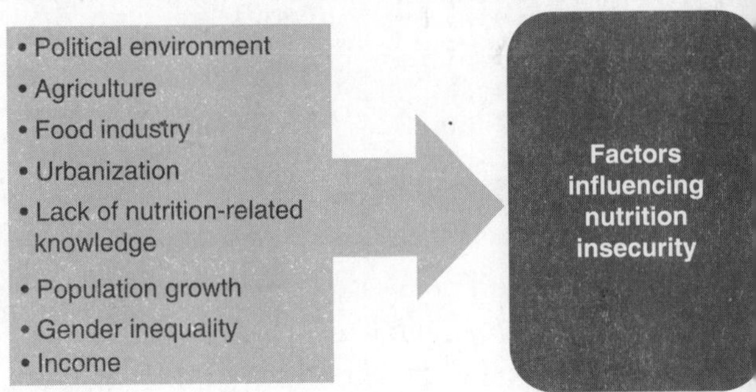

Figure 19.1 Factors influencing nutrition insecurity

19.2 Newer Trends and New Products

New product development is mandatory for the survival of every industry and the same is true in case of food industry. Changes in product composition, menu or ingredients present in a dish are inevitable. Such changes are attributable primarily to research in the field of culinary science, nutrition, medicine, food technology and allied fields. Inter-cultural associations are bringing food around the globe to local market wherein they are modified to suit the preferences of the local customer. About 25 years back when pizza were served to Indian population, they had the original Italian and American taste; however, today we find pizzas with Indian herbs, vegetables and other ingredients to suit the taste of every customer.

Another new trend being observed is the blending of food products manufactured commercially with dishes cooked in the kitchen. The use of functional foods, health foods, nutrition-based products, dietary supplements, alternative sweeteners, prebiotics, probiotics and foods for special needs is the latest trend in food industry. The food service industry uses such food products to prepare appealing and appetizing dishes.

Functional Foods: A food which contains an essential nutrient (usually in enhanced amount), phytochemical or zoonutrient with the objective of providing health benefit is known as a functional food. Functional foods promote optimal health beyond helping the body to meet its basic nutritional needs. Thus, whole, fortified, enriched and enhanced foods are examples of functional foods. Enhanced amount of an essential nutrient refers to the quantity above the levels naturally present in that foodstuff. Several foods have added nutrients such as orange juice with added vitamin C, vitamin E, calcium or zinc. Nutrients added to such foods are referred to as additives.

Phytochemicals literally means *plant chemicals.* They are biologically active compounds or substances present in plants which possess health protective effects, though they are not essential for life. In plants phytochemicals act as natural defence systems; protecting the host plant from infections and microbial invasions as well as providing them colour and characteristic flavour. More than 2000 phytochemicals have been identified and their attributes of health promotion are being explored. The potential health benefits of some phytochemicals and their good food sources are given in Table 19.1.

Table 19.1 Phytochemicals and their potential health benefits

Phytochemical	Food Source	Possible Health Benefit
Allicin	Garlic, onions and chives.	Useful in preventing ulcers (anti-microbial) and treatment of heart diseases (hypo-cholesterolemic).
Carotenoids (including lycopene)	Orangish-red and green-coloured fruits and vegetables (tomatoes, carrots, pumpkin and green leafy vegetables).	Lower the risk of cancer because of antioxidant properties.
Curcumin	Turmeric	Inhibits enzymes which activate carcinogens or reduce blood clot formation (prevent cardiac events).
Beta glucan	Oat bran, oat meal, oat flour, barley, rye, cell wall of baker's yeast, certain seaweeds and mushrooms (reishi, shiitake and maitake).	Lowers blood cholesterol and may facilitate weight reduction.
Flavonoids (flavones, flavanols, isoflavones)	Berries, cherries, purple grapes/eggplant/cabbage, oregano, olives, tea, soy and soy products.	Prevents cancer by inhibiting cell proliferation, scavenge carcinogens and act as an antioxidant.
Geinstein	Soybeans, soy flour, soy milk, tofu and texturized soy protein.	Reduce risk of osteoporosis and cancer of colon, breast, ovary and prostrate.
Lignans	Flaxseed	Reduce risk of cancer of colon, breast, ovary and prostrate.
Resveratrol	Raspberries, strawberries, red wine and grapes.	Useful in the prevention and treatment of heart diseases and inhibit cell multiplication in cancer.
Tannins	Cowpea, lentils, tea and wine.	Acts as an antioxidant and is useful in prevention of cancer and heart diseases.
ω-3 PUFAs (DHA/EPA)	Salmon, tuna, marine and other fish oils, walnuts and flaxseeds.	Reduce the risk of heart diseases, maintain mental and vision function.

Zoonutrients is a latest term used to describe compounds (other than nutrients) present in animal foods which may provide health benefits beyond the capacity of nutrients and energy; for example, certain non-essential lipids present in milk, milk products and seafood have been found to lower the risk of developing heart diseases. Similarly, jelly produced by honey bee is being explored for its potential in lowering incidence of infections.

Phytochemicals and zoonutrients have been extracted or purified from food and are being added to food products and are also sold as dietary supplements. Research has indicated that when a phytochemical is consumed alone, it is less effective as compared to when taken along with other such compounds and nutrients. Thus, increased intake of fruits, vegetables, protein-rich foods, whole grains (especially oats), fish and seafoods as well as fermented products can fulfil the need for functional foods.

Natural Health Products: They are based on the principle of phytotherapy, which is defined as the science of using parts of a plant or plant-derived substances to prevent and/or treat illness. It involves the use of botanicals (leaves, flowers, stems, rhizomes and roots) and herbs. They are either sold as such or added to dishes/food products. The most common example of natural health product sold in India being the 'natural herbal tea' which contains ginger, ashwagandha (Indian ginseng), mulethi and tulsi. The details of some commonly used herbs and botanicals are given in Table 19.2.

Natural products such as mint, lemon, rosemary and paprika contain carotenoids, vitamin C and antioxidants. They are safe for everyday use and popularly used in Indian cookery. Herbs and plant-based supplements should be consumed with caution. Herbs and botanicals listed in GRAS (generally regarded as safe) should be consumed as per prescribed limits.

Dietary Supplements: Health foods, health drinks, vitamin and mineral tablets/capsules are easily available at either the grocery or the medical store. In India, most (not all) can be purchased without a medical prescription. Dietary supplements may contain nutrients in variable amounts and are therefore categorized as follows:

1. Prophylactic or daily use; to supplement the diet for day-to-day inadequacies.
2. Therapeutic use, that is; to supplement the diet during states of severe deficiencies or illness.

Table 19.2 Proposed health benefits and contraindications of some botanicals and herbs

Botanical/Herb	Potential or Proposed Health Benefits	Contraindications
Chamomile	Helpful in insomnia.	Can cause allergic reactions, digestive upset.
Garlic	Moderate reduction in blood pressure, atherogenic lipids and inhibition of blood clot formation.	Safe. >5 cloves may cause heart burn, flatulence and mild dermatitis.
Ginger	Treatment of nausea, motion sickness, hyperemesis and gravidarum. Ease symptoms of arthritis and any kind of inflammation.	Safe. Interferes (mild) with blood pressure and blood-thinning drugs.
Ginkgo biloba	Treatment of heart diseases and cerebrovascular insufficiency. Improves blood circulation.	Interferes with several drugs. Should not be combined with ω-3 fatty acids, ginger, garlic, licorice, ginseng and horse chestnut.
Ginseng	Improves mental and cardiac performance, mood stimulator, antistress and antifatigue agent.	Over stimulation, insomnia, gastrointestinal distress. Not suitable during hypertension, diabetes and pregnancy.
Licorice (mulathi)	Treatment of peptic ulcers, mouth ulcers, mild infections and Crohn's disease.	Very large dose may cause headache, oedema, hypertension and heart failure.
Milk thistle	Protective to liver and treatment of cirrhosis.	Safe. Can cause rejection after organ transplantation in few cases.
Saw palmetto	Promotes prostrate health.	Safe. Should not be combined with ginseng, licorice and red clover.
St. John's wort	Management of mild to moderate depression. Antiviral and antibacterial agent.	Adversely interferes with antidepressants, oral contraceptives and drugs used for treatment of AIDS/heart diseases.

Prophylactic supplements: Multivitamin and mineral supplements along with essential amino acids and fatty acids are more health promotive, as compared to single nutrient supplements. Such supplements should contain at least 20 different nutrients in proportions as per the recommended dietary allowances given by an authentic nutrition advisory organization. Supplements which do not contain nutrients in the desired proportions are generally not helpful in promoting good health. Consumption of such supplements can even be deleterious to good health. Milk-based dietary supplements for young children are most popular in India. Such supplements are useful only if there are nutrient inadequacies in the diet.

Therapeutic supplements: Mega doses of vitamins and minerals are often prescribed as an adjunct or a substitute to medicines (drugs). For example, 10–20 mg niacin works as a vitamin but 500–1000 mg acts as drug to lower blood lipid level. Similarly, high doses of vitamin E are given to retard the progression of neurological degeneration as seen during diabetes. High doses of vitamins and minerals can be toxic, cause deficiency of other nutrients and/or create side effects on health. Such doses should be taken only under strict medical supervision.

GOOD TO KNOW

Who Need Functional Foods, Dietary Supplements and Natural Health Products?

1. Individuals having a dislike for foods rich in a particular nutrient.
2. Those suffering from any kind of food allergy (milk, eggs and wheat) or a disease which inhibits absorption, metabolism or/and storage of nutrients.
3. People consuming diets habitually restricted in certain foods resulting in general deficient intake of nutrients, such as pure vegetarians.
4. Elderly and those with compromised metabolic, gastrointestinal or immune function such as during diabetes, gout, ulcers and constipation, to name a few.
5. Women with special physiological conditions such as pregnancy and lactation.
6. Patients with increased nutrient requirements, as is observed post-operatively.
7. Chronic alcoholics.

Nutritive and Non-nutritive Sweeteners: Recommendations related to moderate consumption of sugars are often misinterpreted as 'no use of sugar' by several consumers. This has led to increased demand and consumption of sweeteners (often referred to as sugar substitutes). Such sweeteners are categorized as nutritive and non-nutritive sweeteners.

Nutritive sweeteners: Substances which provide sweet taste to the food and can be absorbed to yield energy in the body are known as nutritive sweeteners. Sugar, honey, molasses, maple syrup, glucose, dextrose, levulose, xylitol, mannitol, sorbitol, corn syrup, brown rice syrup and lactose are some examples of nutritive sweeteners. They are further categorized as natural sweeteners (honey and maple), refined sweeteners (white/brown sugar, molasses and high fructose corn syrup) and sugar alcohols (xylitol, mannitol and sorbitol). While natural and refined sweeteners provide 4 kcal (average) of energy, sugar alcohols provide only 2 kcal (average) of energy. Thus, a product which contains sugar alcohols is 'sugar free', may be 'low calorie' but not 'calorie free' as compared to a product with same ingredients except sugar added to it instead of sugar alcohol.

Non-nutritive sweeteners: They are several times sweeter than the nutritive sweeteners and provide little or no energy to the body. The amount of such sweeteners present in a single serving is very small and therefore their energy contribution is negligible. Saccharin, aspartame, acesulfame, sucralose, D-tagatose, trehalose and neotame are GRAS (generally regarded as safe) non-nutritive sweeteners. Stevioside has not received GRAS status.

GOOD TO KNOW

The Truth About Saccharin

Saccharin is a non-nutritive sweetener which was discovered in 1879 by Constantine Fahlberg. It is 300 times sweeter than sucrose (table sugar). Research has indicated that very large doses when consumed for several years can cause cancer of the bladder. It is still being used in India though it has been banned by several countries such as USA.

Probiotics and Prebiotics: According to the current definition of FAO/WHO, **probiotics** are 'live micro-organisms which when administered in adequate amounts confer a health benefit to the host'. Lactic acid bifidobacteria and certain yeasts are the microbes, commonly used to manufacture probiotics. Probiotics ar fermented foods and many are traditionally known and prepared at home such as curd, lassi, shrikhand, utta pam, idli, dosa, dhokla, naan and kulcha, to name a few. Commercially prepared probiotic milk is the lates development; it contains live microbes under controlled conditions. Other probiotics being manufactured b the food industry include yogurt and soy yogurt. Probiotics have also been used for the treatment of certai diseases of the gastointestinal system particularly pancreas.

Prebiotics are the non-digestible food ingredients which stimulate the growth and/or activity of beneficia bacteria in digestive system. They are, at times, also referred to as functional food components. They promot gut health and calcium absorption. Commonly used prebiotics are inulin, chitin, fructooligosaccharide (FOS) and poly-dextrose. Whole grains, onions, garlic, honey, leeks and several fortified foods contai prebiotics. Inulin has successfully been added to curd by several leading brands in India. They have bee discussed in Chapter 2 (carbohydrates) of the food science section of this book.

Foods for Special Needs: Apart from the food industry, restaurants, airlines and cafes are preparing an serving food for people with special needs such as diabetes, lactose intolerance and gluten enteropathy Meals for **diabetics** are very popular as its prevalence is rapidly increasing. Such meals provide lo glycemic index foods (Chapter 2). Use of 'multigrain atta' to prepare chapatis, both at household an commercial levels, is one such attempt to provide high-fibre meals.

Special food products and meals are also being made for individuals suffering from **lactose intolerance** (Box 19.1). Use of milk and milk products which contain undigested lactose (khoa, condensed milk, milk powder) is not incorporated in food preparation. Fermentation causes hydrolysis of lactose in milk. Thus, fermented milk, curd and yogurt can be used in meals/dishes. Cheddar, swiss and parmesan cheese contain considerably lesser amount of lactose than milk. Lactose-reduced (70 per cent less lactose), lactose-free (99.9 per cent less lactose), lactose-hydrolyse dairy products can easily be used for cooking meals (icecream, custard, puddings) for individuals suffering fro lactose intolerance. Commercial preparations of lactase enzyme in the form of capsules and chewable tablets are als available. Being proteinaceous, lactase enzyme should not be heated as it would get destroyed.

Box 19.1 What Is Lactose Intolerance?

Lactose intolerance is a condition characterized by diminished presence or complete absence of an enzyme-lactase which is necessary for the digestion of lactose. Lactose is a disaccharide. Milk and milk products are good sources of lactose. Low-lactose or lactose-free diets need to be given to such individuals.

Gluten Enteropathy which is also known as celiac disease or wheat allergy (Box 19.2) is another healt condition for which special foods and meals need to be prepared. Such meals need to be gluten free. The clai 'gluten free' refers to products or meals which contain less than 20ppm of gluten. Wheat, rye and barley are goo ources of gluten. Processed food products such as oup powder, certain sweet meats, confectionaries nd breakfast cereals (including muesli) contain luten. These should not be used in preparing meals or individuals suffering from gluten enteropathy. Corn, rice, potatoes, tapioca, sago, arrowroot, nillets, sorghum, taro, teff, montina, chia seeds, am, colocasia, all legumes/pulses, fruits, vegetables, nilk, milk products, nuts and oilseeds can freely be ısed. Gluten-free flour is available in the market. Gluten-free bread can be made by using rice, sorghum, corn and nuts. Use of ingredients such as eggs, cornstarch, xanthum gum and hydroxypropyl methyl cellulose (HPMC) helps to develop fluffiness in the bread n the absence of gluten.

Box 19.2 Gluten Specific Enteropathy–Need for Gluten-free Products and Meals

Gluten-specific enteropathy is an allergic condition characterized by the inability to digest gluten, a protein present in wheat, barley and rye, resulting in damage to the villi. Villi are structures of the small intestine which help in absorption of nutrients. If gluten is consumed by such individuals, they suffer from diarrhoea and resultant nutritional deficiencies.

Thus, nutritionists develop disease-specific enteral tube feeds, diets and meals for patients with altered nutrient requirements or feeding needs such as galactosemia, maple syrup urine disease, Parkinson's disease and certain neurological conditions, to name a few.

19.3 Critical Evaluation of Nutrition-based Products

Health foods are increasingly in demand. Food products and dishes are often sold through nutrition-based advertisements and health claims. Such information often catches the attention of the customer. However, t is important for the consumer to interpret the health claims and nutrition information correctly. Such nformation on menu cards, labels and advertisements can at times be misleading.

Food Labels: Purchasing a food with proper information on its label is always good (refer to Box 19.3). An important aspect of labels on food products is the information given about their nutritive value. Nutrition labelling helps to

1. know the nutrient content of the product in the containers.
2. compare nutritive value of two similar products.
3. choose the food product most suitable for the customer.
4. understand how reconstitution (if required) or use may influence nutritive value.
5. critically evaluate whether the health claims given on the label are justifiable.

Nutrition Facts Panel: Nutrition-related information given on the label is usually in the form of a nutrition facts panel. It gives the following information:

1. Serving size.
2. Number of servings in the packet.
3. Energy and nutrient content per serving.
4. The percentage contribution of nutrients provided by one serving for a 2000 kcal diet and recommended dietary allowance (such as those given by Indian Council of Medical Research in India or American Dietetic Association in USA). It is usually mentioned as 'per cent daily values'.

Box 19.3 Status of Nutrition Labelling in India

The Ministry of Health and Family Welfare had, on September 19, 2008, notified the Prevention of Food Adulteration Act (5th Amendment) Rules, 2008, mandating packaged food manufacturers to declare on their product labels nutritional information and a mark from the F.P.O and/or Agmark to enable consumers make informed choices while purchasing. Prior to this amendment, disclosure of nutritional information was largely voluntary though many large manufacturers tend to adopt the international practice.

5. Nutrient content is generally given for energy, fats, proteins, carbohydrates and important/relevan micronutrients.
6. Nutrition claims (if any) such as 'fortified' and 'enriched'.
7. Calorie conversion information, that is, calories provided per gram for carbohydrates, fats and proteins
8. Ingredients in descending order of predominance by weight.

Table 19.3 gives a comparative example of nutrition facts panel on two breakfast cereals when reconstitute with equal amount of toned milk.

Health Claims: Any statement or phrase which associates a food or a substance present in a food with a disease or a health-related condition is known as a 'health claim'. 'Cholesterol free', 'good for bones', 'hear friendly' and 'immunity booster' are some of the millions of health claims exhibited by manufacturers on the labels of their products. Tea, coffee, biscuits, bread spreads, breakfast cereals, biscuits and several othe

Table 19.3 Nutrition facts panel: A comparative overview

Product 1: Breakfast Cereal				**Product 2: Breakfast Cereal**			
Serving size:	1cup			Serving size:		1cup	
Servings per container:	18			Servings per container:		18	
Amount per serving	Cereal	With ½ cup toned milk		Amount per serving	Cereal	With ½ cup toned milk	
Calories	104		145	Calories	104		145
Calories from fat	0		0	Calories from fat	.5		5
	Amount	Per cent Daily Value			Amount	per cent Daily Value	
Total fat	0 g	0	0	Total fat	0.5 g	1	1
Saturated fat	0 g	0	0	Saturated fat	0 g	0	0
Trans fat	0 g	–	–	Trans fat	0 g	–	–
Cholesterol	0 mg	0	0	Cholesterol	0 mg	0	0
Total carbohydrate	25 g	8	10	Total carbohydrate	25 g	8	10
Dietary fibre	1 g	4	4	Dietary fibre	5 g	20	20
Sugars	2 g	–	–	Sugars	5 g	–	–
Other carbohydrates	21 g		–	Other carbohydrates	14 g	–	–
Protein	1.2 g			Protein	3 g		
Vitamin A		10	18	Vitamin A		17	20
Vitamin C		10	10	Vitamin C		2	2
Calcium		3	17	Calcium		8	13
Iron		39	40	Iron		42	42
Vitamin D		15	20	Vitamin D		10	15
Thiamin		22	23	Thiamin		26	30
Riboflavin		11	20	Riboflavin		32	35
Cyanocobalamin		34	34	Cyanocobalamin		32	34
* per cent daily values are based on a 2000 kcal diet. our daily values may be higher or lower depending on our calorie needs.				* per cent daily values are based on a 2000 kcal diet. our daily values may be higher or lower depending on our calorie needs.			
Calories per gram: Fat = 9 kcal, carbohydrates = 4 kcal, proteins = 4 kcal.				Calories per gram: Fat = 9 kcal, carbohydrates = 4 kcal, proteins = 4 kcal.			

ACTIVITY

Evaluate the claim 'cholesterol free' on the packet of a sunflower oil.

Evaluation:

1. Plant foods do not inherently contain cholesterol, so sunflower oil any way does not contain this nutrient.
2. The claim gives false belief to the consumer that this oil is safe for heart health and does not give any indication about the quantity to be consumed per day.

Consider the above clai[illegible] bread spread and margarine.

Evaluate the claims:

1. 'Flavonoids are good for health' on a packet of tea.
2. Brand 'z' salad is good for health.

products often have such/similar health claims on their labels. But are these claims true? Are they supported by strong research evidence and have they been approved by some legal agency? Knowledge about nutrition can help to understand the science (if any) behind these claims, but a consumer with inadequate nutrition background is often carried away by false or misleading information.

Some commonly used health claims and their correct interpretations are discussed below.

Health Claim	Interpretation (Per Serving of a Product)
Low fat	≤3 g of fat
Low saturated fat	<1 g of total saturated fatty acids
Low cholesterol	≤20 mg of cholesterol and ≤2gram of total saturated fatty acids
Low sodium	<140 mg of sodium
Low calorie	<40 kcal
Reduced	a nutritionally altered product which contains a particular nutrient 25 per cent lower than that present inherently in a standardized recipe or in its natural/ whole form
High	if the amount of a particular nutrient is 20 per cent or more above the recommended dietary allowance
More	if the amount of a particular nutrient is 10 per cent or more above the recommended dietary allowance
Fat free	<0.5 g of fat
Cholesterol free	< 2 mg of cholesterol and ≤2 g of total saturated fatty acids
Sugar free	<0.5 g of sugar
Calorie free	<5 calories per serving
Fresh	a food which is raw and has never been processed
Healthy	a product which is low fat, low saturated fat, contains <60 mg cholesterol, 360–480 mg sodium and at least 10 per cent or more of vitamin A/C, iron, calcium, protein or fibre

Nutrition information on food labels is not only informative but educative too. While nutritional labelling can help to make healthful choices, fraudulent claims can be quite challenging to good health. Therefore, health or nutrition based claims should be thoroughly evaluated before making a food choice.

The objectives of consuming food are no more limited to satisfaction of hunger or its taste and appearance. Nutrition and health security is receiving greater priority. Newer products are being made with the objective of enhancing the existing health promotional characteristics of food. Nutrition security is the combined effort of the policy makers, farmers, food producers, food industry, hospitality industry, dietetics and medicine. Newer products should be launched only after careful research and their use in food preparation should be carried out wisely. Care must be taken to mention the presence of ingredients in a dish which may not be suitable for a particular category of population (age, physiological state, disease). Food manufacturers have a special responsibility to ensure healthy foods are available in the market. They should reduce the marketing of foods with little or no nutritional value, which are high in sodium, saturated fat and trans fats. Both consumers and producers must be helped to make needed changes in diet. Commercial food systems must be encouraged, and directed where necessary to meet global food and health needs.

To recapitulate what we learnt, let us review the situations given in the beginning of this chapter.

Situation 1: Health surveys help to know the current health status of the target population. Knowing the states of nutrient excess and deficiency would help to chalk out a strategy for a health programme accordingly.

Situation 2: A leading brand of soup bar wanted to develop high-fibre tomato soup without substituting tomatoes with other vegetables. It added chitin to the soup which is a form of dietary fibre.

Situation 3: Nutrition facts on the label help a consumer to find out the product most suitable for their health needs. A variety of breakfast cereals are available in the market such as 'high fibre', 'sugar free' and 'calcium/iron rich'. Health-conscious customers would select products only if it would meet their requirements.

Situation 4: While some processed foods are very healthy, others provide empty calories. The haemoglobin levels of both siblings varied which indicates variations in their choice of food.

Situation 5: Health supplements should be taken with caution, only under the supervision of a nutritionist or a medical expert. Not all family members have similar nutritional status. Vitamin D is fat soluble, a person who is receiving sun exposure may not require supplements.

Know Your Progress

True or False Questions

1. Political environment can influence nutrition insecurity in a country.
2. Geinstein is present in eggplants.
3. Levulose is a nutritive sweetener.
4. Stevioside is enlisted in the GRAS list.
5. If one serving provides less than 40 kcal, it is known as a low-calorie product.

Very Short Answer Questions

1. Describe the following:
   (a) Allicin
   (b) Tannin
   (c) Reduced
   (d) Low sodium
   (e) Nutrition security

Short Answer Questions

1. What are zoonutrients?
2. Enlist any five phytochemicals.
3. What do you understand by therapeutic dosage?
4. Brand 'S' biscuits are good for bones. Is this claim authentic?
5. Flavonoids prevent heart diseases. Brand 'M' tea is a good source of flavonoids. Is this claim correct?

Long Answer Questions

1. Food security is different from nutrition security. Justify this statement giving five examples.
2. Describe the new health/nutrition products recently launched in the market.
3. What is the significance of having nutrition facts panel on the label?
4. What are health claims? Elaborate about correct and misleading health claims.

Beyond the Text

- Read and compare the nutrition facts for five brands of each of the following products:
  1. Breakfast cereals
  2. Biscuits
  3. Noodles
  4. Processed cheese

Part III

Food Safety and Quality

Chapter 20

Introduction to Food Safety

Chapter Objectives

After reading this chapter, the reader should be able to

- have a basic understanding of food safety and the related terms.
- appreciate the importance of food safety in the context of public health and food trade.
- recognize new challenges to the safety of food.
- identify different risk factors associated with food safety.
- understand the relevance of training for developing a food safety culture.
- know the important food safety measures in various activities of hospitality industry.

One of the essential prerequisites for ensuring food quality is food safety. It is an important public health issue in the current scenario where more than 200 diseases are spread through food and nearly 2.2 million people are estimated to die every year due to food-borne and water-borne illnesses. Fortunately, most of these diseases are preventable, provided adequate measures are followed from farm to fork, that is, during primary production, processing, storage, distribution, retailing and catering. The present day food supply has become highly complex and the consumers are demanding a wide variety of foods which are not only convenient and affordable but are safe and wholesome too. Therefore, every food handler and catering professional should be equipped with adequate knowledge and skills related to various aspects of food safety.

THINK TANK

Some Situations Related to Food Safety

Consider the following situations.

- On a hot summer day, a group of friends had chicken curry at a road-side restaurant. By mid night, all of them had to be hospitalized with symptoms of severe diarrhoea, vomiting and abdominal cramps.
- A new recruit in a salad bar was found cutting vegetables using the same knife, chopping board and gloves which were earlier in contact with raw chicken. He was asked to undergo a training course in food safety.
- A particular lot of soft cheese marketed by a dairy company was recalled from the market in a developed country.
- The cooling unit of the food transport van belonging to an established caterer goes out-of-order. He decides to hire another refrigerated van at a higher fare to carry the cooked food to the party venue.

Can you relate these situations to food saftey?

20.1 Defining Key Terms

The terms 'food safety' and 'food quality' are sometimes used interchangeably, but they are rather different from each other. While food safety focuses on hazards that make the food injurious to consumer health, food quality includes attributes that affect consumer acceptability. In contrast to quality, the safety of food is not negotiable. Let us first try to understand the following key terms related to this field.

Food Safety: It is defined as an assurance that the food will not have any adverse effect on the consumer health when it is prepared and/or eaten according to its intended use. The term 'food safety' is also used for a discipline that comprises of sanitation, related sciences and preventive measures to ensure safe and wholesome food supply.

Food Quality: It is a complex characteristic which decides the food's value and/or its acceptability by the consumers. It denotes the degree or extent to which a food fulfils the characteristics desired by the consumer, provided it does not harm the health or the environment. The term includes nutritive value and sensory attributes such as appearance, texture, taste as well as functional properties.

Food Suitability: This term refers to an assurance that food is produced, prepared and packaged according to the standards set by regulatory bodies and is acceptable to the consumer as per its intended use.

Safe Food: It refers to the food which is free from biological, chemical or physical matter capable of causing illness or injury to the consumer. In other words, safe food contains acceptable, tolerable and safe levels of contaminants, adulterants, naturally occurring toxins or any other substance that may make it injurious to health.

Food-borne Illnesses: These are diseases caused by agents entering human body through the ingestion of food (Box 20.1). Though the most common symptoms of food-borne diseases are gastrointestinal in nature, many a times, these illnesses lead to serious lifelong disorders (including neurological, gynaecological or immunological disorders), multiple-organ failure, cancer and death.

Food-borne Zoonotic Diseases: This term refers to infections or diseases that are transmitted from animals to humans through the consumption of contaminated food or drinking water. Many of these pathogens are usually present in the gut of healthy food-producing animals. They enter the human body through the gastrointestinal tract to produce symptoms which range from mild to serious life-threatening conditions.

Food Handler: Any person involved in direct handling of packaged or unpackaged food, related equipment and utensils, or food contact surfaces is known as food handler. He or she is required to follow the hygienic practices for ensuring safety of food.

Contaminant: It refers to any biological or chemical agent, foreign matter or any other substance which is not added intentionally to the food but is present as a result of the production, processing, packaging, transport and holding of food. It may adversely affect food safety or suitability. The term includes bacterial toxins, mycotoxins, seafood toxins, pesticide residues and metallic contaminants.

Contamination: The occurrence or introduction of a contaminant in food or its environment is known as contamination. The sources of contamination are plants, animals, sewage, soil, water and air.

Food Hygiene: It encompasses all conditions and measures required for ensuring

Box 20.1 Categorization of Food-borne Illnesses

Food-borne illnesses are categorized as follows:

- *Food infection* occurs when the disease-causing microorganisms or parasites are present in the food at the time of consumption and grow in the body of the host to cause illness.
- *Food intoxication* occurs when the toxins of bacterial, fungal or chemical origin are present in the food prior to its consumption.

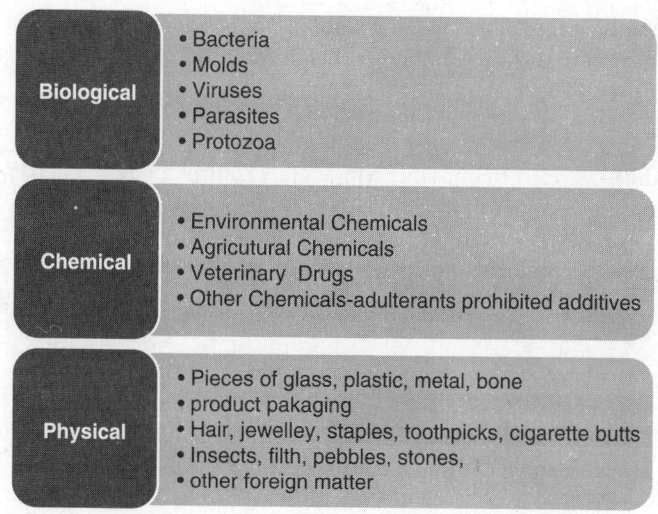

Figure 20.1 Categorization of food safety hazards

food safety and suitability through various stages of food chain such as production, preparation, processing, storage and distribution.

Sanitation: It refers to a comprehensive programme designed to control pathogens in a food establishment. The term includes personal hygiene and environmental sanitation.

HACCP System: Hazard Analysis and Critical Control Point (HACCP) it is a scientific and systematic way of defining, evaluating and controlling hazards to ensure the safety of food. HACCP is a tool to assess hazards and set up control systems which have a preventive approach rather than end-point testing.

Hazard: This term refers to an agent of biological, chemical or physical origin present in the food or a condition of food which can cause an adverse effect on human health (Figure 20.1). The food hazards range from microorganisms (bacteria, mould, viruses), agricultural chemicals (pesticide residues), veterinary drugs residues (of growth promoters, milk enhancers and drugs used in the production of animal products), adulterants and misuse of additives.

Food Safety Management System (FSMS): It is a network of interrelated elements which together ensure that the food does not lead to any adverse effect on human health. It includes Good Hygienic Practices (GHP), HACCP, statutory and regulatory requirements and communication.

Food Safety Objectives (FSO): They specify the maximum concentration of a microbiological or chemical hazard in a food which is considered safe at the time of its consumption. For example, the maximum number of *Clostridium botulinum* at the time of consumption should not be more than 1 in 10^{12} cans.

Good Hygienic Practices (GHP): They include all practices related to the conditions and measures necessary for ensuring the safety and suitability of food at all stages of the food chain.

Potentially Hazardous Food (PHF) and Time/Temperature Control for Safety (TCS) Food: The term 'Potentially hazardous food (PHF)' refers to perishable foods which encourage rapid and progressive growth of pathogenic and toxigenic microorganisms (Box 20.2). 'Time/temperature control for safety (TCS) food' is an equivalent and transitional term (coined in 2005) for PHF. Such foods (Fig. 20.2) need a temperature/time control to limit growth of pathogenic microorganisms or formation of toxin for ensuring their safety.

Risk Analysis: It is a process which includes risk assessment, risk management and risk communication to ensure safe food.

Time/temperature Abuse: It refers to improper temperature control for a prolonged period of time during storage, display and transportation of PHF.

Box 20.2 Examples of Potentially Hazardous Food (PHF)

Milk	Cooked vegetables	Cooked rice	Raw meat, fish, oysters
Soft cheese	Prepared Ready-to-eat (RTE) foods (salads)	Fresh pasta	Smoked/cured fish
Tofu	Cut tomatoes, cut melons, raw seed sprouts	Uncooked pastry and dough products	Smoked/cured RTE meat

The acidity (pH), water activity (a_w), heat treatment and packaging of food are considered to decide whether they are PHF or not. The term does not include foods which do not support microbial growth irrespective of the presence of a microbial, physical or chemical hazard for food safety.

Figure 20.2 Cooked vegetables/rice are potentially hazardous foods

GOOD TO KNOW

Contribution of Various Scientists in Ensuring Food Safety

Food safety is a multidisciplinary science. Experts from the field of medicine, nutrition, veterinary science, agricultural science, microbiology, engineering and law make significant contribution in comprehending and ensuring the safety of food. This fact is illustrated through the following examples:

- **Physicians** → help in understanding the cause, diagnosis and treatment of food-borne illnesses.
- **Nutritionists** → provide advice related to food safety to people particularly the vulnerable individuals.
- **Veterinarians** → recognize relationship between animal health and food-borne diseases.
- **Microbiologists** → identify disease-causing microorganisms in patients and in food. They carry out research to determine the factors affecting survival, proliferation and destruction of these microbes and toxins in food.
- **Agriculture scientists** → determine the relationship among plant diseases, pesticides and food safety.
- **Engineers** → design equipment and systems for ensuring safety of foods.
- **Legislators** → develop food laws at the national and international level.

20.2 Importance of Food Safety

Why should we study about food safety? What is the relevance of this field? All food service professionals should know that there are damaging repercussions of unsafe food. The importance of food safety is listed in Figure 20.3 and discussed in the following paragraphs.

1. **Prevention of Food-borne Illnesses:** Food safety protects the consumer from food-borne illnesses and the associated personal suffering, preventable death and avoidable financial burden. On the other hand, unsafe food can cause food-borne diseases which affect personal health and well-being. They have financial consequences for individuals, families, communities, food businesses and countries. The hospitalization and treatment lead to substantial burden on health-care systems.
2. **Protection of Vulnerable Individuals:** Children, pregnant women, elderly and individuals suffering from other diseases are more susceptible to food-borne diseases and related complications. Safe food protects these individuals and the poor who earn their living from their day-to-day work. Otherwise, the loss of income due to food-borne illnesses in the poor can perpetuate the cycle of poverty.
3. **Boosting of Consumer Confidence:** Safety of food not only protects deception and fraud but also increases consumer confidence in the food producer and the product. Unsafe food, on the contrary, can lead to the loss of trust in the producer and the product, closure of the food business and decline in food trade.
4. **Protection of the Producer:** Safe food supply provides protection to the producer from the financial loss associated with product recall and legal battles. Threat to the safety of food can result in its recall from the market leading to huge financial loss for the producer. Besides, many food companies in the west are facing litigation charges due to the recent increase in number of food-borne illness outbreaks.
5. **Improved Productivity:** Food safety improves productivity and the image of the food establishment. In contrast, an outbreak of food-borne illness can irreversibly damage the brand image of a restaurant or a chain of hotels.
6. **Growth of Food Trade:** Provision of safe food also maintains integrity of food trade and helps in its growth at local, national and global level.

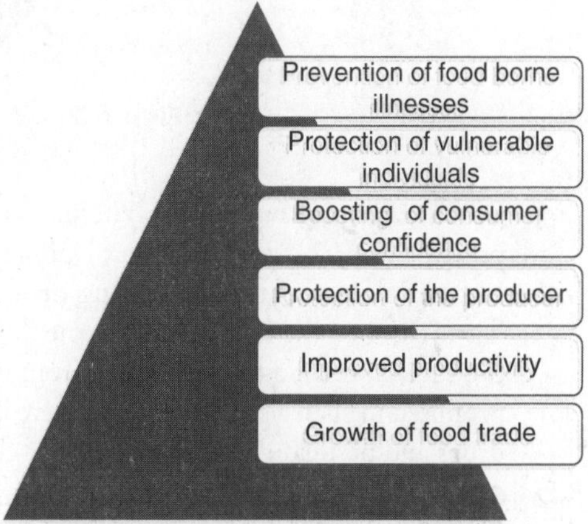

Figure 20.3 Importance of food safety

20.3 New Challenges to Food Safety

The present day food supply is extremely complex and the contamination of food supply by undesirable organisms and chemicals and its effect on the consumer is decided to a large extent by the new challenges to food safety (Figure 20.4).

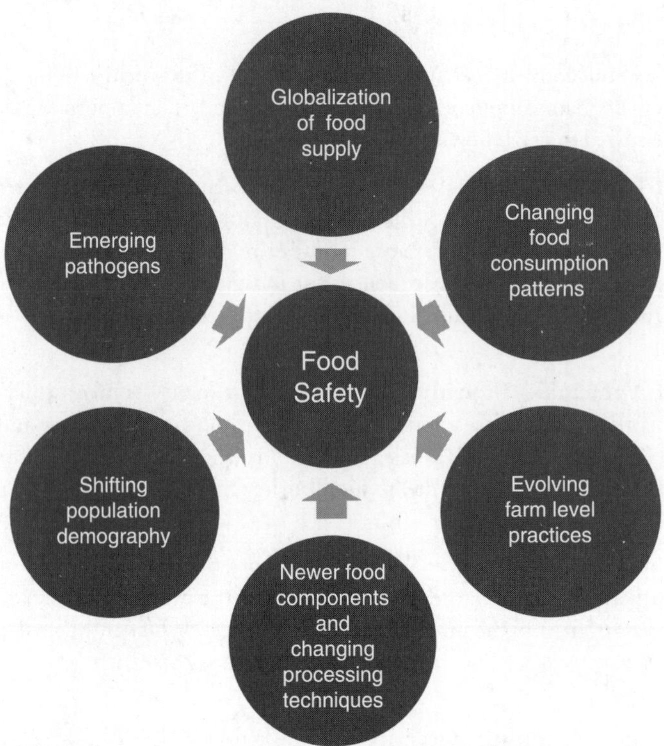

Figure 20.4 New challenges to food safety

1. **Globalization of Food Supply:** Due to globalization, the food grown in one country is being transported and processed in another region and consumed in several other countries. This has resulted in transmission of disease-causing organisms and chemical contaminants to far-off distances and has multiplied the risk of food-borne illnesses. In addition, the pressure on developing countries to produce more food for export has promoted inappropriate agricultural practices. Food safety is compromised in such a scenario where there is higher risk of contamination from mycotoxins, veterinary drugs and pesticide residues, and other chemicals.
2. **Changing Food Consumption Patterns:** Tourism and growing interest in other cultures has predisposed the consumers to try and taste exotic and unusual foods from different parts of the world. They are demanding more variety in food and also calling for foods that are not in season or are not grown in the region of their residence. The trend of eating out has led to an increased number of food establishments catering to a wide range of consumers from streets to five star hotels. Sadly, the food safety education and control have not kept pace with these changes. As a result, food in many of these establishments is not handled or prepared safely and subsequently becomes a threat to food safety.
3. **Evolving Farm-level Practices:** Changing farm practices have created newer ecological place for pathogens and have introduced new chemicals for agriculture. Many of these pathogens survive in the food and cause food-borne illness. On the other hand, the livestock is being attacked by newer

strains of pathogens which have higher ability to cause disease and higher resistance to antibiotics. It is a challenge to deal with these trends.

4. **Newer Food Components and Changing Processing Techniques:** Many of the meals/snacks consumed outside home are exposed to extensive food handling. Being aware of the role of diet in optimal health, the present day consumer is demanding foods that improve quality of life and help in longevity. Some of these may be cold foods (e.g., minimally processed ready-to-eat fruit and vegetable salads) which are not heated prior to consumption. Unless, HACCP is practiced effectively, there is a considerable risk of transmission of food-borne illness from handler to the consumer via such foods.

   Many modern processing and cooking techniques make use of additives (such as flavourings) to create products and dishes. Unless safe and effective additives are used within the prescribed limits, they can become a source of chemical contamination. Newer food components are being extracted from one food and added to another product. Such components should undergo extensive toxicological trials before being launched in the market. One such example is cow pea protein isolate which has recently been extracted at the Indian Agriculture Research Institute (IARI).

   Among the new technologies are genetic engineering, irradiation, and modified-atmosphere packaging which should be used cautiously and only after understanding the potential risks involved. Internationally accepted principles of food safely should be considered along with health benefits, ethical issues, socio-economic factors and environmental considerations.

5. **Shifting Population Demography:** Increased longevity has led to the higher number of elderly who are more at risk of developing adverse health outcomes following food-borne illnesses such as Listeriosis. Other groups which are vulnerable to acquire infections from some pathogens are pregnant women, young children and infants. Being immune-compromised, the Human immunodeficiency virus (HIV) patients are more susceptible to infections resulting from the ingestion of contaminated food.

6. **Emerging Pathogens:** The recent emergence of bacterial strains having increased resistance to antibiotics is a cause of worry. Another example is the meat and bone meal prepared from cattle infected with Bovine Spongiform Encephalitis BSE (refer to Section 30.1 in Chapter 30).

Consumer awareness on fundamental issues related to food safety is very important to ensure safe food during the active chain of supply.

20.4 Factors Affecting Safety of Food

What are the risk factors which hamper the safety of food? Knowledge about these risk factors forms the basis for preventive measures. The risk factors leading to food-borne illnesses can be present in the home-cooked food and the food cooked at a food service establishment. These factors have been identified based on the epidemiological studies, carried out by the Centre for Disease Control and Prevention, and are referred to as food-borne illness risk factors or just risk factors.

Figure 20.5 lists the top five risk factors which are categorized as

1. **contamination factors** (contributing to the transmission of contaminants to the food),
2. **survival factors** (helping in the survival of pathogens) and
3. **proliferating factors** (helping in the growth and replication of pathogens).

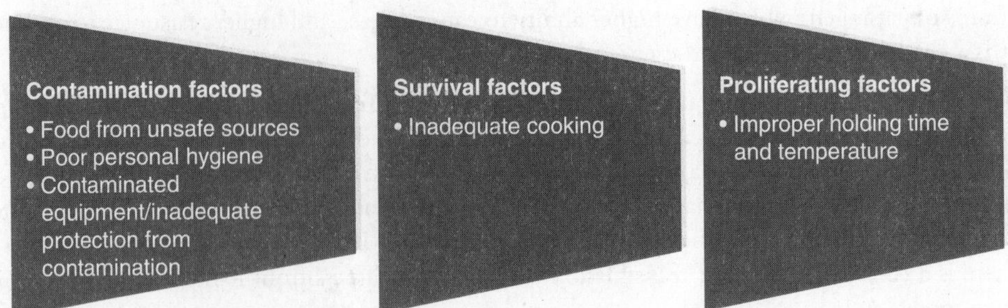

Figure 20.5 Major risk factors for food-borne illnesses

Let us understand how these risk factors affect the safely of food

1. **Food from Unsafe Sources:** Foods obtained from unsafe animal or plant sources are more prone to microbial contamination. Plant foods may be contaminated at the farm level through faulty agricultural practices including the misuse of pesticides. Animal foods, in particular, are at risk of contamination from zoonotic pathogens, veterinary drugs including growth promoters such as arsenal, milk enhancers such as oxytocin and drugs such as antibiotics used in the production of animal products. Fish and shell fish are more likely to be contaminated with bacteria, parasites, algal toxins, viruses and chemicals.
2. **Poor Personnel Health and Hygiene:** If a worker exhibiting symptoms of respiratory or gastrointestinal infection (such as diarrhoea and/or vomiting) continues to handle food, there is a potential risk of food contamination and subsequent transferring of illness to the consumer through its ingestion. The risk is multiplied when the contamination occurs in PHF/TCS foods. Unheated, ready-to-eat foods such as salads are excellent means for the transmission of food-borne infection. In case the infected workers handle huge batches of food, the illness is transferred to a large number of people resulting in an outbreak. Examples of worker-transmitted pathogens are–*Norovirus*, hepatitis virus, *Salmonella, Shigella, Staphylococcus* and *E.coli*.

> **DID YOU KNOW?**
>
> **What Is the Load of Virus Particles Transferred from Hands to Food?**
>
> The number of virus particles
> - present in 1 ml of faeces = 10^{12}
> - left on hand after proper washing = 10^{10}
> - transferred from a bare hand to food = 10^{9}
> - required to make a person sick = 1 – 10

3. **Contaminated Equipment/inadequate Protection from Contamination:** There is ample evidence to show that the food service restaurants responsible for food-borne illnesses do not follow adequate cleaning and sanitizing of utensils, equipment and food-contact surfaces. Moreover, there is cross contamination when raw foods touch the surfaces that were earlier in contact with a raw animal food (Figure 20.6). These practices increase the risk of contamination with pathogenic and toxigenic microorganisms. Environmental contamination and inadequate packaging also hamper food safety. The reasons for these faults include inadequate training of workers, bad design of equipment/surfaces, lack of time allotted for cleaning and sanitizing procedures. Proper training and effective supervision are necessary to control these risk factors.
4. **Inadequate Cooking:** We know that high temperature used in cooking leads to destruction of vegetative forms of disease-causing microorganisms. Since raw animal products are at risk of being contaminated with zoonotic pathogens, special care is needed while cooking eggs, meat, fish and poultry. The guideline

for cooking eggs is at 63°C for 15 seconds. Chicken and turkey should be cooked in a manner that the inner temperature reaches at least 74°C for 15 seconds. For pork, the corresponding temperature should be 63°C for 15 seconds.

However, it is important to note that cooking destroys only the vegetative forms of most pathogens but it does not kill the spores (e.g., of *Clostridium botulinum*) or the preformed toxin (e.g., the toxin formed by *Staphylococcus aureus*). Other measures are needed to prevent the contamination by these spore-forming bacteria (refer to Section 24.2 of Chapter 24).

Figure 20.6 Surface in contact with raw meat can lead to cross contamination

DID YOU KNOW?

Undercooked Eggs Are at Risk of *Salmonella* Infection

Some recipes (e.g., egg nog) use raw eggs and some customers prefer undercooked eggs which are soft and runny in texture. Such eggs are at risk of infection with *Salmonella*. Heating eggs to 63°C for 15 seconds destroys harmful microorganisms including *Salmonella*. Vulnerable individuals such as young children, elderly and immuno-compromised people should specifically be advised to refrain from consuming under-cooked or raw eggs.

5. **Improper Holding Time and Temperature:** As stated earlier, a mild microbial load in PHF/TCS foods is sufficient for rapid and progressive growth of pathogenic and toxigenic microorganisms. Time/temperature abuse is a major risk factor in fast proliferation of bacteria in food. Therefore, PHF/TCS foods should be stored at cold (<4°C) or hot temperature (>60°C). The intermediate temperature represents the danger zone (refer to Figure 24.9 in Chapter 24).

Many a times, chefs prepare or cook dishes ahead of schedule for the convenience of the consumers. Sometimes, leftover foods are also reused. Such practices can also lead to time/temperature abuse. Other factors which can lead to microbial growth and toxin production are out-of-order refrigerating unit, electricity failure, inadequate cooling of cooked food and improper reheating of cooked food. During thawing of frozen PHF/TCS food, insufficient heat transfer makes outer portions thaw faster as compared to inner portions which can again be a risk for food safety.

20.5 Training in Product Safety

In case of hospitality industry, the food safety culture should begin at the top of the organization. Commitment from the corporate leadership, recognition from the management and effective implementation of a well-planned training programme can together reduce the risk associated with safety of food. The chief executive officer (CEO)/president should enforce product safety through the vice presidents, general managers and supervisors right up to the entry-level workers/labour. Training in product safety should be planned and executed keeping the following points in mind:

1. All workers should be empowered to prioritize food safety while working. They should be imparted the necessary knowledge about microorganisms, food-borne illnesses, sources of contamination, causes of food spoilage and principles of preservation. Additionally, they should be trained in sanitation and hygiene practices necessary to ensure food safety.
2. The first-line workers must be instructed and trained in food safety measures related to their work activity, that is, the type of food being handled, its processing/cooking, storage and service. They should be able to recognize deviations and report them to their supervisors.
3. Supervisors/managers should be able to identify critical points and take remedial action.
4. Behavioural science should be employed to ensure that the food safety knowledge is put to actual practice. In other words, human behaviour, group dynamics and organizational culture should also be given due consideration.
5. The effectiveness of training and awareness of food handlers should be evaluated periodically.
6. Routine supervision and checking must be carried out to ensure adherence to food safety practices.
7. The training programme should be reviewed and updated keeping pace with the current developments in the field of food safety.

The hospitality industry must reinforce the messages related to product safety through different means (Figure 20.7).

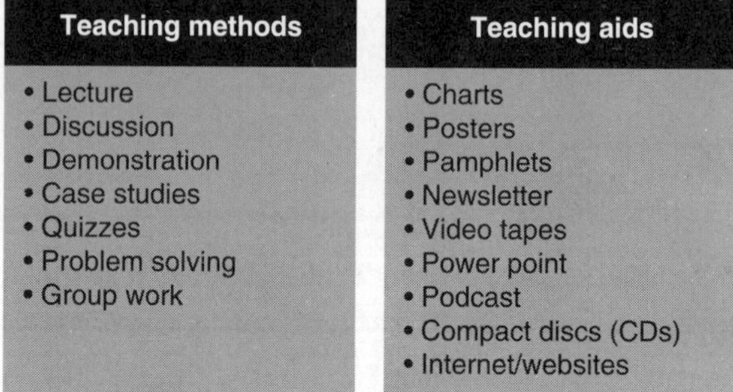

Figure 20.7 Training methods and aids for training in product safety

20.6 Food Safety in Hospitality Industry

Safety of food can be ensured if adequate quality control and food safety measures are followed through all the unit operations in various catering/food service establishments such as follows:

(a) Eating houses
(b) Restaurants and hotels
(c) Snack bars
(d) Canteens (schools, colleges, offices)
(e) Food service at religious points
(f) Tiffin service
(g) Railways catering
(h) Airline catering
(i) Hospital catering

The food business operators must ensure that specific hygiene and sanitary practices are adhered to at the various steps of preparation, processing/cooking, storage and transportation.

Purchase of Raw Material: To minimize microbial, chemical and physical hazards, all raw materials should be procured from reliable and authorized dealers. Meat, fish and poultry should specifically be purchased from licensed outlets. They should be checked visually for any signs of deterioration or foreign body contamination. The raw material should also be tested for the expiry dates, temperature control, adulteration, damage, pests and contamination.

Pre-preparation: Whole fruits and vegetables should be washed properly and only food-grade equipment and utensils should be used. While peeling and/or cutting the fruits and vegetables, care should be taken not to use the same chopping boards, knives and gloves which were earlier in contact with raw meat/fish/poultry.

Figure 20.8 Cooked food should be held at a temperature above 63°C or below 4°C

Cooking/preparation/processing: Adequate time and temperature control is crucial particularly for PHF. The non-vegetarian food should be cooked to an adequate inner temperature to destroy the pathogens and reduce their number to an acceptable level. The vegetarian and non-vegetarian food must be cooked separately. Cooked food should be held at temperature above 63°C or below 4°C to prevent the multiplication of microorganisms (Figure 20.8).

Reheating: The food should be reheated in an oven, steamer, microwave oven or in a cooking range. It should not be reheated on a steam table, in a bain-marie or under a heat lamp. While reheating the inner temperature of food should reach 74°C.

Chilling: PHFs such as semi-/fully cooked dishes, salads, desserts and puddings provide good medium for replication of pathogens. Therefore, they must be chilled at 4°C or below. The refrigerator and display units must have temperature below 4°C.

Transportation: The vehicle used for transporting cooked food should be used clean and should carry nothing other than food. It should have provision to keep hot food hot and cold food cold beyond the danger zone temperature (between 5°C and 63°C). To prevent the multiplication of microorganisms, the time of transport should be minimized. To avoid contamination, the food should be kept covered and the seal of packaged food should be protected. All surplus and unused thawed food should be discarded.

Storage: Raw meat/fish/poultry should be stored away from other foods. The cooked food should be stored above the uncooked food and the vegetarian food above the non-vegetarian food on separate racks of refrigerator. Frozen foods should be stored at −18°C or below. Cooked food should be cooled quickly before being stored in the refrigerator. To minimize wastage and ensure food safety, the packages or the containers should be labelled with the help of stickers to denote the date of manufacturing. To prevent contamination from pests, low moisture foods such as flour and grain pulses should be stored above the ground level in re-sealable containers.

Special Requirements for High-risk Foods: Some foods need special handling since they are at high risk as per HACCP.

1. **Fresh Fruits/vegetables When Cut or Juiced** should be used quickly (Figure 20.9). They can be stored under refrigeration for short duration in sanitized and covered containers. Only potable water should be used for beverages and ice. Juice dispensing machine should be cleaned regularly.
2. **Confectionery Products** should be stored in airtight containers and displayed hygienically. Finished products should be labelled indicating the 'best before date' prior to being refrigerated.

Figure 20.9 Fresh fruits should be used quickly after cutting

3. **Meat, Fish and Poultry** should be purchased in chilled (<5°C) or frozen (-18°C) state from authorized slaughter houses/vendors. They should be washed with potable water and cooked adequately until the core temperature reaches 75°C (for at least 15 seconds). Uncooked meat, fish and poultry should be stored below the cooked ones in the refrigerator. Processing/cooking area should be cleaned and disinfected quickly and the waste should be removed promptly.
4. **Water-based Chutneys and Sauces** should be prepared from clean fruits and vegetables using potable water and permitted additives within the prescribed limits. They should be stored in food grade plastic or glass containers and kept refrigerated when not in use. Perishable chutneys should be consumed immediately. They must be discarded if signs of spoilage in terms of changes in colour/flavour/odour are noted.
5. **Foods with Gravy** should not be kept at room temperature for more than two hours. They should be either refrigerated (<4°C) or kept hot (>63°C) for prolonging their storage life. No water should be added after cooking/reheating/boiling.
6. **Fried Foods** should be prepared in good quality oils/fats. Use of hydrogenated fat (rich in trans fat) and oils which have been reheated several times should be avoided.
7. **Leftover Food** which has been on display should be discarded. PHF/TCS foods such as hollandaise sauce, omelettes and cut melons and tomatoes should not be reused.
8. **Thawing of Frozen Foods** should be carried out in a refrigerator at 5°C or less. Alternatively, microwave or running water should be used for thawing. Thawed food should be labelled again with date and time to denote the beginning of second shelf life.

Worker Hygiene: To ensure food safety, all the workers handling food must be in a good state of health. Many a times, the workers may be asymptomatic carriers of disease-causing organisms or else they may not report their illness. In either case, they may have already caused the damage. Therefore, knowledge about food safety and sanitation must be imparted to all the personnel handling food (refer to Chapter 29). In order to ensure better compliance, they should be explained in simple terms the scientific basis for the specified hygienic practices. The workers should use hand-free methods such as disposable gloves, tongs, spoons or spatula for ready-to-eat foods which should never be handled with bare hands.

Hence, food safety measures must be followed at various unit operations of food service industry. This chapter provided an introduction to the concept of food safety and its relevance in the changing scenario of present times. What makes food unsafe? Where do the contaminants in food come from? How can the beneficial microorganisms be utilized in food preparation and processing? What are the different types of food-borne illnesses? Why does food get spoilt and what are the different principles of food preservation? What is the role of food additives in ensuring food safety and quality? How do food laws and regulations regulate food safety at national and global level? What is the role of risk management in ensuring food safety? How can sanitation and hygiene be maintained in a food establishment? What are the emerging pathogens in the area of food safety? These queries will be addressed in the present section on Food Safety and Quality.

Let us analyse the situations given in the Think Box at the beginning of the chapter.

Situation 1: Investigation revealed that the raw chicken for the curry was obtained from an unauthorized place and the workers handling it had poor personal hygiene. The cooked chicken curry was stored at ambient temperature (30–42°C) for about five hours and was not reheated adequately before being served to the college students. Contamination, survival and proliferation related risk factors lead to the severe food infection.

Situation 2: The new recruit in the healthy sandwich bar was putting the vegetables at risk of cross contamination by using the same knife, chopping board and gloves which were earlier in contact with raw meat/chicken/fish. Therefore, he was asked to undergo a training course in food safety.

Situation 3: The particular lot of cheese was recalled from the market because of the presence of *E.coli* and *Listeria* which can cause serious food-borne illness.

Situation 4: The established caterer knew the importance of controlling temperature and time of the cooked food in ensuring food safety. In order to prevent food-borne illness and the loss of trust of his client, he decides to hire another refrigerated van at double the fare.

Know Your Progress

True or False Questions

1. Perishable foods which encourage rapid and progressive growth of pathogenic and toxigenic microorganisms are known as potentially hazardous foods (PHF).
2. Cooking destroys vegetative cells and spores of pathogens.
3. When raw fruits and vegetables touch the surfaces which were earlier in contact with a raw animal food, the risk of contamination with pathogenic and toxigenic microorganisms increases.
4. Chicken and turkey should be cooked adequately so that the inner temperature reaches at least 63°C for 15 seconds.
5. Human behaviour, group dynamics and organizational culture should also be given due consideration for training in product safety.

Very Short Answer Questions

1. Define the following terms
   - **(a)** Food safety
   - **(b)** Food-borne illnesses
   - **(c)** Time/temperature control
   - **(d)** Good hygienic practices
   - **(e)** Food safety management system

Short Answer Questions

1. What do you understand by the term 'hazard'? Categorize the various food safety hazards.
2. How do the experts from different disciplines contribute to the field of food safety?
3. What is the relevance of safe food for consumers and producers?

4. How has globalization of food supply affected food safety?
5. What points should be kept in mind while planning and executing training in product safety?

Long Answer Questions

1. What are the new challenges in the present day food supply and how do they affect food safety?
2. Explain the top five risk factors identified by Centre for Disease Control and Prevention in context of food safety?
3. Adequate measures must be followed at various unit operations to ensure safe food in hospitality industry. Justify the statement.

Beyond the text

Identify the possible risk factors for food safety at the following food service/catering facilities:

1. School/college canteen
2. Railway catering
3. Tiffin service
4. Street food joint

Chapter 21

Microorganisms in Food

Chapter Objectives

After reading this chapter, the reader should be able to:

- know the relationship between microorganisms and food.
- identify common microorganisms based on their size, shape and structure.
- understand ways through which the microorganisms multiply or increase in number.
- comprehend the role of intrinsic and extrinsic factors in the growth, survival and death of microorganisms.
- learn about food hazards of microbial origin.

Microorganisms are so tiny that they are generally invisible to the naked eye. They can be seen only with the help of a microscope. Their presence on this planet is universal as they are found in almost every part of natural and man-made environment including the food we eat. Modern day consumers are eating out more often than before and expect their food to be wholesome and safe. Therefore, the food service professionals should have adequate knowledge about the characteristics of microorganisms and the factors affecting their growth and development. Such information can be utilized not only to prevent the growth of harmful microorganisms during cooking, processing, storage and service but also to optimize beneficial microbial fermentations.

THINK TANK

Harmful and Useful Interactions Between Microorganisms and Food

Consider the following examples:

- Fuzzy, cottony growth on a slice of bread
- Ropiness in yoghurt
- Brown, soft spots in stored apples
- Bulged can of fish
- Porous texture of bread
- Blue colour in Roquefort cheese
- Clean acid flavour of sauerkraut
- Sourness and effervescence in kefir

Which organisms are responsible for these changes? What are the factors controlling these interactions? Is it possible to prevent the harmful interactions and optimize the useful ones? This chapter will help us find answers to such questions.

21.1 General Characteristics of Microorganisms

The food we eat is not sterile. Microorganisms are present in food partly due to the natural microflora of raw material and additionally due to contamination during harvesting/slaughtering, processing, storage and distribution. The presence of microorganisms (particularly when the microbial load is less) does not usually result in any adverse effect on food or health but many a times it becomes obvious through food spoilage, food-borne disease and beneficial transformation of food (due to fermentation).

In other words, microorganisms can have both deleterious and beneficial effects on the quality of raw and processed food. The accompanying decomposition of food has great significance in maintenance of ecological balance of nature. Moreover, the changes in appearance, taste and texture of food often help us in detecting spoilage and sometimes toxins. Of the vast variety of microorganisms found on earth, a relatively small proportion is harmful to food. The microorganisms of major concern in food include bacteria, fungi (including mold and yeasts), viruses and protozoa. Among these, bacteria form the largest and most important group responsible for food spoilage and food-borne disease. They are present ubiquitously and exhibit rapid growth rate even in conditions where yeasts and mold are unable to replicate and propagate.

21.1.1 Morphological Characteristics

We know that the size of microorganisms is very small. Viruses are the smallest, whereas the bacteria are smaller than yeast and protozoa. On the other hand, mold and parasites are multicellular and can be seen even with unaided eye, though a microscope is required to observe their detailed structure. Various types of microorganisms differ in their size, shape and structure.

Bacteria, the most abundant of all organisms, are unicellular, relatively small in size and have varied shapes. They are found in many morphological forms: spherical (cocci), rod-shaped (bacilli), curved rods (comma) and spiral (spirochetes) as shown in Figure 21.1. The cocci and bacilli can exist as such or form associations (Figure 21.2). The cocci can form clusters (*Staphylococcus*), chains (*Streptococcus*) or tetrad (*Micrococcus*). Among the rods, *E.coli* and *Salmonella* are short, whereas *Clostridium* is large. *Vibrio* is an example of curved rods.

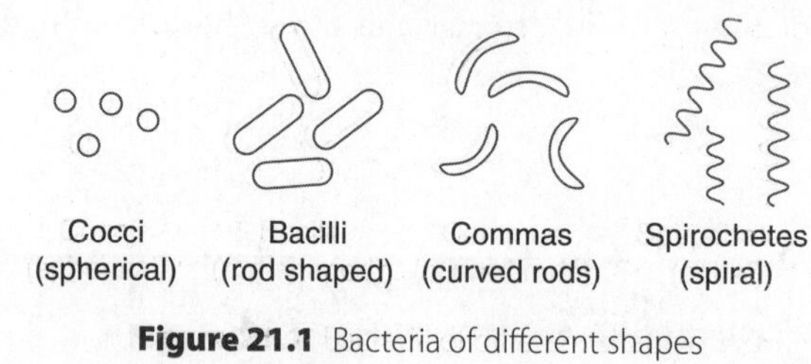

Figure 21.1 Bacteria of different shapes

Coccus Diplococcus Tetrad

Staphylococcus Streptococcus

(a) Cocci

Bacillus

Diplobacillus

Streptobacillus

(b) Bacilli

Figure 21.2 Arrangements of cocci and bacilli

Some bacteria can form endospores (commonly known as spors) in adverse environment (such as nutrition depletion or environmental stress). The spores are metabolically dormant, encased in two hard layers and extremely resistant to heat, radiation and chemicals. The spore-forming bacteria (*Clostridium* and *Bacillus*) are significant in causing food-borne illnesses (refer to Chapter 24 and 25).

Mold commonly grow as a mass of thread-like structures (hyphae) which enlarge and overlap to form visible tufts (mycelium) on food. The hyphae may be growing within the food (submerged) or growing above the food (aerial). They may be vegetative (providing food to the molds) or fertile (involved in production of reproductive parts). Molds are non-motile, filamentous and branched organisms (Figure 21.3). Their size varies from microscopic forms to macroscopic mushrooms. They are multicellular organisms which lack chlorophyll and do not participate in photosynthesis. Nevertheless, they draw on organic sources such as sugars, starch, cellulose and proteins for their growth and energy. Capable of living both on plants and animals, they are generally saprophytes but some are parasitic in nature.

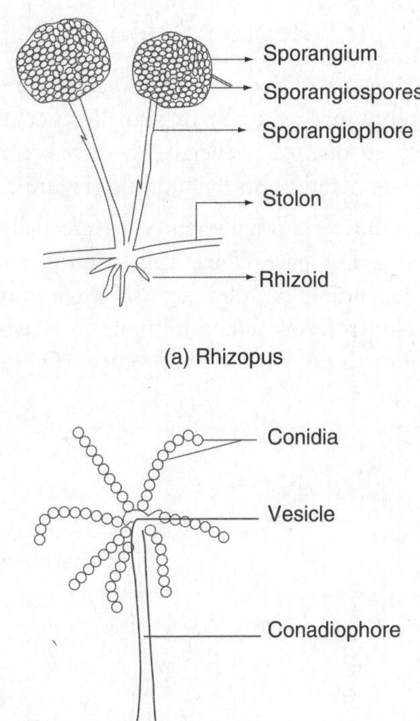

Figure 21.3 Structure of molds

Yeasts are single-celled fungi; oval, spherical or elongated in shape; larger than bacteria and do not form mycelium. The cell wall can have scars, known as bud scars, which indicate the sites of budding. They are non-motile.

Viruses are much smaller than other microbes. They do not have a cellular structure; instead, they have an outer protein layer (known as capsid) which wraps the genetic material – the nucleic acids (Figure 21.4). Viruses do not have any organelles for metabolism or for independent reproduction. They need access to a host to grow and multiply; accordingly, they are termed as 'obligate intracellular parasites'.

Protozoans are the most motile unicellular organisms. Most of them are free living in environment but some are parasitic and need multiple hosts to complete their life cycle. The morphology differs among genera, species and also during certain stages of their life cycle. At one stage, they form cysts which can survive adverse environmental conditions. Generally, it is the cyst stage at which a protozoan is ingested through contaminated food. The stage which causes illness is the active feeding stage (trophozoite). They are more highly organized than bacteria.

Helminths are multicellular organisms much larger than protozoans and have complex biology. The adult form for most helminths is macroscopic and is visible to unaided eye. Nevertheless, they can be microscopic in size at other stages of their life cycle, e.g. egg and larval stages.

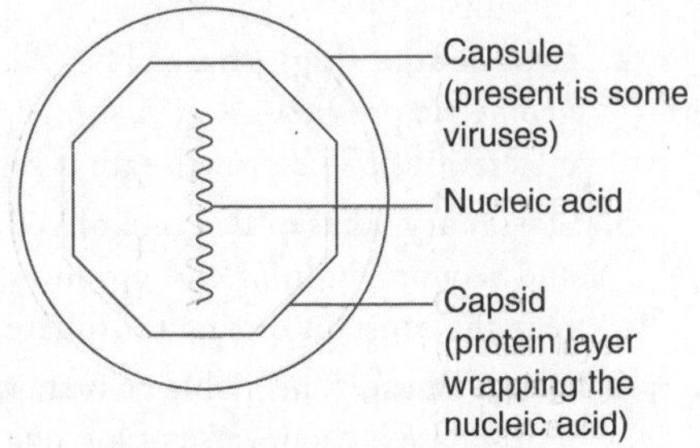

Figure 21.4 Components of a virus

21.1.2 Reproduction

How do different microorganisms replicate? Do they reproduce asexually or sexually? Let us find answers to these queries. Generally, bacteria reproduce by binary fission and yeasts by budding (Figure 21.5).

Bacteria usually reproduce asexually by binary fission at a fast pace. Their cells multiply at a constant rate depending on the extrinsic environmental conditions. Under favourable conditions in a laboratory, bacterial growth can be plotted as a curve (Figure 21.6).

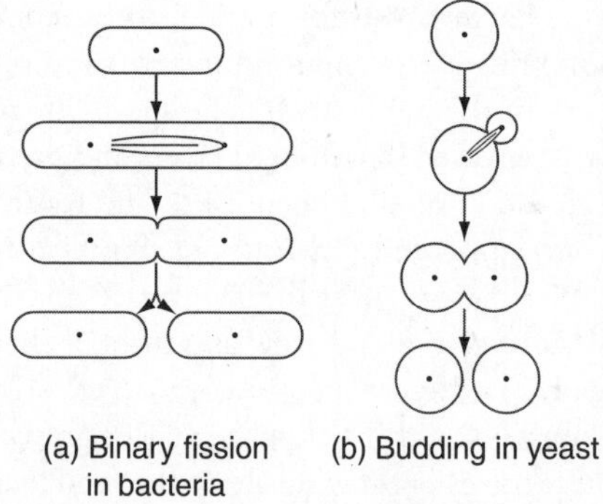

Figure 21.5 Reproduction in bacteria and yeast

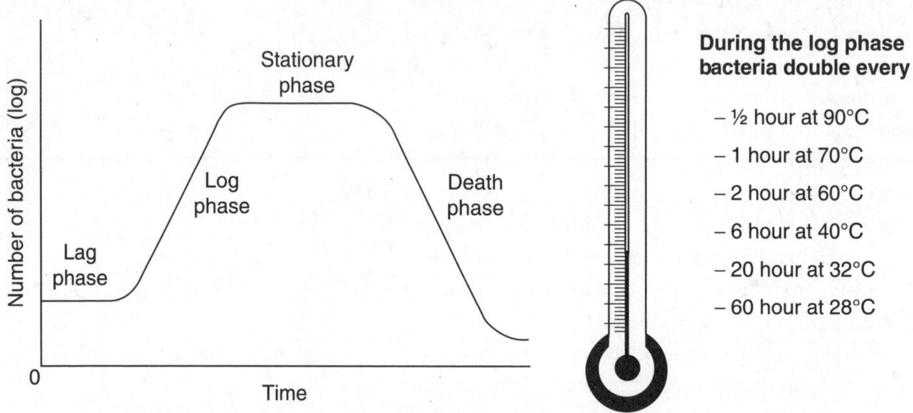

Figure 21.6 Growth curve of bacterial cultures

The characteristic **four phases of growth curve** are as following:

1. **Lag Phase:** The cells adjust to the new environment and there is no increase in their number in this initial phase.
2. **Exponential (log) Phase:** The cells double by binary fission at regular interval. Their growth is by geometric progression 1, 2, 4, 8, 16, … (2^0, 2^1, 2^2, 2^3, 2^4, … 2^n where n represents the number of generations). The growth rate is highest during this phase.
3. **Stationary Phase:** The rate of cell division decreases due to the exhaustion of available nutrients and accumulation of end products. The death rate equals growth rate and the number of bacteria remains constant during this phase.
4. **Death Phase:** The viable cells start dying in an exponential manner and their number decreases. This phase is the reverse of log phase.

Molds can have vegetative or reproductive hyphae. The reproductive hyphae usually extend in the air and form asexual exospores which are either free (conidia) in *Aspergillus* and *Penicillium* or inside a sack (sporangiospores) in *Rhizopus* and *Mucor* (refer to Figure 21.3). The third type arthrospores are formed when hyphae themselves get fragmented. Each of these spores which are released in air for propagation is capable of forming a new mycelium. It is through these asexual or sexual spores that the fungi reproduce. Black bread mold (*Rhizopus*) reproduces asexually by sporangiospores. Sexual reproduction is less common. 'Downy mildews' (*Phycomycetes*) causing blight of potatoes and buckeye rot of tomatoes can produce sexual spores known as oospores.

Yeasts generally reproduce by budding an asexual process in which a bud is formed on the outer surface of the parent cell, the nucleus divides and one nucleus shifts into the enlarging bud. Subsequently, a cell wall forms between the bud and the parent cell. The bud then separates as a new cell (Figure 21.5b). Very few species reproduce by binary fission. Reproduction by sexual spores known as ascospores is uncommon and leads to genetic recombination.

Viruses reproduce by infecting the cells of the host and exploiting their metabolism and resources for their own replication and propagation. They attach themselves to the host cell and release their nuclear matter (DNA or RNA) into it. The virus replicates inside the cell, subsequently the cell bursts and new viruses are released. The viruses are host-specific and cannot grow or multiply outside their host.

Protozoans replicate commonly by asexual reproduction. Free-living protozoans may opt for sexual reproduction in adverse environmental conditions to improve their genetics and survival through mutation and chromosomal crossing over. They resort back to asexual reproduction when the food and environmental conditions become favourable.

Helminths in adult stage are macroscopic and have complex life cycles where multiple intermediate and/or definitive hosts are involved and reproduction is sexual.

21.2 Factors Affecting Microbial Growth in Foods

When microorganisms feed on food to grow and reproduce, they bring alterations (usually undesirable) in its quality. Some microorganisms are beneficial and are added to food as an essential component of preservation to bring about desirable changes in flavour, texture and digestibility. For example, lactic acid bacteria are used for making fermented dairy products and acetic acid bacteria is employed in fermenting alcoholic beverages such as wines (refer to Chapter 23).

When they gain access to food, the microorganisms multiply, use the nutrients and bring enzymatic changes, break down food compounds or synthesize new products and in the process may spoil the food.

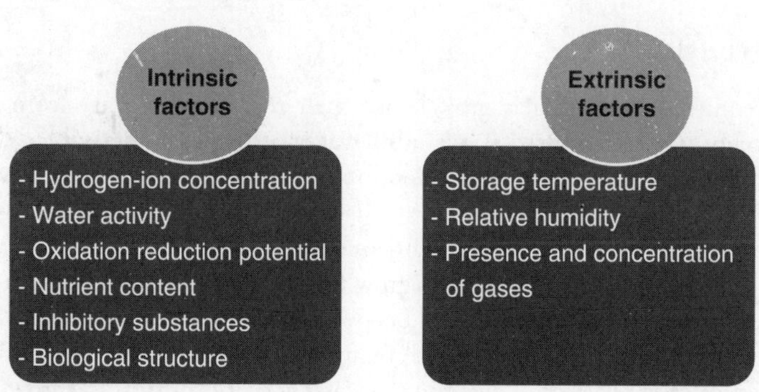

Figure 21.7 Intrinsic and extrinsic factors affecting growth of microorganisms in food

Harmful microorganisms bring undesirable changes in sensory quality of food and/or may lead t food-borne illnesses. Some of them produce toxins in the food, whereas others grow in the body of host t cause illness when ingested. The presence of pathogenic microorganisms in food supply chain is a majo problem from the public health perspective.

In order to understand the principles of food spoilage, processing and preservation, the food servic professionals must understand the factors that affect the growth of microbes. These factors can be groupe as the intrinsic and extrinsic factors (Fig 21.7).

1. *Intrinsic factors* (hydrogen ion concentration, water activity, oxidation reduction potential, nutrient content, inhibitory substances and biological structure) are inherent part of the food.
2. *Extrinsic factors* (storage temperature, relative humidity and presence and concentration of gases) are related to the storage environment which affects both the food and the microorganisms.

21.2.1 Hydrogen-ion Concentration

Hydrogen-ion concentration or the pH of food has great effect on the growth of microorganism Every microbe has a minimum, a maximum and an optimum pH for its growth. Molds can grow ove a wide range of pH. They can even grow in very acidic environment which is otherwise unfavourabl for yeast and bacteria. The yeasts generally prefer acidic pH (nearing 4.0–4.5) and are unable to gro in alkaline medium. Film yeasts grow in acidic foods such as sauerkraut and pickles. Both yeasts an molds tolerate acid better than bacteria which grow best at near neutral pH (6.0–8.0). The acid-formin bacteria prefer moderate acidity, whereas proteolytic bacteria are favoured by alkaline pH. Most fooc are neutral or acidic in nature. Acidic foods such as pickles, soft drinks, fruit juices and fermente drinks are more acidic and have better shelf life. However, yeasts and molds have the ability to spo acidic foods.

The pH of food not only affects the rate of growth of microorganisms but also their rate of surviv during processing and storage. **Buffers**, the compound which resist changes in pH, also have an importa role in food processing. The low buffering action of vegetables permits appreciable increase in acid durir the initial stages of pickle fermentation, thereby inhibiting the growth of undesirable organisms. On th other hand, high buffering power of milk allows adequate growth of lactic acid bacteria and the subseque acid production before yoghurt is finally set.

21.2.2 Water Activity

All microorganisms require water for their growth but their requirements vary from one organism another. This requirement is best expressed as **available water** or the **water activity, a_w**, which is define as the ratio of the vapour pressure of water in the food to the vapour pressure of pure water at a specif temperature.

The optimum water activity for the growth of many microorganisms is between 0.99 and 0.98. Most the bacteria grow at a_w nearing 1.00, and they fail to grow when a_w falls below 0.91. Compared to bacteri the moisture requirements of yeasts and mold are much less. The available water (or the water activity) food can be lowered by adding sugar and salt (which tie up water), by forming hydrophilic colloids (gel or through dehydration or crystallization. Lowering the water activity controls the growth of microorga isms (Table 21.2).

Table 21.2 Water activity of foods and related microbial growth

Water Activity (a_w)	Examples of Foods	Microorganisms Involved
1.00 to 0.95	Meat, poultry, fish, fruits and vegetables, milk, most beverages, canned fruits in syrup, canned vegetables in brine, processed cheese	Gram-negative bacilli, spore-forming bacteria, some yeasts
0.95 to 0.91	Foods with 50 per cent sugar content (low-sugar jams) or 12 per cent salt content (cured meat, pickles), some cheeses (Cheddar, Swiss), tomato paste	Most cocci and bacilli (vegetative cells) including lactobacilli, some mold
0.91 to 0.86	Foods with 65 per cent sugar (jams, syrups, preserves) or 15 per cent salt (dry ham, salami, sausages, dry cheese)	Most yeasts
0.86 to 0.80	Foods with 15–20 per cent moisture (cereals, pulses, cakes, sweetened condensed milk, fruit juice concentrate, chocolate syrup)	Most mold
0.80 to 0.75	Foods with high sugar (molasses, fondants, brittles) or very high salt 26 per cent (salted fish)	Halophilic bacteria
0.75 to 0.65	Foods with less than 10 per cent water (dried apricots, peanuts, dates, figs, rolled oats, dry bread)	Xerophilic mold
0.65 to 0.60	Confectionery, candies, dried fruits (15-20 per cent water), honey	Osmophilic microorganisms (yeasts)
Less than 0.60	Flour, wheat, corn, dried milk, instant coffee, dried egg, spices, some biscuits, chocolate, honey, potato chips	No growth of microorganisms

21.2.3 Oxidation Reduction Potential

Different microorganisms have different oxygen requirement for their optimal growth. The oxidation reduction (OR) potential is the relative measure of oxidizing or reducing capability of the medium and it has significant influence on the growth of microorganisms. Based on their ability to use free oxygen, microorganisms can be categorized as follows:

1. *aerobes* which require free oxygen,
2. *anaerobes* which develop best in the absence of free oxygen and
3. *facultative* which can grow either in the presence or absence of free oxygen.

The bacteria may fall in any of the three categories; whereas, the mold are generally aerobic and the yeasts are usually facultative anaerobic.

The OR potential of the food depends on

1. the particular OR potential of the original food.
2. the resistance to change in OR potential of the food.
3. oxygen tension of the atmosphere around the food.
4. access of atmosphere to the food.

Negative OR potential in fresh plant and animal foods, and their ability to resist any change prevents the growth of aerobes except on their surface. Head-space of a can has low oxygen tension because of the vacuum and inaccessibility to the atmosphere. Hermetic (air tight) and exhausting (removal of air) containers are essential in the canning process to prevent microbial growth and undesirable changes, for example, bulging of can. On the other hand, positive OR potential encourages the growth of aerobes such as bacilli and micrococci, whereas a negative OR potential is suitable for anaerobes, for example, *Clostridium*. Processing decreases the resistance to change the OR potential of food, allows more oxygen to diffuse in and makes the food more prone to spoilage.

GOOD TO KNOW

What Is Oxidation Reduction (OR) Potential?

The oxidation reduction potential is defined as a relative measure of the oxidizing or reducing capacity of the growth medium. In oxidation, the reactant loses electrons, whereas in reduction it gains electrons.

21.2.4 Nutrient Content

Microorganisms need food for energy, growth and accessory substances such as vitamins.

Energy is derived primarily from carbohydrates, particularly the simple sugars. The ability of micro-organisms to utilize different types of simple sugars varies and is an important tool in their identification. For example, *Lactobacillus* grows in milk, whereas many other organisms which cannot use lactose do not grow in milk. Similarly, *Saccharomyces cerevisiae* (baker's yeast) ferments bread dough because of its ability to utilize maltose and helps in achieving a porous texture. Besides carbohydrates, organic compounds such as esters, alcohols and organic acids are also used as energy sources. Meats being rich in proteins and low in carbohydrates are attacked by proteolytic enzymes. In the absence of a better energy source, the microorganisms use hydrolysis products of proteins, peptides and amino acids for energy. Only very few organisms obtain energy from fat sources.

For **growth** purposes, nitrogenous compounds such as proteins, peptides, amino acids and urea are used. Many types of molds, some types of bacteria and few types of yeast are proteolytic in nature. Microorganisms get carbon for their growth from organic compounds and also from carbon di oxide. Minerals needed by these organisms are generally present in the food but sometimes they may be lacking, deficient or unavailable.

Accessory substances such as vitamins needed for the activities of microorganisms also come from the food they grow on. For example, B vitamins are obtained from meat and ascorbic acid from fruits. Another B vitamin, biotin is present in egg white but is unavailable for microorganisms because of being bound to avidin. Therefore, the presence of avidin prevents spoilage of eggs. Processing reduces the level of some of the required accessory substances such as loss of B vitamins and ascorbic acid by heat treatment and drying.

21.2.5 Inhibitory Substances

Inhibitory substances are either present originally, added purposively or unintentionally or else developed through microbial growth or processing methods. Some naturally occurring antimicrobial substances such as lysozyme in egg white, eugenol in cloves, allicin in garlic, thymol in oregano and benzoic acid in cranberries prevent microbial growth. Inhibitory substances may also be produced by microorganisms. For example, alcohol production by wine yeasts prevents competitive organisms from growing; nisin production by *Streptococcus* inhibits growth of *Clostridium* during the curing of cheese. Processing techniques such as heating can also be important in the formation of inhibitory substances.

1.2.6 Biological Structures

ome foods have natural covering such as outer shell of eggs, scales of fish, layers of lard over meat, covering f nut and peels of fruits and vegetables. These biological structures prevent the entry and subsequent ecay by the spoilage organisms. The inner tissues of fresh plants and animals are low in microbial count. eeling, chopping, dressing or comminution not only expose the inner parts but also distribute the invading nicroorganisms. The juices released during these processes provide food for these organisms and hasten heir replication and subsequent spoilage.

1.2.7 Storage Temperature and Relative Humidity

he rate of microbial growth is affected to a large extent by the storage temperature. Different microorganisms ave a minimum and a maximum temperature at which they can grow. Based on the optimum temperature equired for their growth, microorganisms can be divided into three groups:

1. *Thermophiles* (grow best at 55–65°C)
2. *Mesophiles* (need 30–40°C for their optimal growth)
3. *Psychrotropic* (grow at refrigerating temperature of 20–30°C)

reservation methods such as heating, cooling, drying, removal of air and use of preservatives control the rowth of microbes in food. Nevertheless, killing the spores is much more difficult than destroying the egetative bacterial cells. Therefore, all processing times for canned foods are computed by using both he time and the temperature required to kill bacterial spores.

Relative humidity of the storage environment also plays an important role in the growth of micro-rganisms. It also affects a_w within the foods. Dehydrated foods such as milk powder are highly hygroscopic nd can absorb considerable moisture from the surrounding environment when relative humidity is high.

1.2.8 Presence and Concentration of Gases

he most frequently used gas for control of microbial growth is carbon dioxide (CO_2) which is effective gainst many microorganisms including aerobes. On the other hand, presence of oxygen promotes the rowth of microorganisms. In modified gas atmosphere, the concentration of carbon dioxide is increased, vhereas that of oxygen is lowered.

1.3 Food Hazards of Biological Origin 

acteria, yeasts, molds and viruses are significant because they cause food-borne diseases and/or food spoil-ge. In contrast to yeast, many bacteria, some molds and few viruses can lead to food-borne illnesses. Many ypes of bacteria, molds and yeasts are also responsible for food spoilage. The undesirable effects of these nicroorganisms are discussed in the subsequent paragraphs. However, the viruses cannot lead to spoilage ecause they cannot grow on food. It is important to know that some microorganisms have beneficial role n food processing (refer to Chapter 23). For example, yeast (*Saccharomyces cerevisiae*) ferments the dough, ncreases volume, gives porous texture and desirable flavour to bread; mold (*Penicillium roqueforti*) produces lue colour during ripening of Roquefort cheese, and lactic acid bacteria contribute to the desirable flavour n sauerkraut and kefir (refer to Think Tank at the begining of this chapter).

21.3.1 Bacteria

Among the microorganisms present in food, bacteria are the most important group because of their wic variety of species, ability to utilize food nutrients to grow rapidly under varying environmental conditior and survive adverse conditions by forming spores. They are usually free living but can be parasitic. For co venience, bacteria are grouped based on some similarities in their characteristics (Table 21.3). Based on the gram-staining behavior, bacteria are grouped as gram positive (*Staphylococcus, Bacillus, Clostridium, Lister. monocytogenes*) and gram negative (*E.coli, Salmonella, Shighella, Yersinia*). Spore-forming bacteria (*Bacillu Clostridium*) form heat-resistant endospores in adverse environment. Their oxygen and temperatu requirements also vary.

Table 21.3 Groups of bacteria important in food safety

Group	Characteristic Features	Examples
Psychrotropic bacteria	Grow rapidly at refrigeration temperature (≤5°C); e.g., cooked and refrigerated food.	*Pseudomonas, Flavobacterium, Achromobacter, Listeria, Yersinia*
Mesophillic bacteria	Grow at 35–55°C temperature; e.g., cooked food held at serving temperature.	*Bacillus*
Thermophillic bacteria	Grow at high temperature (55°C and above); e.g., canned food.	*Bacillus, Clostridium*
Thermoduric bacteria	Survive heat treatment such as pasteurization of milk.	*Bacillus, Clostridium, Streptococcus, Lactobacillus, Micrococcus*
Osmophillic bacteria	Can grow in substrates of high osmotic pressure (honey and jam).	*Leuconostoc*
Halophilic bacteria	Can grow at high concentration of salt (pickles).	*Pseudomonas, Bacillus, Micrococcus, Pediococcus, Clostridium, Vibrio*
Aciduric bacteria	Can survive at low pH (<4.0); e.g., soft drinks.	*Lactobacillus, Pediococcus, Lactococcus, Enterococcus, Streptococcus*
Proteolytic bacteria	Breakdown proteins; e.g., spoilage of meat.	*Bacillus, Staphylococcus, Clostridium, Pseudomonas, Proteus*
Lipolytic bacteria	Hydrolyse or oxidize lipids; e.g., rancidity of fats.	*Micrococcus, Pseudomonas, Staphylococcus, Alcaligenes,*
Saccharolytic bacteria	Hydrolyze sugars to acids; e.g., honey and syrup.	*Bacillus, Clostridium*
Pectinolytic bacteria	Break down pectin; e.g., softening of fruits.	*Bacillus, Clostridium, Achromobacter*
Lactic acid bacteria	Produce large quantities of lactic acid from carbohydrates; e.g., curdling of milk.	*Lactococcus, Leuconostoc, Pediococcus Lactobacillus, Streptococcus*

(Continue

able 21.3 (Continued)

Group	Characteristic Features	Examples
Acetic acid bacteria	Produce acetic acid; e.g., fermentation of fruit juice.	*Acetobacter aceti*
Gas-forming bacteria	Produce gas (CO_2, H_2, H_2S) from nutrients.	*Leuconostoc, Lactobacillus, Escherichia, Enterobacter, Clostridium, Bacillus*
Slime-forming or rope-forming bacteria	Produce slime or rope; e.g., in milk, curd.	*Leuconostoc, Alcaligenes, Enterobacter, Lactobacillus*
Coliforms	Used as an index of sanitation in food and water.	*Escherichia, Enterobacter, Citrobacter*

everal bacteria are capable of causing food infections and food intoxication through the production of oxins (refer to Chapters 24 and 26). Many lead to food spoilage, for example, *Alcaligens* causes ropiness n milk and yoghurt. Some selected genera of bacteria which lead to food hazards are listed in Table 21.4:

able 21.4 Undesirable effects of different bacteria

Bacteria	Undesirable Effect	Bacteria	Undesirable Effect
Acetobacter	Souring of fruits and vegetables	*Listeria*	Food infection
Alcaligens	Ropiness in milk, yoghurt	*Mycobacterium*	Tuberculosis through raw milk
Bacillus	Food poisoning	*Proteus*	Spoilage of meat, seafood, eggs
Brochotrix	Spoilage of meat and products	*Pseudomonas*	Food spoilage
Campylobacter	Gastroenteritis	*Salmonella*	Food infection
Clostridium	Food poisoning	*Shigella*	Bacillary dysentery
Erwinia	Soft rot in fruits and vegetables	*Staphylococcus*	Food poisoning
Escherichia	Food infection	*Streptococcus*	Sore throat, scarlet fever
Halobacterium	Discoloration on high salt foods	*Vibrio*	Food infection
Lactobacillus	Fermentation of sugars	*Yersinia*	Food infection

Some bacteria play a beneficial role in food fermentation. For example, *Lactobacillus* and *Leuconostoc* are used for sauerkraut and pickle fermentation; whereas, *Acetobacter* is employed for fermenting vinegar.

21.3.2 Mold

Mold are generally aerobes and so they are observed more on the surface than throughout the food. They are more tolerant to sugar or salt as compared to bacteria. They can be easily recognized by their fuzzy, cottony appearance on the food, for example, bread and pickles. Different mold growing on food manifest themselves in different ways, for example, fluffy, compact, velvety, powdery or sticky. The pigments in the mycelium (red, yellow, purple, brown, black) and in the spores (green, blue-green, black, pink, orange, yellow) are characteristic of various types of mold. They exhibit varied heat resistance. Examples of mold include *Aspergillus, Mucor* and *Penicillium*.

An undesirable effect of molds is that they produce mycotoxins which can lead to adverse effects such as allergy, inflammation or irritation of mucous membrane. They are also responsible for spoilage in many kinds of food, for example, *Penicillium expansum* causes brown soft spots in stored apples. Some selected genera of molds leading to food hazards are listed in Table 21.5.

DID YOU KNOW?

Mushrooms Are Fungi

Mushrooms are in reality fungi where the large underground filamentous mycelium develops into a fruiting body (basidiocarp) due to differentiation. They may be classified as both macroscopic fungi and filamentous fungi.

Table 21.5 Undesirable effects of different molds

Mold	Undesirable Effect	Mold	Undesirable Effect
Mucor	'Whiskers' of beef	*Fusarium*	Production of mycotoxins
Rhizopus	Spoilage of bread	*Botrytis*	Spoilage in grapes
Aspergillus	Food spoilage, Production of mycotoxins	*Trichothecium*	Spoilage in apples, peaches cucumbers, cantaloupes
Penicillium	Soft rot of fruits	*Sporotrichum*	White spot on chilled meat

Desirable effects of mold are exhibited in manufacture of certain foods. For example, some kinds of cheese (Roquefort, Camembert) are ripened by *Penicillium roqueforti* and *Penicillium camemberti*, respectively. Oriental foods, for example, soy sauce and miso are prepared by using molds such as *Mucor* and *Aspergillus*.

21.3.3 Yeast and Yeast-like Fungi

Yeasts are fungi which are generally unicellular and oval or spherical in shape. They are non-filamentous and reproduce by budding or fission. Being non-pathogenic, they do not cause food-borne disease. Nevertheless, they can cause food spoilage and many a times are seen as powdery coating on plant foods. **Osmophillic yeast**, for example, *Saccharomyces rouxii* and *S. mellis* grow well in high osmotic pressure, that is, at high concentration of sugar, salt or other solutes. They cause spoilage in dry fruits, concentrated fruit juices, honey, jam, squash and chutneys. **Film yeasts** (*Pichia, Candida and Trichosporan*) grow on the surface of acid food, for example, sauerkraut and pickles. Some selected genera of yeasts resulting in food spoilage are listed in Table 21.6

Desirable yeast fermentations with *Saccharomyces* are carried out in the manufacture of foods such as bread, beer, wines and vinegar. *Saccharomyces cerevisiae* (baker's yeast) is used to ferment the bread dough (refer to Chapter 23).

Table 21.6 Undesirable effects of different yeasts

Yeast	Undesirable Effect	Yeast	Undesirable Effect
Candida	Spoilage of butter and margarine	*Rhodotorula*	Discolouration on meat and sauerkraut
Pichia	Spoilage of beef and wine	*Saccharomyces*	Spoilage of concentrated fruit juices, honey
Trichosporan	Spoilage of chilled beef	*Zygosaccharomyces*	Spoilage of honey, syrup and molasses

21.3.4 Viruses

The viruses are host specific and cannot grow or multiply outside their host. They must have access to a host to grow and multiply; therefore, they are termed as 'obligate intracellular parasites'. The viruses which attack bacteria are known as bacteriophages. Norovirus and Hepatitis A virus can cause food-borne illness (refer to Chapter 24). Nevertheless, they cannot grow on food and therefore do not cause its spoilage.

21.3.5 Parasites

Helminths responsible for food-borne illnesses are *Trichinella, Ascaris, Taenia* and *Anisakis* (refer to Chapter 24). Protozoans significant from food safety perspective are *Giardia, Entamoeba histolytica, Toxoplasma* and *Cryptosporidium parvum*. Generally, it is the cyst stage at which a protozoan is ingested through the contaminated food. The stage which causes illness is the active feeding stage (trophozoite). Some genera of parasites (including protozoans) which result in food infections are listed in Table 21.7.

Table 21.7 Harmful effects of different parasites

Parasites	Harmful Effect	Parasites	Harmful Effect
Trichinella	Trichinosis	*Giardia*	Giardiasis
Ascaris	Ascariasis	*Toxoplasma*	Toxoplasmosis
Taenia	Taeniasis	*Entamoeba histolytica,*	Amoebic dysentery
Anisakis	Anisakiasis	*Cryptosporidium parvum*	Cryptosporidiosis

In this chapter, we learnt about the characteristics of different microorganisms, their interactions with food and also how they are influenced by various intrinsic and extrinsic factors. In modern catering, food is handled by different people at different stages of production, distribution, processing and consumption. As a result, the chances of contamination by different microorganisms are high. A thorough understanding of causes of spoilage, sources of contamination and methods of preservation can go a long way in providing safe food to the clients. Next chapter (Chapter 22) will focus on these important areas. Chapter 23 will highlight the beneficial aspects of microorganisms in food preparation.

Know Your Progress

True or False Questions

1. Viruses have an outer protein layer known as capsid which contains DNA or RNA.
2. Yeasts are multicellular and filamentous organisms having hyphae and mycelium.
3. Anaerobes can grow in either the presence or the absence of free oxygen.
4. Avidin which binds with biotin is present in egg white, makes it unavailable for microorganisms and thereby prevents spoilage of eggs.
5. Osmophilic bacteria can grow in substrates of high osmotic pressure, for example, honey and jam.

Very Short Answer Questions

1. Differentiate between the following:
   (a) Mold and yeasts
   (b) Virus and protozoan
   (c) Anaerobic and facultative organisms
   (d) Psychotropic and thermophilic bacteria
   (e) Lactic acid and acetic acid bacteria

Short Answer Questions

1. Describe the morphological characteristics of different microorganisms found in food.
2. How do bacteria, mold and yeast reproduce?
3. Briefly discuss the role of water activity in microbial growth.
4. What are the nutrient requirements of microorganisms and how are they met?
5. List five groups of bacteria important in food along with their characteristic features and suitable examples.

Long Answer Questions

1. Discuss the various factors affecting the growth of microorganisms.
2. Describe the detrimental and beneficial role of different microorganisms in food.

Beyond the Text

- Search the website 'Microbiology in pictures' www.microbiologyinpictures.com, collect two pictures each of bacteria, yeast and mold. Study their characteristics and prepare a PowerPoint presentation to share with your class.

Chapter 22

Food Spoilage and Preservation

Chapter Objectives

After reading this chapter, the reader should be able to

- know the types and causes of food spoilage.
- identify different sources of microbial contamination.
- recognize signs and symptoms of spoilage in fresh and preserved food.
- comprehend the basic principles of food preservation.
- learn and apply different methods of food preservation.

Milk, meat, fish and poultry, and most fruits and vegetables, spoil rapidly within days, whereas butter, oils, honey, onions and garlic can be protected from spoilage for few weeks or months. Grains, flour, nuts and sugar do not deteriorate in months or years unless they are handled carelessly. For preventing spoilage and ensuring safety of various food items, different methods of preservation are employed. This chapter discusses the types and causes of spoilage in various foods along with main methods of preservation.

THINK TANK

Which of These Foods Will Spoil Faster?

Unrefrigerated raw milk	versus	Pasteurized milk kept under refrigeration
Fresh pineapple	versus	Canned pineapple slices
Fresh lemon juice	versus	Lemon squash
Fresh fenugreek leaves	versus	Dehydrated fenugreek leaves

Surely, the fresh foods in the first column will spoil faster than the preserved ones in the last column. Let us try to understand the reasons for spoilage and the principles of preservation of food.

Food spoilage can be defined as the undesirable changes in appearance, colour, flavour and texture of food which make it unacceptable for human consumption. Sour milk, mouldy bread, overripe fruits and mushy vegetables as well as insect-infested grains are examples of spoilt food. Ingestion of spoilt food is generally not harmful for human health unless disease-causing organisms are also present.

22.1 Types of Spoilage

From the time it is harvested or slaughtered, food is prone to undergo progressive spoilage through chemical reactions, microbial action or a combination of both. Spoilage is characterized by three important reaction viz. putrefaction, fermentation and oxidation. These may occur alone or in combination with each other leading to deterioration in quality of food and making it unfit for consumption.

Putrefaction refers to the decomposition of food which is associated with anaerobic breakdown o proteins and production of foul smell. Generally, it is caused by proteolytic and anaerobic bacteria such as *Clostridium* which produce enzymes to break proteins into simpler compounds (hydrogen sulphide ammonia, mercaptan, indole and amines). Many of these compounds contribute to foul smell, for example, in spoilt meat. Softening of texture and darkening of colour often accompany putrefaction.

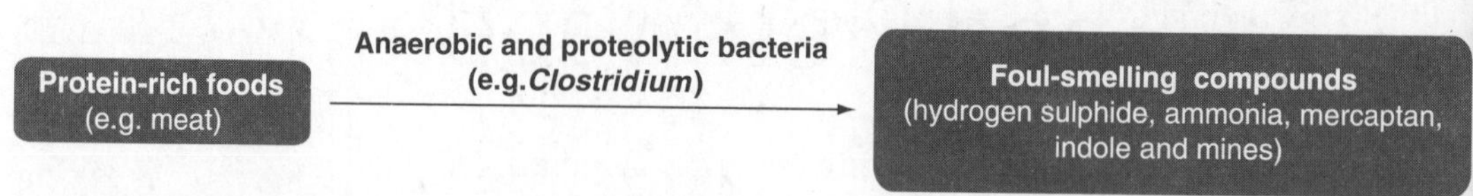

Fermentation is the process of breakdown of carbohydrates by microorganisms into simpler compounds such as organic acids, alcohol and carbon dioxide. The term is generally used for desirable fermentation carried out to prepare wine, beer, cider, bread, cheese, yogurt and sauerkraut (refer to Chapter 23). However, when fermentation is uncontrolled, it can lead to food spoilage, for example, souring of dough, curdling of milk and deterioration of fruit juices. Lactic acid bacteria (*Lactobacillus and Leuconostoc*) and yeasts (*Saccharomyces cerevisiae*) are saccharolytic microorganisms which are responsible for fermentation of simple sugars.

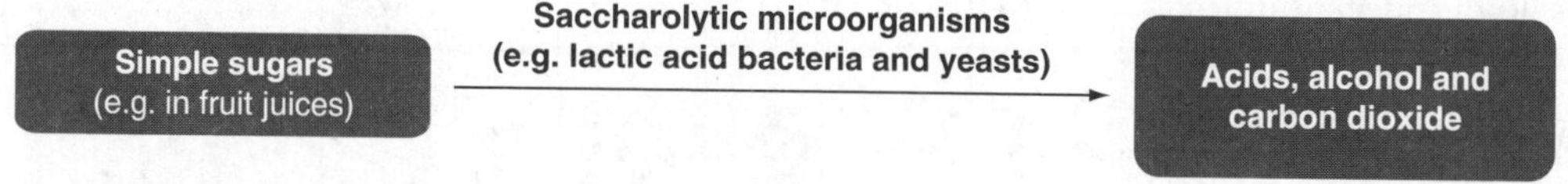

Oxidation was earlier recognized as the addition of oxygen to a compound but now it is defined as the loss of electrons by a molecule, atom or ion. Autoxidation has been observed in almost all foods of plant and animal tissues. This process gets accelerated upon rupturing of tissues as witnessed during primary processing. Mechanisms of oxidative deterioration in fats (refer to Section 3.4 in Chapter 3) and oxidative enzymatic browning in fruits and vegetables (refer to Section 10.1 in Chapter 10) have been discussed in detail in earlier chapters.

22.2 Causes of Spoilage

It is important to know the causes of food spoilage so as to understand and apply the principles and methods of preservation. Food spoils due to physical, enzymatic, chemical, microbiological and other causes (Figure 22.1). Generally, these changes do not occur in isolation. For example, microorganisms and insects can act simultaneously to cause deterioration in cereals.

Physical Causes: Bruising and breaking during harvesting, handling, processing, transportation as well as storage can result in deterioration of food. The cut surfaces of

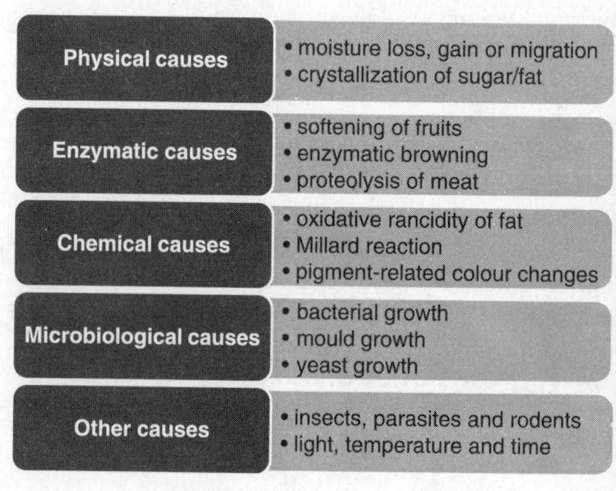

Figure 22.1 Causes of food spoilage

fruits and vegetables act as entry points for the microorganisms and also promote chemical and enzymatic changes. **Loss, gain or migration of moisture, crystallization of sugar or fat, breaking of emulsion** and **syneresis of gel** are other examples of physical changes (Table 22.1).

Table 22.1 Physical changes leading to food spoilage

Physical Change	Foods Commonly Affected
Bruising	Fresh fruits and vegetables
Breaking	Biscuits, potato chips, breakfast cereals
Moisture loss	Green leafy vegetables
Moisture gain	Sugar, salt, – coffee, milk powder, instant mixes
Moisture migration	Bread (staling)
Crystallization of sugar	Jam, syrups, preserves, chocolates
Crystallization of fat	Chocolates
Breaking of emulsion	Salad dressings, e.g., mayonnaise, margarine
Syneresis of gel	Custards, jam

Enzymatic Causes: Produced by living tissues (plants, animals and microorganisms), enzymes are proteinaceous compounds which catalyse chemical reactions. They remain in food even after their harvesting and slaughtering and are capable of causing food spoilage such as post-harvest excessive softening of fruits (by pectinase), browning in cut fruits and vegetables (by phenolase), and post-slaughter proteolysis of meat by protease.

Not all enzymatic changes are objectionable. Some examples of desirable ones include desirable softening of fruits during ripening (by pectinase), browning during processing of tea leaves (by phenolase), and breakdown of starch to maltose (food for yeast in bread making) by amylase.

GOOD TO KNOW

Properties of Enzymes

- Specialized proteins, produced by living things.
- Catalysts (increase the rate of a chemical reaction without themselves undergoing any change).
- Very specific (can act only on one type of substrate).
- Require optimum temperature and pH to act.
- Generally work best at body temperature (37 °C) and near neutral pH (~7).
- Slow down at low temperature, denatured at high temperature.
- Their name generally ends with the suffix 'ase'

Enzymatic action can be controlled by inactivation of enzyme through heat application, exclusion of oxygen, storage at low temperature, lowering the pH or by use of inhibiting chemicals.

Chemical Causes: There are many pure chemical reactions among food constituents which are not catalysed by enzymes.

1. *Oxidation of fats,* particularly the ones that are unsaturated, results in compounds which have undesirable flavours and odours. It causes deterioration not only in oils but also in foods containing oils, for example, fish, flour and fried food. It can be prevented by exclusion of air from food though controlled atmosphere packaging (CAP) and modified atmosphere packaging (MAP), replacement of oxygen by nitrogen or carbon dioxide, or use of reducing substances (refer to Section 3.4 in Chapter 3).

2. *Maillard reaction* is a chemical reaction between the amino group of an amino acid and the carbonyl group of a sugar. This non-enzymatic browning reaction leads to the formation of insoluble brown nitrogenous compounds e.g. (refer to Section 10.2 in Chapter 10). It can be controlled by lowering the cooking temperature, lowering pH, decreasing water activity and storage at low temperature (refer to sub-section 10.2.1 in Chapter 10).
3. *Pigment-related colour changes* in fruits and vegetables green pigment (*chlorophyll*) changing to dull olive-green pheophytin during processing or storage, orange and yellow pigments (*carotenoids*) getting bleached due to oxidation and red pigments (*anthocyanins*) turning blue at higher pH.

Microbiological Causes: Common food spoilage bacteria include *Pseudomonas, Acetobacter, Alcaligenes, Lactobacillus* and *Leuconostoc*. Among the moulds, *Rhizopus, Penicillium* and *Aspergillus* and from yeasts *Saccharomyces* and *Candida* can harm the sensory quality of food (refer to Section 21.3 in Chapter 21). Microorganisms are the most important cause of food spoilage. They can attack and decompose nearly all food constituents leading to fermentation of sugars, hydrolysis of starch and oxidation of fats. Some microbes digest proteins and produce rotten flavour, whereas others produce gas which renders a foamy texture to food.

Insects, Parasites and Rodents: *Insects* grow in foods such as cereal grains, legumes, fruits and vegetables particularly when the environment is warm and humid. They not only result in 5–50 per cent loss in the contaminated food but also make it more prone to microbial attack. Other undesirable changes include the by-products of their activities, for example, holes in grains and clumping of food particles. Insect infestation can be prevented by chemicals such as methyl bromide and ethylene oxide which are used for fumigating grains and spices.
Parasites infect humans through improperly cooked poultry and hogs. Use of raw human excreta as fertilizer for crops and infected water and poor sanitation/hygiene also helps in spreading parasitic infection. High temperature or cooking can kill these organisms. **Rodents** such as rats and mice contaminate food through their excretions (droppings and urine); they also carry pathogenic bacteria which can result in food-borne infections among human beings. Consumption of large quantity of food by rodents can lead to food shortage. Such losses can be prevented by proper pest proofing, sanitation and storage.

Light, Temperature and Time: The rate and extent of food spoilage is greatly affected by light, temperature and time. **Light** promotes light-induced oxidation reactions (e.g. fat rancidity) and has an adverse effect on some nutrients particularly vitamin A, riboflavin and ascorbic acid. **Temperature** plays an important role in preservation. However, in uncontrolled environment, excessive heat can lead to denaturation of proteins and destruction of vitamins. It increases the rate of chemical and enzymatic reactions and can lead to moisture loss. Microbial growth is favoured when the temperature of food is between 4°C and 63°C. Low temperature (freezing) and thawing in uncontrolled conditions can rupture the cellular structure of fruits and vegetables and thus affect their texture adversely. **Time** has a role in spoilage because food quality starts deteriorating soon after its harvest or slaughter. Therefore, animals, fish and poultry should not be slaughtered until required. Immediate cooling of plant produce after harvest and that of meat, fish and poultry after slaughter helps in delaying the progression of food spoilage.

22.3 Sources of Contamination

Microorganisms are present everywhere, for example, soil, water and air (Figure 22.2). However, they are present in negligible number in the inner tissues of food from plant or animal origin.

Plants: Type and number of microorganisms present on the surface depend on the plant itself and its environment. Though the number of microorganisms in the interior of plant foods is nil or very small, sometimes, coliforms, *Pseudomonas, Achromobacter* and yeasts are found in the inner tissues of healthy fruits.

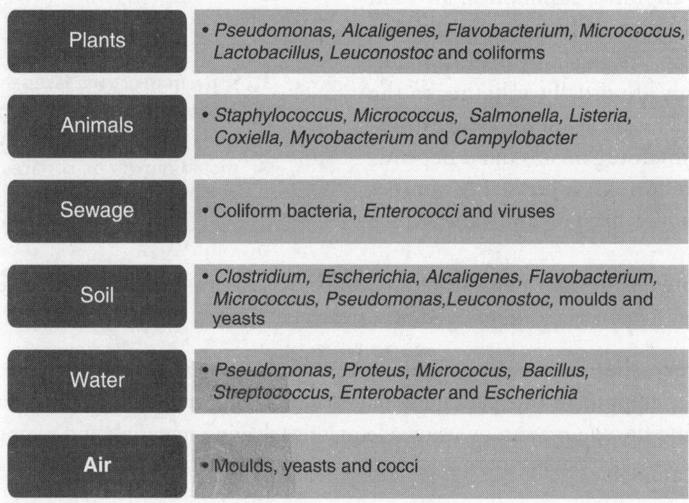

Figure 22.2 Contaminating microorganisms from different sources of food contamination

Animals: In case of animals, the surface flora and, more importantly, the flora of respiratory tract and intestinal tract are the sources of microbial contamination. The hides, hooves and hair of animals, and feathers and feet of birds are also loaded with undesirable microorganisms. Moreover, they release microorganisms through excretions and secretions. When animals are slaughtered and dressed, microorganisms from skin and respiratory tract get transmitted to the carcass and product.

Sewage: Many a times, untreated sewage is used as a fertilizer in food crops. This practice increases the risk of plant foods getting contaminated with disease-causing microorganisms. Use of night soil close to the shore can easily get contaminated with sewage.

Soil: It contains widest variety and largest number of microorganisms which contaminate the plants growing on them and animals roaming over the land. Soil particles are easily blown away by air currents and also carried by flowing water. Different types of moulds, yeasts and bacteria are found in soil. Washing and cleaning are processes that remove soil from the surface of plant and animal foods.

Water: It not only contains its natural flora but is also contaminated with microorganisms from soil, animals and sewage. Surface and stored water in lakes have more number of microorganisms than ground water which gets filtered while passing through several layers of rocks and soil. Water used for foods should be safe from a public health point of view, that is, it should be safe to drink (potable). Equally important is the economic aspect which emphasizes that the taste, odour, colour, clarity and chemical composition. Moreover, the microbial count should meet the standards.

Air: Microorganisms do not grow in air; they merely exist there. The ones which can survive low moisture conditions, live the longest. For example, the mould spores are dry and they exist for long periods in air. *Cocci* are more common in air than *bacilli*. Bacterial spores are uncommon in air that is free from dust. Yeasts are generally present in air. These microorganisms are usually adhering to the suspended solid particles or in the moisture droplets. The number of microorganisms in air depends on several factors. Air currents increase the number of microorganisms while dryness and rain reduce their number.

DID YOU KNOW?

Where Do Microorganisms in Air Come From?

Air does not have a natural flora of its own. The microorganisms present in air come from the following sources:

1. spray from oceans, lakes and streams.
2. moisture droplets from coughing, sneezing and talking.
3. mould growth on surfaces such as walls, ceilings, floor and food.

To remove the microorganisms from air artificially, it can be treated in a number of ways including filtration, washing by water spray, chemical treatment and heat treatment.

Handling and Processing: Beside the natural sources, food may get contaminated during handling and processing through the equipment, the packaging material and the personnel. Such contamination can be minimized by sanitizing the equipment, by using adequate packaging materials and maintaining personnel hygiene.

22.4 Spoilage of Different Products

Though foods are spoilt by many changes such as physical, chemical, enzymatic and microbial, the growth and activity of microorganisms is the most important cause of food spoilage. The type of spoilage varies among different food groups due to their different composition.

22.4.1 Milk and Milk Products

Milk is an excellent medium for growth of spoilage microorganisms because of its high moisture content and almost neutral pH (Table 22.2). Moreover, it is a good source of energy (in the form of milk sugar and milk fat), nitrogenous compounds (proteins, amino acids) and minerals.

Table 22.2 Spoilage in dairy products

Spoilage Symptoms	Microorganisms Involved
Souring of milk	*Streptococcus thermophilus/faecalis* and *lactobacilli*
Foam at the top of milk Gas bubbles suspended in curd Swelling of canned evaporated milk	*Clostridium, Bacillus*
Shrunken curd Separation of whey, bitterness	*Micrococcus, Bacillus, Streptococcus faecalis*
Ropiness in pasteurized milk/curd	*Alcaligenes*
Tallowy odours	*Pseudomonas, Proteus, Alcaligenes, Clostridium, Bacillus*

Liquid milk contains many microorganisms from different sources which result in acid fermentation of milk sugar, gas formation, proteolysis of milk protein and hydrolysis of butterfat. Among these, souring is the most common form of spoilage. Heat-resistant bacteria (*Streptococcus thermophiles/faecalis* and *lactobacilli*) survive through the pasteurization process and cause lactic acid fermentation in milk if it is stored at sufficiently high temperature. Ropiness is observed in milk contaminated with rope-forming *Alcaligenes* post-processing.

Milk products include dried milk powder, evaporated milk, frozen desserts and cream. Dried milk powder does not get spoilt unless it is handled inappropriately; but it can exhibit non-enzymatic browning due to the Maillard reaction. Canned evaporated milk can show signs of swelling by gas-forming anaerobic bacteria . In case of sweetened condensed milk, gas formation by yeasts, thickening by *Micrococci* and mould growth on the surface can be seen. Frozen desserts such as ice creams do not deteriorate as long as they are kept frozen. Cream may develop undesirable flavour due to the growth of bacteria, mould and/or yeast which is transferred to the butter prepared from it.

22.4.2 Cereals and Cereal Products

Cereals and seeds resist spoilage because of their low moisture content (10–12 per cent) and low water activity (<0.60). However, even a slight increase in the moisture content during harvesting, processing and storage permits microbial growth which is carried over to their products but does not usually harm the cereals per se. An important cause of spoilage is insect infestation which can lead to holes in the grains. Besides, stones, filth and dirt may be present. White wheat flour is more resistant to microbial spoilage as compared to other milled wheat products because it is usually bleached (by an oxidizing agent such as benzoyl peroxide) which reduces its microbial load considerably. In cereals, moisture content up to 13 per cent prevents microbial growth, 15 per cent facilitates mould growth, whereas 17 per cent allows mould, yeast and bacterial growth.

Cereal products include pasta, bread and other bakery products. Bread gets contaminated with mould from air (during cooling), handlers, wrappers or slicing blades. Bread mould (*Rhizopus stolonifer*) exhibits itself as white cottony growth with black dots. Rope-forming bacteria do not grow in commercially baked breads because of the added preservatives but may spoil home-baked breads. Staling of bread is a physical change due to the migration of moisture from crumb to the crust. Moist pasta and refrigerated doughs have high moisture content which favours bacterial growth and subsequent spoilage (Table 22.3).

Table 22.3 Spoilage in cereals and cereal products

Spoilage Symptoms/Adverse Effects	Microorganisms Involved
Production of mycotoxins in cereals	Mould (*Aspergillus, Penicillium, Mucor, Rhizopus, Fusarium*)
Cottony growth (white, greenish, purplish brown or black); yellowish pigment on bread	Mould (*Rhizopus, Aspergillus, Penicillium*)
Soft and sticky crumb (ropiness) in home-baked bread	Rope-forming bacteria (*Bacillus mesentericus*)
Swelling of moist macaroni (pasta)	Bacterial growth
Gas formation in refrigerated doughs	Bacteria (*Lactobacillus, Leuconostoc*)

22.4.3 Meat, Fish, Poultry and Eggs

Meat, fish, poultry and eggs have high water activity, are rich sources of proteins, minerals and vitamins and thereby provide ideal medium for the growth of microorganisms.

Meat spoilage can occur in both aerobic and anaerobic conditions. Spoilage caused by bacteria, mould and yeast under aerobic (presence of oxygen) conditions is summarized in Figure 22.3.

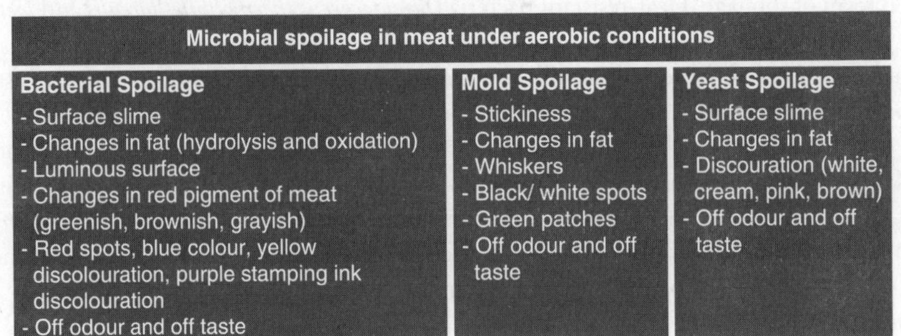

Figure 22.3 Spoilage in meat and meat products (under aerobic conditions)

In the absence of oxygen (anaerobic conditions), for example, in vacuum packed meat, both anaerobic and facultative bacteria grow and spoil meat in the following ways:

1. *Putrefaction* – It is associated with breakdown of proteins and foul smell production.
2. *Souring* – It is related to sour odour and sometimes sour taste caused by acetic acid, fatty acids (butyric acid) or organic acids (lactic acid).

For meat spoilage, though microbial changes are the most important, chemical changes in meat fat and enzymatic changes in meat proteins also play a significant role. For example, autolysis by enzymes is desirable in moderation for meat tenderization but when in excess, it causes souring.

Fish is also spoilt by autolysis, oxidation and bacterial growth but it is more perishable than meat because the autolysis by fish enzymes is faster and the unsaturated fish oils are more prone to oxidation than the meat fat. Lowering pH (e.g rubbing lemon juice) prevents bacterial growth in fish. Common signs of spoilage in fish are off odour, sunken eyes, grey gills, soft flesh and low rigidity. The spoilage of fish can be grouped in four phases (Figure 22.4).

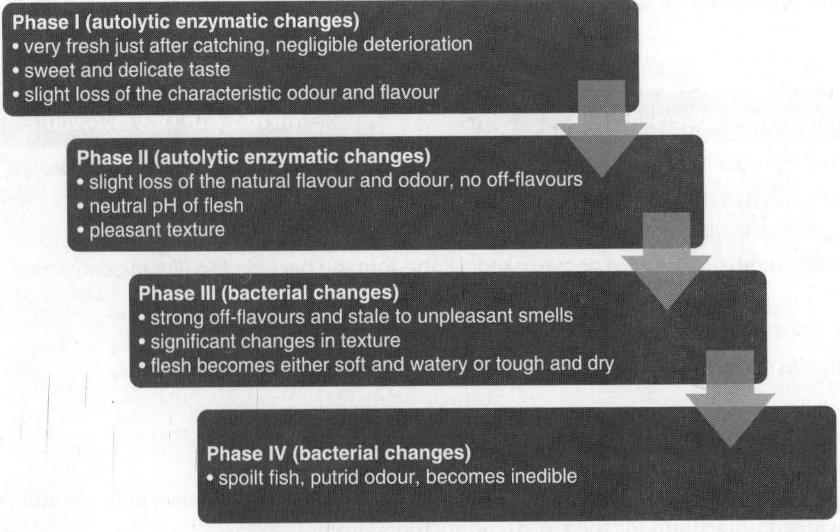

Figure 22.4 Four phases of fish spoilage

Poultry is contaminated with intestinal bacteria which grow on its surface and are the primary cause of its spoilage. Frozen and cut poultry can develop slime and sour flavour due to the growth of *Pseudomonas.*

Eggs are spoilt by physical changes (cracking/leaking shells, loss of bloom, stained shells, moisture loss). Candling (observing intact eggs against strong light) is a technique employed to test defects such as embryonic growth, blood clots, general bloodiness or translucent spots in the yolk. Moisture loss is exhibited by an increase in the size of air cell located at the inner blunt end of the shell. As egg stales, the white becomes thinner, the yolk membrane weakens, the yolk becomes more mobile and pH increase from 7.6 to 9.5. Microorganisms can enter the egg through the pores of its shell. Among them, bacterial spoilage is more common and is exhibited as green, red or black rot caused by *Pseudomonas, Serratia* and *Proteus,* respectively.

22.4.4 Fruits and Vegetables

Fresh vegetables and fruits provide good medium for microbial growth because they are high in moisture and generally rich in carbohydrates. **Vegetables**, being less acidic, are spoilt primarily by moulds (*Penicillium* and *Aspergillus*) and sometimes by bacteria (*Pseudomonas, Clostridium* and *Bacillus*). **Fruits** are more acidic and are attacked by moulds, yeasts (*Saccharomyces* and *Candida*) and aciduric bacteria (lactic acid bacteria and *Acetobacter*). Microbial spoilage in fruits and vegetables is often referred to as 'rot'. Examples of different types of rot are as following:

Bacterial soft rot	Black mould rot	Brown rot
Grey mould rot	Black rot	Sliminess or souring
Blue mould rot	Pink mould rot	Water soft rot
Downy mildew	Green mould rot	Rhizopus soft rot

Loss of moisture can also lead to undesirable changes such as shrivelling in green leafy vegetables. The outer skin protects whole fruits and vegetables, whereas peeling, cutting, shredding, slicing or crushing makes them more susceptible to spoilage. Even physical damage, for example, bruising or breaking leads to undesirable changes in their texture and colour. Enzymatic browning of damaged fruits and vegetables is also undesirable. Pigment-related colour changes are common during processing, for example, frozen green peas turn olive green and red colour of dehydrated carrots gets bleached.

22.4.5 Sugar and Sugar Products

Sugar (sucrose) is a poor medium for growth of microorganisms. However, it is hygroscopic and cakes in the presence of moisture; and thereby loses its organoleptic properties. Liquid forms of sugar such as molasses, syrup and honey and **sugar confectionery** (soft-centred fondant, chocolates and jellies) have water activity below 0.8. They are usually not spoilt by bacteria but can be fermented by some moulds and yeasts (*Saccharomyces, Zygosaccharomyces* and *Candida*).

When humidity is high, moisture condenses on the surface of chocolate which brings out sugar from inside to the surface. Fat crystallization is observed as a whitish, greasy haze on the chocolates. Crystallization of sugar is seen in jams, jellies, marmalades and concentrated syrups.

22.4.6 Canned Products

Heat processing which is fundamental in all canned foods kills practically all microorganisms. Spoilage in these foods occurs primarily due to non-microbial reasons, for example, chemical changes (production of hydrogen leading to hydrogen swell, browning, production of CO_2, corrosion of can) and enzymatic changes (liquefaction, gelation, discolouration). Table 22.4 summarizes the types of spoilage in canned foods.

Table 22.4 Types of spoilage in canned foods

Type of Spoilage	Spoilage Symptoms
Chemical	Hydrogen swell (production of hydrogen), Production of CO_2, Browning (metallic), Corrosion of can
Enzymatic	Liquefaction, Gelation, Discolouration
Microbiological	Flat sour spoilage (flat ends of can and souring) Sulphide stinker spoilage (production of hydrogen sulphide) Swelling of can

Microbial spoilage in canned foods is possible due to the following faults:

1. Insufficient heat processing permits the survival and growth of gas-producing mesophilic microorganisms (*Clostridium* and *Bacillus*).
2. Inadequate cooling after heating favours the growth of thermophilic spore formers which cause spoilage (flat sour spoilage, thermophilic anaerobe spoilage and sulphide stinker spoilage) particularly in low acid foods such as peas and corn.
4. Leaking cans lead to microbial contamination (bacteria, yeast or mould) from outside.

As a rule, puffed can should be discarded without tasting its contents. Sometimes, canned fruits develop black discolouration (ferric tannate) due to the reaction between iron (present in tin plate steel cans) and tannins (present in fruits). Another reason for black colour is the reaction between iron and sulphur dioxide (present in the sugar syrup of canned fruits) forming ferrous sulphide. Prolonged heat processing and/or delayed cooling of canned products result in undesirable changes such as overcooked texture and flavour.

22.4.7 Miscellaneous Foods

A chef deals with many other foods which may exhibit varying spoilage symptoms.

1. ***Fats and oils*** can spoil because of hydrolytic or oxidative rancidity.
2. ***Salad dressings*** are prone to microbial spoilage because of the oil, moisture and other ingredients. Growth of yeasts (*Zygosaccharomyces, Saccharomyces*) and surface moulds, separation of oil and water, fat changes, production of gas and off-flavour are some spoilage symptoms seen in salad dressings.
3. ***Bottled beverages*** are generally carbonated and have low pH which inhibits microbial growth. Nevertheless, cloudiness (by yeast and bacteria) and ropiness (by bacteria) are observed in non-acid beverages.
4. ***Spices*** have low moisture, but during the process of drying, the load of mould spores increases. Treatment with chemicals, irradiation and heat is employed to prevent mould growth in spices.
5. ***Shelled nuts*** can also get contaminated with bacteria and moulds.
6. ***Leftover foods*** Off odour, discolouration if stored for more than 36 hours in refrigerator they should not be used unless cooked again.

ACTIVITY

Help the Chef Choose an Option

A chef has to choose one option for each of the three food products at his disposal. Can you help him?

A. Sealed can of fish bulging from both the ends	Option 1 – throw it without tasting
B. Pack of frozen peas left out of freezer overnight	Option 2 – taste to check the spoilage
C. Left over vegetable curry kept refrigerated for 8 hours	Option 3 – heat the contents and use

22.5 Basic Principles of Food Preservation

Preservation is the process of using appropriate methods for preventing microbial, chemical and enzymatic changes in food. It prevents decay of food, extends its shelf life and ensures safety for human consumption (Figure 22.5).

The basic principles of food preservation are as follows:

1. Prevention or delay of microbial changes
   (a) by keeping out microorganisms (e.g. aseptic processing)
   (b) by removing the microorganisms (e.g. filtration)
   (c) by hindering the growth and activity of microorganisms (e.g. use of low temperature, removal of oxygen, drying or adding preservatives)
   (d) by killing the microorganisms (e.g. use of high temperature, irradiation)
2. Prevention or delay of self-decomposition of food
   (a) by destroying or inactivating endogenous enzymes (e.g. blanching)
   (b) by prevention or delay of purely chemical reactions (e.g. use of antioxidants)
3. Prevention of damage caused by insects, animals and physical causes

Figure 22.5 Some preserved fruit and vegetable products

22.6 Methods of Preservation

Based on the discussed principles, there are several methods of food preservation (Figure 22.6). Many a times, a combination of these methods is used for improving the effectiveness of preservation.

22.6.1 Asepsis

The principle of asepsis aims at preventing the entry of disease-causing and spoilage organisms. The hygienic practices followed during harvesting, handling, washing, cooking and processing help in keeping the microbial load minimum. Low microbial burden minimizes the use of chemical preservatives and the severity of heat treatment during the preservation process. As a result, the food is safer and the nutrients are better retained.

Aseptic packaging is one of the recent techniques of preservation in which aseptic food is placed in an aseptic container under aseptic environment. The packaging is so designed that it maintains aseptic condition until the seal is broken. This technique is used extensively for fruit juices and has several advantages over conventional thermal techniques. The product quality in terms of colour, flavour and nutritive value of the food is excellent; and the production costs are also less. However, the shelf life of aseptically packed foods is shorter than the commercially sterilized foods and this technique is unsuitable for products containing large particles.

Figure 22.6 Different methods of food preservation

22.6.2 Use of High Temperature

Heating may be considered as the foundation of food processing and preservation. It extends the shelf life of food by killing microorganisms through denaturation of proteins and inactivation of enzymes. It is an economical method which is safe too as it does away with the use of chemical preservatives. Moreover, it helps the product to acquire desirable flavour, colour and texture. At the commercial level, blanching, pasteurization and canning are the methods of preservation by heat (for details refer to Section 5.3.1 of Chapter 5).

Blanching is a pre-treatment usually carried out in vegetables prior to dehydration, freezing or canning to inactivate the enzymes (e.g. catalase, peroxidase, polyphenol oxidase). The vegetables are dipped in boiling water for 1–5 minutes. It can be carried out using hot water, steam or microwave and its efficiency is judged by peroxidase test. The rationale behind the selection of peroxidase as an indicator is that peroxidase is a heat-resistant enzyme so if it is inactivated, all other enzymes can be assumed to be destroyed.

Pasteurization is a mild heat treatment which is carried out at a temperature below 100°C followed by immediate cooling to kill the undesirable microorganisms (e.g. *Mycobacterium tuberculosis, Brucella abortus* and *Streptococcus pyogenes* in milk). It is an important heat treatment for processing milk and fruit juices. Since it preserves the food partially, it has to be followed by other preservation techniques such as refrigeration (of milk) and aseptic packaging (of fruit juices). Three processes used for pasteurization are low temperature long time (LTLT), high temperature short time (HTST) and ultra-heat treatment (UHT) pasteurization (refer to Box 5.3 in Chapter 5).

Sterilization is a severe heat treatment at a temperature of 121°C which results in complete destruction of microorganisms and also alters the organoleptic and nutritive quality of food adversely. Therefore, it is advisable to commercially sterilize the foods as is carried out in canning.

DID YOU KNOW?

What Is Phosphatase Test?

The adequacy of pasteurization can be determined by **phosphatase test**. Phosphatase is an enzyme which has a slightly higher heat resistance than *Mycobacterium tuberculosis* (tuberculosis-causing bacterium present in milk). Negative phosphatase test ensures that this bacterium is also destroyed.

Commercial sterilization refers to the degree of sterilization which destroys all disease-causing, toxin-forming and spoilage-causing organisms. For example, most canned and bottled foods are commercially sterile. Heat-resistant spores, if present, do not usually multiply due to the vacuum and high acidity in the canned foods. The main objective of heat processing during canning is to destroy *Clostridium botulinum* and its spores; but the degree and time of heat treatment depends upon the size of can, nature of food and its pH. Heat penetration is better in a smaller can and at a lower pH. For preservation purposes, acidic fruits are processed at a boiling temperature (100°C), whereas vegetables being alkaline need a higher temperature (121°C).

Heat preservation should be used judiciously because it increases the risk of overcooking, cooked flavour, disintegrated texture and the loss of heat-labile nutrients particularly thiamine and ascorbic acid.

22.6.3 Use of Low Temperature

The underlying principle of low temperature preservation is that it decreases the rate of chemical and enzymatic reactions and controls the replication and propagation of microorganisms. It extends the useful storage life of food and also protects the nutritive value, colour, flavour and texture (Table 22.5). Based on the principle of use of low temperature, refrigeration and freezing are the two methods which are used extensively in preservation industry and in catering units.

Refrigeration temperature permits slow microbial activity as it does not usually destroy the microorganisms; but it prevents their replication and thereby delays spoilage. The effect the sensory quality and nutritive value of food is mild. For cold storage, the requirements are controlled low temperature, air circulation, humidity control and modification of gas atmosphere. All food handlers should start refrigeration of foods (e.g milk, fruits, vegetables, meat, fish and poultry) immediately after slaughter or harvest and maintain it throughout transportation and storage to preserve their quality and safety.

Table 22.5 Extension in useful storage life of fresh food at low temperature

Food	Useful Storage Life	
	At 22°C	At 0°C
Fruits	3 weeks	Up to 3 months
Leafy vegetables	1–7 days	Up to 3 weeks
Root vegetables	1–7 weeks	3 months to 1 year
Meat	1 day	1 week to 10 days
Fish	1 day	Up to 1 week
Poultry	1 day	Up to 2 weeks

Freezing preserves by converting water into ice and making it unavailable for microbial and enzymatic activity. As a result, the microbial growth is halted completely, whereas the enzymatic activity is slowed down. Appropriately frozen foods such as vegetables, vegetarian snacks, meat and meat products have a shelf life of many months when packaged properly and kept in frozen storage. Freezing has made available a large variety of convenience foods including the completely cooked frozen foods which can be served just after a single thawing-heating operation. It is wiser to use microwave thawing which is not only faster but also preserves the quality and safety of food being thawed. Three basic methods of commercial freezing are freezing in air, freezing by indirect contact with the refrigerant and freezing by direct immersion in a refrigerating medium. Frozen storage at a temperature of –18°C or less forms an integral part of the freezing process. The chef should know that repeated freezing and thawing can affect the food quality adversely. Moreover, an efficient cold chain during processing, storage, distribution and sale of food should be maintained to avoid resumption of microbial growth and the subsequent food spoilage.

22.6.4 Removal of Moisture

Drying or removal of moisture is possibly the oldest method of food preservation practiced by man. The availability of water in food is expressed in terms of **water activity** (a_w). It is water activity (a_w) rather than total moisture content that affects growth, death and survival of different microorganisms (refer to Box 5.5 in Chapter 5). Traditional processes based on moisture removal include sun drying, dehydration and concentration.

ACTIVITY

Match the Following

Low Moisture Foods (*LMF*, $a_w < 0.60$)	Minimally processed foods (fresh-like fruits and meat products)
Intermediate Moisture Foods (*IMF*, $a_w = 0.60–0.90$)	Milk powder, potato chips, instant mixes
High Moisture Foods (*HMF*, $a_w > 0.90$)	Pickles, jams, marmalades, preserves

Sun drying is still practised in many parts of our country. But it has many limitations. It is a slow process limited to hot and dry climate and is suitable for few fruits (grapes, raw mangoes, figs, dates) and vegetables (potato, peas, leafy vegetables, onion, garlic).

Dehydration leads to almost complete removal of moisture from food under controlled conditions of temperature, humidity and air velocity in special equipment (dehydrators). Drying preserves food, prolongs its shelf life, decreases its bulk and weight as well as the transportation and packaging costs. It has also made available many convenient foods such as instant food mixes. Based on the type of dehydrator used, drying is classified as air convection drying, drum drying, vacuum frying and freeze drying. The sensory attributes of food are preserved better with freeze drying and vacuum drying. Since dried foods are hygroscopic (tendency to absorb moisture), they should be packed properly and kept at dry place.

GOOD TO KNOW

Vacuum Drying and Freeze Drying for Heat Sensitive Foods

Use of vacuum lowers the total pressure allowing the moisture to be removed at a comparatively lower temperature. Freeze drying uses the principle of sublimation where water is converted from its solid form (ice) to gaseous form without coming into the liquid state. The quality of vacuum-dried and freeze-dried products is excellent.

Concentration is employed for liquid foods such as milk, juices and purees. Generally, all fruit juices are concentrated prior to drying. Concentration decreases the moisture content by about 20 per cent but it is not a complete method of preservation and has to be followed by some other preservation technique. For example, evaporated milk is sweetened with sugar and canned.

22.6.5 Use of Preservatives

Preservatives are food additives which are intentionally added to inhibit, retard or arrest the growth of microorganisms and thereby prevent fermentation, acidification or other decomposition of food (refer to Section 25.3.1 of Chapter 25). They can be categorized as follows:

1. **Class I preservatives** (e.g. salt, sugar, citric acid, vinegar and spices) are safe for human consumption and can be added in any amount to the food. High concentration of sugar in jams, jellies and marmalade and that of salt in sauerkraut and pickles is responsible for their preservation.
2. **Class II preservatives** are chemicals (such as benzoic acid and benzoates; sulphur dioxide and sulphites; nitrates and nitrites) which are permitted in foods within the prescribed limits (refer to Table 25.2 in Chapter 25).

The rationale behind using chemical preservatives is to stop the growth and activity of microorganisms by interfering with the cell membranes, the genetic mechanism or the enzyme activity. The use of Class II preservatives is governed by Food Safety and Standards Act, 2006, so they should be used judiciously.

GOOD TO KNOW

Properties of Ideal Preservative

An ideal chemical preservative should have the following properties. It should

1. destroy the microorganisms rather than just inhibiting their growth.
2. be non-toxic and safe for human beings.
3. not get inactivated by the food or its constituents.
4. not bring adverse changes in the appearance, flavour, taste or aroma of the food.
5. be cost effective and economical.
6. not lead to the development of resistant strains of microorganisms.

22.6.6 Modification of pH

Every microorganism grows best at an optimal pH. Its growth and activity is inhibited if the pH is lowered or increased. Since very few food items (e.g. eggs) are edible at a pH exceeding 9, acidification or decreasing pH is used for preserving the food. Acidity can be increased either by adding organic acids or by producing organic acids through fermentation.

Adding **organic acids** such as citric acid (in squashes) and acetic acid (in ketchup) alone cannot achieve the desired safety and stability of food. Therefore, it is combined with other methods of preservation, for example, with addition chemical preservatives such as potassium metabisulphite in squashes and sodium benzoate in ketchup.

Fermentation is defined as the process of breakdown of carbohydrates by microorganisms into simpler compounds such as organic acids, alcohol and carbon dioxide. Alcoholic and acetic acid fermentation are used for preparing and preserving alcoholic beverages. Sauerkraut and pickles are preserved by partial lactic acid fermentation and high concentration of salt (refer to sub-section 23.2.2 in Chapter 23).

22.6.7 Irradiation

Food irradiation works upon the principle that ionizing radiations interact with food material by transferring energy to the electrons and ionizing molecules by creating positive and negative charge (ions). The food (either pre-packaged or bulk) is exposed to ionizing rays such as gamma rays to kill bacteria, insects and sometimes to prevent sprouting. Despite absorption of some radiation, the food itself does not become radioactive but it may lead to some chemical and physical changes. Therefore, the dose of radiation should be such that the preservation of food is achieved without harming its quality. The Codex Alimentarius Commission has recommended a maximum limit of 10 kGy as dose for food irradiation. The advantages and disadvantages of irradiation have already been described in Chapter 5 (refer to sub-section 5.3.5).

GOOD TO KNOW

Methods of Radiation

1. **Radappertization** leads to destruction of nearly all microorganisms through very high dose of radiation (25–45 kGy). It results in adverse changes in the quality attributes of the irradiated food and is, therefore, equated with commercial sterilization.
2. **Radicidation** kills non-spore forming bacteria particularly *Salmonella* but spores of *Clostridium botulinum* are not destroyed. The dose of radiation is low (2–8 kGy), the sensory quality (colour, flavour and texture) of food is not affected adversely and this process is equated with pasteurization of milk.
3. **Radurization** uses low dose of radiation (0.4–10 kGy) to destroy most of the harmful microorganisms. It extends the shelf life of food and is equated with pasteurization.

Despite many developments in the field of food preservation, there is no single procedure that alone can be considered as satisfactory in preventing food spoilage. Many a times, a combination of preservation procedures is used for increasing the effectiveness. When foods are preserved by a combination of procedures, the process is referred to as **'hurdle technology'**. For example, combining low pH with other preservation techniques such as preservatives, water activity, low temperature and/or oxygen availability is gaining popularity for hurdle technology in minimally processed foods. Similarly, preservation of fruit beverages with chemical preservatives is combined with addition of acid to lower pH. The future trends in food preservations include the use of high hydrostatic pressure, pulsed electric field, manothermosonication and aseptic packaging.

In this chapter, we studied that food spoils due to microbial, chemical, physical or other causes including insects, parasites and rodents. Among the food groups, milk, meat, fish, poultry, fruits and vegetables provide good medium for growth of microorganisms, whereas cereals, sugars and canned foods are less prone to spoilage. We learnt that different methods of food preservation are based on some basic principles such as preventing or delaying either the microbial reactions or the self-decomposition of food. The next chapter focuses on fermentation which is the only method of preservation that encourages microbial growth.

Know Your Progress

True or False Questions

1. Loss, gain or migration of moisture in foods is a chemical change.
2. Acid and heat accelerate the conversion of green colour of chlorophyll to olive green.
3. The exposed surfaces of fruits and vegetables harbour more contamination than the inner tissues.
4. Low temperature preserves food by killing microorganisms through denaturation of proteins and inactivation of enzymes.
5. Vacuum drying and freeze drying are used for heat-sensitive foods.

Very Short Answer Questions

1. Differentiate between the following:
   (a) Putrefaction and fermentation
   (b) Class I and Class II preservatives
   (c) Physical and chemical changes of spoilage
   (d) Desirable and undesirable enzymatic changes
   (e) Contamination from soil and contamination from air

Short Answer Questions

1. What are the different sources of food contamination?
2. Compare the spoilage of milk and milk products with that of sugar and sugar products.
3. How do canned foods get spoilt?
4. Describe ways through which cereals and cereal products deteriorate.
5. List the various methods of preservation. Describe preservation by high temperature in detail.

Long Answer Questions

1. What are the various causes of food spoilage?
2. Describe the spoilage of different food groups with particular focus on microbial changes.
3. What are the basic principles of preservation? Discuss the different methods of preserving foods.

Beyond the Text

- Identify the method(s) of preservation used in the following foods:

1. Pineapple jam
2. Kasoori methi
3. Apple juice in a tetrapak
4. Mango pickle
5. Lemon squash
6. Tomato ketchup
7. Milk powder
8. Ice cream

Chapter 23

Beneficial Role of Microorganisms

Chapter Objectives

After reading this chapter, the reader should be able to:

- learn the beneficial role of microorganisms in the fermentation process with particular emphasis on lactic acid bacteria.
- identify different foods preserved by the process of fermentation.
- understand and apply methods of preparing some fermented food products.

A chef in an Italian restaurant uses pizza base, mozzarella cheese, pepperoni and olives, whereas another chef in an oriental restaurant uses soy sauce, tamari sauce, miso and tempeh. Another chef in a South Indian restaurant prepares idli and dosa on a regular basis. A wide range of rum, whiskey, wine and beer are available at the bar in a five star hotel. What do these ingredients, foods and beverages have in common? They are all examples of fermented foods which not only have pleasant flavour, aroma and texture but are wholesome too. They are highly valued for their sensory properties and for their role in cooking and processing.

In contrast to other methods of preservation which prevent microbial growth, fermentation is a process that encourages replication of microorganisms with the objective of bringing desirable changes in the food. It is an energy efficient method of preservation; therefore, its relevance is high in developing countries where advanced equipment for preservation is lacking, particularly in underprivileged areas (FAO, 2011). Many fermented foods, for example, pickles, achar, idli and kimchi are traditionally prepared at household level in different parts of the world. This chapter attempts to explain the beneficial role of microorganisms in food fermentation and the method of preparation of some fermented foods.

THINK TANK

Beneficial microbial changes in foods

Consider the following examples:

- Setting of milk to produce yoghurt.
- Fermentation of dough in bread making.
- Fermentation of rice and black gram batter for preparing idli.
- Curing of cabbage to prepare kimchi.
- Curing of pork to produce salami.
- Fermentation of wheat flour batter to prepare jalebis.

These are desirable microbial changes brought through the fermentation process.

Beneficial microorganisms, if consumed in optimum amounts, have health benefits. They help in preventing the growth and multiplication of harmful microorganisms. For example, lactic acid and the lactic acid bacteria present in fermented milks inhibit the growth of disease-causing organisms in the gut. They are also useful in preservation of food, e.g., some bacteria produce nisin which is used as a preservative in *paneer* and canned *rasgulla* (refer in sub-section 25.3.1 of Chapter 25).

DID YOU KNOW?

Probiotic Yoghurt and Its Health Benefits

Probiotics are live microorganisms which when administered in adequate amounts confer a health benefit to the host. Their health benefits include tolerance to infections, prevention of gut cancer, reduced cholesterol, improved immune system and controlled diarrhoea and allergy. The most commonly consumed probiotic product is yoghurt. The probiotics added to yoghurt are *Lactobacillus acidophilus* and *Bifidobacterium* spp. both of which are resistant to acid and bile salts to survive the stomach passage.

23.1 Fermentation and Role of Lactic Acid Bacteria

In food processing, fermentation is best described as the conversion of carbohydrates to alcohols and carbon dioxide or organic acids by yeasts, bacteria or a combination there of under anaerobic conditions. The following are some examples:

1. Yeasts are used for fermenting bread, beer, wine and distilled liquors (rum and whisky).
2. Bacteria are utilized for preparing fermented milks, fermented vegetables and sausages.
3. Mold help in fermentation of soybeans to produce miso, tamari sauce and soy sauce.
4. Yeast and bacteria are used in the preparation of vinegar and for fermenting cacao beans.
5. Bacteria, yeast and mold are used in the preparation of soy sauce.

Most of the food fermentations are brought about by more than one type of microorganism which either work together or act in a sequence, but their growth is generally initiated by bacteria, followed by yeasts and then moulds. Based on the source of microorganisms, foods can be fermented in three ways as follows:

1. *Natural fermentation* utilizes the natural flora of raw material. For example, the naturally present microorganisms on the leaves of cabbage are used in the preparation of sauerkraut.
2. *Back sloping* makes use of small amount of product from successful fermentation of earlier lot. This method is used for preparing sour dough bread and also for setting of dahi.
3. *Controlled fermentation* utilizes a pure culture of single or mixed species of microorganisms to inoculate the starting material. For example, yeast is used for dough fermentation in bread making.

Lactic Acid Bacteria (*Lactobacillus, Leuconostoc, Pediococcus* and *Streptococcus*) are the most important bacteria in desirable food fermentations. They are responsible for the fermentation of most pickled vegetables, fermented milks, cheese and sausages. They are a diverse group of bacteria which ferment sugars anaerobically (without the need for oxygen) to produce lactic acid. The lactic acid so produced inhibits the growth of undesirable organisms. The lactic acid bacteria can be categorized into two groups as follows:

Homofermenters (*Lactobacillus bulgaricus and Lactobacillus thermophilus*) produce mainly lactic acid. They produce high acidity, for example, in pickled vegetables.

$$\underset{\text{Sugar}}{C_6H_{12}O_6} \rightarrow \underset{\text{Lactic acid}}{2C_3H_6O_3}$$

Heterofermenters (*Lactobacillus fermentum*) produce about 50 per cent lactic acid plus 25 per cent acetic acid and ethyl alcohol plus 25 per cent carbon dioxide. They produce less acidity but impart characteristic flavour to the fermented food.

$$\underset{\text{Sugar}}{C_6H_{12}O_6} \rightarrow \underset{\text{Lactic acid}}{C_3H_6O_3} + \underset{\text{Ethyl alcohol}}{C_2H_5OH} + \underset{\text{Carbon dioxide}}{CO_2}$$

Leuconostoc mesenteroides initiates desirable fermentation in sauerkraut and pickles. It is different from other lactic acid bacteria because it can tolerate high concentrations of salt and sugar (up to 50 per cent sugar).

23.2 Fermentation in Foods

Diverse variety of food is fermented in different parts of the world. It is beyond the scope of this chapter to cover them all. Therefore, some of them are being discussed under the food categories listed in Figure 23.1.

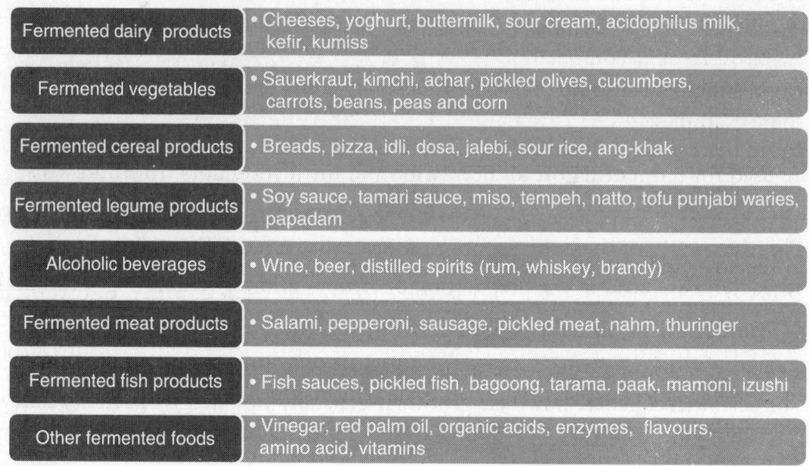

Figure 23.1 Categorization of different fermented foods

23.2.1 Dairy Products

Fermented dairy products comprise of a large variety of fermented milks and cheeses. The **fermented milks** include Bulgarian buttermilk, cultured buttermilk, acidophilus milk, kefir, kumiss, skyr and taette (Figure 23.2). Kefir is prepared from kefir grain, whereas kumiss is made from mare's milk. Skyr is semi-solid and taette is ropy buttermilk. Sour cream and cultured buttermilk use mixed cultures; one strain is for producing lactic acid (*Streptococcus lactis* subsp. *cremoris*) and another is for the aroma (*Leuconostoc mesenteroide* subsp. *cremoris*).

Yoghurt is prepared by fermenting milk with a mixture of *Streptococcus thermophilus* and *Lactobacillus bulgaricus*. Very often, these two bacteria are co-cultured with other lactic acid bacteria (probiotics) for improving the taste or health effects. The acid produced by fermentation causes the milk protein (casein) to coagulate into a semisolid curd. Sometimes, fruits are added to the yogurt after fermentation to prepare fruit yogurt.

Cheese is one of the oldest foods developed nearly 8000 years ago. There are approximately 2000 varieties of cheese prepared by using different milk types, starter cultures, fermentation, ripening microorganisms and maturing conditions. Cheese is primarily prepared by lactic acid fermentation of milk resulting in coagulation of milk proteins to form a curd. The liquid separated is known as the whey. *Streptococcus lactis* is the bacterium used for major cheeses. The curd is pressed and ripened to acquire flavour and texture specific to the cheese. Unripened cheeses (e.g. cottage cheese and cream cheese) are prepared by starter microorganisms similar to those used in buttermilk. They have short keeping quality and must be kept chilled until consumed.

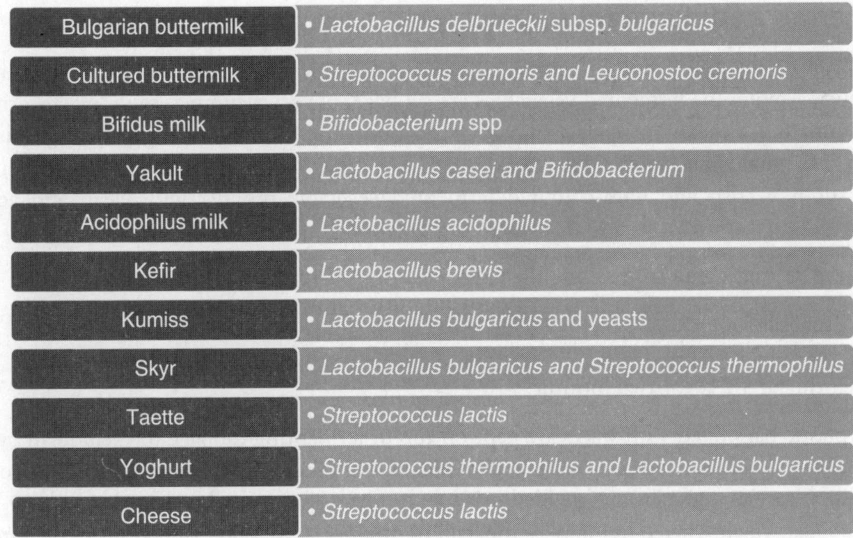

Figure 23.2 Microorganisms responsible for fermented milk products

Ripened cheeses are fermented initially by lactic acid bacteria and thereafter acted upon by enzymes and other microorganisms. In some cheeses, rennet enzyme is also added. The flavour of ripened cheese results from salt, lactic acid, fatty acids, amino acids, aldehydes and ketones. Cheeses are broadly classified as hard, semi-soft and soft cheese.

1. *Soft cheeses* are generally prepared without a starter culture; instead, the natural flora of raw milk is permitted to carry out fermentation, for example, in Mexican queso blanco.
   - Camembert and Brie undergo short fermentation.
   - Feta cheese is pickled in brine.
   - Mozzarella and Cottage cheeses are not ripened at all.
2. *Semi-soft cheeses* are not pressed; they are simply moulded into wheels, are allowed to dry naturally and undergo short maturation process.
   - Brick, Munster and Gauda are ripened principally by bacteria.
   - Limburger, Port du Salut and Trappist are ripened by bacteria and surface microorganisms.
   - Blue, Stilton, Roquefort, Gorgonzola and Wensleydale are ripened principally by blue mold in interior.
3. *Hard cheeses* undergo long maturation process; they generally have highest solids and lowest moisture content. They are pressed to expel the whey and form characteristic wheels or cylinders.
   - Cheddar, Granular and Caciocavallo are ripened by bacteria without eyes.
   - Emmental, Swiss and Gruyere are ripened by bacteria with eyes.
4. *Very hard (grating) cheeses* include Asiago old, Parmesan, Romano, Sapsago and Spalen.
5. *Whey cheeses* comprise of Ricotta, Primost and Mysost.

23.2.2 Fermented Vegetables

Fermentation not only preserves the vegetables but also brings desirable chemical and/or textural changes in the product. Among the fermented vegetables are cabbage, olives, cucumber, cauliflower, carrots, beans, peas, corn, tomatoes, okra and peppers. Cabbage is the most widely fermented vegetable which is known as sauerkraut (in Germany) and kimchi (in Korea).

Sauerkraut is a crisp and light-coloured fermented vegetable product with clean and acid flavour (1.7 per cent acidity and pH 3.4–3.6). Its processing involves the following steps:

1. Washing, trimming and shredding the cabbage.
2. Stacking tightly to exclude air and layering with salt (2–3 per cent by weight) in large containers.
3. Covering with a weight to seal the surface.
4. Fermenting at 18°C for two months.

Nowadays, plastic bags filled with water are kept on top of fermenting cabbage to exclude air and prevent growth of aerobes (mold and surface yeasts). The salt prevents the growth of undesirable microorganisms and results in release of water from shredded cabbage to form a salt solution (brine). Cabbage leaves also release some sugar which is acted upon by lactobacilli to produce lactic acid. Sauerkraut so prepared has low pH and high salt concentration both of which help in its preservation. Usually, no starter culture is added; instead, natural flora of cabbage leaves results in lactic acid fermentation.

Kimchi is also prepared using shredded cabbage and salt, but the fermentation period is short, ranging from 2 to 3 days at 20°C (Figure 23.3a). Other vegetables are also fermented in a similar manner to facilitate lactic acid fermentation.

Pickles such as cucumber pickle may be unfermented, partially fermented or fully fermented. Among the fermented pickles are salt or salt-stock pickles and dill pickles.

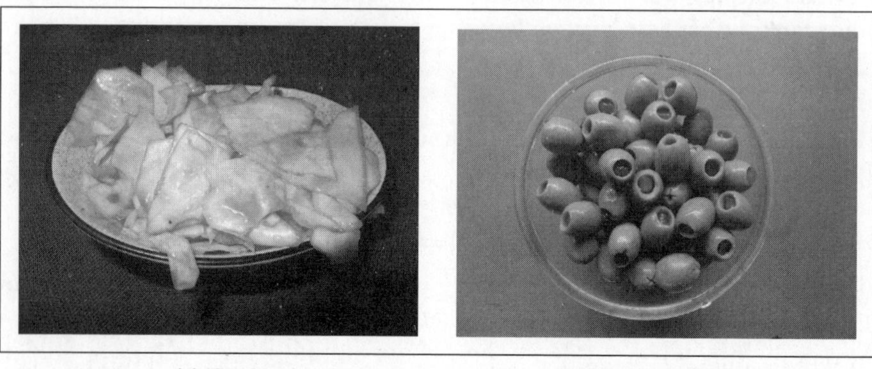

(a) Kimchi (b) Olives with pimento

Fig 23.3 Fermented vegetable products

1. For preparing salt or *salt-stock pickles*, immature cucumbers are washed, kept in barrels and brined. The traditional fermentation is completed in about six to nine weeks depending on the salting method and the temperature used. In case of controlled fermentation, special cultures (*Pediococcus cerevisiae* and *Lactobacillus plantarum*) are used for inoculation followed by active fermentation for 10–14 days.

2. *Dill pickles* are flavoured with dill herb along with other spices. Overnight dill pickles employ slow acid fermentation at low temperature in a rather weak acidified brine. Genuine dill pickles are prepared by using higher concentration of salt along with vinegar to increase the acidity and prevent undesirable fermentation.

Green olives are fruits but their method of fermentation is similar to that of vegetables. Their processing involves the following steps:

1. Harvesting fully developed olives (green or straw yellow in colour), taking care to prevent bruising.
2. Treatment with lye solution until the latter penetrates half to three-fourth of the fruit towards the pit.
3. Washing repeatedly without exposure to air to remove the lye; subsequently,most of the bitterness is removed.
4. Keeping olives in barrels, covering with salt brine and fermenting for 6–10 months.

Olives can be pitted and stuffed with brined pimento before barrelling and brining (Figure 23.3b).

23.2.3 Cereal Products

The fermented cereal products include bread, pizza, idli, dosa, jalebi, sour rice, miso and ang-khak. **Bread making** involves leavening of the dough usually by baker's yeast (*Saccharomyces cerevisiae*). The fermentation process which involves breakdown of simple sugars to ethanol and carbon dioxide serves the following functions in bread making:

1. It leavens (raises the volume of dough) and helps the bread in acquiring a loose and porous (spongy) texture.
2. It conditions the dough by making flour proteins (gluten) more elastic so that the dough is able to hold the carbon dioxide produced by yeast. Proteolytic enzymes produced by yeast and decrease in pH by the acids formed help in conditioning the dough.

Sometimes, dough conditioners such as potassium bromate, potassium iodate, calcium oxide and ammonium sulphate are added to improve the characteristics of dough. Different methods employed for dough mixing are as follows:

1. *Straight dough method* involves mixing all the ingredients and carrying out fermentation for about 30 minutes at about 27°C.
2. In case of *sponge method*, a part of flour is mixed with all the yeast and fermented to the desired maturity before adding the rest of the flour and resuming fermentation.
3. The *sour dough process* uses a portion of dough from the previous batch to inoculate the next batch. Lactic acid bacteria are the major microorganisms involved in sour dough fermentation.
4. In *continuous bread making*, more yeast is added along with other supporting ingredients. The leavening may be accomplished in the baking tins prior to baking.

Bread is baked at a high temperature (nearly 240°C) which kills yeasts, inactivates enzymes, increases the volume and sets the gluten and the structure. The starch gelatinizes and contributes to the structure of baked bread along with gluten. Most of the ethanol evaporates during baking of the bread. The flavour of bread results from the ingredients used, the Maillard reaction and the products of fermentation. Similar process is used for preparing pizza base, kulcha, bhatura and naan.

Idli is a fermented and steamed South Indian food which is soft and spongy. It is prepared from rice and dehulled black gram by the following process:

1. Soaking washed bengal gram dal and rice separately in water overnight at ambient temperature.
2. Grinding dal coarsely and the rice to a smooth paste, mixing them and adding salt (0.8 per cent).
3. Fermenting for 15–24 hours and steaming in small moulds.

Dosa, dhokla and appam are foods fermented in a manner similar to idli (Figure 23.4).

GOOD TO KNOW

Is Eno a Substitute for Yeast?

Eno fruit salt is a product containing sodium bicarbonate, citric acid and sodium carbonate. It is often used to leaven naan, bhatura and dhokla, particularly when yeast is not available or when there is lack of time for fermentation. Similar to fermenting microorganisms, Eno performs the twin function of producing carbon dioxide (for making the product spongy) and of lowering the pH (for making the dough elastic). However, it does not enhance the nutritive value of food.

(a) Idli (b) Dhokla

Figure 23.4 Some fermented cereal/pulse products popular in India

Jalebis are prepared by frying fermented wheat flour in spiral shape and dipping them in sugar syrup.

Ang-khak, also known as Chinese red rice, is used for colouring and flavouring foods such as fish. It is produced by growing *Monascus purpureus* on autoclaved rice (Table 23.1).

Table 23.1 Microorganisms responsible for fermenting cereals

Fermented Cereal Product	Microorganism Involved
Bread	Baker's yeast (*Saccharomyces cerevisiae*)
Idli, dosa	*Leuconostoc mesenteroids, Streptococcus faecalis, Pediococcus cerevisiae*
Ang-khak	*Monascus purpureus*

23.2.4 Legume Products

Among legumes, soybeans are the most often fermented to produce a variety of products such as soy sauce, tamari sauce, miso, tempeh, natto and tofu (Table 23.2).

Soy sauce is a major oriental food which is a brown, salty and tangy sauce. The method of preparation involves the following steps:

1. Moistening a mixture of autoclaved soybean, cracked wheat and wheat bran.
2. Inoculating it with a mold (*Aspergillus oryzae*) and holding at 30°C for 3 days.
3. Soaking with sterile 24 per cent brine.
4. Fermentation for a period of 2.5 months to 1 year.

Table 23.2 Microorganisms responsible for fermentation of legumes

Fermented Legume Product	Microorganism Involved
Soy sauce, Miso	*Aspergillus oryzae*
Tamari sauce	*Aspergillus tamari*
Tempeh	*Rhizopus (R. stolonifer, R. oryzae)*
Natto	*Bacillus natto*
Tofu	*Mucor*

Tamari sauce is similar to soy sauce but is prepared from a mash of soybean and rice which is fermented by another mould (*Aspergillus tamari)* for a shorter period.

Miso is prepared from a starter (koji) which is a culture of a mould (*Aspergillus oryzae*) grown on steamed, polished rice mash until the grains are fully covered but the mould sporulation has not occurred. The process of fermenting miso is as follows:

1. Adding the starter to steamed soybean mash along with salt.
2. Fermenting initially for a week at 28°C and subsequently for 8 weeks at 35°C.
3. Ripening for many weeks at ambient temperature and grinding the product to a paste.

Tempeh is an Indonesian food which is fermented by the following process:

1. Soaking soybeans overnight at 25°C.
2. Removing the seed coat, splitting the beans into halves.
3. Boiling in water for 20 minutes, drying on mats and cooling.
4. Inoculating with a starter from a previous lot of tempeh or with mold spores (*Rhizopus*).
5. Packing the mash into plastic container, hollow tube or rolled in banana leaves.
6. Incubating at 32°C for 20 hours for good growth of mycelium.
7. Slicing thin, dipping in salt water and frying golden brown.

Natto is a thick viscous product that is used as a side dish with soy sauce in Japan. It is slimy in appearance, grey in colour and has a strong and persistent flavour. For its preparation, boiled soybeans are wrapped in rice straw and fermented for 1–2 days until slimy.

Tofu or soybean cheese is a Chinese food which is prepared by following steps:

1. Soaking soybeans and grinding them into a paste and filtering through linen.
2. Curdling the filtrate by magnesium or calcium salts and pressing the curd into blocks.
3. Fermenting the cheese blocks at 14°C for a month to permit the growth of white mould (*Mucor*).
4. Carrying out final ripening in brine or special wine.

Punjabi waries are ball shaped (3–8 cm diameter) spicy condiments used in Indian cooking. They are fermented products prepared from Bengal gram and green gram by following process:

1. Soaking dehulled legumes in water and grinding them to a paste.

2. Mixing spices (asafoetida, cloves, cardamom, caraway, fenugreek, red pepper, ginger and salt).
3. Fermenting at room temperature for 4–10 days by yeast.
4. Forming balls from the paste and drying in the sun.

Papadams are fermented in a manner similar to waries but fenugreek or ginger is not added. These tortilla-like products are circular and thin in shape. They are roasted or fried and are served as accompaniments with North Indian meals (Figure 23.5).

Papadams

Figure 23.5 Traditionally fermented legume products

23.2.5 Meat and Fish Products

Fermented meat products are prepared from beef and pork and more recently from poultry and turkey. Examples of these products are pepperoni, salami, sausages, Italian salami, thuringer and cervelat. Processing of semi-dry sausages involves the following steps:

1. Grinding good quality meat to reduce particle size.
2. Mixing meat, salts, glucose, cure (containing nitrite), spices and starter culture uniformly.
3. Stuffing the mixture in casings.
4. Fermenting at 29.4–43.3°C and 80–90 per cent relative humidity for 8–12 hours.
5. Incubating (ripening) to lower the pH to 5.2–4.6.
6. Heat cooking to raise the internal temperature to 60°C to inactivate the starter culture and the pathogens.
7. Storing at 4.4–10°C for 3–4 days and packing in vacuum packages.

Fermented fish products are prepared by mixing freshly caught small fish and trash fish with high amount of salt. They are then fermented to produce fish sauce and paste which are salty condiments added to food in South East Asia. Fish is also fermented by cutting into strips, cooking it and allowing fermentation by molds (*Aspergillus*).

23.2.6 Alcoholic Beverages

Beer, wine and distilled spirits (rum, whiskey, brandy) are examples of fermented alcoholic beverages. **Beer** and ale are prepared from malted grain because the yeast (*Saccharomyces cerevisiae*) used for their fermentation lacks amylase and cannot utilize starch. Beer processing involves the following steps:

1. *Malting* is a process in which barley is germinated to produce amylase enzyme which converts starch to maltose (a fermentable sugar).
2. *Drying and roasting* the grains stops germination. Short roast gives a light golden colour and mild flavour and is used for lagers. Long roast produces darker colour along with strong, toasted/burnt flavour and is used for strong beer, for example, ale and stout.

3. *Mashing* uses hot water to make soluble the major portion of malt. The insoluble material settles at the bottom and the clear liquid called 'wort' is extracted.
4. *Wort boiling with hops* (dried flowers of hop plant) is carried out to extract resins from hops, to concentrate the wort, to kill the contaminating bacteria and to inactivate the enzymes.
5. *Fermentation* is carried out after cooling the wort to 8–18°C and inoculating it with yeast (*Saccharomyces carlsbergensis*) at 3.3–14°C for 1–2 weeks. During this process, anaerobic fermentation of glucose takes place to produce ethanol and carbon dioxide.
6. *Maturing* or *aging* ensures the development of full flavour of beer through chemical interactions between the secondary metabolites of *Saccharomyces.*
7. *Finishing* is the process of carbonating beer to a CO_2 content of 0.45–0.52 per cent. It is then cooled, clarified and packaged.

Wine is prepared by the fermentation of fruit juice (primarily grape juice). It can also be prepared from berries, dandelions and honey. Grape wines are either red or white. For red wines, black or red grapes are used, and their skin is not removed during fermentation. In case of white wines, white grapes are generally used, and the skins are removed before fermentation. To produce red wines, red grapes are used and the juice is kept in contact with the skins for a short period so that a pink colour develops. Wine processing is carried out as follows:

1. *Harvesting* special varieties of grapes suitable for wine making at a stage when sugar content is optimal. They are crushed to form 'must'.
2. *Primary fermentation* of 'must' by the natural inoculums (natural yeast flora of grapes) or by adding 2–5 per cent wine yeast (*Saccharomyces cerevisiae, Saccharomyces ellipsoideus*) at 24–27°C for 3–5 days (for red wines) and at 10–21°C for 7–14 days (for white wine).
3. *Secondary fermentation* of the fermented juice drawn from the pomace in a storage tank at 21–29°C for 7–11 days so that the remaining sugar is also fermented.
4. *Cooling and filtering* wine and transferring it to wooden tanks completely excluding air and sealing to prevent oxidation. Final ageing may be carried out in bottles for months or years to develop the full aroma of wine.
5. *Clarifying and barrelling or bottling* the wine.

23.2.7 Other Fermented Foods

Vinegar is a condiment which must contain at least 4 per cent acetic acid. Generally, it derives its name from the material it is prepared from, for example, apple vinegar from apples and malt vinegar from malt. It can be prepared from the following:

1. Juices of fruits such as apples, pears, grapes and oranges.
2. Sweet potatoes (after hydrolysing starch to sugar).
3. Malted cereals such as barley, wheat and corn.
4. Sugars, for example, syrups, honey and molasses.
5. Spirits or alcohol.
6. Starchy vegetables, for example, potatoes.

Vinegar is prepared from sugary or starchy foods by two-step fermentation:

1. **Alcoholic Fermentation:** It is the fermentation of sugar to ethyl alcohol in anaerobic (taking place in absence of oxygen) conditions by yeasts (*Saccharomyces cerevisiae*).

$$\underset{\text{Simple sugar}}{C_6H_{12}O_6} \xrightarrow{\text{yeast}} \underset{\text{Ethyl alcohol}}{2C_2H_5OH} + \underset{\text{Carbon dioxide}}{2CO_2}$$

2. **Acetic Acid Fermentation:** It is the oxidation of alcohol to acetic acid in aerobic (taking place in presence of oxygen) conditions by acetic acid bacteria (*Acetobacter aceti*).

$$\underset{\text{Ethyl alcohol}}{C_2H_5OH} + \underset{\text{Oxygen}}{O_2} \xrightarrow{\text{acetic acid bacteria}} \underset{\text{Acetic acid}}{CH_3COOH} + \underset{\text{Water}}{H_2O}$$

Tea leaves and coffee and cocoa beans are fermented during their processing. Fermentation process is als employed to produce flavours (monosodium glutamate), organic acids (lactic, citric and acetic acids), amino acid (lysine and glutamic acid), vitamins (A, C, B_{12}, and riboflavin) and enzymes (amylase, protease and invertase).

This chapter highlighted the beneficial role of microorganisms in food fermentation. We learnt the pro cess of preparing and the microbiology of a wide variety of fermented foods ranging from fermented milk cheese, sauerkraut, pickles, bread, idli, beer and wines as well as distilled liquors. Products fermented from soy, meat and fish were also covered. Though some beneficial microorganisms help in preventing food-born illnesses, the latter are a major public health problem. The next chapter will elaborate the causes, symptom and sources of food-borne diseases along with the preventive measures.

Know Your Progress

True or False Questions

1. Fermentation is a technique which discourages replication of microorganisms with the objective of bringing desirable changes in the food.
2. Lactic acid bacteria are the most important bacteria in desirable food fermentations.
3. Tamari sauce, miso, tempeh and natto are examples of fermented cereal products.
4. *Baker's yeast* or *Saccharomyces cerevisiae* is used for fermentation of bread dough.
5. Vinegar is prepared by lactic acid and acetic acid fermentation.

Very Short Answer Questions

1. Differentiate between the following:
   (a) Hard cheese and soft cheese
   (b) Sauerkraut and kimchi
   (c) Natto and soy sauce
   (d) Alcoholic fermentation and acetic acid fermentation
   (e) Wine and whiskey

hort Answer Questions

1. Define fermentation and give examples of fermented food from different groups.
2. List the different types of cheese.
3. Give examples of fermented vegetables and explain the process of preparing sauerkraut.
4. What is the role of fermentation in bread making?
5. How is vinegar prepared?

ong Answer Questions

1. Describe the process of preparing fermented dairy products and alcoholic beverages.
2. What are the different types of cereal and legume products prepared by fermentation? Explain.

eyond the Text

- Dissolve half tsp of sugar and 1 tsp of flour in 15 ml of warm water. Sprinkle ½ tsp of dry pellet yeast and keep at warm place for 15 minutes to ferment. Take 95 gm of refined flour (maida), add the ferment and 45 ml of water and knead well to form dough. Transfer the dough to a measuring jar and note the initial volume after levelling the surface. Keep covered at 27–30°C for 1 hour and note the final volume.

$$\text{Calculate dough raising capacity}\left(\%\ \text{DRC} = \frac{\text{Final volume}}{\text{Initial volume}} \times 100\right).$$

Chapter 24

Food-borne Diseases

Chapter Objectives

After reading this chapter, the reader should be able to:

- differentiate between food infections and food intoxications.
- know the symptoms and sources of food-borne infections of bacterial, viral and parasitic origin.
- learn about the sources and health consequences of food intoxicants produced by bacteria, fungi (mycotoxins) and chemicals.
- understand and apply preventive measures to control the incidence of food-borne diseases.

Till date more than 250 different food-borne diseases have been identified; some may result in massiv outbreaks but very often these diseases are not reported. World Health Organization (WHO) has define **food-borne diseases** as illnesses either infectious or toxic in nature caused by agents which enter the bod through the ingestion of food. It is important to note that every person is at risk of food-borne illness Since the microorganism or the toxin enters the human body through the gastrointestinal system, nause vomiting, abdominal cramps and diarrhoea are symptoms which are common in many food-borne illnesses Adequate knowledge about the causes, symptoms, food sources and preventive measures relating to food borne diseases is essential for ensuring food safety from 'farm to fork'.

THINK TANK

Food-borne Diseases/Outbreaks

Worldwide, millions of people become ill every year after consuming contaminated food. Consider the following examples:

- Consumption of contaminated ice cream resulted in an outbreak (of salmonellosis) in 1994 in the United States, affecting an estimated 224,000 persons.
- Another outbreak, reported in China in the year 1988, resulted in hepatitis A among 290,000 persons on the consumption of contaminated clams.
- 70 per cent of the 2 million diarrhoeal deaths in developing countries (e.g. India) can be attributed to the ingestion of contaminated food.
- In early 2012, a multistate outbreak of *Salmonella* infection was associated with eating food from a Mexican-style fast food restaurant chain in the United States.

What are the causes of these diseases? What measures should be adopted to prevent these food-borne illnesses? This chapter attempts to answer these queries.

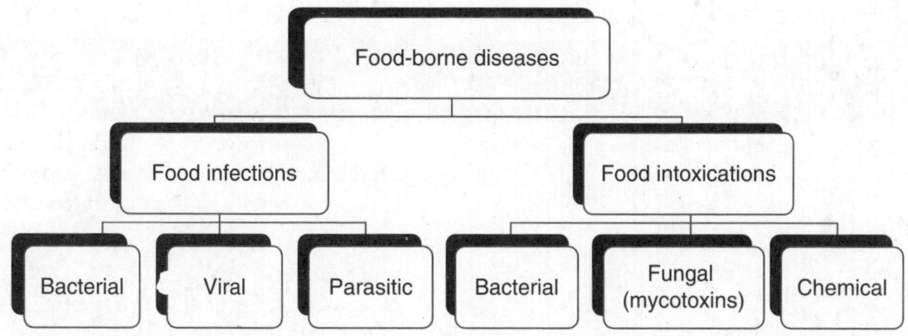

Figure 24.1 Categorization of food-borne diseases

When the disease-causing microorganisms or parasites are present in the food at the time of consumption and grow in the body of the host to cause illness, the condition is known as **food infection** (caused by a variety of bacteria, viruses and parasites). In case of **food intoxication**, the toxins (of bacterial, fungal or chemical origin) are present in the food prior to its consumption (Figure 24.1). Top five pathogens causing food-borne illnesses listed by the Centre for Disease Control and Prevention are presented in Figure 24.2.

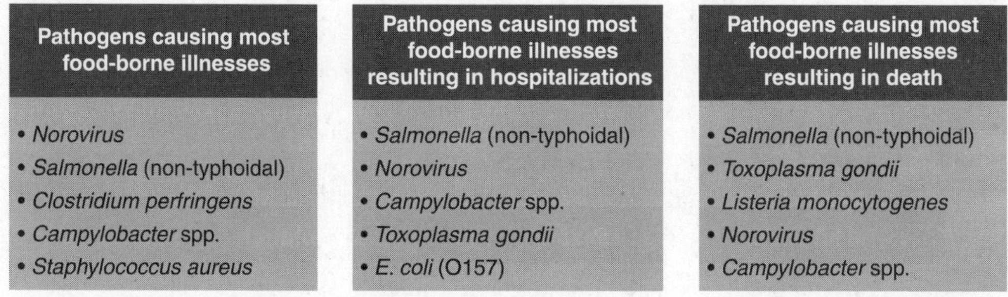

Figure 24.2 Pathogens causing food-borne illnesses

The most common pathogens causing food-borne illnesses are *Norovirus* and *Salmonella*. Food infections caused by *Escherichia coli* (O157) and *Listeria monocytogenes* are among the most serious food-borne diseases in terms of their severe and sometimes fatal health consequences. The threat or outbreaks of food-borne diseases can often force the closure of restaurants or result in recall of the food products.

24.1 Food Infections

Food-borne infections can be caused by bacteria, viruses or parasites. The bacterial food infections include salmonellosis, *Clostridium perfringens* illness, *Escherichia coli* infection, shigellosis, scarlet fever and listeriosis. Campylobacteriosis, cholera and yersiniosis are other food infections caused by bacteria (Table 24.1).

Table 24.1 Food-borne infections of bacterial origin

Food Infection	Causative Organism	Incubation Period	Symptoms	Food Commonly Involved
Salmonellosis	*Salmonella* (non-typhoidal)	6–72 hours	Diarrhoea, fever and abdominal cramps	Undercooked/fresh meat and poultry products and eggs
Clostridium perfringens illness	*Clostridium perfringens*	8–12 hours	Watery diarrhoea and abdominal cramps (usually no fever or vomiting)	Beef, poultry, gravies and dried or pre-cooked foods
Escherichia coli STEC infection	Shiga toxin–producing *Escherichia coli* (STEC O157:H7)	3–4 days	Severe abdominal cramps, haemorrhagic colitis, haemolytic uremic syndrome and diarrhoea (often bloody) and vomiting	Unpasteurized (raw) milk, unpasteurized apple cider and soft cheeses made from raw milk
Shighellosis	*Shighella*	1–2 days	Fever, abdominal cramps, nausea and diarrhoea (bloody stools)	Moist prepared foods, milk, macaroni salads, potato, tuna and turkey
Scarlet fever, septic sore throat	*Streptococcus pyogenes*	1–3 days	Sore throat, pain during swallowing, tonsillitis, chills, vomiting, abdominal pain and sometimes rash	Milk, eggs, egg salad, ice cream, custard and pudding
Listeriosis	*Listeria monocytogenes*	up to 90 days	Fever, muscle aches, sometimes preceded by diarrhoea or other gastrointestinal symptoms	Milk, soft cheese, coleslaw, cook-chill chicken and ready-to-eat meats (hot dog, deli meat)
Campylobacteriosis	*Campylobacter jejuni*	2–7 days	Acute abdominal cramps, fever and inflammatory diarrhoea	Poultry, red meat and raw milk
Vibrio parahaemolyticus infection	*Vibrio parahaemolyticus*	2–48 hours	Abdominal pain, diarrhoea (with blood and mucus) and mild fever	Raw salt water fish, shellfish and fish products
Cholera	*Vibrio cholera* O1 or O139	2–3 days	Profuse watery diarrhoea, vomiting and leg cramps	Contaminated water, raw shellfish, uncooked fruits and vegetables
Yersiniosis	*Yesinia enterocolitica*	1–11 days	Diarrhoea, low grade fever and abdominal cramps	Raw milk, pork and other meats

24.1.1 Salmonellosis

Salmonellosis, one of the most common bacterial food infections, is caused by the ingestion of *Salmonella,* which are gram negative, facultative, non-spore-forming, anaerobic bacteria. They can grow over a wide range of temperature (7–45°C), pH (4.1–9.0) and a_w (0.93–0.95). The optimum temperature required for their growth is 37°C. These rods are tolerant to bile salts and are found primarily in the gastrointestinal system of infected humans and other animals (including birds) and their faeces through which they find a route of transmission to food products.

> **GOOD TO KNOW**
>
> **New Nomenclature of *Salmonella***
>
> Serotyping is used to define different bacterial species beyond the subspecies level. As per the new naming convention, the genus, species and sub species names are in italics but the serotype name is not italicized; it starts with a capital letter, for example, *S. enterica* ssp. *enterica* ser. Typhimurium. For convenience, only genus and serotype can be used; for example, *Salmonella* Typhimurium or *S.* Typhi.

Food sources of *Salmonella* are undercooked/fresh meat and poultry products including eggs, sausages and cured meats. The risk of contamination is higher when these foods are held unrefrigerated for long periods. Raw milk and raw milk cheese, salad leaf products, chocolate and processed foods (such as orange juice, peanut butter and bakery products) are among the other food sources. The **symptoms** of salmonellosis include nausea, vomiting, abdominal pain, diarrhoea and low grade fever. The severity and intensity varies for this infection which may prove fatal in 10–15 per cent of untreated patients. It can be **prevented** by (i) avoiding food contamination from sources such as infected human beings and other animals, carriers and *salmonellae* contaminated food items; (ii) destroying the bacteria with heat processes; (iii) controlling its growth by the use of low temperature and (iv) maintaining adequate sanitation and hygiene.

> **GOOD TO KNOW**
>
> **Typhoid-causing Salmonella**
>
> While non-typhoidal *salmonella* causes Salmonellosis, *S.* Typhi and S. Paratyphi A, B and C are responsible for typhoid and parathyroid fever (enteric fever accompanied by headache, loss of appetite, abdominal pain and diarrhoea). Constipation is more characteristic of *S.* Typhi infection.

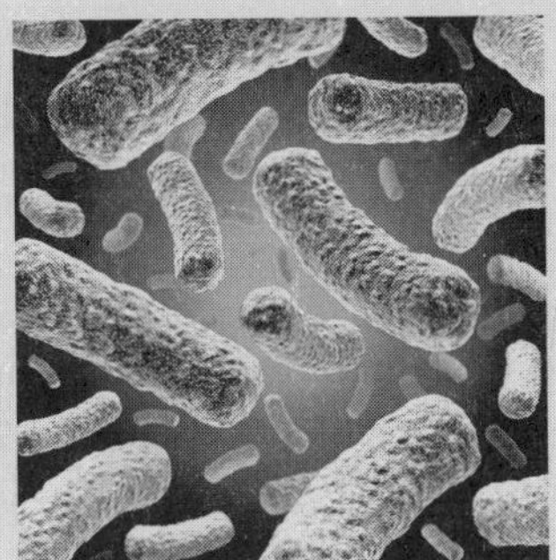

Source: Shutterstock.

Figure 24.3 Salmonella Typhimurium

24.1.2 *Clostridium Perfringens* Illness

Clostridium perfringens is a gram positive, anaerobic, rod-shaped, spore-forming bacterium with sub-terminal oval spores. The optimum temperature needed for its growth is 43–47°C but it can grow at a temperature range of 15–55°C, pH range of 5.0–9.0 and a_w of 0.95–0.97. Though Types B, C, D and E can cause illness, Type A is the most common cause of food-borne illness and gas gangrene. Common salt (6 per cent) and sodium nitrate (present in cured meats) have an inhibiting action on this bacterium.

The **sources** of *Clostridium perfringens* are raw foods, soil, sewage and animal faeces. Poor hygiene during food handling and preparing meat dishes in advance of eating and keeping them unrefrigerated substantially increase the risk of contamination. The **symptoms** (nausea, acute abdominal pain, diarrhoea and flatulence) appear after the ingestion of millions of viable cells of these bacteria which release a protein enterotoxin during sporulation in the intestines. The **preventive measures** includ (i) adequate and fast cooling of cooked foods particularly meat and meat products; (ii) keeping cooked foo hot; (iii) reheating of leftover food and (iv) following hygienic practices in food preparation and handling.

GOOD TO KNOW

Risk of *Clostridium perfringens* Illness in Bulk Catered Food

When foods are prepared in large quantities and kept warm (at 20–60°C) for a long time before being consumed, the risk of *Clostridium perfringens* infection increases significantly (Figure 24.4). Many outbreaks of this food infection have been reported in institutions such as hospitals, school cafeterias, prisons and nursing homes, or at events and parties with catered food. However, the toxin can be inactivated by heating at 60° C for 10 minutes.

Figure 24.4 Bulk catered food: a risk for *Clostridium perfringens* illness

24.1.3 *Escherichia Coli* Infection

E. coli generally considered as a normal flora of the alimentary tract of human and many animals, has been recognized as a pathogen in recent years as it can result in diarrhoeal diseases. Some strains of *E. coli* produce an enterotoxin and cause cholera-like symptoms such as rice-water stools, vomiting, dehydration and shock in human beings. For the symptoms to appear, the ingested food such as raw milk and undercooked beef needs to be highly contaminated. *E. coli* grows at temperature ranging from 10°C to 40°C, the optimum being 37°C. It can grow at a wide pH ranging from 4.0 to 8.5 but the optimum pH is between 7.0 and 7.5. There are six types of *E. coli*:

GOOD TO KNOW?

'Shiga Toxin-Producing' *E. coli* (STEC)

Enterohaemorrhagic *E. coli* (EHEC) commonly known as 'Shiga toxin-producing' *E. coli* (STEC O157:H7) is the most susceptible cause for outbreaks related to beef consumption (in North America, United Kingdom and Japan). It has also been detected in dairy cattle and the beef marketed in Kolkata, India, but no outbreaks have been reported (Figure 24.5).

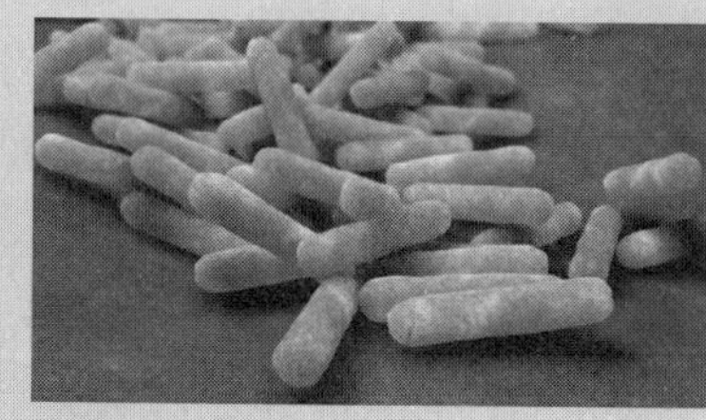

Source: Shutterstock.

Figure 24.5 Escherichia coli O157:H7

1. Enteropathogenic *E. coli* (EPEC) – do not generally produce any enterotoxins
2. Enterotoxigenic *E. coli* (ETEC) – can produce one or more enterotoxins
3. *Enteroinvasive E. coli* (EIEC) – produce cytotoxin and result in dysentery-like syndrome
4. Enteroaggregative *E. coli* (EAEC) }
5. Diffusely adherent *E. coli* (DAEC) } – are more important for less-developed countries like India where they can cause diarrhea among infants and pre-school children
6. Enterohaemorrhagic *E. coli* (EHEC) – are most susceptible for outbreaks

E.coli is a heat-sensitive bacterium, and it can be destroyed through proper cooking or pasteurization. The **preventive measures** include (i) chilling the food quickly in small amounts (ii) protecting and treating water (iii) following hygienic practices for the preparation of food (iv) maintaining personal hygiene and (v) disposing the sewage properly.

24.1.4 Shigellosis

Shigellosis refers to enteric infections caused by the members of genus *Shigella* such as *S. sonnei, S. flexneri, S. dysenteriae* and *S. boydii*. These bacteria are somewhat heat-resistant and can grow even in saline medium. They can grow at temperature ranging from 10°C to 40°C and need an optimum temperature of 37°C for their growth.

Shigella releases a lipopolysaccharide endotoxin and causes bacillary dysentery (shigellosis). The infective dose for the symptoms to appear is very low. The **symptoms** may be mild to severe and include abdomen cramps, fever, chills, nausea, frequent liquid stools which often contain blood, mucus or pus, dehydration, headache and exhaustion. The primary **food sources** are moist foods, mixed foods, milk, potato, macaroni salads, tuna and turkey. Food worker hygiene and sanitation are extremely important for **preventing** this infection. It is possible to prevent shigellosis by (i) cooking foods properly; (ii) adopting hygienic practices for food preparation; (iii) personnel hygiene and adequate sewage disposal and (iv) chilling foods rapidly in small amounts. The water must be protected from contamination and the vectors of contamination (flies, rodents, dogs and cats) must be controlled strictly.

DID YOU KNOW?

What Is Shiga Toxin?

Shiga toxin is an exotoxin produced by some strains of *shigella*. This cytotoxin binds to and inactivates 60S ribosomes of the host cell; halting protein synthesis and killing the cell.

24.1.5 Scarlet Fever and Septic Sore Throat

Streptococcus pyogenes can cause scarlet fever and septic sore throat. It grows in **foods** such as milk, eggs and their products such as ice cream, egg salad, custard and pudding. The **symptoms** are sore throat, pain during swallowing, tonsillitis, high fever, vomiting and malaise; sometimes, rash is also observed. In addition, there may be chills, vomiting and abdominal pain. The **preventive measures** include cooking the food thoroughly, chilling the food quickly in small amounts and following hygienic practices. Milk must be pasteurized and the workers suffering from respiratory disease or having skin wounds should be kept away from food handling.

24.1.6 Listeriosis

Listeriosis is a serious infection caused by *Listeria monocytogenes* which is a gram positive, facultative anaerobe in coccoidal rod shape. It is catalase positive and oxidase negative. It is capable of growing at a wide range of temperatures (0–42°C) and its optimum growth takes place between 30°C and 35°C. The minimum pH required is 5.5. It can survive refrigeration temperature and high concentration of salt.

The **symptoms** of this serious public health problem include fever, headache, vomiting, diarrhoea or other gastrointestinal symptoms. Nearly all infected persons have 'invasive' infection, which spreads beyond the gastrointestinal tract to mucosal epithelium and lymphatics of the intestine. Among the **food sources** of *L. monocytogenes* are soft cheese, raw milk, raw seafood coleslaw and cook-chill chicken. The **preventive measures** include adequate processing of food products particularly the ready-to-eat (RTE) food products such as hamburger and deli products.

24.1.7 Campylobacteriosis

Campylobacter is a rather newly identified food-borne bacterium which is a gram negative spiral rod. The most prevalent species is *C. jejuni* and the disease enterocolitis caused by it is the most common cause of diarrhoea in the developed countries. However, the disease is less acute in developing countries where it results in asymptomatic excretion. Campylobacteriosis is significant because it can lead to a neurological disorder, Guillain-Barre syndrome, and result in paralysis of respiratory muscles and death.

DID YOU KNOW?

Listeriosis can lead to Serious Health Consequences in Pregnancy and Old Age

Listeria monocytogenes is capable of crossing materno-fetal barrier and blood-brain barrier and thereby leads to complications related to pregnancy (abortion, still birth and premature birth) and the nervous system (meningitis). Pregnant women, elderly and immune-compromised persons are more at risk of developing complications.

24.1.8 Food-borne Viral Infections

A characteristic feature of viruses is that they cannot grow or multiply outside their host. They are 25–30 nm in size and are made up of nucleic acids Deoxy Ribonucleic Acid (DNA) or Ribonucleic Acid (RNA) packaged in a protein capsid. Though the viruses lack the molecular components or resources required for their own growth and reproduction, they are the most abundant microorganisms. It is due to their capability of infecting the host cells and utilizing their metabolic machinery and resources for their own growth and multiplication. Different viruses can cause gastrointestinal disease in human beings (Table 24.2). They enter the human body via the oral route, usually through as a consequence of faecal contamination of water or food such as shellfish; milk and other beverages; and specific foods, for example, potato salad.

Table 24.2 Food-borne viral infections

Viral Infections	Causative Virus	Incubation Period	Symptoms	Foods Commonly Involved
Gastroenteritis	Norovirus	1–2 days	Nausea, vomiting, watery large volume diarrhoea	Undercooked shellfish, ready-to-eat food touched by infected worker, salads, sandwich
Gastroenteritis	Human Rotavirus (HRV)	1–3 days	Vomiting, watery diarrhoea, mild fever	Ready-to-eat food touched by infected worker
Hepatitis A	Hepatitis A virus (HAV)	30 days	Diarrhoea, dark urine, jaundice, flu like symptoms	Shellfish grown in contaminated water, food contaminated after contact with infected worker

Norovirus: Norovirus is the official genus name for the group of viruses previously known as 'Norwalk-like viruses' (named after an outbreak in Norwalk, Ohio). The most common food-borne virus norovirus (Figure 24.6) belongs to genus *Norovirus* and family *Caliciviridae*. The gastroenteritis caused by noroviruses is mild and self-limiting and sometimes referred to as stomach flu. They are transmitted through contaminated food or water, from person to person and by touching contaminated surfaces.

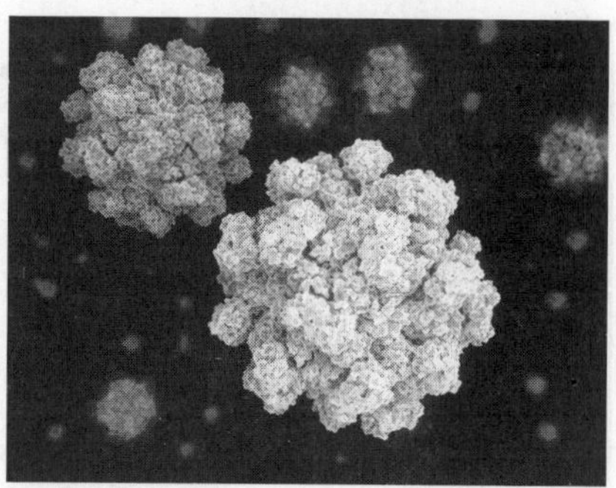

Source: Shutterstock.

Figure 24.6 Norovirus is the most common pathogen causing food-borne illnesses

Human Rotavirus: It is more common in less developed countries and spreads primarily through person-to-person contact and sometimes through water and food sources. It causes diarrhoeal disease and can lead to severe dehydration among infants and young **children**.

Hepatitis A Virus: Hepatitis A virus is responsible for major outbreaks of hepatitis frequently caused by contaminated water and food in developed world. It is less common in lesser developed countries where most people are infected with the virus during childhood and develop immunity by the time they are adults. Food sources include shell fish, fruits, salads and milk. The incubation can be as long as 6 weeks before **symptoms** of grave liver damage appear. Jaundice is apparent in nearly half of the cases; besides, there is loss of appetite and gastro-intestinal disturbances. Infectious hepatitis can be **prevented** by (i) cooking the shellfish properly; (ii) examining the personnel handing food on a regular basis; (iii) heating milk and foods adequately; (iv) disinfecting water and (v) maintaining cleanliness of food handlers.

DID YOU KNOW?

How does Sewage Water Cause Viral Contamination of Shellfish?

Shellfish are generally grown in water close to the shore which is easily contaminated with sewage. Moreover, they feed by filtering the water so the concentration of bacteria and virus in shell fish tends to be high. Putting them in uncontaminated water helps in removing bacteria within two days but virus decontamination takes much longer.

24.1.9 Infections Due to Food-borne Parasites

Food-borne parasitic infections enter human body primarily through ingestion of raw or insufficiently cooked beef or pork (Table 24.3). Animal parasites cannot grow or multiply in the food or the environment. To complete their life cycle, they need access to animal and human hosts which is accomplished when food is contaminated with their eggs or larvae.

Trichinosis is a parasitic infection caused by *Trichinella spiralis*. The initial symptoms include diarrhoea, nausea, abdominal pain and, sometimes, vomiting. Gradually, muscle swelling, fever, difficulty in breathing and swollen eyelids and other symptoms may also appear. Later, neurological and cardiac complications and death are possible.

Taeniasis is caused primarily by beef tapeworm (*Taenias saginata*) or pork tapeworm (*Taenia solium*) through their live larvae present in muscle tissues of undercooked beef or pork respectively. The mature tapeworm grows in human intestine to produce symptoms such as abdominal pain, anaemia and discomfort. Infection by fish tapeworm (*Diphyllobothrium latum*) does not produce any symptoms but they may result in anaemia when infection is severe. To prevent the entry of live larvae of tapeworms in human body, beef, pork and fish must be thoroughly cooked and processed under strict inspection.

Table 24.3 Food-borne parasitic infections

Food-borne Parasitic Infections	Causative Parasite	Food Commonly Involved
Trichinosis	*Trichinella spiralis*	Uncooked or insufficiently cooked pork
Taeniasis	*Taenia saginata*	Uncooked or insufficiently cooked beef
	Taenia solium	Uncooked or insufficiently cooked pork
Ascariasis	*Ascaris lumbricoides*	Raw food grown in soil contaminated with human faeces
Anisakiasis	*Anisakis simplex*	Uncooked or insufficiently cooked fish
Cryptosporidiosis	*Cryptosporidium*	Drinking water and recreational water
Amoebic dysentery (amebiasis)	*Entamoeba histolytica*	Contaminated moist food
Toxoplasmosis	*Toxoplasma*	Contaminated meat, fish, poultry fruits and vegetables

Anisakiasis, caused by a nematode *Anisakis simplex*, has an incubation period of several days. Consumption of raw or insufficiently cooked fish can cause this infection which results in irritation of throat and digestive tract. It can be prevented by proper cooking of fish.

Besides *Entamoeba histolytica, Toxoplasma, Cryptosporidium and Giardia* are protozoa important in food safety. *Entamoeba histolytica* causes **amoebic dysentery (amoebiasis)** which is an endemic diarrhoeal infection spread through oral–faecal route in many parts of lesser developed countries. **Toxoplasmosis,** caused by *Toxoplasma*, is generally without symptoms but it can cause influenza-like illness. Immuno-compromised individuals and pregnant women are prone to this infection which is also the most common cause of cerebral infection among AIDS patients.

24.2 Food Intoxications

As stated earlier, food intoxication (or food poisoning) is caused by harmful toxins present in the food before it is consumed; the sources of the toxins may be bacterial, fungal or chemical in nature. Some food intoxications of bacterial origin, their causative organism, symptoms and food sources are listed in Table 24.4.

Table 24.4 Bacterial food intoxications

Bacterial Food Intoxication	Causative Bacterium	Incubation Period	Symptoms	Foods Commonly Involved
Botulism	*Clostridium botulinum*	12–36 hours	Vomiting, diarrhoea, fatigue, double vision, difficulty in swallowing and speaking	Inadequately processed low-acid home-canned foods (string beans, sweet corn, spinach, asparagus), contaminated canned food, meat, fish
Staphylococcus food intoxication	*Staphylococcus aureus*	2–6 hours	Salivation, nausea, vomiting, abdominal cramps and diarrhoea	Custard and cream filled bakery products, ham, puddings, pies, salads, salad dressings, meat and products

24.2.1 Botulism

Botulism is a disease caused by *Clostridium botulinum* which is a rod-shaped, spore-forming anaerobe motile with peritrichous flagella. It produces eight types of neurotoxins (A, B, C_1, C_2, D, E, F and G) of which Type A, B and E cause botulism in human beings. It needs near neutral pH for its growth and does not usually grow at pH of 4.5 or below. This bacterium can grow at a temperature ranging from 15°C to 48°C; the optimum temperature for the production of toxin is 35°C. Compared to other spore-forming bacteria, the heat-resistance of *C. botulinum* is extremely high.

The **food sources** are meat, fish, canned fruits and vegetables particularly the ones which are low or medium acid foods processed at home. The **symptoms** of botulism initiate with acute digestive disturbance, followed by nausea and vomiting and diarrhoea, fatigue, dizziness and headache. Thereafter, signs of constipation, double vision, dry mouth and difficulty in swallowing and speaking may appear. It results in paralysis of involuntary muscles which may lead to failure of respiratory system or heart. The disease can result in death in 1–7 days.

The single most important measure to prevent botulism is adequate canning of foods. It can, further, be controlled by (i) using approved heat processes for canning; (ii) rejecting swollen or spoilt canned foods; (iii) refusing to taste suspected food; (iv) avoiding cooked food held at room temperature and (v) boiling suspected food for a minimum of 15 minutes.

24.2.2 Staphylococcus Food Intoxication

Staphylococcus aureus is an important pathogen capable of causing food poisoning in humans. It can produce enterotoxins (Types A, B, C_1, C_2, C_3, D, E, G–I) which are heat stable, pH stable and resistant to proteolytic enzymes. Due to these characteristics, this toxin is not destroyed by the cooking process or during the digestive process. The nasal passages of humans, infected wounds or boils are generally loaded with this bacterium which is a gram positive, facultative anaerobe. The enterotoxin producing cultures of this organism are facultative but they grow better in the presence of oxygen than in its absence. They are fairly tolerant to salt, nitrites and sugars. For growth, this bacterium requires a temperature ranging from 4°C to 46°C, pH ranging 4.8 to 8.0 and a water activity (a_w) of 0.85.

The **food sources** associated with staphylococcal food poisoning include custard- and cream-filled bakery products, milk and milk products and salads. Other foods such as meat and meat products, fish and fish products, ham and poultry can also result in this poisoning. The **symptoms** which appear within 2–6 hours of consumption include salivation followed by nausea, vomiting, abdominal cramps and diarrhoea which last for nearly 24 hours. Other possible signs are headache, muscular cramping, sweating, chills, prostration, weak pulse, shock and shallow respiration. The illness is self-limiting, the duration is short, recovery is complete and the risk of death is minimal. **Prevention** of staphylococcal food poisoning can be achieved by preventing contamination of food with this bacterium, controlling its growth and by killing it. Therefore, effective sanitation, refrigeration and thermal treatment should be employed during food processing and storage.

24.2.3 Mycotoxins

'Mycotoxin' is a term used to describe toxins produced by fungi. They are secondary metabolites having no role in the growth and development of fungi but can lead to adverse health outcomes in humans. These toxins contaminate the food either in the field or at home. Storage of food in high humidity and warm temperature in many less-developed countries facilitates the fungal growth and subsequent production of mycotoxins. In contrast to microorganisms, the mycotoxins are not destroyed by cooking or freezing and remain in the food until it is consumed. Some of the important mycotoxins along with their food sources and health outcomes are listed in Table 24.5.

Table 24.5 Causative fungi and food sources of important mycotoxins

Mycotoxins	Causative Fungi	Food Sources	Health Concerns
Aflatoxins	*Aspergillus flavus* *Aspergillus parasiticus*	Barley, corn, millet, oats, peanuts, rice, soybeans	Aflatoxicosis, carcinogenic
Ochratoxin A	*Aspergillus ochraceus*	Corn, wheat, barley, bread, eggs, peanuts, coffee, oats, dried beans	Nephrotoxic, hepatotoxic, teratogenic, carcinogenic
Patulin	*Penicillium expansum*	Apple sap, apple cider, apple juice, wheat, sausages and ham	Genotoxic and probably carcinogenic
Ergot alkaloids	*Claviceps purpurea*	Rye, other grains	Ergotism
Yellow rice toxins (e.g. citrinin)	*Penicillium*	Mouldy rice, mouldy bread, cured ham, wheat, rye, oats	Nephrotoxic, hepatotoxic, paralytic effects, respiratory disorders, carcinogenic
Fumonism	*Fusarium* spp.	Corn, sorghum, rice, other grains	Carcinogenic, hepatotoxic
Roquefortine	*Penicillium roqueforti*	Blue cheese, Roquefort cheese	Toxic to mice
Trichothecenes	*Fusarium* spp.	Corn, wheat, barley, oats, rye	Hepatotoxic, nephrotoxic
Luteoskyrin	*Penicillium islandicum*	Mould, rice flour	Carcinogenic

Aflatoxin is produced by *Aspergillus* in many foods such as cassava, wheat, barley, millet, sorghum, oats, rice, soybeans and spaghetti. Besides, peanut meal, peanut butter, cottonseed meal, soybean meal, cowpeas, peas, sesame, raisins, sweet potatoes and cocoa are other sources. Since peanut and cottonseed products are imported into many parts of the world, a guideline for aflatoxin level (30 parts per billion) has been established. **Ochratoxin A** is a potent nephrotoxin, produced by *Aspergillus* and *Penicillium*. It is found in wheat, barley, white beans, bread, eggs and peanuts; and can lead to kidney failure in animals such as rats, chicks and ducklings. **Patulin** has carcinogenic and mutagenic properties and can be fatal in mice and rats. It is found in apple products such as apple sap, apple cider, apple juice and is produced by *Penicillium* and *Aspergillus*. This mycotoxin can result in brain oedema, lung haemorrhage and capillary damage of the liver, kidney and spleen.

24.2.4 Rickettsial Diseases

Rickettsias are a form of life that closely resembles bacteria but they cannot grow outside the living cells. They are generally transmitted to human body through the bite of a louse or tick. *Coxiella burnett* are important from the perspective of food safety as they can infect cows, are excreted in their milk and thereby infect the human bodies to cause Q fever. Antibiotics can be used effectively to treat the rickettsial diseases.

24.2.5 Seafood Intoxication

Seafood intoxication is an area of concern in many parts of the world. The most important type of thi intoxication is the **ciguatera,** also known as fish poisoning. It generally occurs on the consumption of bigge fish which have fed on smaller fish surviving on toxic algae or plants. Another type of marine intoxication is the **shellfish poisoning** which can be caused by the shellfish (e.g. clams and mussels) feeding on toxic algae such as dinoflagellates (*Gonyaulax catenella* and *G. tamarensis*). Based on the different symptoms produced, the shellfish intoxication is categorized as paralytic shellfish poisoning (PSP), diarrhoetic shellfish poisoning (DSP), neurotoxic shellfish poisoning (NSP) and amnesic shellfish poisoning (ASP).

24.2.6 Poisoning by Chemicals

The contamination of food by chemicals is a worldwide public health concern. It may result through environmental pollution, the intentional use of various chemicals such as agrochemicals and veterinary drugs or other chemicals associated with food processing (Figure 24.7).

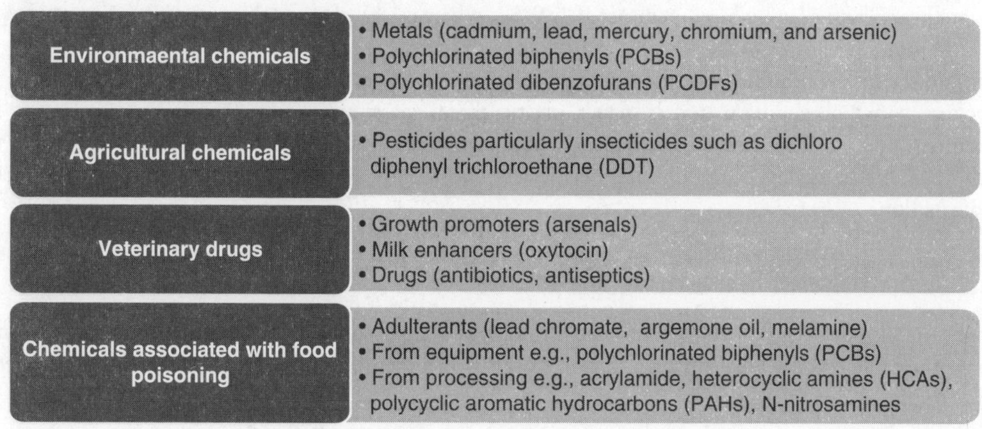

Figure 24.7 Chemical contaminants of food

Environmental chemicals have a long history of persistence and bio-accumulation in human body. Metals (lead, mercury and arsenic); polycyclic aromatic hydrocarbons (PAHs); polychlorinated biphenyls (PCBs) and polychlorinated dibenzofurans (PCDFs) fall under this category. The PAHs, PCDFs and PCBs may be carcinogenic and mutagenic. Recently another category of chemicals called persistent organic pollutants (POPs) have gained importance; they are found in solvents, pharmaceuticals and flame retardants.

Agricultural chemicals such as pesticides, herbicides, insecticides, rodenticides and fungicides are chemicals which can leave harmful residues on foods. Chlorinated hydrocarbon insecticides such as dichlorodiphenyltrizchloroethane (DDT) have the ability to bioaccumulate; therefore, organophosphates and carbamates have gained preference these days. **Veterinary drugs** (antibiotics and antiseptics) used for ensuring better health of animals can also leave harmful residues in the animal food.

Chemicals associated with food processing include both food additives and adulterants.

> **GOOD TO KNOW**
>
> **Grilling and Smoking Produce Cancer-causing Compounds**
>
> Outdoor grilling is a trendy method of cooking which imparts a wonderful flavour to meat. Nevertheless, high temperature grilling results in the development of two cancer-causing compounds, heterocyclic amines (HCAs) and polycyclic aromatic hydrocarbons (PAHs), in flesh foods. HCAs form when the meat is grilled on a flame or a very hot surface. PAHs, which are constituents of wood smoke, tend to contaminate foods such as fish, ham and cheese during smoking. Their levels are higher when processing techniques such as barbecuing and charcoal broiling are employed.

Food additives are approved substances added intentionally to improve the shelf life and acceptability of food but their use in amounts exceeding the prescribed limit is detrimental to food safety (refer to Chapter 25). Food adulterants are unapproved and illegal substances which can have harmful health consequences (refer to Chapter 26).

Figure 24.8 High temperature cooking: A risk for chemical poisoning

Some toxic chemicals, for example, acrylamide, heterocyclic amines (HCAs) and polycyclic aromatic hydrocarbons (PAHs) can be produced during cooking and processing (Figure 24.8). Acrylamide is a neurotoxin produced when foods containing high amount of carbohydrates and amino acids are heated at a very high temperature (>120°C), whereas N-nitrosamines are powerful carcinogens associated with cured meats and texturized proteins. The food-contact surfaces (such as cookers, pans), packaging materials (paper, plastic, metal, glass) and counter tops also contribute to food contamination with chemicals, many of which pose significant health hazards. The PCBs which are used widely in industry as dielectric fluids, as plasticizers and as fluids for heat transfer can also lead to food contamination.

24.3 Control and Preventive Measures

Food-borne illnesses can be prevented through measures which reduce or prevent contamination from the farm to the table. World Health Organization (WHO) has identified **Five Keys to Safer Food** based on solid scientific evidence.

1. **Keep Clean:** The disease-causing microorganisms, present in soil, water, animals and people, find way to food through hands, wiping clothes, chopping boards and utensils. Washing hands before handling food and after using toilet; cleaning and sanitizing all food contact surfaces and providing protection from insects and pests are some measures which will ensure food safety. It is also important to use all chemicals judiciously and carefully.
2. **Separate Raw and Cooked Food:** Microorganisms are naturally present in the intestines of healthy animals and can contaminate the carcass of meat and poultry by contact during slaughter. Uncooked food, particularly meat, fish and poultry, may contain harmful microorganisms which can get transferred to other foods during the process of food preparation and storage. Therefore, raw meat, fish and poultry should be kept separate from other foods by using separate equipment and containers.
3. **Cook Food Thoroughly:** Microorganisms are killed at high temperature. Heating the food to 70 °C destroys viruses, parasites and bacteria in 30 seconds. The only exception is *Clostridium* which produces heat-resistant spores and is killed only at temperatures above 100°C. So, all canned foods must be processed at high temperature to prevent this bacterium from growing. As a rule, all food should be cooked thoroughly, particularly meat, eggs, poultry and sea food. Bacteria are generally present on the surface of meat but in case of minced meat, they are present in the whole lot, so it needs adequate heating. Large joints of meat and whole poultry also need more attention. Cooked food should be boiled while reheating to reach the desired temperature.

4. **Keep Food at Safe Temperatures:** Microorganisms multiply at a fast pace when the food is kept at a temperature between 4°C and 63°C which presents the danger zone. Therefore, all cooked food should be refrigerated at a temperature below 5°C and hot food should be kept piping hot at 60°C. Two food-borne bacteria, *Listeria monocytogenes* and *Yersinia enterocolitica* are capable of growing at the low temperature of refrigerator. Other microorganisms can also grow at refrigeration temperature (though at a slower pace). Therefore, food should not be stored for too long even in the refrigerator, and the frozen food should be thawed quickly.

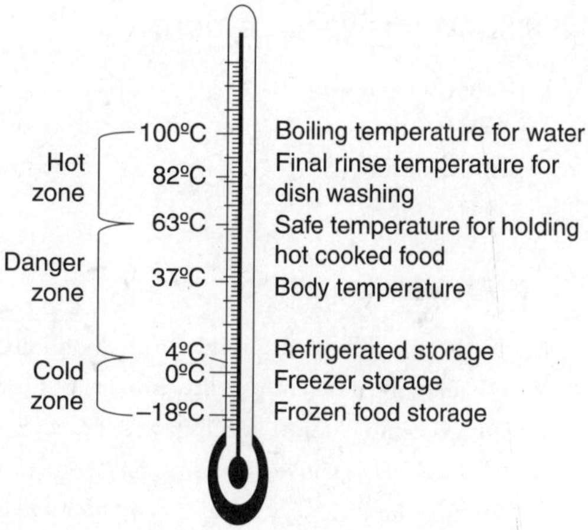

Figure 24.9 Temperature control: A requirement for prevention of food-borne disease

5. **Use Safe Water and Raw Materials:** Microorganisms may be present in the raw material including the water and ice used for food preparation. Bruised or mouldy food may contain harmful toxins. Therefore, only safe potable water; fresh and wholesome raw material should be used. Fruits and vegetables should be washed especially when they are served raw. All packaged food should be checked for the best before date.

In this chapter, we learnt that food-borne diseases are common but preventable public health problems. Food infections are caused by bacteria such as *Salmonella, Clostridium perfringens, E. coli, Shigella, Streptococcus pyogenes* and *Listeria monocytogenes.* Food-borne viruses particularly Hepatitis A virus, rotavirus and noroviruses can cause gastrointestinal disease in human beings. Besides, parasitic and protozoan infections are significant in food safety. This chapter also highlighted the health concerns related to toxins produced by bacteria (*Staphylococcus aureus and Clostridium botulinum*); mycotoxins (aflatoxins, ochratoxin and patulin) produced by fungi and chemicals contaminating the food during production, processing and storage. Last but not least, we studied the basic measures to control and prevent food-borne illnesses to ensure food safety. Many a times, the chefs and processors add substances other than the basic food to improve the latter's quality including the organoleptic attributes. These substances termed as 'additives' are covered in the next chapter.

Know Your Progress

True or False Questions

1. The toxin produced by *Clostridium botulinum* is one of the most toxic substances known.
2. When foods are prepared in large quantities and kept warm for a long time before being consumed, the risk of *Clostridium perfringens* infection increases significantly.
3. *Taenias saginata* is found as live larvae in muscle tissues of uncooked or insufficiently cooked pork.
4. Hepatitis A virus is responsible for major outbreaks caused by contaminated water and food in lesser developed countries.
5. Ochratoxin A is nephrotoxic, hepatotoxic and carcinogenic in nature.

Very Short Answer Questions

1. Define the following terms
   (a) Mycotoxins
   (b) Botulism
   (c) Listeriosis
   (d) Norovirus
   (e) Salmonellosis

Short Answer Questions

1. Discuss the role of viruses in causing food infections.
2. What are the different parasites important for food safety?
3. Compare and contrast botulism with *Staphylococcus* food intoxication.
4. What are the causative organisms, food sources and health concerns of mycotoxins in food poisoning?
5. Describe the Five Keys to Safer Food identified by World Health Organization.

Long Answer Questions

1. Explain the causes, symptoms and food sources of food infections of bacterial origin.
2. What are the different types of food intoxications? Discuss.

Beyond the Text

- A freshly recruited student in a restaurant kitchen is asked to decide the fate of the following dishes/product:
  1. Chicken salad left unrefrigerated overnight
  2. Tuna sashimi prepared with tuna from an unknown source
  3. Can of meat that is puffed
  4. Makhani dal kept warm for over six hours

  Can you help him take a decision? Justify your suggestions in the broader view of food safety.

Chapter 25

Food Additives

Chapter Objectives

After reading this chapter, the reader should be able to

- justify the use of additives in food processing.
- comprehend the meaning of terms related to food additives.
- know the different types of additives and their specific functions in food processing.
- select the appropriate additive for enhancing quality and shelf life of foods.
- use additives with caution to ensure food safety.

Imagine the texture of a cake baked without baking powder or the shelf life of a squash without preservative, or visualize a candy without any colour and flavour. Surely, these products will be low in consumer acceptability. Baking powder, preservatives, colouring and flavouring agents are examples of additives which perform different functions such as raising the volume, preventing spoilage and improving the appearance and flavour of food for better consumer acceptance. Owing to their benefits relating to aesthetics, convenience and safety, the additives have become an integral part of present day processed foods. The only way consumers can to do away with these additives is that they should spend long hours in cooking, accept less attractive products/dishes or withstand increased risks of food spoilage.

THINK TANK

Additives in Food Products

Consider the following substances added to the respective foods:

- Baking powder to cookies
- Sodium nitrite to ham
- Sunset yellow to iced candy
- Pectin powder to pineapple jam
- Sucralose to yoghurt
- Potassium meta bi sulphite to lime juice cordial
- Ajinomoto to chilli garlic noodles
- Nisin to canned rasgulla
- Potassium ferrocyanide to 'free-flowing' table salt
- Aspartame to diet soda

What is the contribution of these additives?

25.1 Definition and Benefits of Food Additives

Food additive, in simple terms, can be defined as a substance or a mixture of substances, other than the basic food stuff, which is present in food as a result of any aspect of production, processing, storage or packaging; the term does not include chance contamination.

GOOD TO KNOW

Definition of Food Additive

'Food additive' means any substance not normally consumed as a food by itself or used as a typical ingredient of the food; whether or not it has nutritive value, the intentional addition of which to food for a technological (including organoleptic) purpose in the manufacture, processing, preparation, treatment, packing, packaging, transport or holding of such food results, or may be reasonably expected to result (directly or indirectly), in it or its by-products becoming a component of or otherwise affecting the characteristics of such food but does not include 'contaminants' or substances added to food for maintaining or improving nutritional qualities.

Source: FSS Act, Rules and Regulations, 2011.

As the definition states, food additives are used intentionally and are expected to become a component of the food to which they are added. During its journey from 'farm to fork', the food is subject to varying environmental conditions such as temperature change, oxidation and exposure to microbes. Addition of additives helps in preserving the appeal, wholesomeness and safety of food during this process. There are many other benefits of using additives; some of which are at times taken for granted (Figure 25.1).

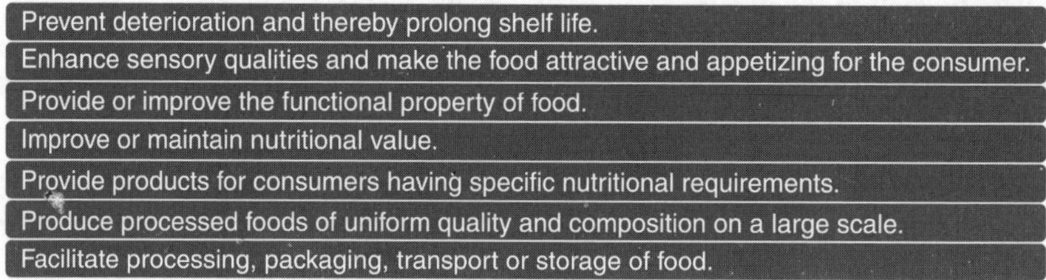

Figure 25.1 Benefits of food additives

Despite the diverse benefits they provide, food additives cannot be used indiscriminately. The addition of only permitted additives (within the prescribed limits) is allowed in foods; their use is prohibited in the following conditions:

1. in infant foods and foods for patients.
2. to disguise deterioration or poor quality of food.
3. to bring about desirable changes without using improved techniques of processing.
4. to make the food appealing without considering the harm to its nutritive value and safety.

25.2 Important Terms

Before we discuss various types of additives and their application in processed foods, let us try to understand the meaning of some terms relevant in this context.

Acceptable Daily Intake: Acceptable Daily Intake (ADI) is a measure of the quantity of a particular additive in food which can be consumed on a daily basis over a lifetime without appreciable health risk to human body. It is expressed on a body-weight basis (milligram of additive per kilogram of body weight) per day but it does not refer to a level of toxicity. The maximum level of additive (with no demonstrable toxic effect) is decided based on the currently available data on animal studies; it is referred to as 'no-observed-adverse-effect level' (NOAEL).

Generally Recognized as Safe (GRAS): Additives may be naturally occurring or synthetic. 'Generally Recognized as Safe' is a designation given to those food additives which are approved by experts as safe based on their extensive history of use in food before 1958 or supported by scientific evidence. Presently, GRAS status has been granted to over 2500 additives including salt, sugar, spices, vinegar, baking soda, citric acid, monosodium glutamate and nisin (refer to box in sub-section 25.3.1).

Delaney Clause: Named after James Delaney, this clause bans the use of any food substance found to be carcinogenic in humans or experimental animals. It governs the regulation of pesticide residues in processed foods and recommends that the tolerance levels be based on only carcinogenic risk and that not on the benefits of the pesticide.

'INS Number': Additives generally have lengthy chemical names, so the Codex Alimentarius Commission has developed an International Numbering System (INS) to identify each food additive by a short code. Another advantage of using 'INS number' is that it approves the safety of additive used in food. This system has been adopted in India under the Food Safety and Standards (Food Products Standards and Food Additives) Regulations, 2011. Label of processed food may list the additive by its 'INS number' or by its generally known name. For example, 100–199 refer to colour additives, 200–299 denote preservatives, 300–399 imply antioxidants, whereas 400–499 represent emulsifiers, stabilizers, thickeners and gelling agents. On similar lines, the European Union (EU) has developed an 'E Number' to identify each food additive. The E numbers generally correspond to INS numbers for the same additive.

GOOD TO KNOW

INS numbers of Common Food Additives

Preservatives		*Antioxidants*		*Emulsifiers, Stabilizers, Thickeners and Gelling Agents*	
200	Sorbic acid	300	Ascorbic acid (L)	322	Lecithins
210	Benzoic acid	301	Sodium ascorbate	407	Carrageenan
220	Sulphur dioxide	306	Mixed tocopherols	412	Guar gum
224	Potassium metabisulphite	310	Propyl gallate	413	Tragacanth
234	Nisin	316	Sodium isoascorbate	414	Acacia gum
251	Sodium nitrate	319	Tertiary-butyl hydroquinone (TBHQ)	415	Xanthan gum
252	Potassium nitrate	320	Butylated hydroxyanisole (BHA)	440	Pectins
280	Propionic acid	321	Butylated hydroxytoluene (BHT)	460	Cellulose
282	Calcium propionate	392	Extracts of rosemary	466	Carboxy methyl cellulose

Colours		*Sweeteners*		*Others*	
100	Curcumin and turmeric	420	Sorbitol and sorbitol group	260	Acetic acid glacial
102	Tartrazine	950	Acesulfame K	290	Carbon dioxide
110	Sunset Yellow	951	Aspartame	296	Malic acid
123	Amaranth	954	Saccharin (and Na, K and Ca salts)	330	Citric acid
124	Ponceau 4R	955	Sucralose	422	Glycerol
127	Erythrosine	967	Xylitol	948	Oxygen

GMP for Additive Use: As per the Food Safety and Standards (Food Products Standards and Food Additives) Regulations, 2011(refer to Section 27.2.1 in Chapter 27), 'Good Manufacturing Practices (GMP) for the use of food additive' mean that

1. the quantity of additive added to the food is limited to the lowest possible level; necessary to achieve the desired effect.
2. the quantity of additive becomes a component of food and is not intended to achieve any physical or other technical effect in the food itself (e.g. use of citric acid in preserve, squash and syrup).
3. the additive is handled in the same way as a food ingredient (e.g. use of pectin in jam, jelly and marmalade).

25.3 Types of Additives

Food processors and culinary chefs use a large variety of food additives which can be classified in to different groups such as preservatives, antioxidants, sweeteners, colouring agents, flavouring agents and stabilizers (Table 25.1). A single additive may perform many functions so it can be listed under more than one category.

Table 25.1 Types of food additives, their functions and applications

Additive	Functions	Applications	Examples
Preservatives	Prevent food spoilage from microorganisms	Fruit beverages, baked goods, cured meats	Sodium benzoate, calcium propionate, citric acid, sodium nitrite, calcium sorbate
Antioxidants	Delay rancidity and prevent changes in colour, flavour or texture	Oils and margarines, cereals, dressings, fried and baked snack foods	Butylated hydroxyanisole (BHA) and butylated hydroxytoluene (BHT), tocopherols (vitamin E)
Sweeteners	Add sweetness with or without the extra calories	Beverages, baked goods, confectionery, table-top sugar, substitutes, many processed foods	Sucrose (sugar), glucose, fructose, sorbitol, mannitol, corn syrup, aspartame, acesulfame-K, sucralose
Colouring agents	Correct natural variations in colour; enhance colours that occur naturally; provide colour to colourless and 'fun' foods	Sweetmeats, candies, snack foods, margarine, cheese, soft drinks, fruit beverages, jams, jellies, gelatins, pudding and pie fillings	Tartrazine, sunset yellow, ponceau 4R, indigo carmine, fast green, annatto extract, beta-carotene, caramel, saffron
Flavouring agents	Impart, modify or enhance the flavour	Squashes, jam, preserves, ice cream, desserts	Herbs, spices, citral, vanillin, allyl hexanoate
Flavour enhancers	Enhance flavours already present in foods	Many processed foods	Monosodium glutamate, hydrolyzed soy protein

(Continued)

Table 25.1 (Continued)

Emulsifiers	Prevent separation and keep emulsified products stable	Salad dressings, chocolate, table spread, bread spread, margarine, frozen desserts	Soy lecithin, mono- and di glycerides, egg yolks, sorbitan monostearate
Stabilizers and thickeners	Produce uniform texture, improve 'mouth-feel'	Frozen desserts, dairy products, cakes, dry soup powders, dressings, sauces	Gelatin, chitin, inulin, pectin, guar gum, carrageenan, xanthan gum, whey
Leavening agents	Promote rising of baked goods	Cakes, biscuits and other bakery products	Baking powder, mono calcium phosphate, calcium carbonate
Dough Conditioners	Produce more stable dough	Breads and other baked goods, frozen doughs	Ammonium chloride, potassium bromate
Anti-caking agents	Keep powdered foods free-flowing, prevent moisture absorption	Salt, baking powder, confectioner's sugar	Calcium silicate, ammonium calcium silicate, potassium ferrocyanide
Sequestrants	Bind with trace metals to prevent oxidative changes in colour and flavor	Dairy products, meat products, fats and oils	Citric acid, malic acid, ethylene di amine tetra acetic acid (EDTA) and poly phosphates
Humectants	Retain moisture	Soft candies	Glycerin, sorbitol
Nutrients	Replace micronutrients lost in processing (enrichment), add nutrients lacking in the diet (fortification)	Flour, breads, cereals, rice, margarine, salt, milk, fruit beverages, energy bars, instant breakfast drinks	Thiamin, riboflavin, niacin, folate, beta carotene, potassium iodide, iron or ferrous sulfate, tocopherols, ascorbic acid, vitamin D
Firming Agents	Maintain crispness and firmness	Processed fruits and vegetables	Calcium chloride, calcium lactate
Fat replacers	Provide expected texture and a creamy 'mouth-feel' in reduced-fat foods	Baked goods, dressings, frozen desserts, confections, cake mixes, dairy products	Olestra, cellulose gel, carrageenan, modified food starch, guar gum, xanthan gum, whey protein concentrate

25.3.1 Preservatives

Preservatives are intentionally added to inhibit, retard or arrest the growth of microorganisms (e.g. bacteria, yeast, mould) and thereby prevent fermentation, acidification or other decomposition of food. FSS Act, Rules and Regulations, 2011 have categorized them as Class I and II preservatives (Figure 25.2).

Table 25.2 lists the commonly used preservatives along with their active compounds and application. The sulphites release sulphur dioxide which distributes itself uniformly in the product and is, therefore, preferred over benzoic acid. Due to its bleaching property, sulphites are not added to coloured products such as tomato ketchup where benzoic acid is a better option. Nitrites impart colour and flavour to cured meat and more importantly they inhibit the growth of *Clostridium botulinum* – a food poisoning bacterium which causes botulism (refer to sub-section 24.2.1 in Chapter 24).

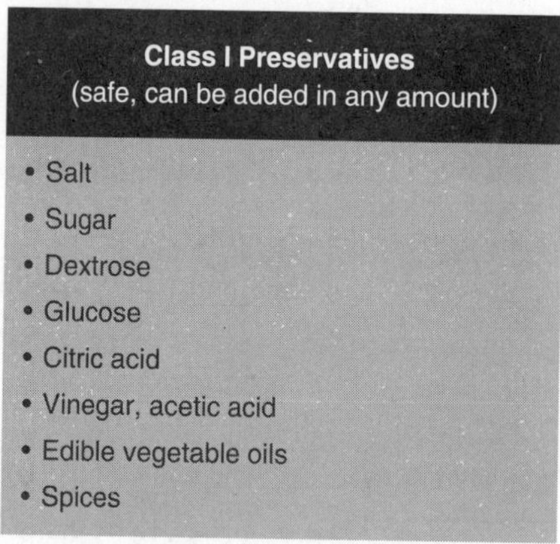
Class I Preservatives
(safe, can be added in any amount)

- Salt
- Sugar
- Dextrose
- Glucose
- Citric acid
- Vinegar, acetic acid
- Edible vegetable oils
- Spices

Class II Preservatives
(chemicals, allowed only within the prescribed limits)

- Benzoic acid and benzoate
- Sulphurous acid and sulfites
- Nitrates and nitrites
- Sorbic acid and its sodium, calcium and potassium salts
- Propionic acid and its calcium and sodium salts
- Nisin

Figure 25.2 Categorization of preservatives

Table 25.2 Common preservatives used in processed foods

Preservatives and their E Number	Active Compound	Microorganism Affected	Application in Food
Sulphur di oxide and sulphites (220–228)	Sulphur dioxide	All microbes	Fruit beverages (squashes, cordial), wines and fermented foods
Calcium propionate (282)	Propionic acid	Moulds	Breads, flour confectionery (cakes)
Sorbic acid and its calcium/ potassium salts (200–203)	Sorbic acid	Moulds; bacteria (*Bacillus mesentericus*)	Salad dressing, cakes, cheese, cider, wine, instant mixes, prunes, preserved chapattis
Benzoic acid and its salts (210–219)	Benzoic acid	Yeast and moulds	Tomato ketchup, soft drinks
Nitrates and nitrites (249–252)	Nitrite	*Clostridium botulinum*	Cured meat products (ham, bacon, frankfurters)

Preservatives and antioxidants both delay spoilage but the latter aim to prevent oxidative changes, where as the preservatives are used primarily to prevent microbial changes. Ideally, the preservatives should be affordable, non-toxic and safe for human consumption; and they should not affect adversely the sensory quality of food (refer to box in sub-section 22.6.5 of Chapter 22). They are added to many products including baked foods, fermented foods, wine, cheese, cured meats, fruit juices, margarine and processed fruits and vegetable products.

DID YOU KNOW

What Is Nisin?

Nisin is commercially the most important bacteriocin produced by a bacterium (*Lactococcus lactis* subsp. *lactis*). Due to its ability to kill or inhibit the growth of other bacteria (such as *Lactobacilli*), this polypeptide is used as a preservative in *paneer*, canned *rasgulla* and pre-packaged coconut water. FAO has granted nisin a GRAS status.

25.3.2 Antioxidants

Oxygen is one the important factors which affects stability of vegetable oils. It may be present in the oil or the head space of the package or it may gain entry by permeation through the package. Oxidation results in rancid flavour not only in the oil but also in fish, meat and poultry and the food prepared/fried in vegetable oils (refer to Chapter 3). ***Antioxidants*** are additives which prevent oxidative deterioration of food such as rancidity in edible fats and vegetable oils and salad dressings and also inhibit enzymatic browning of cut fruits and vegetables (refer to Chapter 10). They can be categorized into following two groups:

1. *Naturally occurring antioxidants*, e.g., ascorbic acid (INS300), tocopherols (INS306–309), citric acid (INS300) and some phenolic compounds.
2. *Synthetic antioxidants*, e.g., butylated hydroxyanisole (BHA or INS320) and butylated hydroxytoluene (BHT or INS321), propyl gallate (PG or INS310), sodium isoascorbate (added to cured meats) and sulphur dioxide (used in fruits and vegetables).

Removal of natural antioxidants (such as tocopherol) during the process of refining makes the oils more prone to rancidity. Therefore, antioxidants are added to prevent oxidative changes and to prolong the shelf life of refined vegetable oils.

25.3.3 Sweeteners

The additives which contribute to the sweetness of food and beverages are known as *sweeteners*. Sucrose (table sugar) is the most commonly used *natural sweetener*. Fructose is twice as sweet as sucrose and does not crystallize due to its high solubility in water. Invert sugar is a 50:50 mixture of glucose and fructose and is produced by enzymatic or acidic hydrolysis of sucrose. It is used to prevent crystallization of sugar in some foods. Other examples of natural sugars are lactose (milk sugar), maltose (malt sugar), honey, molasses, corn syrup, high-fructose corn syrup and maple sugar syrup. *Artificial sweeteners* include intense sweeteners and bulk sweeteners and they find many uses in food processing (Figure 25.3).

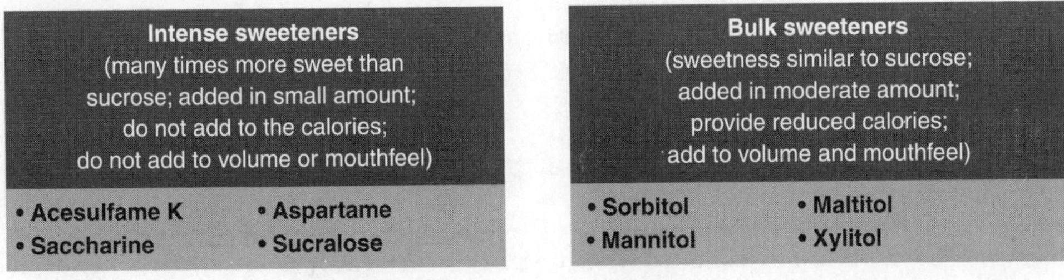

Figure 25.3 Intense and bulk sweeteners

1. *Intense sweeteners* (non-nutritive sweeteners) have sweetness many times that of sucrose. The sweeteners permitted by FSS Act include acesulfame K (INS950), aspartame (INS951) and saccharin (INS954) which are 130–200 times, 200 times and 300–500 times sweeter than sugar respectively. Since they do not add to the calories, these additives are highly beneficial in low-calorie and special dietary products for the diabetics where they provide the sweetness but do not appreciably increase the blood glucose levels. Aspartame is unstable in heat and should not be consumed by phenylketonuria

patients. Sucralose (INS955) is heat-stable, 400–800 times sweeter than sucrose and is highly soluble in water. Saccharin is the oldest artificial sweetener and is sometimes not acceptable because of its bitter after-taste.

2. *Bulk sweeteners* (nutritive sweeteners) have same/lower sweetness as sucrose but provide reduced calories (2.4 kcal/g) as compared to sucrose (4 kcal/g). They include sorbitol (INS420) and maltitol (INS965) which are added to energy-reduced foods and additionally provide volume and mouth feel. Sorbitol and malitol are less sweet than sucrose according to sweetness index (refer to Chapter 2).

25.3.4 Food Colours

Colour of food is an important attribute for consumer acceptability because people 'eat with their eyes' and their palates. *Colouring agents* enhance the colour and thereby the visual appeal of a wide variety of foods such as bakery products, snack foods, ice cream, beverages, jams, jellies, puddings and desserts in many ways. For example, orange colour is added to an orange-flavoured beverage to lend credibility to the flavour and to improve the consumer acceptance. The colours are added to cakes and confectionery products purely for visual decoration, and they brighten the fun foods.

Natural colouring agents permitted by FSS Act include carotene, carotenoids, chlorophyll, riboflavin, caramel, cucumin/turmeric, annatto and saffron. However, most of the natural pigments are unstable during processing and storage conditions; so they may lose colour, whereas the flavour, texture, safety and nutritive value are still intact.

GOOD TO KNOW

Functions of Food Colours

- Enhance natural colours which are of lighter tone than generally associated with that food
- Compensate colour loss due to exposure to heat, light, air, moisture and storage
- Amend natural or seasonal variations in colour and ensure batch uniformity in production
- Provide colour to colourless and 'fun'foods

Synthetic or man-made colouring agents are more intense and less expensive than natural colouring agent. They have better colouring ability, uniformity and stability and are needed in very small quantities. Synthetic food colours permitted by FSS Act are listed as follows:

Red	Ponceau 4R, Carmoisine, Erythrosine	Yellow	Tartrazine, Sunset Yellow
Green	Fast Green	Blue	Indigo Carmine, Brilliant Blue

Under the FSS Act, the use of permitted synthetic food colours is allowed in only the following foods:

1. Ice-cream, milk lollies, frozen desserts, flavoured milk, yoghurt and ice cream mix powder.
2. Biscuits, biscuit wafers, pastries, cakes, confectionery, thread candies, sweets, and savouries.
3. Peas, strawberries and cherries in cans, preserved/processed papaya, canned tomato juice, fruit syrup, fruit squash/crush/cordial, jam, jelly, marmalade and candied/crystallized/glazed fruit.
4. Non-alcoholic carbonated and non-carbonated ready-to-serve synthetic beverages including synthetic syrups, sherbet, fruit drinks, synthetic soft-drink concentrates, fruit bar and fruit beverages.

5. Custard powder. **6.** Jelly crystal and ice candy.

7. Flavour emulsion and flavour paste for use in carbonated or non-carbonated beverages.

Generally, the **maximum permitted limit** for use of permitted synthetic food colour is 100 parts per million (ppm) of the final quantity of food/beverage. The colour used should be pure and free from impurities.

25.3.5 Flavouring Agents and Flavour Enhancers

Flavouring agents are added to foods to impart, modify or enhance the flavour (odour or taste or both). Forming the largest single group of additives, they find many uses in processed foods such as dairy food, spices and herb blends, sea food, fruit/vegetable products and wine (Figure 25.4). They do not include substances (such as sugar, salt or vinegar) which are exclusively sweet, salty or sour in taste.

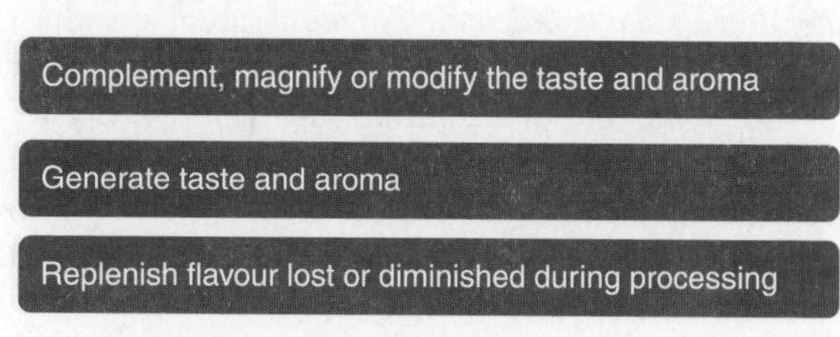

Figure 25.4 Functions of flavouring agents

They comprise of flavouring substances (natural or synthetic) and flavours produced as a consequence of thermal process or smoking and at times contain non-flavouring food ingredients. The chemicals responsible for flavours include alcohols, esters, aldehydes, ketones and protein hydrolysates.

Natural flavours and Natural flavouring substances mean flavour preparations and single substance, respectively. They are extracted from plants, herbs and spices by physical processes. They include essential oils, oleoresins, herbs and spices (refer to Chapter 9).

Nature-identical flavouring substances are substances which are or obtained synthetically or chemically isolated from aromatic raw materials. Chemically, they are identical to substances present in natural products.

Artificial flavouring substances are substances which have not been identified in natural products.

Some examples of synthetic flavouring agents are as follows:

Isoamyl valerate	⟶	Banana	Ethyl vanillin	⟶	Vanilla
Allyl hexanoate	⟶	Pineapple	2-trans-6-cis nonadienal	⟶	Cucumber
Methyl anthranilate	⟶	Grape	Benzaldehyde	⟶	Almond
Benzoyle butyrate	⟶	Raspberry	Diacetyl	⟶	Buttery
Citral, Lemonene	⟶	Lemon	2-isobutyl thiazole	⟶	Fresh tomato
Nootkatone	⟶	Grapefruit	2-methoxy-3isobutyl pyrazine	⟶	Green pepper

There is restriction on the use of some flavouring agents such as caumarine and hydro caumarine.

While adding flavouring agents, following precautions must be taken:

1. Use within safe levels of intake.
2. Do not compromise on the purity level.
3. Do not mislead the consumer about the nature or quality of food.
4. Employ good manufacturing practices (GMPs) which include limiting the quantity used in food to the lowest level necessary to accomplish the desired flavouring effect.
5. Limit the non-flavouring food ingredients to the lowest level required to ensure the safety and quality of the flavourings, facilitate their storage and ease of use.

Flavour enhancers intensify or enhance the flavour of other ingredients in food. Monosodium glutamate (MSG or INS621), the best known example of flavour enhancer, is a GRAS additive and is added primarily to savoury products and oriental dishes. There are reports to suggest that MSG can cause brain damage in mice so its use is prohibited in infant foods.

25.3.6 Emulsifying and Stabilizing Agents

Emulsifying and stabilizing agents mean substances which are capable of facilitating a uniform dispersion of oils and fats in aqueous media or vice versa, and/or stabilizing such emulsions. They are widely used in the production of chocolate, margarine, bread, ice cream, ice confectionery and flour confectionery. The *emulsifying agents* help in the formation of a stable emulsion, for example, lecithin, mono- and diglycerides, egg yolks and sorbitan monostearate (Figure 25.5).

Figure 25.5 Lecithin present in egg yolk is an important emulsifying agent in food preparation

The *stabilizing agents* help in maintaining an emulsion after it has formed and in producing uniform texture and consistency. The latter include alginates (from kelp), carrageenan (a seaweed derivative), cellulose, gums (arabic, guar), pectin, dextrins, hydrocolloids, gels and starches. Based on their properties, stabilizers find many uses in food production (Table 25.3).

Table 25.3 Application of stabilizing and thickening agents as per their properties

Property	Application in Food Products	Property	Application in Food Products
Binding agent	Formed meat (sausages)	Thickening agent	Sauces, gravies
Coating agent	Confectionery	Whipping agent	Icing, toppings
Gelling agent	Jam, jelly, puddings	Crystallization inhibitor	Ice cream, sugar syrups
Stabilizer	Salad dressings	Clouding agent	Fruit juices
Suspending agent	Chocolate milk		

Stabilizers and thickeners perform many functions. They:

1. protect emulsions and prevent volatile oils from escaping.
2. provide consistency and texture to many foods.
3. absorb and hold water and thereby stabilize, gel or thicken foods.
4. increase the viscosity of final product and prevent the formation of ice crystals in ice creams.
5. enhance visual appeal and mouthfeel.

25.3.7 Leavening Agents

Leavening agents raise the volume of batters and doughs by releasing carbon dioxide. During baking, the release of carbon dioxide coupled with the incorporated air (through sifting of flour; beating and creaming of batters and doughs) and water vapour (formed by evaporation of moisture present) contributes to the desired porous and light texture of bakery products (bread, biscuits and cakes).

Chemical leavening agents such as sodium bicarbonate (baking soda) and ammonium bicarbonate (used n cookies) react with an acid to produce carbon dioxide. When used alone, baking soda reacts quickly with eat to produce this gas which may escape from unbaked batter. The excessive use of baking soda may result n soapy flavour, excessively brown coarse textured product.

Baking powder, a leavening agent used in bakery products, is a mixture of sodium bicarbonate, dry acid nd inert corn starch filler. The starch filler prevents sodium bicarbonate from prematurely reacting with cid. Earlier, tartarate and phosphate were used as dry acid but now sodium aluminium sulphate (SAS) hosphate is used.

DID YOU KNOW

What Is Double-acting Baking Powder?

Double-acting baking powder consists of two acid ingredients: one fast acting and another slow acting. The fast acting part (e.g. calcium acid phosphate) releases a good share of their leavening gas in a relatively short time after mixing and continues to release gas rather rapidly while the batter is on the bench to keep the batter smooth flowing. The slow acting ingredient (e.g. sodium aluminium sulphate, SAS) does not release the most of their leavening gas until the batter is heated in the oven. It liberates the gas uniformly during baking to hold the raised batter until set. Double-acting baking powder is commonly used as a leavening agent in many bakery products (Figure 25.6)

Figure 25.6 Baking powder: A leavening agent for cakes

5.3.8 Dough Conditioners

ough conditioners or dough improvers modify starch and protein (gluten) of flour in bread making. As a result, ough handling and baking quality in terms of increased volume and uniform grain improves considerably. mmonium chloride, potassium bromate and diammonium phosphate are examples of dough improvers. SS Act (2006) permits use of sodium bisulphate and sodium metabisulphate as dough conditioners in breads nd biscuits.

5.3.9 Anti-caking Agents

nti-caking agents are anhydrous substances which pick up moisture without themselves becoming wet. hey are added to prevent lumping or caking in fine powders such as dry mixes, salt and baking powder keep them 'free flowing'. Carbonates and phosphates of calcium and magnesium; silicates of calcium, agnesium and aluminium; and calcium potassium or sodium ferrocyanide are among the examples of ti-caking agents permitted in India.

5.3.10 Sequestrants

nder FSS Act, Rules and Regulations, *sequestrants* are defined as substances which prevent adverse effects of etals catalysing the oxidative breakdown of foods forming chelates thus inhibiting decolourization, off-taste and ncidity. They are also known as *chelating agents* or metal *scavengers* (refer to Chapter 3). Besides, they catalyse undesirable enzymatic browning (refer to sub-section 10.5.1 in Chapter 10). Examples of sequestrants are citri acid, malic acid, ethylene diamine tetra acetic acid (EDTA) and poly phosphates. They are added to dairy prod ucts, meat products and fats and oils to stabilize colour and flavour and to prevent the loss of vitamins.

GOOD TO KNOW

Applications of Acid Regulators in Food Processing

Acids (e.g. citric acid and potassium acid tartarate) and bases or alkaline substances (sodium hydroxide, sodium bicarbonate) are additives which find many applications in processing of food.

Acid/Base	Applications in Food Processing
Citric acid	canned foods (to lower pH and help in sterilization at less severe temperature)
Potassium acid tartarate	fondant and fudge (to inhibit crystallization of sugar)
Sodium hydroxide	ripe olives (to remove bitterness and develop a darker colour) pretzels (to alter proteins and starch, smoothen surface and develop deep brown colour) soy (to solubilize the protein) fruits and vegetables (to aid in peeling)
Sodium bicarbonate	peanut brittle (to make the texture porous and colour darker brown)
Potassium carbonate	dark chocolate (to enhance browning)

Details of other additives such as humectants, nutritive additives, fat replacers, anticaking agents, antifoamir agents (which reduce foams, e.g., during the boiling of jam) and firming agents are beyond the scope of this bool

25.4 Safety of Food Additives

Consumers often perceive additives as complex chemical compounds having long and unfamili names. Sometimes, they are apprehensive about the safety issues concerning these substances. It important to understand that every additive undergoes a careful analysis before being approved for u in foods. At international level, Joint FAO/WHO (Food and Agriculture Organization/World Heal Organization) Expert Committee on Food Additives (JECFA) evaluates the safety of food additive Prevention of Food Adulteration (PFA) Act, 1954 and the recently introduced Food Safety and Sta dard Act (2006) are national food law governing the use of food additives in India (refer to Chapter 27

GOOD TO KNOW

What Is JECFA?

The Joint FAO/WHO Expert Committee on Food Additives (JECFA) is an international expert scientific committee that evaluates safety of food additives and contaminants and naturally occurring toxicants and residues of veterinary drugs in food. In context of food additives, it

1. explains principles for evaluating their safety and risks;
2. carries out toxicological evaluations and determines acceptable daily intakes (ADIs);
3. evaluates the execution, quality and applicability of analytical methods;
4. brings out specifications related to purity of food additives; and
5. measures exposure of people to chemical substances in food.

Due to strict regulation and extensive testing, food additives can be considered to be safe for human beings. However, there are some safety related issues concerning few additives. Use of colours has been associated with increasing hyperactivity, whereas additives such as colouring agents, preservatives (*sulphites, benzoic acid*), artificial sweeteners (*aspartame*) and antioxidants (*butylated hydroxyanisole, BHA,* and butylated hydroxytoluene, BHT) have been reported to trigger generalized allergic reactions and/or specific symptoms such as hives, urticaria and asthma. It has been found that an oral dose of flavour enhancer *mono sodium glutamate* (MSG) of 3 g in the absence of food can result in 'MSG symptom complex' encompassing symptoms such as burning sensation/numbness of the back of the neck, forearms and chest, facial tightness, chest pain, headache, nausea and bronchospasm (in asthma patients). Therefore, MSG is prohibited in foods for infants and pregnant women. *Nitrates,* added to cured meats, are definite potential hazard to human health because they react with the natural breakdown products of proteins to form cancer-causing nitrosamines. Use of saccharine in massive doses can cause cancer. Addition of *base/alkali* can result in loss of nutrients (thiamine) and racemization of amino acids. Therefore, additives should be used judiciously.

In this chapter, we learnt about different food additives grouped as preservatives, antioxidants, sweeteners, colouring agents, flavouring substances and stabilizers. Synthetic colourants and flavouring agents are more convenient options as compared to natural ones as they are more intense and less expensive, provide better uniformity and stability and are needed in small quantities. Use of chemical preservatives is permitted only in restricted amount to prolong the shelf life of processed foods. We discussed that intense sweeteners provide sweet sensation to the taste buds but do not appreciably increase blood glucose levels. The benefits of using emulsifiers, stabilizers, dough conditioners, leavening agents and sequetrants were also deliberated.

Additives help immensely in making available flavourful, nutritious, safe, convenient, colourful and affordable foods to meet consumer expectations throughout the year. National and international regulations ensure that each of these additives is safe at its intended level of use before it is added to the food. Food processors and culinary experts should also resist the indiscriminate addition of additives to their preparations. The general criteria for their use should be that the additives perform a useful purpose, are safe and do not mislead the consumer. The next chapter focuses on food contaminants and adulterants.

Know Your Progress

True or False Questions

1. Antioxidants are substances which combine with trace amounts of unwanted metals so that they are not available for hastening deteriorative reactions.
2. Anti-caking agents keep powdered foods free-flowing by preventing moisture absorption.
3. Propionates are preservatives which inhibit the growth of *Clostridium botulinum* in cured meat products.
4. Butylated hydroxyanisole (BHA) and butylated hydroxyltoluene (BHT) are examples of emulsifiers.
5. Alginates, cellulose, gums and pectin contribute to the thickness and stability of foods.

Very Short Answer Questions

1. Give the full form of JECFA.
2. What is double-acting baking powder?
3. How does a dough conditioner help in bread making?
4. List four natural colouring agents.
5. Differentiate between flavouring agents and flavour enhancers.

Short Answer Questions

1. What are the benefits of food additives?
2. List the conditions where the use of additives is prohibited.
3. Define the following terms.
   (a) Acceptable Daily Intake
   (b) GRAS additives
   (c) Delaney Clause
   (d) E Number
4. Describe four artificial sweeteners used as food additives.
5. Discuss briefly the role of four chemical preservatives.

Long Answer Questions

1. Describe the functions of colouring agents, flavouring substances and leavening agents in food processing citing suitable examples.
2. What are the benefits and applications of emulsifiers, stabilizers, thickeners and anti-caking agents?

Beyond the Text

- Visit a self-service store and pickup 10 packs of food products (such as cakes, biscuits, snacks, chocolates, jams, pickles and fruit beverages). Read carefully the ingredients listed on the label, identify the food additives and categorize them as preservatives, colouring agents, flavourings and emulsifiers. Refer to the additives listed in Box 25.1 and identify the category they belong to.

Chapter 26

Food Contaminants and Adulterants

Chapter Objectives

After reading this chapter, the reader should be able to

- differentiate between contaminants and adulterants.
- recognize food contaminants from various sources.
- minimize contamination during preparation, processing and storage of food.
- identify adulterants commonly associated with food.
- learn and apply various methods of detecting food adulteration.

n chapter 24, we studied that food intoxication is aused by harmful toxins or chemicals present in ood before it is consumed. One example of bacterial oxin is the *Staphylococcal* enterotoxin which led to ood poisoning among thousands of people through ontamination of reconstituted milk in Osaka, Japan, n the year 2000. Consider another example where ne consumption of mustard oil adulterated with rgemone oil led to several cases of epidemic dropsy in Delhi, India, in 1990s. Food contaminants and adulterants form a significant group of substances which an lead to adverse health outcomes. Adequate knowledge about contaminants and adulterants is essential or the hospitality professionals so that they can adopt ppropriate measures to ensure food safety and quality.

THINK TANK

Contaminants and Adulterants in Various Foods

Consider the following examples:

- *Staphylococcal* enterotoxin in reconstituted milk
- Aflatoxin in maize
- Polycyclic aromatic hydrocarbons in smoked fish
- Argemone oil in mustard oil
- Metanil yellow in turmeric powder
- Methanol in liquor
- Melanin in milk
- Vanaspati in pure ghee

What do these substances have in common? These are examples of contaminants and adulterants in foods.

6.1 Difference Between Contaminants and Adulterants

ood Safety and Standards Act defines **food contamination** as the presence of harmful chemicals and microrganisms in food which can cause consumer illness (FSS Act, Rules and Regulations, 2011). A **contaminant** neans any substance, whether or not added to food, which is present in such food as a result of the production ncluding operations carried out in crop husbandry, animal husbandry or veterinary medicine), manufacture, processing, preparation, treatment, packing, packaging, transport or holding of such food, or as a result of environmental contamination and does not include insect fragments, rodent hairs and other extraneous matter. The term comprises of bacterial toxins, mycotoxins, seafood toxins, pesticide residues, veterinary drug residues and chemical contaminants from packaging and processing (Figure 26.1).

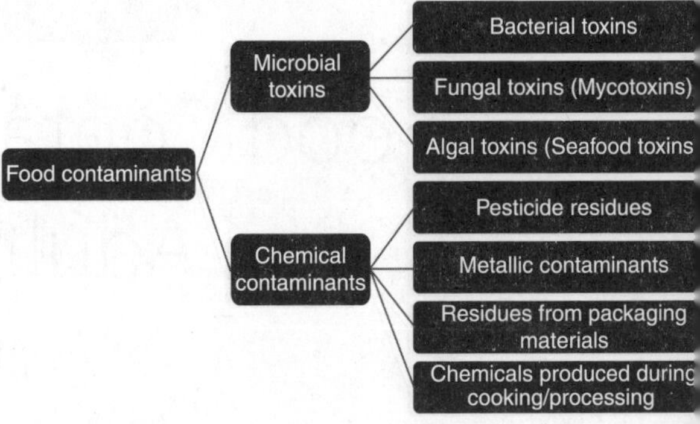

Figure 26.1 Categorization of food contaminants

On the other hand, **adulteration** includes not only the intentional addition, substitution or abstraction of substances which adversely affect the nature, substances and quality of foods, but also their incidental contamination during harvesting, storage, processing, transpo and distribution. **Adulterant** means any material which is or could be employed for making the food unsaf sub-standard or mis-branded or containing extraneous matter (FSS Act, 2006).

GOOD TO KNOW

Definition of Adulteration

An article of food shall be deemed to be adulterated:

1. if the article sold by a vendor is not of the nature, substance or quality demanded by the purchaser and is to his prejudice, or is not of the nature, substance or quality which it purports or represents to be;
2. if the article contains any other substance which affects, or if the article is so processed as to affect, injuriously the nature, substance or quality thereof;
3. if any inferior or cheap substance has been substituted wholly or in part for the article so as to affect, injuriously the nature, substance or quality thereof;
4. if any constituent of the article has been wholly or in part abstracted so as to affect, injuriously the nature, substance or quality thereof;
5. if the article has been prepared, packed or kept under insanitary conditions whereby it has become contaminated or injurious to health;
6. if the article consists of wholly or in part of any filthy, putrid, rotten, decomposed or diseased animal or vegetable substance or is insect-infected or is otherwise unfit for human consumption;
7. if the article is obtained from a diseased animal;
8. if the article contains any poisonous or other ingredient which renders it injurious to health;
9. if the container of the article is composed of poisonous or deleterious substance which renders its contents injurious to health;
10. if any colouring matter other than that prescribed in respect thereof is present in the article, or if the amount of prescribed colouring matter which is present in the article are not within the prescribed limits of variability;
11. if the article contains any prohibited preservatives or permitted preservatives in excess of the prescribed limits;
12. if the quality or the purity of the article falls below the prescribed standard or its constituents are present in quantities not within the prescribed limits of variability, which renders it injurious to health;
13. if the quality or the purity of the article falls below the prescribed standard or its constituents are present in quantities not within the prescribed limits of variability, which does not render it injurious to health.

Source: PFA Act, 1954.

26.2 Types of Contaminants

Some toxicants are naturally present in certain foods. For example, fluoride present in drinking water can lead to fluorosis characterized by mottling of teeth and skeletal and neurological disorders. Oxalic acid (in spinach) and gossypol (in cottonseed flour and cake) can cause renal calculi and cancer, respectively. Another example is solanine (in potatoes) which is the cause of solanine poisoning leading to vomiting, diarrhoea and abdominal pain. Cryogenic compounds found in bitter almonds can also cause gastrointestinal disturbances. Khesari dal (*Lathyrus sativus*) contains beta-N-oxalyl amino-L-alanine (L-BOAA)–a toxin which can lead to a crippling disease known as lathyrism.

Food contaminants can be broadly categorized based on their origin as microbial toxins (bacterial, fungal and algal) and chemical contaminants (pesticide residues, metallic contaminants, residues from packaging materials and chemicals produced during cooking/processing). The categories have been listed in Figure 26.1.

26.2.1 Bacterial Toxins

We know that bacterial toxins produced by *Staphylococcus aureus* and *Clostridium botulinum* lead to food intoxication (Table 26.1).

***Staphylococcus* enterotoxins** produced by *Staphylococcus aureus* are one of the major toxins resulting in food-borne disease. The intoxication is not very serious from clinical point of view but the abrupt onset of symptoms such as vomiting, diarrhoea and cramps and low blood pressure often incapacitates the individuals. Therefore, some of *staphylococcus* enterotoxins are considered as potential agents of bio-terrorism.

The bacterium generally enters the food (meat, fish, egg-based dishes) through infected workers (nasal passages, cuts or sores) or through cross-contamination. Inadequate temperature control following the contamination helps the bacteria to multiply and produce the enterotoxins. These proteinaceous and water-soluble toxins can tolerate boiling temperature for nearly 30 minutes and are not destroyed by the proteolytic enzymes. In other words, they can survive cooking and digestion processes. The bacterium, *Staphylococcus aureus*, also has high tolerance to salt and changes in pH but can be destroyed easily by heat. Therefore, the preventive strategies should focus on (i) preventing contamination with *staphylococci*; (ii) inhibiting its growth and destroying the bacterium before it produces the enterotoxin; and (iii) controlling the production of toxin.

Botulinum toxins are produced by spore-forming *Clostridium botulinum*. Among them, only Type A, B, E and F affect humans and Type A is the most toxic. The symptoms which appear 12–36 hours after the ingestion of contaminated food include nausea, vomiting and diarrhoea. Subsequently, these neurotoxins lead to muscle weakness, blurred vision and difficulty in speaking. If untreated, it may cause respiratory failure and death.

Table 26.1 Health effects of bacterial toxins

Bacterial Toxin	Bacteria Involved	Foods Commonly Involved	Health Effects
Staphylococcus aureus enterotoxin	*Staphylococcus aureus*	Custard/cream filled products, salads, meat and meat products	Salivation, vomiting, abdominal cramps, diarrhoea, thirst
Botulinum toxins	*Clostridium botulinum*	Defectively canned low or medium acid foods, meats	Botulism (double vision, respiratory failure, death)
Bacillus cereus enterotoxin	*Bacillus cereus*	Cereal products, custard, puddings	Nausea, vomiting, abdomen pain, diarrhoea
Scombrotoxins	Marine bacteria	Fish (tuna, mackerel)	Allergy-like symptoms

The products associated with botulism generally include the home-processed foods. The risk of botulism from fresh foods whether they are eaten raw or cooked is very little. *Clostridium botulinum* toxin is formed only when its spores germinate under optimal environment (anaerobic condition, warm temperature, high moisture, nutrients and pH above 4.6). In contrast to staphylococcus enterotoxins, the botulinum toxin is destroyed at cooking temperature (of 80°C for 10 minutes). Therefore, the preventive measures are to ensure adequate canning of foods (the most important measure) and to heat the food thoroughly before consumption.

DID YOU KNOW?

***Clostridium botulinum* Produces the Most Lethal Toxin**

The toxin produced by *Clostridium botulinum* is one of the most toxic substances known and the lethal dose for a human is as low as 10^{-8} g. The toxin binds to the nerve endings at the nerve-muscle junctions blocking the release of acetylcholine and resulting in cranial neuropathy and descending flaccid paralysis.
Source: Aldsworth, Dodd and Waites, 2009.

***Bacillus cereus* enterotoxin,** or cereulide, is a preformed toxin which is heat stable and causes emetic syndrome. It is produced at temperature ranging from 12°C to 37°C primarily in cereal products. The incubation period is short and the symptoms appear in half to six hours after the ingestion of contaminated food. Low temperature storage can prevent the growth of *Bacillus cereus*.

Bacterial-produced histamines (Scombrotoxins) are produced when marine bacteria metabolize histidine (an amino acid) present in the muscle of fish (such as tuna and mackerel) kept at 45°C for many hours. The ingestion of contaminated fish leads to allergy-like symptoms including rashes, sweating, dizziness, nausea, diarrhoea and mouth swelling. These toxins are not destroyed by heat; therefore, the fish must be refrigerated from the time of catching until its consumption to prevent the growth of bacteria.

GOOD TO KNOW

First Aid for Food Poisoning

A person who exhibits symptoms of food poisoning should be advised rest and given plenty of liquids. Call for emergency medical assistance if

- the symptoms are severe, e.g., watery diarrhoea which turns very bloody within 24 hours.
- the affected person belongs to a high-risk group (older adults, infants, young children and immune-compromised persons).
- botulism poisoning is suspected; it requires immediate medical attention.

26.2.2 Mycotoxins

Mycotoxins, produced by fungi (moulds), are important from public health viewpoint because of their carcinogenic, mutagenic and teratogenic effects on animals. Since fungi grow better at high humidity and temperature, the crops grown in tropical and subtropical regions are more prone to mycotoxin contamination. Besides agriculture, fungal growth is common during post harvest handing. The mycotoxins so produced are not destroyed by heat or low temperature. Therefore, it is very important to prevent the growth of fungi and the subsequent production of mycotoxins in food for which good agricultural and manufacturing practices must be followed.

Among the hundreds of mycotoxins identified, **aflatoxin** is the most common and the most important toxicant. It is produced by some strains of *Aspergillus flavus* and *Aspergillus parasiticus* in many foods including peanuts, maize, cotton seed, rice, wheat and soybean. Aflatoxins can also contaminate animal feeds and thereby find way into the food consumed by human beings. For example, Aflatoxin B present in cow's feed is metabolized and excreted as aflatoxin M. Other important mycotoxins, for example, ochratoxin A, patulin, furmonism and trichothecenes have been described in chapter 24. The preventive measures include minimizing contamination with toxgenic fungi; controlling storage conditions (low moisture and low temperature) and using fungicides.

26.2.3 Seafood Toxins

Marine bio-toxins or seafood toxins are significant from public health perspective in many parts of the world. The most important seafood toxins are produced by algae such as dinoflagellates. Tropical or subtropical fish (primarily coral fish) feeding on these algae, and bigger fish consuming the contaminated coral fish can lead to fish poisoning (ciguatera). Another type of food intoxication is reported on consumption of shellfish (usually clams or mussels and sometimes scallops and oysters) feeding on toxic algae. The symptoms may be diarrhoeal, paralytic or neurological in nature. Accordingly, the shellfish intoxication is grouped as paralytic shellfish poisoning (PSP), diarrhoeic shellfish poisoning (DSP), neurotoxic shellfish poisoning (NSP) and amnesic shellfish poisoning (ASP).

26.2.4 Pesticides Residues

Since pesticides are used widely at agricultural, industrial and household level, their residues find way to foods as contaminants. Based on their target pests, the pesticides can be categorized as insecticides (insects), herbicides (weeds), rodenticides (rodents), miticides (mites), nematocides (nematodes) and fungicides (fungi).

Foods are most often contaminated with insecticides. There are two main classes of organic insecticides: the chlorinated hydrocarbons and organophosphorous insecticides. The chlorinated hydrocarbons (such as dichlorodi phenyl trichloroethane (DDT)) tend to persist in the environment and bio-accumulate in humans and so they are no longer used in most developed countries. However, DDT is still used in India primarily for control of mosquitoes. The second category of organic insecticides, the organophosphates, is water soluble and volatile. Due to their lower risk of food contamination, they are the most frequently used chemicals as insecticides. **Veterinary drugs** such as growth promoters (arsenal), milk enhancers (oxytoxin) and antibiotics and antiseptics can also leave harmful residues in animal foods.

> **GOOD TO KNOW**
>
> **Chemicals Used for Artificial Ripening of Fruits and Vegetables Can Be Toxic**
>
> Calcium carbide is a chemical reportedly being used for artificial ripening of fruits such as mangoes in India. Commonly known as 'masala', it is carcinogenic in nature and its use is banned legally.

26.2.5 Metallic Contaminants

Metals can find way to food either through industrial/environmental contamination or through contact with equipment and containers. Table 26.2 lists the adverse health effects of heavy metals (arsenic, cadmium, lead, mercury, methyl mercury, tin and zinc) present in food.

Arsenic is present in some wood preservatives, pesticides (lead arsenate), rodenticides (arsenic trioxide) and medicinal drugs. It is found in high concentration in seafood (fish, shellfish and crustaceans), rice and mushrooms and sometimes in poultry which has been fed fish meal containing arsenic. Its presence in drinking water is also a matter of health concern.

Cadmium is a naturally occurring contaminant which is often present in some volcanic soils. It is also an industrial pollutant and component of certain fertilizers. Plant foods such as vegetables, cereals and tobacco uptake cadmium from the soil they grow in. Shellfish, crustaceans and fungi are natural accumulators of this metal. Massive exposure to cadmium can result in severe joint pain; whereas, long-term low-level consumption leads to renal failure.

Table 26.2 Health effects of some metallic contaminants

Metal	Foods Commonly Involved	Health Effects
Arsenic	Fruits sprayed with lead arsenate and seafood	Dizziness, cramps, chills, paralysis and death
Cadmium	Fruit juices and soft drinks in contact with cadmium plated ware	Excessive salivation, liver and kidney damage, pain in joints and gastritis
Lead	Water and processed food	Foot drop, anaemia, constipation and brain damage
Mercury	Grains treated with mercury fungicide, fish contaminated with methyl mercury	Paralysis and brain damage
Tin	Foods in tin plate steel cans	Colic, vomiting and photophobia
Zinc	Foods in contact with galvanized iron	Dizziness and vomiting
Barium	Contamination of food with rat poison (barium carbonate)	Forceful peristalsis, convulsions and muscular twitching
Cobalt	Beer and water	Heart failure
Copper	Acid food in contact with tarnished copper vessels	Diarrhoea, vomiting and abdominal pain

Lead contaminates water, food and beverages in different ways. Water, food and beverages can get contaminated through the lead pipes, lead-coated water tanks/equipment or lead-soldered cans. Fruits and vegetables grown near big cities or busy roads are often laden on their surface with lead deposited from the exhaust of motor vehicles. Lead is also a component of lead chromate–a yellow colouring which is a common food adulterant. It can lead to anaemia, abortion, paralysis, foot drop and mental retardation.

Mercury, a naturally occurring metallic element, can contaminate food through industrial or environmental factors which elevate its level beyond the safe limit. Methyl mercury is the most toxic form of mercury which is found in aquatic organisms particularly the large predatory fish. It can have serious effects on the nervous system of foetus, infants and children.

Tin, used in tin-plate steel cans, also finds use in solders and in alloys including dental amalgams. It is present in small amounts in fresh meat, cereals and vegetables, and in foods stored in unlacquered cans. Consumption of food contaminated with tin can lead to colic, vomiting and photophobia.

ACTIVITY

Safe Limit of Metal Residues in Food

Refer to Food Safety and Standards (Contaminants, Toxins and Residues) Regulation, 2011 of Food Standards and Safety Authority of India (http://www.fssai.gov.in) and find out the maximum permissible limits of the following metals:

- Arsenic in milk
- Cadmium in infant foods
- Lead in fruit and vegetable juices
- Tin in Jam, jelly and marmalade
- Copper in hard boiled confectionery
- Zinc in ready-to-drink beverages

26.2.6 Residues from Packaging Material

Earlier metal and glass were the main food packaging materials. With recent developments in the field treated paper and plastics are being used extensively for food packaging.

Metal containers such as tin-plate steel cans have, many a times, been the cause of contamination with iron, tin and lead. Use of non-lead soldering and lacquering of interior surfaces of cans has lowered the incidence of such food intoxication. However, the metal alloys, used in recent years, may contaminate acidic food at high temperatures to cause cadmium intoxication.

Glass is considered to be the most inert material yet metals such as lead may migrate from crystal glassware to alcoholic beverages and citrus juices. Lead and chromium have also been reported to leach from ceramic containers.

Paper is generally coated or laminated to improve its packaging properties. This involves the use of many additives and chemical processes which can be a source of contamination from metals such as zinc, aluminium, manganese and barium.

Plastics are being used widely for food packaging in the present times and this development has created new problems of food contamination. The plastics may contain many chemicals as plasticizers, catalysts, antioxidants, colorants, stabilizers and light absorbers. Some of these compounds get migrated to the food packaged in plastic materials. For example, poly vinyl chloride (PVC) is a type of plastic which has been found to contaminate food with vinyl chloride (a carcinogen). There are also issues of some cancer-causing chemicals leaching into food from plastic packaging during the process of microwave heating.

The Food Safety and Standards (Packaging and Labeling) Regulations (2011) have specified that the containers made of plastic materials (e.g., polyethylene, PVC, polypropylene, styrene polymers and ionomer resins) in contact with food should conform to the Indian Standards Specifications. Harmful chemical such as acrylamide, heterocyclic amines (HCAs) and polycyclic hydrocarbons (PAHs) are produced during cooking and processing are also included as food contaminants (refer to Chapter 24).

26.3 Common Adulterants in Food

Adulteration of food has been rather rampant in India. It is an evil practice adopted by traders to maximize their profits at the cost of food quality. Common food adulterants are listed in Table 26.3.

Do you know that food additives, if not used judiciously, can also fall under the category of adulterants? Most of the additives are chemicals which are permitted in foods only within the prescribed limits. **Use of prohibited additives** is a form of adulteration which can lead to grave health consequences such as the following:

1. Malachite green and metanil yellow (colours) can cause mental retardation and cancer.
2. Coumarin and dihydrocoumarin (flavourings) are blood anticoagulant.
3. Talc (anticaking agent) can be adsorbed in particulate form and may lead to cancer.
4. Brominated vegetable oils (emulsifiers) can cause skin and neurological disorders.

The following **permitted additives when used beyond the safe limit** can cause harm to the consumers:

1. Nitrates and nitrites (preservatives) can be carcinogenic.
2. Monosodium glutamate (flavour enhancer) can lead to mental retardation in infants.
3. Cyclamates (sweeteners) can cause cancer.
4. Butylated hydroxyanisole, BHA, and butylated hydroxytoluene, BHT (antioxidants), can lead to allergy and liver damage.

The health effects of common adulterants are summarized in Figure 26.2.

Table 26.3 Common food adulterants

Foodstuffs	Common Adulterants
Cereals	
Wheat, rice	Stones, marble chips, sand and other filth
Wheat flour, refined flour	Talc, sand , grit, chalk powder, gluten deficiency, foreign starch
Pulses	
Whole bengal gram, red gram	Whole Khesari dhal (*Lathyrus sativus*)
Bengal gram dhal, red gram dhal	Khesari dhal (*Lathyrus sativus*)
Bengal gram flour	Powdered khesari dhal, maize/peas flour, starch coloured yellow with dye
Milk and Milk Products	
Milk liquid	Water, fat deficiency, addition of oil, starch, coal tar dyes, urea, formaldehyde
Milk powder, condensed milk	Excess moisture, starch , fat deficiency
Cream	Fat deficiency, other fats
Curds	Water, fat deficiency
Ice cream	Non-permitted colours and sweeteners, fat deficiency, other fats
Khoya, paneer	Starch, paper, fat deficiency
Ghee, butter	Hydrogenated fat, animal body fat, starch, rancid stuff
Vegetable Oils and Fats	
Margarine	Excess moisture, fat deficiency, animal body fat
Vanaspati	High melting and rancid fat, animal body fat
Vegetable oils	Argemone oil, orthotricresyl phosphate (TCP), oil deficiency, cheaper oils
Spices and Condiments	
Mango powder	Heavy insect infestation, sand
Asafoetida	Foreign resins and gums, deficiency in alcoholic extracts, sand, starch, talc, sand
Black pepper	Dried papaya seeds, berries, skin of black gram
Chillies	Coated with colour Sudan I
Red chilly powder	Red colour, starch, powdered bran, brick powder
Cinnamon	Cassia bark
Cloves, cardamom	Exhausted spices, imitation samples
Coriander powder	Foreign starch, excess stalky matter, bran or sawdust coloured green
Coriander whole	Foreign seeds, green colour, foreign matter
Cumin seeds	Artificial jeera from mud and earth, foreign seeds
Mixed spices	Foreign starch, sand, grit, exhausted spices, excess common salt, coal tar dyes
Mustard seeds	Argemone seeds
Saffron	Added colour, excess moisture, foreign matter, dried tendrils of maize cob

Table 26.3 (Continued)

Foodstuffs	Common Adulterants
Turmeric powder	Yellow earth, coloured starch or talc, lead chromate, metanil yellow
Whole turmeric	Coating of lead chromate
Non-alcoholic Beverages	
Soft drinks	Metallic contamination, microbial contamination, non-permitted colour, flavour, sweetener or permitted additive beyond the prescribed limits
Coffee powder	Exhausted coffee powder, chicory, date or tamarind seed powder
Tea	Exhausted tea leaves, foreign leaves, added colour, sawdust, cashew husk
Alcoholic Beverages	Fuel oil, methanol, mineral oil,excess water, metallic contamination
Fruit and Vegetable Products	Deficiency in fruit/vegetable content, low total soluble solids, non-permitted colour, flavour, sweetener and preservatives, synthetic products labelled natural

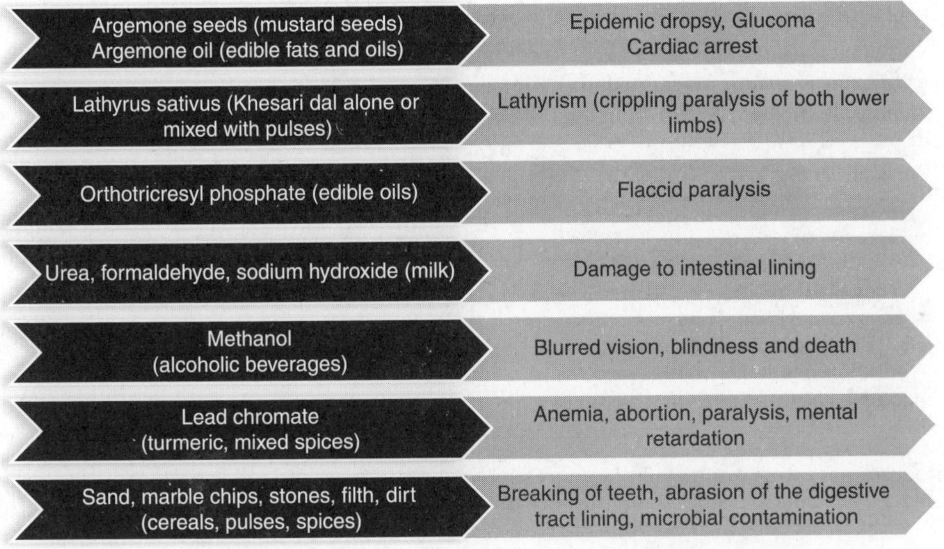

Figure 26.2 Health effects of common adulterants

26.4 Method of Detection of Food Adulterants

Food adulteration is an unethical practice which is followed by many sub-standard food outlets. It not nly leads to adverse health consequences but also undermines the credibility of the catering unit. As a responsible professional, it is the duty of a chef to ensure that the ingredients used by him or her are pure and unadulterated. It is possible to detect presence of adulterants in food by simple tests (Table 26.4).

Table 26.4 Simple tests for detecting adulterants in food

Food Stuff	Adulterant	Method of Detection
Fats and oils	Argemone oil	Take 5 ml of oil in a test tube. Add 5 ml of concentrated nitric acid and shake carefully. Orange yellow to reddish brown colour indicates the presence of argemone oil.
Milk	Water	Place a drop of milk on a polished surface (e.g. a ceramic tile) and slant it. The pure milk flows slowly leaving behind a white trail, whereas adulterated milk flows fast without leaving any mark.
Milk	Starch	Take some sample and add few drops of iodine solution. Blue colour indicates the presence of starch.
Pure ghee, butter	Vanaspati	Take one teaspoon of melted fat in a test tube, add quantity of HCl and a pinch of sugar. Shake and keep for five minutes. Appearance of crimson colour in the acid layer indicates the presence of vanaspati.
Ghee, *paneer*, *khoa*, milk powder	Coal tar dyes	Take one teaspoon of sample in a test tube and 5 ml of dilute H_2SO_4 or concentrated HCl. Shake well. Pink colour (with H_2SO_4) or crimson colour (with HCl) indicate the presence of coal tar dyes.
Khoa, chhana	Starch	Boil a small quantity of sample with water and cool it. Add few drops of Iodine solution. Appearance of bluish black colour indicates the presence of starch.
Rabbari	Blotting paper	Take one teaspoon of sample in a test tube. Add 3 ml each of HCl and distilled water. Stir with a glass rod, remove the rod and examine it carefully. Presence of fine fibres on the rod indicates the presence of blotting paper.
Ghee	Mashed boiled potatoes	Add few drops of Iodine solution to the sample. Appearance of bluish black colour indicates the presence of starch.
Black pepper	Dried papaya seeds	The papaya seeds can be separated as they are shrunken, oval and greenish brown/brownish black in colour.
Mustard seeds	Argemone seeds	Unlike mustard seeds that are smooth, the argemone seeds have a grainy and rough surface and are black in colour. Besides, they are white from inside but mustard seeds have a yellow interior.
Turmeric powder, sweetmeats, spices	Metanil yellow	Extract colour in lukewarm water in a test tube and add few drops of concentrated HCl. Instant appearance of pink colour persisting on dilution with water indicates the presence of metanil yellow.
Turmeric powder, salt	Chalk powder	Mix a teaspoon of sample with water in a test tube. Add few drops of concentrated HCl. Effervescence indicates the presence of chalk.

Table 26.4 (Continued)

Food Stuff	Adulterant	Method of Detection
Chilli powder	Brick powder	Take a teaspoon of sample in a beaker. Fill the beaker to half with water. Stir and feel the matter at the bottom. Grittiness indicates the presence of brick powder.
Chilly powder	Water soluble colours	Sprinkle a small amount of sample on the surface of water in a beaker. Artificial colour starts descending immediately.
Asafoetida	Foreign resin	Pure sample burns like aromatic camphor.
Cinnamon	Cassia bark	Unlike cinnamon bark which is thin and can be rolled, the cassia bark has several layers between the rough outer and smooth inner most layer.
Green chillies and green vegetables	Malachite green	Soak a piece of cotton in paraffin and rub over the outer green surface of the sample. Green colour on the cotton indicated adulteration with malachite green.
Saffron	Dried tendrils of maize cob	Artificial saffron breaks easier and leaches all the colour sooner than pure saffron
Silver foil	Aluminium foil	Add two drops of concentrated nitric acid. The silver foil will dissolve completely but the aluminium foil will remain undissolved.
Bengal gram, Red gram whole/ washed, *besan*	*Khesari dal*	Put the sample in dilute hydrochloric acid. Development of pink colour indicates the presence of *khesari dal*.
Wheat flour, refined flour, semolina	Iron filings	Move a magnet through the sample to separate the iron filings

This chapter focussed on different contaminants and adulterants found in food and highlighted their implications on human health. In order to prevent food from contamination and adulteration, the food service professionals must follow best practices such as Good Manufacturing Practices (GMP), employ Hazard Analysis Critical Control Point (HACCP) approach, reduce environmental pollution and apply appropriate control measures in production, processing, packaging and storage. They should also abide by the related laws and thereby ensure safe food supply to their clients. Chapters 27, 28 and 29 will describe some of these measures.

GOOD TO KNOW

Lactometer helps to measure the specific gravity and hence the presence of water in milk (Fig 26.3). It can be used at household and commercial level at the time of purchase of milk. Normal specific gravity of standardized milk is 0.32.

Figure 26.3 Lactometer

Know Your Progress

True or False Questions

1. *Staphylococcus* enterotoxin and mycotoxins are destroyed at usual cooking temperature.
2. Ensuring adequate canning is the most important measure to prevent botulism.
3. Methyl mercury is a toxin found in large predatory fish which can have serious health effects on the nervous system of foetus, infants and children.
4. Alcoholic beverages are adulterated with fuel oil, methanol and mineral oil.
5. Appearance of bluish black colour on addition of iodine solution to milk indicates that the sample is pure.

Very Short Answer Questions

1. Differentiate between contaminants and adulterants.
2. List four natural toxins present in foods along with their health effects.
3. What are the common adulterants in chilly powder, asafoetida, cumin seeds and turmeric powder?
4. How would you test the presence of argemone oil in edible oils?
5. Write the test for detecting vanaspati in pure ghee.

Short Answer Questions

1. Discuss the source and health effects of various bacterial toxins.
2. List the different metals which contaminate food and discuss their health hazards.
3. How do residues from packaging materials contaminate food?
4. Describe the role of pesticides and mycotoxins in food contamination.
5. Intentional adulteration leads to adverse health effects. Justify the statement.

Long Answer Questions

1. What are the different types of food contaminants? Discuss their origin and the health consequences.
2. List the common adulterants present in various types of food.
3. Describe the different methods of detection for food adulterants.

Beyond the Text

- Collect five samples each of the following food items from the market/street restaurants and test for the presence of adulterants listed against them:

  1. Vegetable oil – argemone oil
  2. Pure ghee – vanaspati
  3. Milk – starch
  4. Turmeric powder – metanil yellow

Chapter 27

Food Laws and Regulations

Chapter Objectives

After reading this chapter, the reader should be able to:

- understand the meaning of act, law, rule, regulation and standard.
- identify the mandatory/voluntary food laws, standards and regulations in India.
- get an overview of the comprehensive Food Safety and Standards Act.
- appreciate the role of Codex and ISO in harmonizing international food standards.
- recognize the significance of global and national regulatory agencies in this field.
- learn about legislation related to consumer protection.

Every consumer has a right to safe food. The responsibility of ensuring food safety rests jointly with all the main players – international organizations, government authorities, food business operators and the consumers. Global organizations play an important role in setting harmonious standards for food trade which is expanding beyond the boundaries of different nations. Government, at the national level, sets up food laws through the legislative process and lays down rules and regulations for their effective implementation. Food industry, at its end, must exhibit self-discipline and follow food standards and codes of practice set by national and international agencies. Finally, the consumers must be aware of their rights and should learn to assert themselves individually or collectively to demand food that is wholesome and safe.

THINK TANK

Queries from a Prospective Food Business Operator

An entrepreneur is keen to start a medium scale food business. Can you help him find answers to his queries?

- Is it essential to obtain a licence before starting a food business?
- Should the weights and measures be verified before use?
- Is there a restriction on the use of colouring and flavouring agents in foods?
- Are ISO standards of any relevance for food preparation and processing?

The answer to all these questions is 'yes' and detailed specifications can be found in the respective food laws and standards.

27.1 Introduction to Food Acts, Laws and Standards

Food-related acts, laws, rules, regulations and standards are required to meet the changing needs of food trade, to maintain the overall quality of food supply and to protect the consumers and the producers from food-related frauds. Food acts, orders and laws are usually formulated by the government, which make it mandatory for the food business operator to obtain a working licence. On the other hand, food standards may be mandatory or voluntary in nature. They are set either at the national or at the international level and are generally not developed by the national government. Box 27.1 gives the definition of important terms in this field.

Box 27.1 Important Terms Related to Food Laws and Regulations

- **Act** is a written ordinance (order) passed by the parliament.
- **Law** is the system of rules which a particular country recognizes as regulating the actions of its members and which it may enforce by the imposition of penalties.
- **Rule** is a statement explaining what someone can or cannot do in a particular system.
- **Regulation** is a principle, rule, or law designed to control or govern conduct.
- **Standard** is a document that provides requirements, specifications, guidelines or characteristics which can be used consistently to ensure that materials, products, processes and services are fit for their purpose.
- **Food standard** has been defined as a set of criteria that a food must meet if it is to be suitable for human consumption, such as the source, composition, appearance, freshness, permissible additives, and maximum bacterial content, Bender, 2005.

27.2 National Food Safety Acts, Laws and Standards

Over the last six decades, the Government of India has formulated multiple food laws, standard-setting agencies and enforcement bodies for different sectors of food. However, changing consumer demands, laws at national/international levels, scientific advancements and challenging business environment necessitates the regular upgrade of existing food laws. To promote innovations in food business, to encourage the investors and include the component of consumer education, the government of India has introduced the Food Safety and Standards Act (FSS Act). This recently formulated act is modern in approach and will gradually replace many of the earlier mandatory laws (Box 27.2).

Box 27.2 FSS Act Will Replace Many of the Earlier Laws

The Food Safety and Standards Act will gradually repeal the following eight laws:

- Prevention of Food Adulteration Act, 1954.
- Fruit Products Order, 1955.
- Meat Food Products Order, 1973.
- Milk and Milk Products Order, 1992.
- Vegetable Oil Products (Control) Order, 1947.
- Edible Oils Packaging (Regulation) Order, 1998.
- Solvent Extracted Oil, De oiled Meal, and Edible Flour (Control) Order, 1967.
- Essential Commodities Act, 1955, relating to food.

The standards, safety requirements and other provisions of the repealed acts and orders will remain in force only until the new rules and regulations are put in place under the FSS Act.

Source: MoFP, 2012.

27.2.1 Food Safety and Standard Act

The FSS Act, 2006, is a mandatory law which has been devised to meet the changing requirements of food business. This act aims to ensure food safety through standards based on scientific evidence. The focus is on the end consumers and the food industry, whereas the role of government is minimized. The characteristics of this contemporary and comprehensive legislation differ from the earlier laws. The **main features of FSS Act** are listed in Figure 27.1 and are discussed as following:

1. **Integrated Line of Command:** Under FSS Act, the Food Safety and Standards Authority of India has been established as the single national regulatory agency which forms an integrated line of command to control food safety. Earlier laws had multi-level and multi-departmental control.

2. **Focus on Strategic Issues:** In contrast to earlier laws, the FSS Act focuses on intended issues (e.g. international trade) and also on foods which are new to the world (e.g. novel foods, genetically modified foods, functional foods, caffeinated drinks).
3. **Licensing Through a Single Window:** Under the act, the licence is provided through the Commissioner of Food Safety at the state level. Earlier, it was granted by central agencies under various acts and orders.
4. **Promotion of Self-compliance:** It promotes self-compliance through Food Management Safety Systems. Earlier, laws had an inadequate regulatory regime enforced primarily through the government.

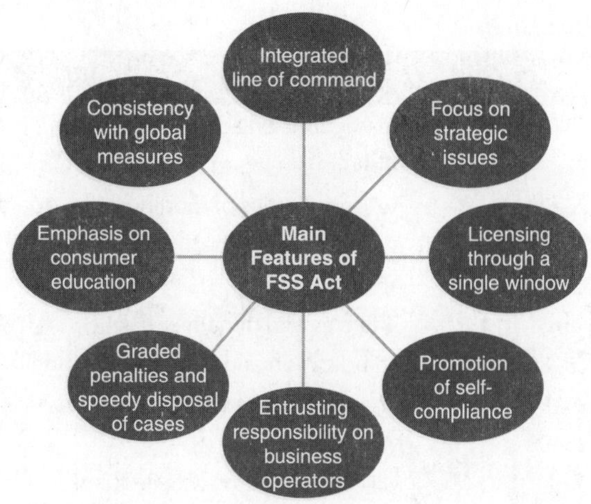

Figure 27.1 Main features of food safety and standards act

5. **Entrusting Responsibility on Business Operators:** Under the FSS Act, the food business operators have to ensure that the food (processed, manufactured, imported or distributed) complies with the domestic food laws.
6. **Graded Penalties and Speedy Disposal of Cases:** In contrast to earlier laws, the FSS Act imposes graded penalties as per the gravity of the offence (civil penalties for minor violations and punishment for serious offences) and has mechanism for speedy disposal of cases.
7. **Emphasis on Consumer Education:** It emphasizes on educating the consumers about food safety to enable them to make informed choices. This component was completely missing in earlier laws.
8. **Consistency with Global Measures:** The act provides consistency between national and global food policy measures. The earlier laws were not in line with international standards.

Description of FSS Act

The FSS Act is based on international legislation and Codex Alimentarius. Accordingly, it has nearly hundred clauses which systematically describe the following areas:

Clause	Description
1–3	Short title, extent, commencement and definitions
4–17	Food Safety and Standards Authority of India–composition and functions
18	General principles of food safety
19	Use of food additive or processing aid
20	Contaminants, naturally occurring toxic substances, heavy metals etc.
21	Pesticides, veterinary drug residues, antibiotic residues and microbial counts
22	Genetically modified, organic, functional and propriety foods
23	Packaging and labelling of foods

(Continued)

(Continued)

Clause	Description
24	Restrictions on advertisements and prohibition of unfair practices
25	Provisions relating to import
26–28	Responsibilities of food business operator and food recall procedures
29–42	Enforcement of the act
43–47	Analysis of food
48–67	Offences and penalties
68–80	Adjudication and Food Safety Appellate tribunal
81–84	Finance, accounts, audit and reports
85–98	Miscellaneous
99	Milk and Milk Products Regulations (MMPR, 09)
100	Amendments to the Infant Milk Substitutes, Feeding Bottles and Infant Foods (regulations of production, supply and distribution) Act, 1992

Rules and Regulations (related to FSS Act)

The food safety and standards related rules and regulations came into effect in the year 2011. These **rules** give explanation for the following:

1. Enforcement structure for FSS Act.
2. Procedure for seizure of articles by the food safety officer.
3. Subsequent sampling and analysis.
4. Mechanism of holding an inquiry.
5. Procedure of appeal to the appellate tribunal.

On the other hand, the **regulations** focus on licensing and registration of food businesses, food products standards and food additives, packaging and labelling, contaminants, toxins and residues, and laboratory and sample analysis (Figure 27.2).

Figure 27.2 Food Safety and Standards Regulations, 2011

It is beyond the scope of this chapter to describe all these regulations. However, highlights from two regulations relating to licensing of food businesses and the food products standards, respectively, are being discussed for the benefit of hospitality professionals. First are the Food Safety and Standards (Licensing and Registration of Food Businesses) Regulations, 2011, which cover four important schedules: (1) list of food business falling under the purview of Central Licensing Authority, (2) application for registration/renewal of registration under the FSS Act, (3) fee for grant/renewal of licence registration/licence fee per annum in rupees and (4) general hygiene and sanitary practices to be followed by food business operators.

Of these, the fourth schedule is particularly important for **maintaining sanitation and hygiene in a food establishment**. It focuses on the concept of good hygienic practices (GHPs) and has four parts (Figure 27. 3). It specifies the General Requirements on Hygiene and Sanitary Practices to be followed by all food business operators applying for licence (refer to Section 29.2 of Chapter 29) including the petty food business operators. It also states separately the Specific Hygiene and Sanitary Practices to be followed by all food business operators engaged in milk and milk production, meat and meat production, or catering/food service establishments.

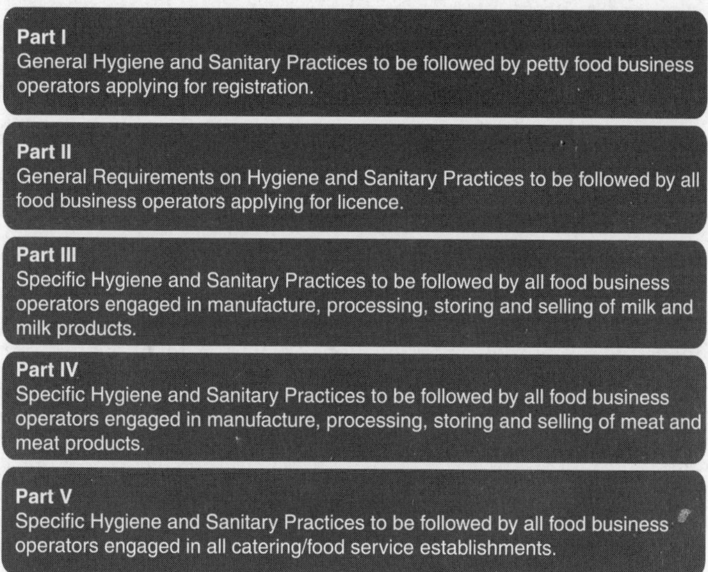

Figure 27.3 FSS regulations for general hygienic practices for food business operators

List of Food Articles

The FSS Act covers a long and exhaustive list of food articles. The standards for these food articles are specified in detail in the Food Safety and Standards (Food Products Standards and Food Additives) Regulations, 2011, under the following categories:

- Dairy products and analogues
- Fats, oils and fat emulsions
- Fruit and vegetable products
- Cereals and cereal product
- Meat and meat products
- Fish and fish products
- Sweets and confectionery
- Sweetening agents including honey
- Salt, spices, condiments and related products
- Beverages
- Other food products and ingredients
- Propriety food
- Irradiation on food
- Food additives

As stated earlier, the FSS Act, 2006, will eventually repeal many of the earlier food laws which will remain in force only until the new rules and regulations are put in place. The following laws and regulations are being discussed briefly in the subsequent paragraphs:

1. Prevention of Food Adulteration Act, 1954.
2. Fruit Products Order, 1955.
3. Meat Food Products Order, 1973.
4. Milk and Milk Products Regulations, 2009 (which replace Milk and Milk Products Order, 1992).
5. Legal Metrology Act, 2009 (which repeals and replaces the Standard of Weights and Measures Act, 1976).

While all these laws are mandatory, the use of BIS and Agmark standards is purely voluntary in nature.

27.2.2 Prevention of Food Adulteration Act

The Prevention of Food Adulteration (PFA) Act was promulgated in 1954 for preventing adulteration of food and the PFA Rules were incorporated in 1955.

1. PFA is a mandatory law.
2. Both primary and processed foods, whether domestic or imported, are covered under this act which also defines food adulteration (refer to Box 26.1 in Chapter 26).
3. It aims at protecting the consumer from foods that are impure, unsafe or falsely labelled.
4. The PFA Act includes food standards, general procedures for food sampling and analysis, powers of authorized officers, penalties and other food-related matter.
5. It also specifies parameters relating to food additives, preservatives, colouring matters, packing and labelling of foods and prohibition and regulations of sale of certain foods.

Central Committee of Food Standards (CCFS) was setup by central government under the Ministry of Health and Family Welfare. It comprises of members from different regions of the country and amends PFA laws from time to time (Box 27.3). However, the responsibility of implementation of PFA Act and Rules rests with state governments and union territories; so each of them has formed its own structure/organization for the purpose.

Box 27.3 Important Amendments in PFA

- Several amendments have been carried out in PFA, the most recent being the Prevention of Food Adulteration (7th Amendment) Rules, 2006, which include
  - mandatory GMP/GHP requirements.
  - food recall procedure.
  - genetically modified (GM) food labelling.
  - contemporary methods of analysis.
- Another important amendment for consumer protection came out in 1998 which enforced the mentioning of 'best before date' on the package of food articles as per international regulations.

27.2.3 Legal Metrology Act

The Legal Metrology (LM) Act, 2009, repeals and replaces the Standard of Weights and Measures Act, 1976, and the Standards of Weights and Measures (Enforcement) Act, 1985.

1. It prohibits the use, manufacture or import of any non-standard weight, measure or numeral.
2. Under the act, only standard weight or measure should be used (Box 27.4). They should use only standard units, should be verified and re-verified periodically and should conform to the physical characteristics, configuration, construction details, materials, equipment, performance and tolerance as per provisions of the LM (General) Rules, 2011.
3. The use of non-standard units in any document, quotation, invoice, packages and advertisements is prohibited.
4. Every manufacturer, dealer and repairer of weights and measures is required to obtain a licence.
5. All weights and measures require verification and stamping before use.
6. Selling, delivering or using any un-verified weight or measure is a punishable offence. The act also gives specifications for declaration of weights and measures on the pre-packaged commodities.

Box 27.4 What Are Standard Weights or Measures?

Standard Weights or Measures are defined as the weights or measures which conform to the standard in accordance with metric system based on international system and have the physical characteristics, configuration, construction, material, equipment, performance and tolerance as per the General Rules.

Source: Legal Metrology Act, 2009

27.2.4 Fruit Products Order

The Fruit Products Order (FPO) was promulgated under Section 3 of the Essential Commodities Act, 1955, and is administered by the Ministry of Food Processing Industries, Government of India.

1. It is a mandatory law which lays down specifications for the manufacture of fruit and vegetable products under sanitary and hygienic conditions and with quality standards as laid down in the order.
2. Under this law, it is essential for all manufacturers of fruit and vegetable products such as jams, pickles and squashes (including some non-fruit products such as non-fruit vinegar, syrup and sweetened aerated water) to obtain a licence.
3. No fruit and vegetable products can be sold in the market without the FPO mark.
4. The order lays minimum requirements for hygienic production and quality standards in the following areas:

Location of the factory	Personnel hygiene	Water quality
Sanitary and hygienic conditions of premises	Facilities for quality control (including the technical staff)	Product standards
Machinery and equipment	Maximum limits for additives and metallic contaminants	

5. The order also specifies requirements for packing, marking, labelling and sealing of containers of fruit products.
6. The label should not contain any false or misleading statement, claim, design or device.

The advisory body for FPO is the Central Fruit Product Advisory Committee which comprises of representatives from the Central Food Technology Research Institute, Bureau of Indian Standards, fruit and vegetable processing industry and consumer organizations.

27.2.5 Meat Food Products Order

Meat Food Products Order (MFPO), established in the year 1973, is presently being implemented by Ministry of Food Processing industries (after it was transferred from the Directorate of Marketing Inspection, Ministry of Agriculture, in 2004).

1. The order aims to regulate production and sale of meat and meat products (related to sheep, hog, goat, cow, buffalo, fish and poultry) through licensing of manufacturers.
2. It specifies sanitary and hygienic conditions along with strict quality control at all stages of production of meat, fish and poultry food products.
3. MFPO applies to all manufacturers/processors engaged in the business of manufacturing, packing, repacking, relabelling meat food products meant for sale.
4. Exempt from licensing are the manufacturers who are engaged in production of meat products for on-the-spot consumption, for example, restaurant, hotel, snack bar, eating house or similar establishments.

27.2.6 Milk and Milk Products Regulations

Milk and Milk Products Order (MMPO), 1992, was promulgated by the Department of Animal Husbandry, Dairying and Fisheries under Section 3 of the Essential Commodities Act, 1955. It has now been included as Milk and Milk Products Regulations (MMPR, 2009) under Section 99 of the FSS Act, 2006.

1. Its objective is to sustain and improve the liquid milk supply of desired quality for the general public and also to regulate the production, processing and distribution of milk and milk products.
2. Under this order, registration is required by any person/dairy plant handling more than 10,000 litres milk/day or 500 metric tonnes of milk solids/year.
3. It specifies sanitary and hygienic conditions, and food safety and quality measures.
4. The task of inspection of registered dairy units is assigned on rotation basis to National Productivity Council and Export Inspection Council of India.

27.2.7 Indian Standards

The Bureau of Indian Standards (BIS) looks after standardization of processed foods. BIS is involved in the formulation of Indian standards for processed foods and the implementation of standards through promotion and through voluntary and third-party certification systems.

1. BIS are voluntary standards which have been formulated for most of the processed foods.
2. They specify the permitted raw materials, their quality characteristics, hygienic conditions for manufacturing and also the packaging and labelling requirements.
3. The manufacturers complying with standards laid down by the BIS can obtain exhibit 'ISI' mark on the packages of their products.
4. BIS is the licensing authority for manufacture of packaged mineral water in India.
5. It has also identified certain items for compulsory certification, for example, food colours and additives, milk powder and condensed milk.

27.2.8 Agmark Standards

Agricultural Produce (Grading and Marketing) Act, 1937, is one of the oldest acts and is operated by the Directorate of Marketing and Inspection. 'Agmark' standards are prescribed under this act. A person or a body of persons desirous of being authorized to grade and mark an article as per the act can apply to the

Agricultural Marketing Adviser (or any other officer of the central or state government authorized by the Agricultural Marketing Adviser). The concerned authority duly verifies the credentials of the applicant, inspection of the premises, laboratory and processing units before issuing the certificate of authorization.

1. Agmark standards are grade standards prescribed for all agricultural and allied commodities included in the schedule.
2. They are purely voluntary in nature.
3. Under the act, the article should be packed in a manner using specified packaging material in specified package sizes by weight or number as prescribed for the said article.
4. After obtaining the certificate of authorization, the Agmark logo can be exhibited on the package of the food.
5. The foods covered in the act include vegetable oils, ghee, butter, eggs, cereals, pulses, jaggery, honey and sugar.
6. Failure to conform to the standards can result in downgrading or removal of the authorized mark from the article.

27.3 International Standards

Codex Standards and ISO Standards laid down, respectively, by Codex Alimentarius Commission and International Organization for Standardization address the problems of global food trade. Being a member of WTO, it is binding for India to match its national standards to the Codex Standards.

27.3.1 Codex Standards

An increasing number of developing countries are becoming involved with Codex to compete better in the sophisticated world market and also to enhance food safety for their own population. The exporters gain knowledge about the expected standards by the importers, whereas the importers are safeguarded from poor quality shipments (Box 27.5).

Box 27.5 What is Codex?

Codex Alimentarius Commission (commonly referred to as Codex) is the body established by Food and Agriculture Organization (FAO) and World Health Organization (WHO) in 1963 to implement their Joint FAO/WHO Food Standards Programme. Over the years, Codex has evolved as an open, transparent and inclusive way to meet the emerging challenges in food trade. Deriving knowledge from best available scientific evidence, it has contributed to international harmonization of food standards.

Benefits of Codex are as follows. It

- *develops* international food standards, guidelines and codes of practice to protect the health of consumers and ensure fair practices in the food trade.
- *promotes* coordination of all food standards work undertaken by international governmental and non-governmental organizations.
- *assembles* scientists, experts, government regulators, international consumers and industry organizations.
- *encourages and assists* participation from developing countries through finance and training.

Codex is an intergovernmental body whose purpose is to develop international food standards to protect the health of consumers and to facilitate the trade of food. It sets international standards on foods and other texts which can be recommended to governments for acceptance. This collection of food standards, known as the **Codex Alimentarius** or the **Codex Standards**, has become an international reference point not only for the consumers, food producers and processors but also for the national food control agencies and the international food trade. These standards are based on scientific evidence and the expert advice provided by ad-hoc expert consultations, by joint experts or by committees set up under FAO and WHO, such as:

1. Joint FAO/WHO Expert Committee on Food Additives (JECFA),
2. Joint FAO/WHO Meeting on Pesticide Residues (JMPR) and
3. Joint FAO/WHO Expert Meeting on Microbiological Risk Assessment (JEMRA).

27.3.2 ISO Standards

The **International Organization for Standardization (ISO)** is an international federation of national standards body comprising of nearly 163 members, one in each country. As the world's largest developer of voluntary international standards, it provides benefits for business, government and society.

ISO standards ensure quality, ecology, safety, efficiency and effectiveness at an economical cost. They assist in trade, extend knowledge and share technological advances and good management practices. Their benefits are many as they:

1. create confidence in the products we eat and drink by ensuring that the world uses the same recipe for maintaining food quality, safety and efficiency.
2. bring benefits for all participants in the supply chain (e.g. agricultural producers, food manufacturers, laboratory analysts, regulators and consumers).
3. provide a platform for all stakeholders to share best practices, encourage state-of-the-art technology and ensure food safety and quality.

DID YOU KNOW?

ISO Provides Standards for Cooking/Serving Food

- **ISO 3591** determines the size of wine-tasting glasses.
- **ISO 7304** provides cooking times for spaghetti.
- **ISO 3103** explains how to make a perfect cup of tea.

Source: ISO, 2012.

GOOD TO KNOW

Food Safety Checklist for Small Businesses

'ISO 22000 Food Safety Management Systems – an easy-to-use checklist for small business – Are you ready?' is a product consisting of a combined handbook and CD-ROM which is meant for small- and medium-sized enterprises (SME) in the food sector in both developed and developing countries. It aims to benefit them from the advantages of implementing a food safety management system (ISO 22000:2005).

Source: ISO, 2012.

27.4 Regulatory Agencies

As a result of globalization, the national borders are softening, and it is becoming increasingly important for nations to be aware of food standards and practices which are internationally accepted and harmonized. Equipped with such knowledge, they can address food-related global challenges effectively. Many international and national organizations are playing significant role in regulating the food standards in the interest of the consumer. Food Safety and Standards Authority of India (FSSAI) is a recently formulated and important regulatory body at the national level.

27.4.1 Food Safety and Standards Authority of India

The regulatory body for FSS Act is the Food Safety and Standards Authority of India (FSSAI). Since the act emphasizes on a single reference point for all matters relating to food safety and standards, FSSAI has been established as a new national regulatory body. Its function is to develop science-based food standards and to regulate and monitor the manufacturing, processing, storage, distribution, sale and import of food so as to ensure the supply of safe and wholesome food for human consumption. The administrative ministry for the implementation of FSSAI is the Ministry of Health and Family Welfare, government of India and the chairperson is in the rank of secretary to the government of India. FSSAI performs its functions through an implementation structure.

Duties and functions of FSSAI are as follows:

1. It *regulates and monitors* the manufacture, processing, distribution, sale and import of food for ensuring its safety and wholesomeness.
2. FSSAI is involved in *framing regulations* relating to
   (a) food standards and guidelines along with specifying suitable system for their enforcement.
   (b) limits for food additives and contaminants (including pesticide residues, metal contaminants, mycotoxins, antibiotics) and irradiation.
   (c) mechanisms and guidelines for accreditation of bodies engaged in certification of food safety management system for food businesses.
   (d) procedures and guidelines for accreditation of laboratories and notification of the accredited laboratories.
   (e) food-labelling standards (including claims on health, nutrition and special dietary uses).
   (f) procedures for undertaking risk analysis, risk assessment, risk communication and risk management.
3. Another function of the authority is to *make available information and support based on the scientific evidence* to government, food operators and consumers through the following measures:
   (a) providing scientific advice and technical support to central government and state governments in framing policy and rules in areas relating directly or indirectly to food safety and nutrition.
   (b) collecting, collating, analysing and summarizing relevant scientific data about food consumption, incidence and prevalence of biological risk, food contaminants and their residues, identification of emerging risks and introduction of rapid alert system.
   (c) creating an information network across the country so that the public, consumers, and *panchayats* receive rapid, reliable and objective information about food safety and related issues.
   (d) providing training to persons who are involved or intend to get involved in food businesses.
   (e) contributing to the development of international technical standards for food, sanitary and phyto-sanitary standards.
   (f) promoting general awareness about food safety and standards.
4. It *publicizes* the opinion of scientific committee, results of scientific studies and report of annual activities.
5. It *directs* the Commissioner of Food Safety on matters relating to food safety and standards.
6. However, it does not disclose any information which is confidential.

The rules and regulations about the procedures for business transactions, meetings and the service conditions of the office bearers of FSSAI have been formulated recently.

GOOD TO KNOW

Rules and Regulations that Govern FSSAI

- FSSAI (salary, allowances and other conditions of service of chairperson and members) Rules, 2008.
- FSSAI (procedure for transaction of business of the Central Advisory Committee) Regulations, 2010.
- FSSAI (transaction of business at its meetings) Regulations, 2010.
- FSSAI (procedure for transaction scientific committee and scientific panels) Regulations, 2010.

27.4.2 The Export Inspection Council

The Export Inspection Council (EIC) of India was set up under Section 3 of the Export (Quality Control and Inspection) Act, 1963. It aims to enhance export trade through quality control and inspection. EIC is an advisory body to the central government, which is empowered under the act to

1. notify commodities which will be subject to quality control and/or inspection prior to export,
2. establish quality standards for the notified commodities and
3. specify the quality control measures for these commodities.

27.4.3 World Health Organization

World Health Organization (WHO) has a mission to assist member states to strengthen their programmes for improving food safety along the entire food production chain from production to consumption. Recognizing food safety as an essential public health problem, it has developed 'WHO global strategy for food safety: safer food for better health' to reduce the burden of food-borne diseases. WHO works in close association with the Food and Agriculture Organization (FAO) and other international organizations. WHO Department of Food Safety and Zoonoses provides leadership in international efforts to decrease the burden of diseases from food and animals.

The International Food Safety Authorities Network (INFOSAN), a joint initiative between WHO and FAO is a global network including 177 member states. It publishes information notes in the six official WHO languages, to make its members aware of food safety issues through summaries. It is also developing guidance in many areas related to food safety emergency response and management.

27.4.4 Food and Agriculture Organization

FAO not only works at the global level to set norms and standards, guidelines, protocols and codes of practice but also involves itself in operational field activities. These two areas of work are interdependent and mutually reinforcing. The field activities are supported by normative resources, whereas its normative work is supported by lessons learned in the field.

FAO also helps the nations in capacity building so that they can adopt the accepted norms and standards into their own national policies and legal frameworks. Such efforts promote sustainable agriculture and food safety. Assistance is also provided to introduce internationally accepted best practices and principles into the legislation of the concerned countries. These include principles of good governance, which are significant for sustainable agriculture. FAO and WHO implemented their Joint FAO/WHO Food Standards Programme in 1963. To prevent food safety emergencies, FAO Emergency Prevention System for Food Safety (EMPRES Food Safety) unit has been established which works with FAO members and other partners.

27.4.5 World Trade Organization

World Trade Organization (WTO) is the only international organization in the world which deals with the rules of trade between countries. It provides many benefits (Figure 27.4).

The aim of WTO agreements is to help producers of goods and services, exporters and importers to carry out their business. These agreements are negotiated and signed by most of the world's trading nations and ratified in their parliaments. Following two agreements are highly significant for food safety and quality issues:

1. Sanitary and Phytosanitary Measures (SPS) Agreement is a separate agreement on food safety and animal and plant health standards under WTO. It covers all relevant laws, rules and regulations; testing, inspection, certification and approval procedures; and packaging and labelling requirements which are directly related to food safety. The member countries are permitted to set their own standards, provided the regulations are based on science. SPS agreement aims to ensure that the consumers are being supplied with safe food that is 'safe' by the standards considered appropriate by the concerned nation. Simultaneously, it ensures that strict health and safety regulations are not used as justification to protect domestic producers. The agreement has provisions on control, inspection and approval procedures and it complements technical barriers to trade.

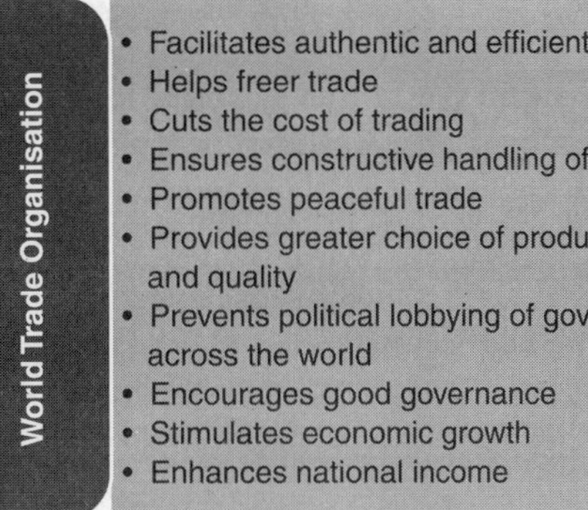

Figure 27.4 Benefits of World Trade Organization

2. Technical Barriers to Trade (TBT) Agreement aims to ensure that regulations, standards, testing and certification procedures minimize unwarranted barriers. This agreement by WTO includes a code of good practice for governments and non-governmental/industry bodies to help them prepare, adopt and apply voluntary standards. TBT agreement emphasizes that fair and equitable methods are used to decide whether a product conforms to relevant standards but it dissuades the procedures that may give unfair advantage to domestically produced goods.

27.5 Consumer Protection Act

Despite the existence of various laws to protect the consumers, they are exploited by well-organized sectors of manufacturers, traders and service providers who use their manipulative skills and marketing knowledge to this effect. Besides, the disposal of cases in the civil courts takes much longer than anticipated.

1. Consumer Protection Act (CPA) was enacted in 1986 as a comprehensive legislation covering all goods and services in the interest of the consumers.
2. The act aims to provide simple, speedy and low-cost redressal to consumer grievances through three-tier redressal machinery.
3. The Consumer Protection (Amendment) Act 2002 was brought with the objective of facilitating faster disposal of complaints, empowering redressal agencies, restructuring the procedures and expanding the scope of the act to make it more functional and effective. It also aimed at strengthening the consumer movement at the grass-root level.
4. In contrast to other laws which impose punitive penalties, CPA is compensatory in nature.
5. It specifies the setting up of Consumer Protection Councils at central, state and district levels (three-tier). The rights of consumers under the CPA are listed in Figure 27.5.

Right to be protected against the marketing of goods and services which are hazardous to life and property.

- **Right to be informed** about the quality, quantity, potency, purity, standard and price of goods or services, as the case may be so as to protect the consumer against unfair trade practices.

Right to be assured for, wherever possible, access to a variety of goods and services at competitive prices.

- **Right to be heard** and to be assured that consumer's interests will receive due consideration at appropriate forums.

Right to seek redressal against unfair trade practices or restrictive trade practices or unscrupulous exploitation of consumers.

- **Right to consumer education.**

Figure 27.5 Rights of consumers under the Consumer Protection Act

This chapter covered various food laws and standards being implemented in India with special focus on the recent FSS Act which will repeal many of the earlier food laws, for example, PFA, FPO and MMPO. We learnt about the importance of international standards such as Codex Alimentarius and ISO standards in supporting global food trade. The significance of regulatory agencies such as FSSAI, FAO, WHO, WTO and EIC was also deliberated. Last but not the least; we learnt the role of CPA in safeguarding the rights of consumers.

Know Your Progress

True or False Questions

1. Food standards, rules and regulations are based on food acts.
2. Agmark and BIS standards are mandatory food standards.
3. The Prevention of Food Adulteration Act, 1954, and the Fruit Products Order, 1955, will be repealed by Food Safety and Standards Act, 2006.
4. The foods covered in the Agricultural Produce (Grading and Marketing) Act include vegetable oils, ghee, butter, eggs, cereals, pulses, jaggery, honey and sugar.
5. The Consumer Protection Act, 2002, aims to provide simple, speedy and low-cost redressal to consumer grievances through a single-tier redressal machinery.

Very Short Answer Questions

1. Differentiate between SPS Agreement and TBT Agreement.
2. Enumerate the six regulations for food safety and standards released in 2011.
3. What are standard weights or measures?
4. Describe briefly Milk and Milk Products Regulations.
5. What is the role of Legal Metrology Act?

Short Answer Questions

1. What are the main features of Food Safety and Standards Act, 2006?
2. List the different functions of Food Safety and Standards Authority of India.
3. How does Consumer Protection Act protect the rights of the consumers?
4. Describe two food standards which are voluntary.
5. Give full form of the following abbreviations:

(a) BIS	(b) CAC	(c) EIC	(d) FAO	(e) WTO
(f) FSSAI	(g) FPO	(h) ISO	(i) WHO	(j) PFA

Long Answer Questions

1. What are the characteristic features of Food Safety and Standards Act, 2006? Highlight the important components of the act.
2. Many international and national organizations are playing significant role in regulating the food standards in the interest of the consumer. Justify the statement.
3. Describe the role of international standards in meeting the emerging challenges of global food trade.

Beyond the Text

- Refer to the Food Safety and Standards Act, 2006, and find out the following information:

    1. Responsibilities of the food business operator.
    2. Liabilities of manufacturers, packers, wholesalers, distributors and sellers.
    3. Food recall procedures.

Chapter 28

Quality Management

Chapter Objectives

After reading this chapter, the reader should be able to

- understand the concept of quality.
- know the various approaches (methods) which can help in improving quality.
- recognize the importance of quality management systems.
- identify good manufacturing practices while preparing food.
- assess risks related to food safety.
- realize the significance of HACCP system and learn to prepare a HACCP plan.

THINK TANK

Identify what went wrong in the situations below.

- Two bakers were preparing chocolate biscuits. The biscuit cutter of one was well framed whereas that of the other baker was slightly dented.
- A chef was preparing purée for gravy when suddenly there was short-circuit in the plug of blender. An electrician promptly corrected the fault. When the dish was served to the guest, there was a piece of electricity tape.
- The shelf life of ready-to-eat 'Dal Bati' varied from two days to one month between different lots.
- A doughnut manufacturing company measures the diameter and height of each doughnut. It discards doughnuts which are not meeting its criteria.

So far we have read about the safety aspects of food. However, it is important to know whether a safe food is also of 'high quality'. Consider this example. A pineapple pudding prepared by a chef was physically, chemically and microbiologically safe. However, the customers did not appreciate the dish because it had watery consistency near the pineapple pieces. In other words, the dish was 'safe' but not of 'desired or high quality'.

The word 'quality' has been described variably or differently by experts. It can be defined as 'the degree or the extent to which a set of inherent characteristics fulfils the requirements' as desired by the customer or consumer of the product or a service. The concept of quality is increasingly becoming popular in the hospitality, catering and food processing industry; both in India and across the world.

.8.1 Characteristics of Quality

Quality has the following characteristics:

1. The concept of quality is obligatory and not mandatory.
2. It helps customers or the users of a product or service to distinguish one from another.
3. Quality is based on the needs and feedback of customers.
4. Maintenance of quality is an 'on-going process' and not a 'one-time' process.
5. This concept helps to achieve higher customer satisfaction.
6. Higher quality products and services facilitate 'brand positioning' in the market and hence ensure a competitive edge over the competitors.
7. Though maintenance of quality based standards may increase expenditure initially, they prove to be profitable in the long run.
8. Maintenance of quality should be the objective of every employee and the management.
9. Quality oriented food safety systems ensure sustained social and economic development by enhancing opportunities for national and international trade.

Quality can be maintained and higher levels can be attained by following a systematic plan of action. Various methods have been propounded by management experts to help food industry improve and maintain the quality and safety of its products and services. The salient ones include Quality Control (QC), Quality Assurance (QA), Total Quality Management (TQM), Quality Management System (QMS), Good Manufacturing Practices (GMP) and Hazard Analysis Critical Control Point (HACCP). These have been discussed briefly in this chapter with relevance to the food industry or more specifically the catering and hospitality establishments (refer to Figure 28.1). Some important terms related to quality have been highlighted in Box 28.1.

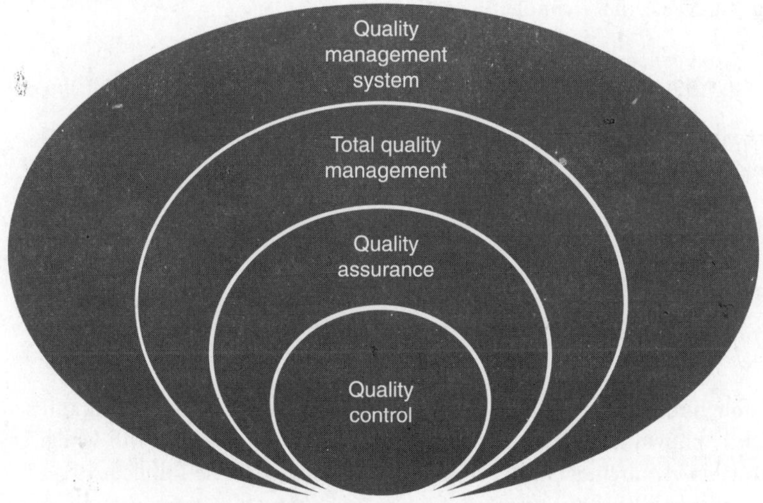

Figure 28.1 Various approaches to management of quality

Box 28.1 Some Quality-related Terms

- **Quality Control** (QC) is the on-going effort to maintain the integrity of a process to ensure the reliability of achieving an outcome.
- **Quality Assurance** (QA) is the planned or systematic action necessary to provide enough confidence that a product or service will satisfy the given requirements.
- **Quality Improvement** can be distinguished from QC in that quality improvement is the purposeful change of a process to improve the reliability of achieving an outcome.
- **Quality Planning** is a part of the quality management system. Quality planning refers to a group of systematic activities which specify the necessary operational processes and resources required to fulfil the objectives of quality improvement.
- **Quality Assessment** is a means of determining the quality of results generated during the evaluation of QC and QA. It helps in identifying the effectiveness of QC and QA. It can be carried out through self-assessment of employees and by an external organization.

28.2 Quality Control

A doughnut manufacturing company is particular about the quality of its products and therefore, it adheres to the following measures:

1. *Checks the quality of ingredients before use and rejects them if they do not meet the specifications and*
2. *Rejects doughnuts which do not conform to its laid down specifications.*

This is known as Quality Control (QC). It works on the principle of 'detection and fire fighting'. QC is part of the QMS which emphasizes on products or processes which improve the reliability of achieving product or service of desired characteristics (quality). Screening, grading, sorting, checking, inspecting and thereafter adopting corrective measures to make the product conform to specified standards is also known as QC. The concept of QC prevents the sub-standard service or product from reaching the customer, but, at times, may result in heavy losses to the manufacturer such as food service operator or the doughnut manufacturing company in the earlier example.

28.3 Quality Assurance

The doughnut manufacturing company soon realized that QC measures were not enough to reduce the manufacture of doughnuts not conforming to desired standards. Therefore, it adopted the following measures:

1. *Ensured that all ingredients, utensils, equipment and packaging materials/serving cutleries conformed to the desired standards.*
2. *Developed a standardized recipe.*
3. *Ensured process control (control of the activities associated with preparation, sale, distribution and storage of doughnuts).*

The number of doughnuts not conforming to the desired standards reduced significantly.

This approach is known as Quality Assurance (QA). QA is a part of QMS which ensures to a great extent that the quality requirements for a service or a product shall be fulfilled. It is different from QC. QC checks the dish once it has been prepared. On the other hand, QA creates conditions which ensure th preparation of a dish which would conform to the desired specifications. QA is a process-oriented approach which aims at 'doing it right the first time'. Therefore, QA for doughnuts would be targeted towards the following:

1. Raw material: providing specifications to suppliers, accepting raw material (ingredients) conforming to the specifications and ensuring the selection of correct cooking/baking equipment.
2. Process: following standardized recipes; in terms of both ingredients and process (such as duration of beating, force applied for beating, ladle used etc.).
3. Finished product: coordinating between preparation and sale, storage, distribution and service of the product, and taking feedback from customers.

Thus, QA refers to the systematic activities implemented in a QMS so that the quality requirements of a food or a service are fulfilled. It involves systematic measurement and comparison with standards, monitoring of processes and feedback; all the three confer error prevention.

28.4 Total Quality Management

The doughnut manufacturing company soon realized that it could not do away with the measures adopted under the QC and QA approach. It also understood that the discussed approaches would be required on a daily basis, several times a day, whenever the doughnuts would be prepared. So, it undertook the following policy decisions:

1. *Allocated quality related responsibilities to each employee.*
2. *Ensured team approach for better coordination between employees.*
3. *Regularly measured performance of each employee, team and department.*
4. *Ensured that the process of doughnut manufacture remained the same every time.*
5. *Discussed customer feedback with suppliers of ingredients.*

Rejection of doughnuts reduced; there was further improvement in quality, higher customer satisfaction, improved brand loyalty and increased profit within a period of one year.

This approach is known as Total Quality Management (TQM). TQM is an integrative philosophy of management which involves the application of quantitative methods and human resources to improve the quality of raw material, processes applied in the preparation of a product (dish/food) or service supplied to the consumer or customer. In catering establishments, 'quantitative methods' refer to the ingredients used, time, money and fuel required for preparing a dish. 'Human resources' refer to the involvement of management and employees at all levels (chef to dish washer and sweeper). TQM is a continuous on-going process which needs to be evaluated and modified from time to time.

Objectives of TQM: It is targeted towards meeting the following prime objectives.

1. **Achieving high Productivity with Minimum Production Costs** by the following means:
   (a) Increasing speed of food production.
   (b) Reducing food production time.
   (c) Minimizing the cost of labour and other overheads.
   (d) Judicious utilization of ingredients and other necessary resources.
   (e) Preferring light weight, easy to handle, easy to clean and efficient equipment, labour saving devices and holding and serving utensils/crockery.

2. **Developing a system involving the approach of 'right first time' and 'zero defects'** by the following means:
   (a) Using high quality (physically, chemically and microbiologically safe) raw materials, equipment, utensils and serving crockery/cutlery.
   (b) Employing the process of preparation, skill and equipment which would provide the highest possible efficiency.
   (c) Producing safe and wholesome food.
3. **Encouraging the involvement of all staff members to contribute towards continuous improvement** of the following:
   (a) Quality and process control achieved so far.
   (b) Efficiency of food production and judicious or alternative utilization of resources.
   (c) Work culture and work environment for each member of the staff and management across all levels.
4. **Evaluating the quality management system** through the following means:
   (a) Taking feedback of customers and sharing with management, staff and suppliers of raw material and equipment.
   (b) Keeping an eye on the profit–loss records of the catering establishments.
   (c) Rewarding and motivating the key players.

These objectives of TQM can be met by taking the following steps:

1. Strategic planning is the first step. Developing a detailed plan of action helps to ensure that the catering unit shall be able to meet the objectives of zero defects and improved quality.
2. Managing the quality of ingredients, equipment, utensils and crockery by continuously interacting with the suppliers and updating oneself about the newer innovations in food industry.
3. Process management such as ensuring smooth work flow and efficient and timely food preparation.
4. Ensuring customer satisfaction by taking their feedback regularly. Involving the customers can also be useful.
5. Assessing performance, promoting team work, delegating responsibility to each employee, empowering them to take relevant decisions and involving employees in food production related decisions.
6. Facilitating periodical medical check-ups, training of employees and management, and rewarding efficient contributors, help to make TQM an inbuilt management approach.
7. Discussing failures and identifying flaws in the entire process/policy, so that timely necessary changes can be made.

By following these steps, not only a food service unit would be able to maintain quality but also aspire for higher quality. This approach would help it to create and sustain its reputation. Thus, TQM is based on the premise that the quality of products and processes is the responsibility of everyone involved with the creation or consumption of the products or services offered by an organization, requiring the involvement of management, workforce, suppliers and customers, to meet or exceed customer expectations.

28.5 Quality Management System

The doughnut manufacturing company soon began to encounter difficulties related to coordination in various activities across various departments, especially the ones involved in purchase and food preparation. It was also aspiring to meet higher standards of quality so that its brand position could be maintained in the market. Therefore, it hired a consultant to build a quality management system for its organization which included the following:

1. *Mandatory compliance of all employees to laid-down specifications.*
2. *Identifying team leaders.*
3. *Success and failure identification system.*
4. *Identifying factual solutions for each problem within a time limit and resources available.*
5. *Greater focus on customers.*
6. *Ensuring healthier relationships with suppliers and employees.*

The approach adopted by the doughnut manufacturing company is known as Quality Management System (QMS). 'Quality management' refers to all the coordinated activities and processes which direct and control an organization with regards to its quality. QMS ensures that quality management becomes an inherent feature which is adopted as a continuous on-going process. The fundamental focus areas of Food Quality Management System include the following:

1. **Commitment of the Management:** Top management people should make a food safety policy for its employees and ensure its compliance. The steps being taken for achieving higher quality must be clearly explained to the immediate subordinate and the messages should be communicated according to the work profile of each employee. The management should also be prepared for failures and untoward emergency situations. Decisions to be taken and steps to be implemented during situations of system failure must be clearly stated in the quality policy document. Top management people must periodically review its quality policy.
2. **Management of Resources:** While aspiring for higher quality, it is important to know the available resources such as finance, time, manpower and infrastructure. The work environment of employees should also be considered. Judicious utilization of resources is a must for maintaining balance between optimal production and cost incurred.
3. **Planning and Realization of Safe Quality Products:** Awareness generation and delegation of responsibility pertaining to safety and quality of food is a must for each employee in every food service unit; be it a dhaba, mess or a five star hotel. **Prerequisite programmes** must be developed to ensure that the basic conditions and activities which are necessary to maintain a hygienic environment for the production, handling, storage and distribution of safe and higher quality foods are available in the organization. There should also be a study of all the processes taking place in the organization to enable hazard analysis. Based on the analysis of hazards, a suitable Hazard Analysis Critical Control Point (HACCP, discussed latter in this chapter) plan should be developed, effectively implemented and reviewed periodically. Traceability of 'not-safe' or 'sub-standard' at the earliest possible stage is also necessary. It is also essential that the organization has a stringent control over non-compliance.
4. **Validation, Verification and Improvement of the Food Quality Management System:** Each organization has its own QMS. It must be ensured that the QMS being adopted is valid, standardized and error-free and would not be a liability for the organization in the long run. It is always advisable to take opinion from experts on the system being adopted.

Several organizations provide guidelines on the development of QMS. Certain organizations also provide accreditation to catering establishments/food industries such as the International Standards Organization (ISO), the British Retail Consortium (BRC), International Food Standard (IFS) and Safe Quality Food Scheme (SQF 2000) (refer to Box 28.2).

Box 28.2 ISO and Food Quality Management System

ISO 22000: 2005 'Food Safety Management Systems – requirements for any organization in the food chain' has been developed by International Standard Organization to meet the specific needs of the food industry including catering establishments. The fundamental concepts discussed in this standard include food safety, HACCP (Hazard analysis critical control point), prerequisite programme (PRP) and Operational PRP or operational pre-requisite programme.

28.6 Good Manufacturing Practices

Good Manufacturing Practices (GMP) is typically a group of QC and QA procedures. In India, adoption of GMP is obligatory but in certain countries such as the United States of America, it is mandatory. Nowadays, the concept of GMP has been replaced by cGMP which means Current Good Manufacturing Practices. During the processing or preparation of food, there are several variables such as the inherent moisture content in tomatoes being used in tomato ketchup. A policy document on GMP cannot take care of such variations. Therefore, the concept of cGMP is replacing GMP. All organizations involved with the pre-preparation, preparation (manufacture), packing, handling and storage of food must follow GMP regulations. The areas covered under cGMP include personnel, building and facility design, sanitary operations, equipment and utensils, and production and process control. It ensures prevention of food contamination, production of safe food and testing of food to ensure its safety and quality.

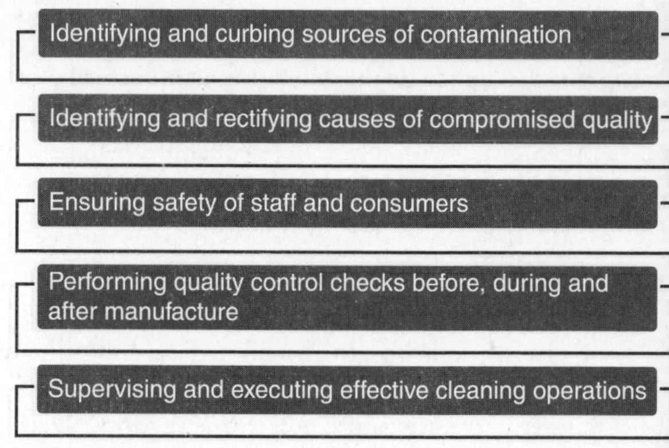

Figure 28.2 Components of good manufacturing practices

The various components which need to be adopted to ensure GMP include (Figure 28.2):

1. **Identifying and Curbing Sources of Contamination:** Building, general area of catering establishments, people, equipment, ingredients, utensils and crockery are the potential sources of contamination. Contamination reduces the chances of quality maintenance. Contaminated work environment reduces work efficiency, creates dissatisfaction among employees and reduces their motivation to help management achieve its goals. Some of the points which should be taken care of include the following:

   (a) Employees

      (i) Communicable diseases spread easily through dirty, improperly washed hands. Good, clean hand-washing facilities comprising of continuous flow of hot and cold water, non-scented disinfectant containing soap in dispensers (not tablets) and preferably towels should be provided.

      (ii) Employers should provide clean overalls, hair coverings and, if feasible, disposable gloves to all working people in the food production areas. Employees should not wear any loose jewellery, rings, nail polish or any other cosmetic in their hands. They should not carry any loose objects which may accidentally drop in the food being manufactured. They should not be allowed to brush hair in the work area.

(iii) Smoking should be strictly restricted and talking should be avoided.

(iv) Sick or medical leaves should be provided, if the employee is suffering from infectious diseases. Nasal and throat excreta can seriously hamper food safety. Cuts and sores on arms/hands can also be a source of contamination.

(b) Equipment, utensils and work environment

(i) Raw food, fresh food, minimally processed food, preserved food, manufactured foods and formulated foods all have variable shelf life and should be stored/handled accordingly.

(ii) All raw materials including ingredients should be carefully removed from sealed containers or boxes, cartons, bottles and polybags in a separate ingredient collection area. Utmost care should be taken that no part of the containers, labels and strings accidently get dropped into the raw materials (ingredients) being used for food preparation.

(iii) All equipment and devices coming in contact with food surface should be of good quality material, preferably metal. Glass containers should be strictly avoided because upon their breakage, the splinters spread over large areas.

(iv) Office equipment should not be brought into the food pre-preparation and preparation area.

(v) At the end of each food preparation schedule (such as breakfast/lunch/tea/dinner), all equipment and utensils should be immediately and thoroughly cleaned with hot water and a suitable sanitizing or cleaning agent should be used.

(vi) All equipment and cooking utensils should be checked for defects and repaired promptly. Repair should be carried out in a separate room and not in the food preparation area. Repair of fixtures should be carried out only when food production is over or all food has been removed from the place where the fault has occurred. Thorough cleaning after repair is a must.

(vii) All food production establishments must take suitable measures to keep away insects, pests, reptiles, animals and birds.

2. **Identifying and Rectifying Causes of Compromised Quality:** Quality may get compromised during food handling, preparation, storage and distribution. All employees should be trained to ensure conformance to quality related policy matters. The following points should be kept in mind:

(a) Give skill-specific tasks according to the specialization of each employee.

(b) Ensure that efficient and correct store-keeping procedures are followed.

(c) Standardized recipes should be followed each time a food is prepared.

(d) Handling and distribution of food should be such that there is no physical damage to the structure of food, nor do the inherent characteristics (flavour, colour, juiciness) get compromised.

(e) Keep the staff motivated by developing good relationships with them.

3. **Ensuring Safety of Staff and Consumers:** Employers must provide a safe and comfortable environment for work to all their employees. The staff must in turn ensure that the consumer is safe within the premises of the catering unit. They must feel responsible for providing safe wholesome and good quality food to their consumers. It is the duty of all employees to maintain the facilities and

promptly report any inadequacies or faults to their immediate superiors and also help in the repair process. To ensure customer and employee safety, the following points should be kept in mind:

(a) Wet, greasy, slippery floors should be promptly cleaned. The floors should be smooth and non-slippery.

(b) Food display and service areas should not have any displays, furniture or fixtures with sharp or broken corners; no fixtures should be loosely hung from the surface; and there should not be any objectionable odour and smoke. Precautions should be taken that during the display, handling, serving or packing, the food does not become contaminated or otherwise unsafe for human consumption.

(c) Employees must receive insurance pertaining to accidents within the work premises. This motivates them to attain higher standards of work.

(d) Facilities for providing first aid to customers and employees must be readily available in case of miss-happening or accident.

(e) Safety masks and gloves should be provided to employees involved with dish-washing and clearing of waste.

(f) Clean toilet facilities must be provided separately for employees and customers.

(g) Food service and work area should be well ventilated and have ambient temperature (preferably between 20°C and 30°C).

(h) Equipment, fans and air conditioners must be provided with safety switches/fuse.

4. **Performing Quality Control Checks Before, During and After Manufacture:** Inspection of work area, ingredients, equipment, utensils, crockery and process control is the backbone for all quality aspiring operations. It is a teamwork which involves the management, suppliers, purchase department, food production and service team, food analyst, food safety officer and the housekeeping department. All team members must be aware of the specification agreed upon in the contract for a particular purchase. A physical examination of the supplies before downloading from the vehicle and thereafter metering, weighing or counting of supplies is also a must. Good store-keeping ensures safety and preservation of quality for all supplies. Inspection of pre-preparation and processing area, dish-washing area and store is necessary to maintain safety and quality of food. For example, spot checking of pack weights, appearance, labelling, sealing, number of broken cookies after specific storage duration and presence of extraneous or other substances in or on food pack can help to identify faults in a bakery.

5. **Supervising and Executing Effective Cleaning Operations:** Cleaning operations in India are labour intensive. They reduce profit of an organization, and yet these operations are significant in determining the reputation of food establishment. Cleaning operations should not be restricted to customer toilets or food display areas; they should be performed in the entire food establishment including office area. Some of the important points which can be helpful include the following:

(a) Adopting and executing HACCP programme.

(b) Developing a functional roaster for all cleaning operations with duty list. This should be supplemented with frequent inspections (daily or several times a day) followed by prompt reporting.

(c) Employing a food safety officer or a hygiene officer to supervise all cleaning and sanitizing operations.

(d) Developing cleaning and maintenance contracts with other companies, if the same is not feasible within the organization.

(e) Providing good quality, easy to use, effective cleaning agents and aids to workers. Ensuring that no cleaning material/aid comes in the proximity of food.

So far, we have discussed about the various aspects of GMP, including HACCP. In the forthcoming section, we will deliberate on the concept of HACCP and its significance in QC and safety.

28.7 Hazard Analysis Critical Control Point System

The inception of HACCP occurred during a space food project of NASA and the US Army in collaboration with the Pillsbury Company. It was developed with the objective of making space food microbiologically safe. Today, HACCP is widely used as a system to prevent physical, biological and chemical hazards which can compromise the safety of food meant for human consumption. HACCP is a system which identifies, evaluates and controls hazards which are significant for food safety. It helps to enlist the potential risks associated with food production, identify causes (hazards) and assess their effects before deploying an effective control mechanism. Some important terms associated with HACCP are given in Box 28.3.

Box 28.3 Important HACCP-related Terms

Control Point (CP): A step involving preventive or controlling actions in the food preparation process as part of good manufacturing practices, regulations or product reputation. They do not affect safety of food but influence its quality significantly.

Critical Control Point (CCP): It is an operation (practice or process or location or ingredient) at which if a preventive or control measure is taken, it would eliminate or prevent or at least minimize the occurrence of a hazard(s).

Critical Limit: A quantitative or qualitative value which differentiates an acceptable norm from an unacceptable one.

Corrective Action: It is a specific prompt action which needs to be taken whenever there is any deviation from the critical limits for critical control points.

Verification: It is a systematic review of the monitoring records to determine whether the HACCP plan is valid and effective for a particular product or service.

The HACCP system relies heavily upon the assessment of risk because all subsequent actions depend upon this step. If the risk of a hazard is identified and understood properly, the corrective actions to be taken for their prevention also prove to be successful.

28.7.1 Risk Analysis

The concept of risk analysis is closely associated with the objective of ensuring food safety. In context of food safety, the word 'risk' refers to the probability of having an adverse effect on health and severity of the effect due to the presence of a 'hazard' in a particular food. In the field of food and health, the word 'risk' is used when an attempt is made to identify, evaluate and control factors associated with compromised food safety or adverse effects on health as a consequence to food handling or consumption. The process involved in identifying, evaluating and controlling hazards is known as 'risk analysis'. The process of risk analysis comprises of three steps viz. risk assessment, risk management and risk communication (refer to Figure 28.3).

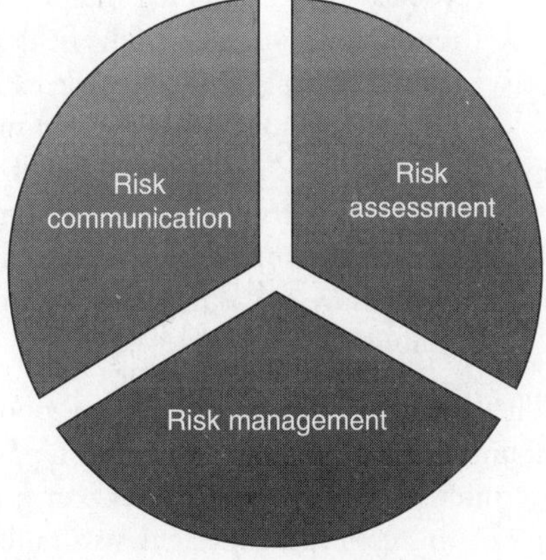

Figure 28.3 Steps involved in risk analysis

1. **Risk Assessment:** It is the first step of risk analysis process. The objective of risk assessment is to quantitatively predict the possibility of harm caused by a hazard with least amount of uncertainty. Risk assessment comprises of four steps.
   (a) Identification of hazard.
   (b) Assessment of the effect of hazard (in variable doses) on health.
   (c) Identification of population most prone to exposure to the hazard.
   (d) Risk characterization, that is, expressing the overall magnitude of risk (problem).
2. **Risk Management:** Based on the results of risk assessment, appropriate control and prevention measures are adopted under the step of risk management. While short-listing the preventive or control measures, the interest of consumers should receive first preference. Risk management process should be outlined and implemented in consultation with management, suppliers and other experts in the field. After implementation of corrective actions, the entire process should be reviewed and documented.
3. **Risk Communication:** It is a multidimensional process which involves an exchange of information pertaining to risk assessment and opinions on risk management. Information related to risk should be communicated to the consumers, all employees associated with handling and preparation of food, professionals from the fields of health food and law, and policy makers. Information related to risk can be communicated through messages given on labels and packs, advertisements, community outreach programmes, education and training of employees and regulatory enforcement communications.

The status of risk analysis is very poor in our country and is often not undertaken particularly by micro-enterprises due to lack of knowledge about this concept. Risk analysis should be carried out by well-qualified professionals such as food safety officers or food analysts in consultation with top management people of a catering or food service unit.

THINK TANK

Participants of a salad competition who consumed salads containing cottage cheese suffered from diarrhoea and vomiting. An enquiry was carried out to identify the problem. There was presence of *E. coli* bacteria in the cheese samples. It was found that workers involved with cheese setting did not wash their hands after using toilet. Can you identify the step of risk analysis which has been discussed here?

THINK TANK

The CEO of a well-reputed frozen meat chain read about salmonellosis outbreak in a daily newspaper. He immediately instructed his subordinates to

- check the storage conditions of raw meat,
- send a sample of raw meat for microbiological assessment and
- ask for a food safety certification from the supplier.

Which step of risk analysis has been performed above?

THINK TANK

A fruit juice company gets the following statement printed on its pack 'Do not buy or consume, if the pack is puffed'. Which step of risk analysis process has the company followed?

The unsold bread loaves indicated mould growth (*aspergillus flavus*). The company immediately called back all the sold loaves. It also advertised a warning in the daily newspaper for the consumers instructing them not to consume bread with a particular lot number. Which step(s) of risk analysis were performed in the above situation?

28.7.2 Principles of HACCP System

CODEX Alimentarius Commission has laid down the following seven principles of HACCP (Figure 28.4).

1. **Conduct Hazard Analysis:** Hazard analysis is a two-step process; hazard identification and hazard evaluation. Hazard identification involves determining the biological, chemical and physical agents or the process(s) which may compromise food safety. This can best be done by the help of a flow diagram which presents the sequential step-by-step process involved with food handling, preparation, service and storage. Hazard evaluation helps to assess the likelihood of occurrence and the possible consequences (risk) of the hazard(s).
2. **Identify Critical Control Points:** The flow diagram helps to identify the possible points at which a hazard gets introduced or created during the food processing chain. It is, then, important to sort out control points (CPs) from critical control points (CCPs). If a potential risk for a hazard gets eliminated by adopting a 'prerequisite programme', it is considered to be a control point. However, the control points which are critical to managing the safety of a product are the critical control points. Yet, identifying critical control points does not mean that safety precautions are not required for control points. Necessary precautions must be taken to curb hazards for both CPs and CCPs. CCPs should be carefully monitored, and recordkeeping of the monitoring efforts is necessary to ensure that no lapses occur with hazard control measures.
3. **Establish Critical Limits for Control Point Criteria:** It refers to critical limits for each critical control point. It is the minimum and maximum value for parameters at each CCP which must be controlled to prevent or eliminate or reduce a hazard (within an acceptable level) which may compromise safety of food. For example, the time and temperature of cooking kidney beans to denature trypsin inhibitors or the maximum permissible amount of a colouring agent added to an ice cream or the smoking temperature of oil being used for frying.

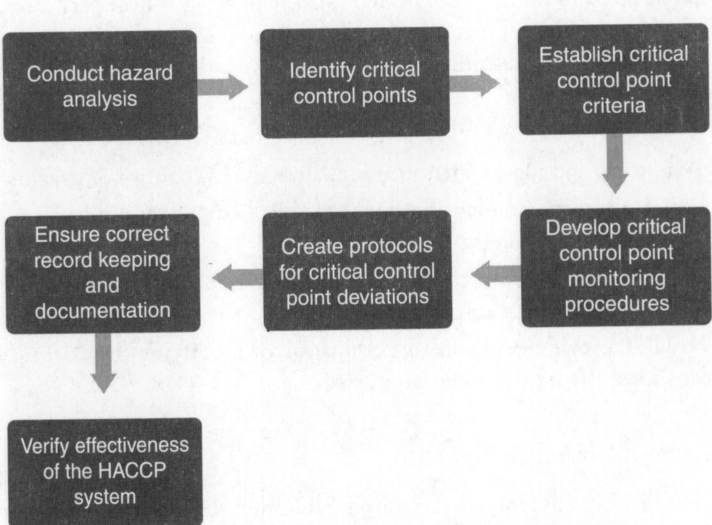

Figure 28.4 Principles of hazard analysis critical control point system

4. **Develop Critical Control Point Monitoring Procedures:** Monitoring involves repetitive measurements of certain parameters or observations in a process. The measurements should be quantitative as far as possible. For example, if the critical limit for raw food storage temperature is 20–25°C, the monitoring protocol would include recording the store temperature after every one hour. Curvature and sponginess of doughnut is also a quantitative monitoring data.
5. **Create Protocols for Critical Control Point Deviations:** Deviations from standard protocols can be minimized but cannot be completely eliminated in practical situations. The purpose of identifying critical control points, establishing critical limits and monitoring procedures is to take corrective actions as and when required so that unsafe food does not reach the consumers. The HACCP system protocol must mention the corrective actions, if there are deviations from the standards. Such corrective actions should be known to the concerned employee; they should be easy to understand and apply. For example, if the cream to be filled in cream rolls has become sour, the corrective actions would be to discard it and use fresh cream. Adding sugar to the sour cream or ignoring the stale status of cream would be a hazard and not a corrective action.
6. **Ensure Correct Record Keeping and Documentation:** Recordkeeping is often disliked and, at times, even manipulated by certain employees. HACCP plan must clearly state the procedures required for documenting the corrective actions. Correct records help in finding out the frequency of occurrence of a hazard, the cause(s) and success of corrective actions. If the corrective actions are not successful, they need to be modified. Records help to evaluate the entire HACCP plan.
7. **Verify Effectiveness of the HACCP System:** HACCP plans should be routinely verified. It helps in understanding the effectiveness of the system for each product. There are three types of verification processes: validation, on-going verification and reassessment.
   (a) Validation helps in testing the suitability of a particular HACCP plan for a particular product or process. It also helps to ensure that correct CCPs have been shortlisted for subsequent actions.
   (b) On-going verification involves testing of HACCP plan when the process of, say, preparing a dish is under progress.
   (c) Reassessment helps to update the HACCP plan. Updating is necessary, because over a period of time, the composition of ingredients may change, newer equipment may have replaced the older ones, and menu or customer profile might have changed and so on.

28.7.3 Benefits of HACCP System

1. It facilitates systematic and logical identification, analysis and control of hazards.
2. The flow diagram approach to identification of CCPs helps in minimizing mistakes or failures which often occur due to oversight.
3. Information on critical limits and corrective actions provide a direction for more efficient work. It ensures supply of safe food to employees.
4. Records on HACCP provide guidelines for evaluating or modifying the existing plan and developing newer plans for different or similar products.

28.7.4 Limitations of HACCP System

1. Implementation of HACCP requires the adoption of other programmes such as cGMP, QMS and TQM. For example, if a CCP is dish-washing, cGMP or GMP would be required to ensure that the dishes are washed by 3-sink method.

2. Many a times, there can be difference of opinion with respect to CPs and CCPs.
3. It is time consuming.
4. Recordkeeping is expensive.
5. Well-qualified and trained individuals are required to develop, implement and maintain the HACCP system.

It is important to note that HACCP plans are industry specific and different for each product, including varieties of products. Box 28.4 briefly gives an outline of an HACCP plan for breaded chicken. In addition

Box 28.4 A Brief Outline of HACCP PLAN: Breaded Chicken

Ingredients	**Method of preparation**
12 chicken strips 1 egg, beaten 1 cup buttermilk 1.5 teaspoon garlic powder 1 cup all-purpose flour 1 cup seasoned bread crumbs 1 teaspoon salt 1 teaspoon baking powder Oil for frying	• Place chicken strips in a flat, wide surface bowl. • In a small bowl, mix beaten egg, buttermilk and garlic powder. • Pour mixture over the chicken strips. Seal with aluminium foil and refrigerate for two to four hours. • Mix the flour, bread crumbs, salt and baking powder in another flat, wide surface bowl. • Remove chicken from refrigerator and drain, discarding the buttermilk mixture. • Place chicken strips on the flour mixture and lightly move to coat all over. • Heat oil in a large heavy bottom skillet. • Carefully place breaded chicken strips in hot oil. • Fry until golden brown. Drain on an absorbent or brown paper.
Control points *(only salient points listed)*	• Quality of ingredients (microbial quality of chicken, eggs, buttermilk and other ingredients, fresh or stale eggs, moisture content and size of bread crumbs and type of skillet). • Process control (storage conditions of ingredients, sealing with aluminium foil, duration of marination and number of times chicken strips are turned in the skillet).
Critical control points *(only salient points listed)*	• Microbiological quality of chicken, buttermilk and eggs. • Storage conditions of ingredients. • Power failure.
Critical limits *(only salient points listed)*	• Microbial load of chicken, buttermilk and eggs. • Temperature at which chicken was stored. • Temperature of oil used for frying. Turnover rate of frying oil.
Monitoring procedures *(only salient points listed)*	• Type of pathogens present and their content. • Temperature fluctuations (if any) in the cold store. • Oil heated above or below smoking temperature. Quality of oil being used, duration and frequency of heating oil.
Corrective action *(only salient points listed)*	• Uninterrupted power supply ensured to cold store and refrigerator. • Microbial load of chicken discussed with suppliers. • Improved storage conditions of ingredients. • Heating oil below smoking temperature.
Recordkeeping and documentation *(only salient points listed)*	• Change of supplier (if any). • Reduction of microbial load in ingredients (if any). • Uninterrupted power supply (achieved or not). • Feedback of customers.
Verification *(only salient points listed)*	• Did the corrective measures improve the safety of breaded chicken? • Did the above HACCP plan improve customer satisfaction and brand loyalty?

to the above methods of quality management, there are other supportive approaches which can be adopted such as Good Hygiene Practices (GHP) and Sanitation Standard Operating Procedures (SSOP).

28.8 Relevance of Microbial Standards for Food Safety

The major hazards in food which compromise its safety include physical, chemical and microbiological. Physical damage to food is easy to ascertain but the same is not true in case of chemical and microbiological hazards. Chemical hazards include presence of pesticides, heavy metals, radionuclides and additives beyond permissible limits. Microbiological hazards include infections and intoxications caused by bacteria, fungi, viruses and protozoa.

Food testing is the fundamental tool which provides quantitative data on food safety. Different tests are available to assess chemical and microbiological hazards. Food should be tested prior to leaving the farm or the processing unit. It should be tested before the food reaches the customers. Testing of food helps to identify sources of contamination and hence prevent food-borne illness outbreaks.

Significance or scope of microbial standards:

1. Authentic microbial standards would reduce microbial contamination and control microbial growth. They would thus help in reducing the episodes of food poisoning and intoxication.
2. The tests required for identifying a pathogen and its load in a particular food are often time consuming. It is not always possible to wait for two to three days for the microbial load report of a dish or a food product, because the shelf life of the dish is very short. Think about a chef waiting for microbial load report of *paneer pasanda*!
3. Not all food service units and catering establishments have a well-equipped laboratory. They often have to rely on outside laboratories for food analysis. Interpretation of reports is not always easy and necessitates the expert advice of a microbiologist.
4. Interpretation of the report to develop/implement corrective actions or preventive measures should be easy for a chef or any employee associated with handling of food.
5. They would help to assess shelf life of the product.

Microbial standards must have the following characteristics:

1. They should be accessible to each employee associated with handling of food.
2. The standards must be easy to understand and apply in practical situations.
3. Microbial standards should be need based, that is, best suited to the food processing capacity of an organization.
4. They must clearly indicate the time–temperature relationship for each pathogen.
5. Incubation period should be clearly mentioned for each microbe.
6. The effect of environmental factors such as water activity, pH, temperature and time must be clearly stated for each microorganism.
7. Corrective and preventive measures should be clearly indicated for each microorganisms and these should be food specific.

Only valid and standardized microbial standards can ensure food safety. They should preferably be recommended by specific organizations such as ISO. Adhering to national or international level microbial standards helps in greater approval of food trade and hence a better economy.

Know Your Progress

True or False Questions

1. Quality assurance works on the principle of detection and fire fighting.
2. Risk communication is based on the results of risk assessment.
3. HACCP plan is product specific.
4. Microbial standards can replace the need for TQM.
5. Adherence to GMP is obligatory in India.

Very Short Answer Questions

1. Describe the following:
   (a) Quality planning
   (b) Risk
   (c) Hazard
   (d) Verification
   (e) Risk analysis

Short Answer Questions

1. What is risk communication?
2. What is the difference between control points and critical control points?
3. Enumerate the various characteristics of quality.
4. What are the limitations of HACCP system?
5. Why is the commitment of top management people essential for the success of a Quality Management System?

Long Answer Questions

1. Describe the concept of total quality management in detail. How is TQM different from Quality Management System?
2. Discuss the various components of GMP.
3. Give the principles of HACCP. Develop an HACCP plan for the preparation of a pineapple ice cream.

Beyond the Text

- Prepare the following for your next Quantity Food Production practical
  1. HACCP plan for each dish to be cooked. Begin from the store in your college.
  2. Identify at least 10 activities which adhere to the norms of GMP.
  3. Enlist at least 10 points which could be improved or where corrective actions could be taken.

Chapter 29

Hygiene and Sanitation in Food Sector

Chapter Objectives

After reading this chapter, the reader should be able to

- understand the general principles of food hygiene.
- learn the basic requirements for designing food establishment, equipment and facilities.
- appreciate the role of operational control, transportation, product information and consumer awareness in ensuring food hygiene and safety.
- gain knowledge about cleaning, sanitizing and pest control measures.
- apply good hygiene practices in food production.
- comprehend the role of water quality and waste disposal in food sanitation and hygiene.

We studied in Chapter 24 that food-borne diseases lead to gastro-intestinal and other symptoms and may be fatal in some cases. Moreover, the outbreaks of these diseases can adversely affect consumer confidence, as well as the reputation and profits of food business. Awareness about the principles of food hygiene and their application in good hygienic practices can go a long way in preventing food-borne diseases and ensuring food safety in food service, catering, as well as other food businesses.

THINK TANK

Lapses in Food Hygiene

Consider the following examples in the context of food safety:

- A trolley boy, whistling while carrying meat out of a cold store room, forgets to close the door.
- The salad bar tender suffering from dysentery is refused sick bay leave. While preparing salad for the buffet, he visits the toilet frequently to ease himself but fails to wash his hands properly.
- Same set of knives is being used for cutting fruits and raw meat in a food service unit.

Can you pinpoint the food safety risks in the above situations? These lapses can lead to contamination of food which if ingested can result in food-borne illnesses. Adequate sanitary and hygienic measures are necessary to prevent such mistakes.

29.1 General Principles of Food Hygiene

Minimal hygienic practices and procedures that have been used for many years to produce safe food and which are considered best by the practitioners and experts of that operation are known as Good Hygiene Practices (GHP) (Box 29.1). They form the basis for Hazard Analysis and Critical Control Point (HACCP) and are significant even in modern times.

Box 29.1 Definitions of Important Terms

Food Hygiene has been defined as all the conditions and measures required for ensuring food safety and suitability at all stages of the food chain.
General Principles of Food Hygiene (formulated by Codex), draw attention to the fundamental hygienic controls necessary for food from the primary production through the processing until it is consumed.
Good Hygienie Practices (GHP) refer to conditions and measures required for ensuring the safety and suitability of food at all stages of the food chain.
Sanitation generally refers to the provision of facilities and services for the safe disposal of human urine and faeces. The word 'sanitation' also refers to the maintenance of hygienic conditions, through services such as garbage collection and waste water disposal.

We studied about FSS Act, Rules and Regulations in Chapter 27. Under the Food Safety and Standards Licensing and Registration of Food Businesses) Regulations, 2011, the fourth schedule stresses on anitation and hygiene in a food establishment. t lists the 'General Hygienic and Sanitary Practices to be followed by Food Business Operators applying for license in India'. These practices are similar to Codex-formulated General Principles of Food Hygiene (Figure 29.1).

Knowledge about principles of food hygiene s equally important for the government, industry (primary producers, manufacturers, processors, food service operators and retailers) and the consumers. The food service operators can apply hese principles to

1. provide food which is safe and suitable for consumption.
2. ensure that consumers have clear and easily understood information.
3. enable the consumers to protect their food from contamination and food-borne pathogens.
4. maintain confidence in food traded/served at various levels.

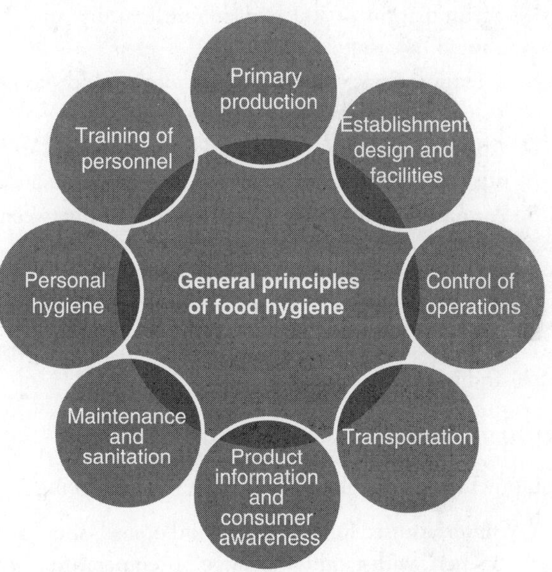

Figure 29.1 General principles of food hygiene

29.1.1 Primary Production

Primary production comprises of steps in the food chain up to and including harvesting, slaughter, milking nd fishing. Segregation of unsafe food and other material as well as prevention of contamination is necessary t this stage to ensure food safety at later stages of food chain.

29.1.2 Establishment Design and Facilities

Food establishments should normally be located away from environmentally polluted areas, industrial ctivities and pest-prone areas which pose a serious threat to food safety. The location should be at such places where solid/liquid wastes can be removed effectively.

Layout and Design

While designing the layout, following points should be given due consideration:

1. **Floors, Walls and Ceiling** of the processing/cooking area should be in a sound condition so that accumulation of dirt, condensation and growth of moulds is minimized. The materials and finishes used should be impervious and easy to clean with no flaking paint or plaster. Concrete flooring is unhygienic as it can absorb water and other constituents to permit growth of microbes underneath the surface.
2. **Doors, Covers and Panels** should prevent accumulation of soil and contamination, and should be easy to clean.
3. **Adequate Space for Storage** of food and for keeping the equipment should be provided away from urinals, drains and waste disposal areas.
4. **Appropriate Placement of Fixtures, Pipes and Ducts** should be ensured so that they do not suspend over the processing area to contaminate the food or the contact surfaces.
5. **Ample Working Space and Wide Aisles** between the equipment and wall should be provided for ensuring smooth and unobstructed movement.
6. **Partitioned Sections** for different operations should be designed to avoid cross contamination. The washing area should be away from the processing area.
7. **Effective Screens or Other Barriers** should be installed to ensure protection from rodents, animals and insects.

ACTIVITY

Challenge 1. Can you help a food business operator to choose a durable and hygienic floor finish for the kitchen of his restaurant from the following options?

(i) vitrified ceramic tiles (ii) concrete
(iii) stainless steel (iv) granite

Challenge 2. A chef specializing in fruit mocktails has to select suitable containers for preparing the beverage. Which of the following materials are appropriate for the task?

(i) enamel ware (ii) glass (iii) stainless steel
(iv) aluminium (v) brass

Equipment

Equipment and containers should meet the following requirements:

1. **Materials** used for equipment and utensils should resist abrasion, corrosion and tainting. They should be inert, withstand wide range of temperature, clean and disinfect easily and have a long life (other than single-use disposables). Food contact surfaces should be made of non-toxic materials. Stainless steel, aluminium and glass are suitable materials for cooking, storing and serving food.
2. **Cleanliness, Sanitation and Maintenance** should be carried out for all equipment and utensils used for food preparation and serving.
3. **A Cover, Lid or Clean Gauze Net** should be provided for the equipment and utensils used for storage/serving so as to prevent the food from dirt, flies, dust or insects (Figure 29.2).
4. **Location** of equipment should be such that it allows easy cleaning and maintenance and also ensures GHPs inside the premises.
5. **Food Control and Monitoring Equipment** should ensure that the temperatures and other conditions necessary for food safety and suitability are rapidly achieved and maintained.
6. **Containers for Waste and Inedible Substances** (including hazardous substances and cleaning agents) should be distinctly identifiable, suitably constructed and, wherever appropriate, made of impervious material. They should be lockable to prevent malicious or accidental contamination of food.

Figure 29.2 Hygienic utensils and covered containers help to prevent contamination of food

Facilities

Hygienic practices are required for facilities such as water supply, drainage and waste disposal, cleaning, personal hygiene, temperature control, air quality, lighting and storage (Figure 29.3).

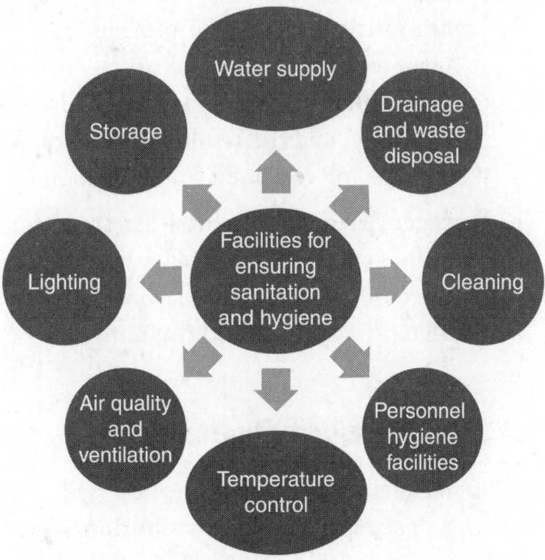

Figure 29.3 Facilities for ensuring sanitation and hygiene in a food establishment

1. **Water Supply:** For food processing and cooking, there should be adequate supply of potable water along with its storage and distribution. The pipes carrying non-potable water should be distinctly separate from the pipes of potable water.
2. **Drainage and Waste Disposal:** The plumbing should ensure holding and supplying of water to the required locations in the food establishment. It should also aid in the removal of sewage and liquid waste. Self-drainage of all surfaces and pipelines of equipment would prevent contamination as well as microbial growth.
3. **Cleaning:** Proper facilities for cleaning food, utensils and equipment should be provided along with adequate supply of hot and cold potable water and steam (if required). Sinks for washing raw foods should not be used for washing utensils or for any other purpose.
4. **Personnel Hygiene Facilities:** In order to ensure personal hygiene and to avoid food contamination, adequate personnel hygiene facilities should be available. These include wash basins and a supply of hot and cold water for proper washing and drying of hands, hygienically designed separate lavatories for men and women, and changing facilities for the personnel. These facilities should be so designed that they do not open directly into food processing or storage areas. Doors should be self-closing and signs should be posted instructing the employees to wash their hands with soap. Water, soap, tissue and towels should be made available with easily cleanable waste receptacles (Box 29.2).
5. **Temperature Control:** Adequate facilities should be provided for heating, cooling, cooking, refrigerating and freezing foods. There should be provision for storing refrigerated/frozen foods and controlling food temperatures particularly for potentially hazardous foods (PHFs).

6. **Air Quality and Ventilation:** Ventilation systems should be designed in a manner that air does not flow from contaminated areas to clean areas. They should be maintained and cleaned at regular intervals. Adequate ventilation must be allowed for exhaust of undesirable fumes and vapours and for preventing air-borne contamination.
7. **Lighting:** It is necessary to provide adequate lighting, natural or artificial, at different storage, processing and washing areas. Light intensity should be suitable to the nature of the operation and the resulting colour of lighting should not be misleading. Lighting fixtures should be protected to prevent food contamination by breakages.
8. **Storage:** Adequate facilities for the storage of food, ingredients and non-food chemicals such as cleaning materials, lubricants and fuels should be provided. These should permit cleaning and maintenance; avoid pest harbouring; enable food protection from contamination and minimize the deterioration of food (by temperature and humidity control).

Box 29.2 Facilities for Personnel Hygiene

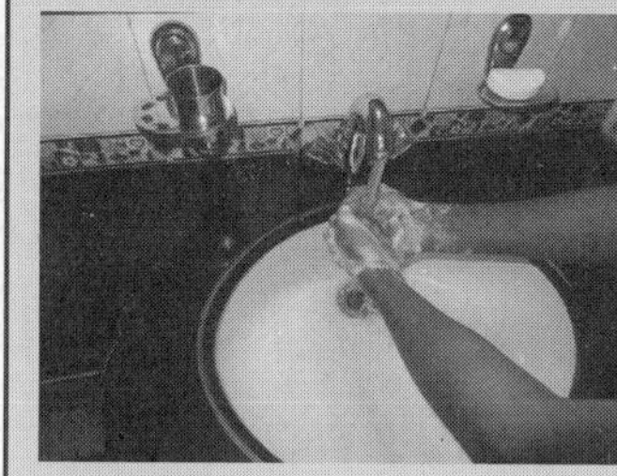

Figure 29.4 Washing hands to ensure hygiene

- Proper wash basins
- Hot and cold water supply
- Sanitary provisions (soap, tissue and towels)
- Waste receptacles (easily cleanable)
- Hygienic lavatories (separate for men and women)
- Changing facilities

29.1.3 Control of Operations

Food operations should be controlled right from the procurement of raw material through its storage, processing, packaging and distribution/service to prevent the food from harmful or undesirable microorganisms or their toxins. HACCP should be used to control food hazards.

1. The **raw materials or ingredients** should be inspected and sorted at procurement level and, if necessary, laboratory tests should be carried out to ensure their quality and suitability. Control of operations during **storage and processing** is important for food hygiene (Figure 29.5).
2. The **hygiene control system** should ensure time and temperature control (through thermometer, automatic temperature regulator and alarm system) in freezer and in cold storage compartments and at specific steps of processing to prevent contamination of food.
3. **Packaging materials** should protect product damage and contamination and should carry proper labelling including the manufacturing and 'best before dates'. Packaging materials should be non-toxic and should not affect the safety and suitability of food when it is stored and used as per the specifications.
4. **Managers and supervisors** should be adequately aware of food hygiene principles and practices so that they can judge potential risks, take suitable preventive and corrective action through effective monitoring and supervision. They should take immediate steps so that complete and rapid recall and destruction of any implicated lot is possible.
5. **Documentation** should be done to improve the effectiveness of the food safety control system.

Procurement of raw material

- Raw materials and ingredients should be free of pathogens, chemical contaminants, toxicants and decomposed substances.
- Use only permitted additives and ingredients.
- Physically check all raw materials.
- Purchase in quantities as per the storage capacity.
- Check best before date of pack-aged food.
- Receive high risk food (e.g., meat, milk) at a temperature <5°C and frozen food at ≤18°C.

Storage of raw material and cooked/processed food

- Prevent contamination and pest access, and ensure cleaning and maintemance.
- Store meat, milk, cooked food under refrigeration.
- Keep frozen food in frozen storage.
- Segregate raw, processed rejected recalled products (mark them distinctly).
- Practice First in First Out (FIFO).
- Use non-toxic containers, keep them above the floor level and away from wall.
- Maintain stock register.

Food processing/ preparation

- Maintain strict time and temperature control for food which is perishable.
- Thaw only the required quantity of frozen food.
- Do not refreeze the food once thawed.
- Provide proper outlets for smoke, e.g., chimneys and exhaust fans.

Figure 29.5 Hygiene control during procurement, storage and processing of food

29.1.4 Transportation

Despite adequate hygiene control measures taken earlier in the food chain, the food may get contaminated or spoilt during transport. Therefore, type of conveyance or containers used for transportation should be designed to ensure that they

1. do not lead to contamination of food or the packaging.
2. can be cleaned and disinfected when required.
3. keep cooked food to be served hot at a temperature ≥60°C and cooked food to be served cold at <5°C.
4. allow effective separation of different foods or foods from non-food items.
5. provide effective protection from contamination, including dust and fumes.
6. minimize time required for transportation.
7. protect food from damage and potential sources of contamination.
8. effectively maintain temperature, humidity, atmosphere and other conditions to control the growth of pathogenic or spoilage microorganisms and the production of toxins in food.

29.1.5 Product Information and Consumer Awareness

Appropriate **information** should be given on the products so that the next person in the food chain can handle, store, process, prepare and display the product safely and correctly. The lot or batch should be easily identifiable and recalled if necessary**. Consumer awareness** about food hygiene will ensure that they comprehend the importance of product information, make informed choices and prevent contamination and growth or survival of food-borne pathogens by correct storage, preparation and usage. Consumer information, particularly on food labels should be clearly distinguishable from the information for industry or trade users.

29.1.6 Maintenance and Sanitation

The food establishments and equipment should be maintained to facilitate all sanitation procedures and to function as intended particularly at the critical steps including pest control and waste management (Box 29.3).

Cleaning is defined as the process of removing soil, food residue, dirt, deposits, slime and microbes from the surface by physical, chemical or mechanical means. It cannot remove most of the microorganisms. The term 'soil' includes all undesirable materials including food residue, spillage, packaging contamination and environmental dust and dirt.

Cleaning agents such as detergents are generally mixed with other components such as water, surfactants, inorganic and organic acids, inorganic alkalis and/or sequestrants for improving their efficiency. Strong alkali helps in saponifying the fats; weaker alkalis act as bulking agents; surfactants increase the ability for wetting, dispersing and rinsing; and sequestrants bind the hard water ions.

Box 29.3 What Is the Meaning of the Term 'Clean'?

In food processing/cooking, a surface is said to be

- *physically clean* when a surface is visibly clean and free from soil or residue.
- *chemically clean* when the cleaning agents are rinsed off and when all deposits have been removed.
- *microbiologically clean* when the number and type of microorganisms is reduced to an acceptable level.

Cleaning procedures are an integral part of the sanitation plan in a food processing unit. Floors, walls, ceiling and drains are cleaned generally during a shut-down period by either dry cleaning or wet cleaning. Dry cleaning comprises of mechanical removal of soil by sweeping, wiping, brushing and vacuum cleaning. Among these, vacuum cleaning is the most preferable option as it removes dust without spreading.

Wet cleaning of equipment can be carried out in two ways: **clean in place (CIP) and clean out of place (COP)**. The CIP method is extremely beneficial for processing operations such as aseptic processing where interior surfaces of the equipment (e.g, tanks and pipes) cannot be reached for cleaning. On the other hand COP is useful for pieces of equipment that need to be disassembled for adequate cleaning (Figure 29.6).

Sanitizing is the process employing chemical or physical agents to a surface so as to reduce the microbial count to a level which is safe for the consumer. Though sanitizing wipes out vegetative cells, it does not necessarily

Clean in place (CIP)	Clean out of place (COP)
• Draining off the food • Switching the system to 'clean' mode • Pre-rinse with cold water to remove the residual food • Circulation of hot detergent to wash off the soil followed by another water rinse • Sanitization with a suitable chemical followed by yet another water rinse	• Removing food and the open contatiners and dismantling • Dry cleaning to remove dust, soil and other debris from the equipment • Pre-rinse of the equipment and area • Cleaning with detergent, hot/cold water and scrubbing the equipment and its components in COP tanks/ vessels followed by rinse in potable water • Applying disinfectant (if necessary) followed by a final rinse . • Reassembling the equipment and allowing it to drain and dry.

Figure 29.6 Clean in place (CIP) and clean out of place (COP): A comparison

destroy the microbial spores. Cleaning reduces microbial load yet many viable microorganisms remain on the surface. Therefore, cleaning process is followed by sanitizing to destroy the viable microorganisms in a manner that they do not grow on the surface during the production period (Figure 29.7). To achieve this objective, sufficient contact time between the sanitizer and the microorganisms is necessary.

Figure 29.7 Basic sequence of cleaning and sanitizing the equipment

High temperature is a good sanitizer because it has high penetration, is non-corrosive and leaves no residue. Milk processing equipment is an enclosed system that uses steam and hot water for sanitizing. Chemical sanitizers are more suitable and cost-effective for open surfaces such as conveyor belts. Commonly used sanitizers include chlorine-releasing compounds, quaternary ammonium compounds, iodophors, bromine compounds, acid sanitizers, ozone and peroxides.The containers for cleaning agents and other dangerous chemicals should be easy to identify and be stored separately to prevent accidental contamination of food.

Pest Control

Pests create a major hazard to the safety and suitability of food (refer to Section 22.2.5 in Chapter 22). It is important to create an environment which is not conducive to pest infestation so that the need for pesticides can be minimized. It can be ensured through good sanitation, inspection of incoming materials and good monitoring and control, and an effective and comprehensive pest control programme.

1. The food establishment building should be well maintained to prevent pest entry (by sealing all cracks, gap and holes) and their breeding sites. Drain covers should be firmly in place to prevent rodent entry. Use of wire mesh screens on open windows, doors and ventilators prevents pest entry.
2. Potential food sources should be stored in pest-proof containers and/or stacked above the ground and away from walls. All the left overs in production and the rejected food should be removed as soon as possible.
3. Sanitation should be maintained both inside and outside food premises.
4. Garbage should be stored in covered containers (which are preferably pest proof).
5. Regular monitoring should be carried out to find out for evidence of infestation.
6. Whenever a pest infestation is found, it should be dealt with immediately.
7. Treatment with chemical, physical or biological agents should be carried out without posing a threat to the safety or suitability of food.
8. A documented pest control program should preferably be maintained and the Pest Control Operator should be licensed, insured and certified. Only permissible pesticides should be used.

29.1.7 Personnel Hygiene

Since every human is a potential carrier of disease-causing and spoilage microorganisms, disease control is important for food safety. Food handlers should maintain a high degree of personal cleanliness.

Health Status: Medical check-up at recruiting and routine medical examination ensures that the personnel handing the food are free from infections. Any person who is a carrier of or is suffering from a communicable disease should be sent on leave and not allowed to work in the food processing/serving area. The workers who are affected with boils, sores, infected wounds or any other condition should also be kept away. If permitted to continue working, their cuts and wounds should be covered with suitable waterproof dressings.

Personnel Cleanliness and Sanitary Practices: In every food processing unit, it should be mandatory for all the food handlers to adopt the following sanitary practices:

1. Wearing special protective outer clothing to maintain a high degree of personal cleanliness.
2. Removing all jewellery (particularly from the hands) and watches.
3. Maintaining the gloves (if used) clean, sanitary and intact; changing disposable gloves regularly.
4. Wearing appropriate head gear (e.g. cap, headband and hairnet) and footwear.
5. Keeping nails short, clean and without paint.
6. Washing the hands thoroughly before initiating the work, immediately after using the toilet, after each break from the work area or whenever their hands become soiled or contaminated.
7. Using the assigned hand-washing area.
8. Drying hands with clean towels, disposable paper towel or under air dryer.
9. Refraining from eating food, drinking beverages, smoking, consuming tobacco or storing personnel belongings in work areas.
10. Following all important precautions to ensure cleanliness and sanitation of food.

Box 29.4 Sanitary Practices for Personnel Hygiene

Wash hands immediately

- before working with food
- after using toilet
- after handling rubbish/waste
- after smoking, coughing, sneezing, using tissue, eating and/or drinking
- after touching hair, scalp mouth or other body part
- after using cell phone
- before putting on the gloves

Ask food handlers about

- last illness with diarrhea/vomiting
- skin trouble anywhere on the body
- discharges from eye, ear, mouth
- digestion problems
- medical history of typhoid/paratyphoid

Wear special dress code

- head cover
- beard cover
- no jewellery
- clean protective clothing
- clean shoes

The dress code applies to everyone who enters the food handling area

Gloves should be changed as often as hand wash.
Hand sanitizers are not a substitute for hand wash.

29.1.8 Training

Food hygiene training should be imparted to all food handlers and supervisors. They should be aware of the risks associated with poor personal hygiene and insanitary practices and also know their role and responsibility in protecting food from contamination and spoilage.

1. Food handlers should be trained in food safety measures which correspond to their work activity: the nature of food, its handling, preparation, processing/cooking, packaging, storage and service.
2. Supervisory and managerial level personnel should have proper education or experience or both to maintain plant sanitation. They should be able to identify sanitary hazards and take corrective action at the earliest.
3. Periodic assessment should be carried out for evaluating the effectiveness of training and awareness of food handlers.
4. Routine supervision is necessary to ensure that food safety practices are being followed.
5. The training programme must be reviewed and updated keeping pace with the current developments in the field of food safety.

29.2 Good Hygiene Practices in Catering and Food Service

The Food Safety and Standards (Licensing and Registration of Food Businesses) Regulations, 2011, state the Good Hygiene Practices (GHPs) which must be followed by food business operators engaged in catering/food service establishments. These GHPs are listed for various operations such as cleaning, raw material handling, cooking, chilling and cross-contamination.

29.2.1 Cleaning

Food area and equipment should be cleaned between different tasks. The surface of work area should be thoroughly cleaned (especially when someone spills food/water/beverage).

1. **A systematic cleaning schedule** along with instructions should put up by the food business operator. It should be strictly followed by the food handlers. It should include what needs to be cleaned along with the frequency and the way of cleaning.
2. The **cleaning instructions** should state the cleaning agents, their amount, the required dilution and the duration for which they should be left in contact with the surface. For dish washing, three sink method is generally employed (Figure 29.9).

29.2.2 Raw Material Handling and Preparation

The raw materials should be procured from a reliable source and checked for visible signs of spoilage and physical contamination. Raw sauces and pastes should be stored in covered food grade containers, kept in refrigerator and checked periodically for fungal growth and/or spoilage.

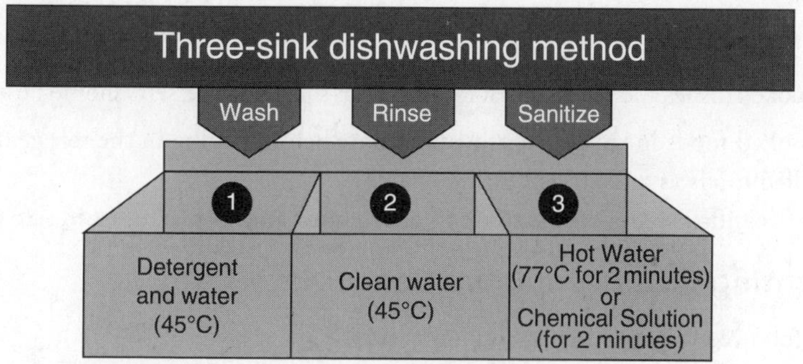

Figure 29.9 Cleaning and sanitizing by three-sink dishwashing method

For preparation of fruits and vegetables, the following practices should be adopted:

1. Select fruits and vegetables which have been protected from cross-contamination.
2. Wash them with potable water prior to cutting (Figure 29.10).
3. Use clean equipment and containers for pre-preparation (peeling, squeezing or cutting).
4. Keep previously prepared fruits and vegetables covered in hygienic food grade containers under suitable temperature in a refrigerator.

Figure 29.10 Wash vegetables with potable wate

While preparing meat, fish and poultry, the following precautions should be taken:

1. Separate raw and processed meat from other foods and surfaces.
2. Avoid cross-contamination by using separate cutting boards, containers and knives for these foods.
3. Wash hands thoroughly after handling raw meat, fish or poultry.
4. Cook these foods thoroughly.
5. Clean the used surfaces with antimicrobial cleaning agent, rinse with water and sanitize.
6. Thaw frozen meat products by no other method than thawing in refrigerator (at ≤5°C), under running water or in the microwave.

29.2.3 Cooking

Preparation, processing and cooking should ensure elimination of hazards and prevent re-contamination.

1. Segregate vegetarian food from non-vegetarian products.
2. Keep food hot during cooking and reheating.
3. Avoid re-using of cooking oil.
4. Hold hot food at 60°C, cool to 21°C within two hours, to 5°C in four hours. Reheat the refrigerated food adequately before consuming.
5. Reheat food quickly in oven/steamer/microwave to ≥74°C (if microwave reheating is used, keep covered for five minutes for the temperature to equalize.
6. Discard frying oil after it has been used thrice.

29.2.4 Chilling

Semi-cooked or cooked dishes and ready-to-eat foods such as salads and desserts should be kept in refrigerato

1. Cool the cooked foods in the shortest possible time before storing in the refrigerator.
2. Process chilled foods as quickly as possible.
3. Maintain the condition and temperature of refrigerator and display units in safe range.

29.2.5 Preventing Cross-contamination

GHPs should be followed to prevent cross contamination.

1. Separate raw meat, fish and poultry from cooked/ready-to-eat foods.
2. Wash hands after handling raw meat, fish and poultry.

3. Clean work surfaces, chopping boards and equipment thoroughly before initiating food preparation and after use.
4. Use separate knives for raw fruits/vegetables and for raw meat/fish/poultry.
5. Train the staff to prevent cross contamination.

The GHPs to be followed for personnel hygiene and transportation and storage of food have already been lescribed in Section 29.1.

29.3 Safety Aspects of Processing Water

ls stated earlier, the water used for processing purposes must be potable. Let us try to understand the term water quality', and how it can be ensured in a food processing/cooking establishment.

29.3.1 Water Quality Standards

Water quality' is a comprehensive term which encompasses microbial, chemical, radiological and sensory aspects Figure 29.11).

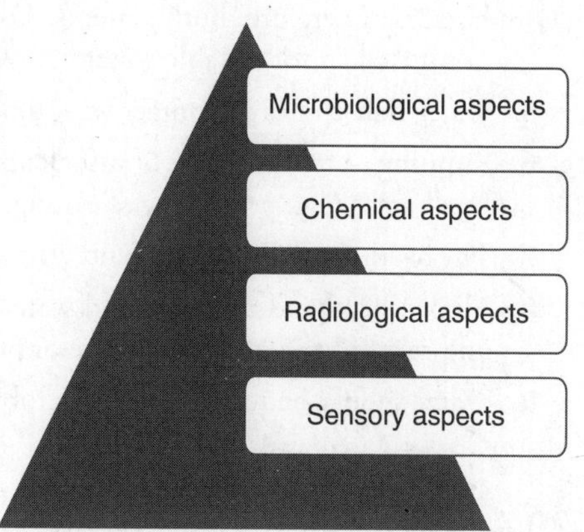

Figure 29.11 Various aspects of water quality

licrobial Aspects: Water is prone to microbial contamiation through public water supplies or faults in internal lumbing systems (cross-contamination of potable water vith sewage system or with non-potable water). Microbial uality of water is generally tested by the presence of *E.coli* vhich is an indicator of faecal contamination. It is better to arry out more frequent microbial examination by a simple est than less frequent testing by a complex test. Chemial disinfectants (e.g. chlorine) are used in water treatment o reduce the load of harmful microorganisms. Though it esults in the formation of chemical by-products, the efficacy of water disinfection should not be compromised.

hemical Aspects: Water is exposed to chemical hazards from natural sources (rocks, soil), industrial ources (solid waste, fuel leaks), agricultural activities (fertilizers, pesticides), water treatment (coagulants) nd materials in contact with water (sanitizers, coolants, heating fuels, oils). The presence of inorganic and rganic chemicals in water is tested by chromatography, spectrophotometry and volumetric and colorimetric nethods.

adiological Aspects: The exposure of drinking water to radionuclides is very small under normal ircumstances. Whenever required, gross alpha and gross beta radiation activity is tested.

ensory Aspects: Taste, odour and appearance of water may be affected by its microbial, chemical and hysical constituents. Though many of these constituents have no direct health effects, their presence may ead to rejection of water by the consumers who have diverse ability to detect them by taste or odour. o ensure better consumer acceptability, it is important that the water is free of objectionable tastes and dours.

29.3.2 Water Supply

A preventive approach from catchment area and source to its use (for processing/consumption) should be adopted for the safety of potable water. The following considerations are important to ensure hygienic water supply:

1. Plumbing design should prevent cross-contamination of potable water to ensure supply of safe water in the establishment. There should be minimum dead ends and other areas of stagnation in the water supply.
2. Sufficient quantities of water should be available to the required locations throughout the food establishment.
3. Back flow of water should be prevented.
4. Separate pipe lines for potable and non-potable water **(dual piped water supply system)** should be provided. Accidental cross-contamination between the two supplies can pose serious health hazards. Therefore, both systems should be labelled conspicuously so that non-potable supply is not mistaken for potable water.
5. Water that comes in contact with food or food-contact surfaces should be of potable quality.
6. Running water should be of sufficient pressure, and at suitable temperature, and should be provided in all areas (e.g. processing, cleaning, and personnel hygiene).
7. The overhead and underground water tanks should be properly covered to prevent contamination.
8. All overhead and underground water tanks should be in good condition, be cleaned periodically and marked to indicate the date of preceding cleaning.
9. Water should be tested for its microbial, chemical and sensory quality.

29.4 Waste Water and Waste Disposal

Adequate disposal of garbage and other food waste forms the basis of food sanitation. Food preparation and processing units produce large volumes of wastes including trimmings and leftover portions of cooked and uncooked foods, packaging materials, waste water from wash and rinse operations. They should be disposed in a manner that they neither act as medium for growth of harmful microorganisms nor act as source of food for insects and pests. They should not be a public health annoyance for consumers and community.

1. There should be a separate garbage disposal point away from the processing/cooking area in the food establishment.
2. Garbage should be collected and stored at a marked place.
3. The waste and garbage collection area should be large enough to carry peak loads and should not contribute to contamination.
4. Containers for waste should remain covered and should be in a state to prevent pest access. They should be lined with plastic bags and cleaned and sanitized regularly.
5. Garbage should be removed daily to the garbage disposal point.

6. Single-use and disposable items are very convenient to reduce cleaning efforts but they increase the volume of solid waste. Because of environmental concerns, recycling and bio-degradation of such containers is being promoted.
7. Sewage should be disposed into an adequate sewage system conforming to local sanitary laws.

Nearly three-fourth of food waste is water. The garbage disposal units can be effective measures for eliminating waste from a food establishment in a sanitary method but their energy cost and water requirements are very high.

We learnt that premises, equipment and facilities of a food establishment should be located, designed and constructed not only to minimize contamination, protect pest access and harbouring but also to permit appropriate maintenance, cleaning and sanitizing. All the surfaces and materials, particularly those coming in contact with food, should be non-toxic, durable (except the ones that are disposable) and easy to maintain and clean. Besides facilities for temperature, humidity and other controls, personnel hygiene, safe water supply and waste disposal are needed to ensure hygiene and sanitation. Periodic monitoring of sanitation systems would ensure better effectiveness of food safety.

Know Your Progress

True or False Questions

1. The General Principles of Food Hygiene recommend HACCP-based approach as a means for improving food safety.
2. Gloves are cleaner than hands so they need not be changed as often as hand wash.
3. Only the workers who enter food-handling area require special dress code.
4. Commonly used sanitizers include chlorine-releasing compounds, quaternary ammonium compounds, iodophors, bromine compounds, acid sanitizers, ozone and peroxides.
5. Microbial quality of water is generally tested by the presence of *E.coli* which is an indicator of faecal contamination.

Very Short Answer Questions

1. Define Food hygiene, GHP and Sanitation.
2. State the eight principles of food hygiene.
3. What is the role of product information and consumer awareness in preventing food safety?
4. List the hygienic control measures during transport of food.
5. What are the different measures for pest control in a food establishment?

Short Answer Questions

1. Differentiate between cleaning and sanitizing. Explain the basic sequence of cleaning and sanitizing in context of CIP and COP
2. State 10 golden rules of personnel hygiene.
3. What are the different aspects of water quality?
4. Discuss the safety aspects of water supply in a food industry.
5. How would you ensure adequate waste water and waste disposal?

Long Answer Questions

1. How can the design of layout, facilities and equipment contribute to food hygiene and sanitation in a food establishment?
2. List the general principles of food hygiene. Explain the role of maintenance and sanitation along with that of personnel hygiene in food safety.
3. What 'Good Hygiene Practices' should be followed by food business operators engaged in catering/food service establishments?

Beyond the Text

- Plan a half-day workshop to impart training in personnel hygiene to the food handlers in a middle level restaurant. Incorporate audio-visual aids and demonstrations to make the training interactive and interesting.

Chapter 30

Recent Concerns

Chapter Objectives

After reading this chapter, the reader should be able to:

- enumerate the emerging pathogens of concern to the food industry.
- learn about genetically modified foods along with their potential risks to human health and concerns for the environment.
- appreciate the significance of appropriate information on food label.
- know the characteristics of the latest packaging materials and related technology.

In the last ten chapters, we deliberated on devierse aspects of food safety ranging from contamination and adulteration to food-borne illnesses. This chapter will focus on some recent concerns relatd to safety of food such as emerging pathogens, genetically modified foods, labeling and packaging.

30.1 Emerging Pathogens

Emerging pathogens are the organisms which are responsible for the newly appearing infectious diseases. Some important pathogens in this category are listed in Table 30.1. They include pathogens that may have existed before with lower incidence of disease. *When the incidence of disease caused by such pathogens increases or the geographical area affected by them expands over the preceding two decades, they are included in the category of emerging pathogens.* In case the pathogens controlled in previous years emerge again, the process is known as re-emergence.

Table 30.1 Some emerging and re-emerging pathogens and the foods commonly involved

Food-borne pathogen	Food commonly involved
Salmonella (non-typhoidal)	Undercooked/fresh meat and poultry products, eggs
Shigatoxin-producing *Escherichia coli* (STEC O157:H7)	Unpasteurized (raw) milk, unpasteurized apple cider, soft cheeses made from raw milk
Shigella	Moist prepared foods, milk, macaroni salads, potato, tuna, turkey
Listeria monocytogenes	Milk, soft cheese, coleslaw and cook-chill chicken, ready-to-eat meats (hot dog, deli meat)
Vibrio parahaemolyticus	Raw salt water fish, shellfish, fish products
Vibrio cholera O1 or O139	Contaminated water, raw shellfish, uncooked fruits, vegetables
Yersinia enterocolitica	Raw milk, pork, other meats

(continued)

Table 30.1 (Continued)

Food-borne pathogen	Food commonly involved
Campylobacter jejuni	Poultry, red meat, raw milk
Cryptosporidium spp	Drinking water, recreational water
Toxoplasma gondii	Contaminated meat, fish, poultry fruits, vegetables
Cyclosporacayetanensis	Fresh imported raspberries, lettuce, fresh basil
Norovirus	Undercooked shellfish, ready-to-eat food touched by an infected worker, salads, sandwich

Food-borne emerging pathogens and illness caused by them are affected by many factors such as the environment, food production, processing and distribution, alterations in microorganisms, and demographic and lifestyle changes in the population (Figure 30.1). Moreover, globalization of food supply can lead to unintentional introduction of pathogens in new geographical areas (refer to Section 20.3 in Chapter 20).

Most of the pathogens are zoonotic in their origin and many of them are associated with the emerging pathogens. It is hypothesized that some species of disease-causing microorganisms were earlier restricted to animals. Over the years, they have evolved and are now infecting human beings. Another concern is that many newly identified pathogens have higher ability to cause disease and have better resistance to antibiotics.

Among the emerging pathogens are prions, the proteinaceous infectious particles which are the smallest infectious agents known. They do not carry any genetic material, resist high temperature and chemicals and are extremely difficult to destroy.

Bovine Spongiform Encephalopathy (BSE) is a disease that affects cattle. Commonly known as Mad Cow disease, it belongs to a group of disease known as Transmissible Spongiform Encephalopathies (TSEs). These diseases result in the appearance of massive amounts of structurally abnormal proteins (prions) in the brain and nerve cells leading to neurological disorders exhibited through poor coordination of movements and mental retardation.

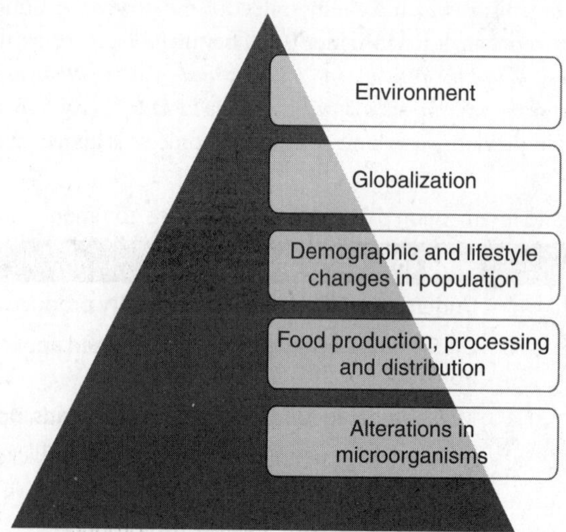

Figure 30.1 Factors affecting the emergence of food-borne pathogens

GOOD TO KNOW?

What Are Prions?

Prions are proteinaceous and infectious particles which are capable of causing TSEs in humans and animals. Size wise, they are smaller to viruses and, therefore, are the smallest infectious agents known.

TSEs are rare in humans. They include Creutzfeldt Jakob Disease (CJD), variant CJD, fatal familial insomnia and Kuru. These are progressive generative diseases of nervous system with serious implications. Though no tests are available to diagnose the disease before the classical neurological symptoms appear, BSE can be detected though appropriate tests in slaughtered animals. Till date, no medicine is available to cure the disease and no vaccine has been developed for its prevention. These illnesses are significant because animal origin material including that from cattle is used for manufacture of medicinal products and vaccines.

30.2 Genetically Modified Foods

Another concern related to safety of food is associated with the Genetically Modified (GM) foods. Let us first understand the meaning of some terms used in this context.

Genetic modification is a special set of technology that changes genetic material of organisms such as animals, plants and bacteria.

Genetically modified organisms (GMOs) are those organisms in which the genetic material (deoxyribonucleic acid, DNA) has been altered in a way that does not occur naturally.

Genetically modified (GM) foods are foods in which the genetic material (DNA) has been altered in a way that does not occur naturally. For example, BT brinjal has been developed by introducing a gene from a different organism.

Box 30.1 How Do GM Foods Provide Crop Protection?

Insect resistance is achieved by inserting into the food plant a gene from the bacterium *Bacillus thuringiensis* (BT). This gene is responsible for the production of a safe toxin that is presently used as an insecticide in agriculture. The GM crops, therefore, require the insecticides in lower amount particularly in areas where pest pressure is high.

Virus resistance is attained through the introduction of a gene from some viruses that cause diseases in plant. As a result, the susceptibility of plant to diseases caused by such viruses decreases.

Benefits of GM foods extend to the producer or consumer in the following ways:

1. They provide improved yield and thereby increase the profit.
2. The crop protection is also better due to increased resistance to plant disease caused by insects or virus (Box 30.1).
3. The future trend could also aim at changing the nutrient content of food, lowering its allergenic potential, or increasing the efficiency of food production systems.

At the same time, GM foods pose some potential risks to human health and concerns for the environment (Figure 30.2).

Risks to human health	Concerns for the environment
• toxiciity • tendency to provoke allergic reactions (allergenicity) • gene transfer (to the cells of human body or the bacteria in gastro-intestinal tract) • movement of genes from GM plants into conventional crops (out crossing)	• escape and introduction of gene into the wild population • persistence of the gene after harvesting • susceptibility of non-target organisms • stability of the gene • loss of biodiversity • increased use of chemicals in agriculture

Figure 30.2 Potential risks and concerns related to GM foods

30.2.1 Safety of GM Foods

Different types of genes are inserted in different ways to develop GM foods. Therefore, it is not advisable to generalize the safety of all GM foods. Instead, the individual GM foods and their safety should be assessed on a case-by-case basis. It is important to follow the international and national guidelines before allowing the GM foods in the market.

At the same time, it is important to understand that the GM foods available in the global market have cleared the risk assessment and therefore, they do not pose any risk for human health. Moreover, there has been no adverse effect on human health due to the consumption of the approved GM foods by general population in their respective countries. There is a well-publicized controversy related to introduction of Bt brinjal in India (Box 30.2). The risk assessments carried out continuously as per the Codex principles and also post market monitoring should be the guiding force for assessing the safety of GM foods.

Box 30.2 What is Bt Brinjal and the Related Controversy?

Bt Brinjal is a transgenic brinjal developed by inserting a gene *cry1Ac* from the soil bacterium *Bacillus thuringiensis* into brinjal. It provides the brinjal plant resistance against insects such as the Brinjal Fruit and Shoot Borer (*Leucinodes orbonalis*) and Fruit Borer (*Helicoverpa armigera*).

The controversy related to introduction of Bt Brinjal in India has been making news. The supporters argue that Bt Brinjal will be advantageous to small farmers because of its insect resistance, better yield, more cost-effectiveness and minimal environmental impact. In contrast, concerns about its possible adverse impact on human health and biosafety, livelihoods and biodiversity have been raised.

The decision is pending. Though the Ministry of Environment and Forests has approved and recommended the environmental release of Bt Brinjal in India, the Minister of State for Environment and Forests has taken cognizance of the strong views raised for and against its release and has called for public consultations across the country before a final decision on this issue is taken.

30.3 Food Labelling

In the context food safety, product information particularly that given on the label is very important (refer to sub-section 29.1.5 of Chapter 29).

A food label can be described as a piece of information given on paper, plastic, cloth, metal or any other material attached to or printed on a food product. A good label is simple, easy to read and understand.

A label not only helps to identify a product it also acts as a means of advertisement and can help provide a competitive edge over other similar products in the market. Food labelling is compulsory and must adhere to the specifications given by the Food Safety and Standards Regulations, 2011.

General Requirements of a Label:

1. Every pre-packaged and packaged food must have a label.
2. Information given on the label should be in English or Hindi. Other languages can also be used but as a means of translation of the information already given in the first two languages.
3. No false, deceptive, misleading or erroneous claims should be given on the label.
4. Label should be attached or applied to the container in such a manner that it does get detached during handling from the manufacturer to the customer.
5. Contents on the label should be clear, prominent, indelible and readily legible by the consumer under normal conditions of purchase and use.
6. If the container is covered by a wrapper, either the wrapper should have all the contents of the label or the contents of the label on the container should be readily legible through the outer wrapper.

Specifications of Labelling on Pre-packaged Foods:
In addition to meeting the above specifications, a label must have the following information:

1. Name of the food which includes the trade or brand name and description of the food in the container.
2. List of all ingredients in descending order according to their amount.
3. Nutrition information or nutrition facts should be given per 100 gram or 100 ml or per serving of the product. Nutrition information must support the health claim given (if any) on the product such as the content of *trans fat* or *MUFA* content of a product. Products which contain hydrogenated fats must mention on label '*hydrogenated vegetable fats or bakery shortening used – contains trans fats*'. Please refer Section 19.3 of Chapter 19 for details on nutrition facts panel and health claims.
4. Declaration regarding the presence of vegetarian or non-vegetarian ingredients used in the product must be in the form of a symbol. A brown–colour filled circle inside a white box with brown outline and a green–colour filled circle inside a white box with green outline should be displayed for foods with animal or vegetarian origin, respectively.
5. Use of additives such as colours, flavours, acidity regulators, anticaking agents, antioxidants, emulsifiers and firming agents must be clearly specified.
6. Name of the manufacturer, manufacturing unit(s), packaging/bottling unit(s), marketing agency/organization as the case may be must be mentioned on the label.
7. The net quantity by weight or volume or number, as the case may be, should be given. In case of liquid foods, weight of the drained food (without liquid) should be written such as in case of *rasgullas* or *gulab jamuns*. The weight should exclude the weight of container or wrappers.
8. The lot/code or batch number which can help in tracing a food during transportation or handling should be given on the label.
9. The date of manufacture or packing or the 'best before' or 'use by' date must specify the date, month and year.
10. In case of imported ingredient(s) or food, the country of origin needs to be mentioned on label.
11. Instructions for use should give details about the reconstitution (if applicable), cooking or serving details which can ensure best use of the product.

Display of Information on Label:
The information discussed in the previous list should be presented on the label as per the following specifications:

1. Legible, simple, plain and unambiguous.
2. Conspicuous so that customers need not search for the desired content.
3. The discussed information should be given on the principal display (central) panel of the container. Forty per cent, 20 per cent and 10–20 per cent of the total surface area of containers should be used for labelling in case the containers are rectangular, oval/cylindrical or of any other shape, respectively.
4. The size of principal display panel and numerals used should be in accordance with the specifications of FSS regulations, 2011.

In addition to these, there are specific requirements and restrictions laid down by regulatory agencies. For example, all infant foods or infant milk substitutes must mention (in capital letters) 'IMPORTANT NOTICE: MOTHERS MILK IS BEST FOR YOUR BABY'. Infant food containers cannot use the claims 'full protein food', 'energy food', 'health food' or 'complete food'. Labels of oils and fats cannot mention expressions such as 'cholesterol fighter', 'anti-cholesterol' or 'saturated fat free' to name a few.

30.4 Newer Trends in Packaging and Technology

Packaging is the process, art or style of packing. The process of packing and the package itself influence the quality (characteristics) and safety of food products. In recent years, residues from packaging materials have created new problems of food contamination (refer to sub-section 26.2.6 in Chapter 26). Packaging performs the following purposes:

1. **Physical protection** from shock, compression, pests, insects and external environment.
2. **Barrier** to microorganisms, air, pollutants, selective permeability to oxygen and other gases within and outside the product (by employing modified atmosphere packaging).
3. **Containment**, that is, holding the food or beverage in a defined container or manner.
4. **Convenience** during handling, stacking, display, sale, opening, use and reuse.
5. **Portion control** especially in single serving packages or if bulk packs have smaller units which helps to control the amount of food being distributed or consumed.
6. **Marketing** through special features (shape, colour, pouring features etc.) of containers, packaging process (UHT, untouched by hands etc.) or food labels which can promote brand positioning of a food product.
7. **Security** is facilitated by using tamper evident packs or anti-theft devices such as dye packs, RFID tags or electronic article surveillance tags which can be activated or detected by devices at exit points and need specialized tools to deactivate.

30.4.1 Packaging

A wide variety of packaging materials are used by the food industry. They can be grouped as follows:

1. **Flexible** such as paper (laminated/waxed), thin plastic films and edible films/coatings.
2. **Semi rigid** includes aluminium foils, paper boards, formed plastics etc.
3. **Rigid** such as glass, metal and thick plastics.

Table 30.2 Characteristics of salient packaging material

Package	Advantages	Disadvantages	Typical use
Glass	Does not react with food, transparent, odour free, hygienic, excellent barrier to oxygen/moisture.	Heavy, easily breakable.	Squashes, jams, jellies, pickles, marmalades etc.
Metal	Physically strong, durable, stable to heat/cold, excellent barrier to oxygen/moisture.	Corrode/rust/dent easily, difficult to open/break seal.	Canned food products, carbonated drinks etc.
Paper	Light weight, easy to print, handle, economical, flexible, can be coated/laminated.	Easy to tear/break, poor barrier to physical damage, air, dust and moisture.	Breads, low moisture sweet meats, snacks etc.
Plastic and other polymers	Light weight, variety of shapes/colours, excellent barrier to physical damage, air, dust and moisture etc.	Low resistance to heat, development of off-flavour, non-biodegradable.	Butter, ice cream, margarine, shortenings, sugar, chips etc.

Each of these materials has several advantages and disadvantages (refer to Table 30.2). Packages should be selected on the basis of their functional requirements, cost and customer demands. Usually, a single packaging material is not able to fulfil all the packaging requirements within economic limitations. Therefore, a combination of two or more packaging material is used. A typical breakfast cereal pack has an inner aluminium bag to provide protection from moisture/odour/microbes etc., whereas the outer paperboard box provides physical protection.

Some of the latest innovations in the field of packaging include the following:

1. **Polymer-based Packages:** Extensive advancements have been made in the development of several polymers such as polyethylene (PE), polystyrene (PS), polypropylene (PP), polyethylene terephthalate (PET), polyvinyl chloride (PVC), polyvinylidene chloride (PVDC), ethylene vinyl alcohol copolymer (EVOH) and several other polymers which provide variable characteristics. For example, PVDC is used as coating or film on food, has excellent oxygen and moisture barrier capacity, is grease resistant and provides tamper-proof sealing upon heating. EVOH has the property of heat resistance, selective permeability and is being used for microwave preparation of ready-to-cook foods.
2. **Protein- and Starch-based Edible Films:** These films have the following properties:
   (a) They are biodegradable.
   (b) They inhibit growth of three major food-borne bacteria viz. *E.Coli, Salmonella and Listeria monocytogenes.*
   (c) They can carry nutrients, antioxidants, flavours, colours and other functional ingredients on the food to which they are applied.
   (d) They are currently being applied on vegetables, fruits, meats, seafood, cereals, nuts and frozen snacks (pizza) but at an elementary level.
3. **Self-heating and Self-cooling Packages:** Self-heating packages such as those required for soup, tea and coffee and self-cooling containers as required for soft drinks, beer etc. have been developed.
4. **Thermochromic Labelling:** It is the latest innovation wherein a thermochromic ink dot is used to indicate that the product is at correct serving temperature following refrigeration or microwave heating. It has been used by a leading fast food outlet for efficient home delivery of its products.

Some new trends related to the packaging process are as following:

1. **Controlled/modified Atmosphere and Vacuum Packaging:** The use of controlled atmosphere packaging (CAP), modified atmosphere packaging (MAP) and vacuum packaging is increasing extensively to retain sensory characteristics of food (especially freshness) and prolong its shelf life. **CAP** refers to alterations made in the normal composition of air surrounding the food during its preparation or preservation process. **MAP** involves changes in the composition of air within the package but the change is not constant as it continues to get modified as per the respiration of food and permeation of gases through the package. **Vacuum packaging** involves removal of oxygen from the air surrounding the food and its replacement with nitrogen.
2. **Active Packaging:** It is a form of 'smart packaging'. It is being used for extending the shelf life of semi-perishable foods such as cheese, beer, wine, fruit juice and mayonnaise. It exhibits the following properties during storage:

-binding oxygen	- anti microbial	- bind off odours
-release flavours	- adjust temperature	- absorb moisture

3. **Intelligent Packaging:** It is also a type of 'smart packaging'. It is often referred to as a 'packaging which senses and informs'. It responds to the changing environment within and outside the package and communicates to the end-user about the status of the food product. It gives indication regarding tampering, time-temperature history, microbial spoilage, exposure to light and physical shocks of the product during handling and transportation. *Intelligent tamper-proof packaging* involves a 'self-bruising' closure on a bottle or jar which indicates that an attempt has been made to open it. Intelligent labelling and printing are currently being developed and would be capable of communicating directly to the customer via thin film devices which will provide sound or visual information in response to touch, motion or other scanning devices. Voice activated safety and disposal instructions would also be given by such packages.
4. **Radiofrequency Identification (RFID) Technology:** It is being applied to food packages for controlling the supply chain. RFID tags or bar codes give information regarding the content, weight, location and number of times the product has gone through the distribution channel. It helps manufacturers to call-back food with sub-standard quality, thereby ensuring safety of the consumers.

Innovations and advancements in the field of packaging and labelling are being made at a fast pace with the objective of maintaining sensory qualities and at the same time keeping the food safe. However, it also the responsibility of the consumer to keep oneself well informed and take necessary precautions while purchasing, preparing and before consuming food.

Know Your Progress

True or False Questions

1. Bovine Spongiform Encephalopathy (BSE) is a disease that affects cattle and is commonly known as Mad Cow disease.
2. The GM foods do not pose any potential risks to human health or for the environment.
3. RFID technology helps to maintain taste of food.
4. Labelling of food is governed by Food Safety and Standard Regulations, 2011.

Very Short Answer Questions

1. Define the following terms:
   (a) Prions
   (b) GM foods
   (c) TSEs
   (d) RFID
   (e) MAP

Short Answer Questions

1. What do you understand by the term emerging and re-emerging pathogens?
2. List the benefits of GM foods.
3. What is a food label?
4. Enumerate the general requirements for a food label.
5. How is active packaging different from intelligent packaging?

Long Answer Questions

1. Write the pros and cons of using genetically modified foods. Discuss the controversy related to Bt brinjal.
2. What are the mandatory requirements for a food label?
3. Discuss the innovations and advancements that have been made in the field of packaging.

Beyond the text

- Search the net to identify some GM foods which have cleared the risk assessment and are available in the global market.

References

Adams, M. R. and Moss, M. O. *Food Microbiology.* Cambridge: Royal Society of Chemistry Publishing, 2008.

Aldsworth, T., Dodd, C. E. R. and Waites, W. Food Microbiology. In: Campbell-Platt, G. ed., *Food Science and Technology*. UK: Wiley-Blackwell, 2009, 115–164.

Aldsworth, T., Dodd, C. E. R. and Waites, W. Food Microbiology. In: Campbell-Platt, G. ed., *Food Science and Technology*. Oxford: Wiley-Blackwell, 2009, 115–164.

Bender, D. A. *A Dictionary of Food and Nutrition*. New York: Oxford University Press, 2005.

Blitz, H. D. and Grosch, W. *Food Chemistry*. Berlin: Springer, 1999.

Bosy-Westphal, A. et al. Effect of Organ and Tissue Masses on Resting Energy Expenditure in Underweight, Normal Weight and Obese Adults. *International Journal of Obesity Related Metabolic Disorders,* 2004, 28(1), 72–79.

Branen, A. L., Davidson, P. M., Salminen, S. and Thorngate, J. H. *Food Additives,* second edition. New York: Marcel Dekker, 2002.

Campbell-Platt, G. *Food Science and Technology,* first edition, Oxford: Wiley-Blackwell, 2009.

Casimir, C. A. and Min, D. B. *Food Lipids: Chemistry, Nutrition and Biotechnology*. New York: Marcel and Dekker Inc., 2002.

CDC. http://www.cdc.gov/foodborneburden/2011-foodborne-estimates.html (accessed on 20 April 2012).

Chadha, D. S. ed. *Prevention of Food Adulteration Act and Rules: A son 31.05.2008*. New Delhi: Confederation of Indian Industry, 2008.

Chandrashekhar, U. *Food Science and Applications in Indian Cookery*. New Delhi: Phoenix Publishing House, 2002.

Codex Alimentarius Commission. *Codex Alimentarius Commission Procedural Manual,* 19th edition. Rome Joint FAO/WHO Programme, 2010.

Committee on Dietary Reference Intakes. *Dietary Reference Intakes for Energy, Carbohydrate, Fiber, Fat, Fatty Acids, Cholesterol, Protein and Amino Acids.* Washington, DC: National Academies Press, 2005 and 2010

Crisosto, C. H., Johnson, R. S., Day, K. R., Beede, B. and Andris, H. L. Contaminants and Injury Induced Inking on Peaches and Nectarines. *California Agriculture* 1999, 53(1), 19–23.

Damodaran, S., Parkin, K. L. and Fennema, O. R. *Fennema's Food Chemistry*, fourth edition. CRC Press CRC Press Taylor & Francis Group. Boca Raton, London, New York, 2007.

Dauthy, M. E. *Fruit and Vegetable Processing*. FAO Agricultural Services, Bulletin No. 119. Rome: FAO 1995.

deMan, J. M. *Principles of Food Chemistry,* third edition. Maryland, USA: Springer, 1999.

Dendy, A. V. D. and Dobraszczyk, B. J. *Cereal and Cereal Products – Chemistry and Technology*. Gaithersburg, Maryland, ASPEN Publications, 2001.

Deshpnade, S. S., Salunkhe, D. K., Oyewole, O. B., Azam-Ali, S., Battcock, M. and Bressani, R. *Fermented Grains, Legumes, Seeds and Nuts: A Global Perspective*. FAO Agricultural Services, Bulletin No. 142. Rome: Food and Agriculture Organization, 2000.

Donato, K. A. National Health Education Programs to Promote Healthy Eating and Physical Activity. *Nutrition Reviews,* 2006, 64, S65–S70.

EIC. Export Inspection Council.http://www.eicindia.gov.in/eic/about-main.htm (accessed on 9 October 2012

Eliasson, A. C. *Carbohydrates in Food.* New York: Marcel and Dekker, 1996.

FAO. *Fermented Cereals—A Global Perspective.* FAO Agricultural Services, Bulletin No. 138. Rome: Food and Agriculture Organization, 1999.

FAO. *Fermented Fruits and Vegetables—A Global Perspective.* FAO Agricultural Services, Bulletin No. 134. Rome: Food and Agriculture Organization, 1998.

FAO and WHO. *Codex Alimentarius – Food Hygiene- Basic Texts,* fourth edition. Rome: Food and Agriculture Organization, 2009.

FAO and WHO. *Compendium of Food Additive Specifications.* Joint FAO/WHO Expert Committee on Food Additives, Monograph No. 11. Food and Agriculture Organization of the United States, 2011.

FAO and WHO. *Understanding Codex Alimentarius.* Rome: Food and Agriculture Organization, 2005.

Faridi, H. and Faubion, M. *Dough Rheology and Baked Product Texture.* New York: Van Nostrand Reinhold, 1990.

FDA. FDA Report on the Occurrence of Foodborne Illness Risk Factors in Selected Institutional Food-service, Restaurant, and Retail Food Store Facility Types, Maryland: Food and Drug Administration National Retail Food Team, 2009.

Fieldhouse, P. *Food and Nutrition: Customs and Culture,* second edition, London: Chapman and Hall, 1995.

Fox, P. F. *Advanced Dairy Chemistry. Part I Proteins.* London: Elsevier Applied Science, 1992, 339–367.

Fox, P. F. *Developments in Dairy Chemistry. Vol. 4: Functional Milk Proteins.* London: Elsevier Applied Science, 1989.

Friedman, M. Food Browning and Its Prevention: An Overview. *Journal of Agricultural and Food Chemistry* 1996, 44(3), 631–653.

FSSA. *The Food Safety and Standards Act Along with Rules and Regulations, 2011.* Delhi: Commercial Law Publishers (India) Pvt. Ltd, 2011.

FSSAI. Guidelines for Food Safety (Medium to Small Food Establishments). Food Safety and Standards Authority of India. http://www.foodsafety.gov/Portals/0/Pdf/DOC_GUIDELINES.pdf (accessed on 20 January 2013).

Gallagher, D., Heymsfield S.B., Heom, Jebb, S.A., Murgatroyd, P.R. and Sakamoto Y. Healthy Percentage Body Fat Ranges: An Approach for Developing Guidelines Based on Body Mass Index. Bethesda, MD: *American Journal of Clinical Nutrition,* 2000, 72, 694–701.

Gerrish, C. J. and Mennella, J. A. Flavor Variety Enhances Food Acceptance in Formula Fad Infants. *American Journal of Clinical Nutrition,* 2001, 73(6), 1080–1085.

Giese, J. Fats Oils and Fat Replacers. *Journal of Food Technology,* 1996, 4, 78.

Glicksman, M. Hydrocolloids and the Search for the 'Oily Grail'. *Food Technology* 1991, 45(10), 94–99.

GoI. *Report of the Working Group on Consumer Protection.* Twelfth Plan (2012–17) Vol II. Government of India: Ministry of Consumer Affairs, Food and Public Distribution, 2012.

GoI. *The Food Safety and Standards Act 2006.* http://www.fssai.gov.in/portals/0/pdf/food-act.pdf (accessed on 15 April 2012).

Gopalan, C., Sastri, B. V. R. and Balasubramanian, S. C. *Nutritive Value of Indian Foods* Hyderabad: National Institute of Nutrition, Indian Council of Medical Research, 2011.

Guyton, A. C. and Hall, J. E. *Textbook of Medical Physiology,* 11th edition. Amsterdam: Elsevier, 2005.

Hall, R. L. Food Flavors: Benefits and Problems. *Food Technology,* 1968, 22, 1388–1392.

Harris, R. S. General Discussion on the Stability of Nutrients. In: Karmes, E. and Harris, R. S. eds. *Nutritional Evaluation of Food Processing*, third edition. New York: AVI/Van Nostrand Reinhold, 1988.

Hasenhuetti, G. L. and Hartel, R. W. eds. *Food Emulsifiers and Their Applications.* New York: Springer, 2008.

Hedberg, C. W., Smith, S. J., Kirkland, E., Radke, V., Jones, T. F., Selman, C. A., and the EHS-NET Working Group. Systematic Environmental Evaluations to Identify Food Safety Differences between Outbreak and Nonoutbreak Restaurants. *Journal of Food Protection,* 2006, 69(11), 2697–2702.

Heldman, D. R. IFT and the Food Science Profession. *Food Technology* October, 2006, 60(10), 1.

Hill, J. O., Thompson, H. and Wyatt, H. Weight Maintenance: What's Missing? *Journal of the American Dietetic Association,* 2005, 105, S63–S66.

IFT. *Bacteria Associated With Food-borne Diseases.* Scientific Status Summary. Chicago: Institute of Food Technologists, 2004.

Indian Council of Medical Research. Nutrient Requirements and Recommended Dietary Allowances for Indians. A Report of the Expert Group of the Indian Council of Medical Research. New Delhi: ICMR 2010.

Institute of Medicine, Food and Nutrition Board. *Dietary Reference Intakes for Water, Potassium, Sodium, Chloride and Sulfate.* Washington, DC: National Academies Press, 2004.

Institute of Medicine. *Dietary Reference Intakes for Vitamin C, Vitamin E, Selenium and Carotenoids.* Washington, DC: National Academy Press, 2000.

International Food Information Council (IFIC). *Food Ingredients and Colors.* http://www.fda.gov/food/foodingredientspackaging/ucm094211.htm#foodadd (accessed on 16 April 2012).

ISO. *ISO and Food– Quality and Safety fromFarm to Fork.* Switzerland: ISO Central Secretariat, 2012.

James, W. P. The Fundamental Drivers of the Obesity Epidemic. *Obesity Reviews,* 2008, 9, S6–S13.

Karleskind, A. *Oil and Fats Manual.* Vols 1 and 2. Secaucus, NJ: Lavoisier Publishing, 1996.

Kleiner, S. M. Water: An Essential but Overlooked Nutrient. *Journal of the American Dietetic Association* 1999(2), 200–206.

Knechtges, P. L. *Food Safety: Theory and Practice.* Burlington, MA: Jones and Barlette Learning, 2012.

Lawless, H. T. and Heymann, H. *Sensory Evaluation of Food: Principles and Practices,* second edition. USA: Springer, 2010.

Lichtenstein, A. H. et al. Diet and Lifestyle Recommendations Revision 2006: A Scientific Statement from the American Heart Association Nutrition Committee. *Circulation,* 2006(114), 82–96.

Lindsay, R. C. Flavors. In: Fennema, O. R. ed. *Food Chemistry,* third edition. New York: Marcel Dekker Inc., 1996, 95–156.

Lopez, M. A. A. and Martos, F. C. Iron Availability: An Updated Review. *International Journal of Food Science and Nutrition* 2004, 55, 597–606.

Lusher, A. 2011. http://www.dailymail.co.uk/news/article-2036058/Tea-victory-Twinings-saves-original-300-year-old-Earl-Grey.html.

Mahan, L. K., Escott-stump, S. and Janice L. Raymond. *Kraue's Food and the Nutrition Care Process,* 2012, 13th edition. St. Louis, Missouri: Elsevier Saunders.

Mahindru, S. N. *Food Additives: Characteristics, Detection and Estimation.* New Delhi: APH Publishing Corporation, 2008.

Manay, S. and Shadakshraswamy, M. *Food Facts and Principles.* New Delhi: New Age International Publishers, 1997.

Marshall, E. and Majia, D. *Traditional Fermented Food and Beverages for Improved Livelihood*. FAO Diversification Booklet, No. 21. Rome: Food and Agriculture Organization, 2011.

McClements, D. J. *Food Emulsions- Principles, Practices and Techniques*, second edition. New York: CRC Press, 2005.

McClements, D. J. Lipid-Based Emulsions and Emulsifiers. In: Akoh, C. C. and Min, D. B. eds. Foods Lipids–*Chemistry, Nutrition and Biotechnology*. Boca Raton, Florida: CRC Press Taylor & Francis Group, 2008, 64–97.

Mejia, C., Mc Entire, J. C., Keener, K., Muth, M. K., Nganje, W., Stinson, T. and Jensen, H. Traceability (Product Tracing) in Food Systems: An IFT Report Submitted to the FDA, Volume 2: Cost Considerations and Implications. *Comprehensive Reviews in Food* Science *and Safety*, 2009(9), 160.

Mermelstein, N. H. Sensory Developments and Challenges. *Food Technology*, 2010, 64(7), 102–112.

Mertens, B. and Knorr, D. Developments of Non-thermal Processes of Food Preservation. *Food Technology*, 1992, XLVI(5), 124–133.

MoFPI. Ministry of Food Processing Industries.http://mofpi.nic.in/ContentPage.aspx?CategoryId=147 (accessed on 13 October 2012)

Montville, T. J. and Mathews, K. R. *Food Microbiology*, second edition. Washington, DC: ASM Press, 2008.

Mooney, K. and Walbourn, L. When College Students Reject Food: Not Just a Matter of Taste. *Appetite* 2001, 36(1), 41–50.

Mottram, D. S. The Maillard Reaction: Source of Flavour in Thermally Processed Foods. In: Berger, R. G. ed., *Flavors and Fragrances: Chemistry, Bioprocessing and Sustainability*. New York: Springer, 2007, 269–284.

Mountney, G. J. and Gould, *W. A. Practical Food Microbiology and Technology*. New York: AVI, 1988.

Nickerson, J. T. R. and Ronsivalli, L. J. *Elementary Food Science*. Connecticut: AVI Publishing Company, Inc, 1980, 59–81.

Paine F. A. and Paine H. Y. A Handbook of Food Packaging, second edition. London: Blackie Academic and Professional, 1992.

Parker, R. *Introduction to Food Science*. USA: Delmar, Thomson Learning, 2003.

Philips, R. D. and Finley, J. W. *Protein Quality and the Effects of Processing*: Marcel Dekker, 1989.

Podmore, J. Applications of Edible Oils. In: Hamm, W. and Hamilton, R. J. eds. *Edible Oil Processing*. Sheffield: Sheffield Academic Press, 2000.

Ray, B. and Bhunia, A. *Fundamental Food Microbiology*. New York: CRC, 2011.

Russin, T. A. Getting Creative with Hydrocolloids. *Food Technology*, 2010, 64(7), 58–65.

Sharma, B., Malhotra, S., Bhatia, V. and Rathee, M. Epidemic Dropsy in India. *Postgraduate Medical Journal*, 1999, 75(889), 657–661.

Shils, M. E., Shike, M., Ross, C. A., Caballero B. and Cousins R. J. *Modern Nutrition in Health and Disease*, 10th edition. Philadelphia, PA: Lippincott Williams and Wilkins, 2006.

Singh, R. P. and Anderson, B. A. The Major Types of Food Spoilage: An Overview. In: Steel, R. ed. *Understanding and Measuring the Shelf life of Food*. Cambridge: Woodhead Publishing Limited, 2004, 3–23.

Singh, R. P. and Heldman, D. R. *Introduction to Food Engineering*, fourth edition. San Diego, CA: Academic Press, 2009.

Sommer, A., West, K. P. Jr. *Vitamin A Deficiency: Health, Survival, and Vision*. New York: Oxford University Press, 1996.

Stone, H. and Bleibaum, R. *Sensory Evaluation. In:* Campbell-Platt, G. ed. Food Science and Technology. United Kingdom: Wiley-Blackwell, 2009, 323–340.

Stone, H. and Sidel, J. L. *Sensory Evaluation Practices*, third edition. California: Academic Press, 2004.

Swaminathan, M. *Food Science, Chemistry and Experimental Foods*. Bangalore: The Bangalore Printing and Publishing Co., 1990.

Turner, R. E., Degnan, F. H. and Archer, D. L. Label Claims for Foods and Supplements: A Review of the Regulations. *Nutrition in Clinical Practice,* 2005(20), 21–32.

Uauy, R., Hertrampf, E. and Dangour, A. D. Food Based Dietary Guidelines for Healthier Populations: International Considerations. In: Shils, M. E., Shike, M., Ross, A. C., Caballero, B. and Cousins, R. J. *Modern Nutrition in Health and Disease*, 10th edition. Philadelphia: Lippincott Williams and Wilkins, 2006.

US Department of Health and Human Services and US Department of Agriculture. *Dietary Guidelines for Americans, 2005,* sixth edition. Washington, DC: US Government Printing Office, 2005.

Vaclavik, V. A. and Christian, E. W. *Essentials of Food Science*, third editon. New York: Springer, 2008.

Villamiel, M., del Castillo, M. D. and Corzo, N. Browning Reactions. In: Hui, Y. H. ed. *Food Biochemistry and Food Processing.* UK: Blackwell Publishing, 2006, 71–102.

Walstra, P. Dispersion Systems: Basic Considerations. In: Fennema, O. R. ed. *Food Chemistry*, third edition. NewYork: Marcel Dekker Inc., 1996, 95–156.

Whitmire, S. J. Water, Electrolytes and Acid-Base Balance (Chapter 6). In: Mahan, K. L. and Escott-Stump, S. eds. *Krause's Food Nutrition and Diet Therapy,* 11th edition, St. Louis, Missouri: Saunders.

WHO, Wholesomeness of Irradiated Food: Summaries of Data Considered by the Joint FAO/IAEA/WHO Expert Committee on the Wholesomeness of Irradiated Food, EHE/81.24, Geneva, 1980.

WHO. *Guidelines for Drinking Water Quality,* fourth edition. Geneva: World Health Organization, 2011.

WHO. Sanitation. http://www.who.int/topics/sanitation/en/ (accessed on 2 October 2012)

Wilcock, A. M., Pun, J. K. and Aung, M. Consumer Attitudes, Knowledge and Behaviour–A Review of Food Safety Issues. *Trends– Food Science and Technology*, 2004(15), 56–66.

Willet, W. C. Diet and Cancer: An Evolving Picture. *JAMA,* 2005(293), 233–234.

World Health Organization. Diet, Nutrition and the Prevention of Chronic Diseases: A Report of a Joint WHO/FAO Consultation. Geneva: WHO Technical Report Series (916), 2003.

World Health Organization: Energy and Protein Requirements. Report of a Joint FAO/WHO/UNU Expert Consultation, Technical Report Series 724. Geneva, WHO, 1985.

Yapp, C. and Fairman, R. Factors Affecting Food Safety Compliance within Small and Medium-Sized Enterprises: Implications for Regulatory and Enforcement Strategies. *Food Control,* 2006(17), 42–51.

Websites

Chapter 1

Association for the Study of Food and Society: www.food-culture.org
Central Food Technology Research Institute: www.cftri.com
European Federation of Food Science and Technology: www.effost.org
International Food Policy Research Institute: www.ifpri.org
International Society of Food Applications of Nano scale Sciences: www.ISFANS.org
International Society of Food Engineering: www.iufost.org/isfe
International Society for Nutraceuticals and Functional Foods: www.functionalfoodscenter.net
International Union of Food Science and Technology: www.iufost.org

Chapter 2

International Food Information Council: www.IFCI.org
International Commission for Uniform Methods of Sugar Analysis: www.ICUMSA.org

Chapter 3

Journal of American Oil Chemists Society: www.aocs.org
Journal of Lipid Research: www.jlr.org

Chapter 4

International Food Information Council: www.IFCI.org

Chapter 5

American society of Heating, Refrigerating and Air-conditioning Engineers: www.ASHRAE.org
Journal of Food Science and Technology: www.onlinelibrary.wiley.com

Chapter 6

Sensoryspectrum: http://www.sensoryspectrum.com
Institute of Food Technologists: http://www.ift.org

Chapter 7

European Food Emulsifier Manufacturer's Association: www.emulsifiers.org
Food Additive and Ingredient Association: www.faia.org.uk

Chapter 8

International Union of Pure and Applied Chemistry: http://old.iupac.org/reports/2001/colloid_2001
Harvard School of Engineering and Applied Sciences: http://seas.harvard.edu/cooking

Chapter 9

Tea Board of India: http://teaboard.gov.in
Bell Flavors & Fragrances: http://www.bellff.com
The International Organization of the Flavour Industry: www.iofi.org

Chapter 10

Department of Food Science and Technology, Ohio State University: http://fst.osu.edu/
Food Info Net: http://www.food-info.net/uk/

Chapter 11

American Dietetic Association: www.eatright.org
American Medical Association: www.ama-assn.org
Healthy People 2010 Initiative: www.healthypeople.gov
Indian Dietetic Association: www.ida.org
International Food Information Council: www.ific.org
Mayo Clinic Nutrition Information: www.mayoclinic.com
United States Department of Agriculture: www.usda.gov
World Health Organization: www.who.int/home

Chapter 12

Centre for Disease Control and Prevention: www.cdc.org
Nutrition Foundation of India: www.nfi.org

Chapter 13

American Dietetic Association: www.eatright.org
Food and Agriculture Organization of the United Nations: www.fao.org
Bread for the World: www.bread.org

Chapter 14

National Osteoporosis Foundation. Fast Facts: http://www.nof.org/node/40
World Health Organization. Micronutrient Deficiencies: http://www.who.int/nutrition/topics

Chapter 15

International Bottled Water Association: www.bottledwater.org
Merck manual for healthcare professionals: www.merck.com

Chapter 16

Dietary reference intake reports: www.nap.edu
Canada's food guide: www.hc-sc.gc.ca

Chapter 17

United States Department of Agriculture: www.Myplate.gov
The Canadian Council of Food and Nutrition: www.ccfn.ca

Chapter 18

International Food Information council: www.ific.org

Chapter 19

Center for hunger, poverty and nutrition policy: www.hunger.tufts.edu
H.J Heinz Company: www.heinz.com
Kraft Foods: www.Kraftfoods.com
Kellogg Company: www.Kelloggs.com
The Dannon Company: www.dannon.com

Chapter 20

Codex Alimentarius: www.codexalimentarius.org
Food Code: http://www.fda.gov/Food/FoodSafety/RetailFoodProtection/FoodCode/default.htm
Food Safety and Standards Authority of India: http://www.fssai.gov.in
FoodSafety.gov: http://www.foodsafety.gov

Chapter 21

USDA Food Safety and Inspection Service: www.fsis.usda.gov
Microbiology in pictures: www.microbiologyinpictures.com

Chapter 22

National Center for Home Food Preservation: http://nchfp.uga.edu
Oregon State University Extension Services: http://extension.oregonstate.edu
Penn State Extension: http://extension.psu.edu

Chapter 23

Wild Fermentation: http://www.wildfermentation.com/
Dakota Bakers' Yeast: http://www.dakotayeast.com/
Sauerkraut.com: http://www.sauerkraut.com/

Chapter 24

Centres for Disease Control and Prevention (CDC): www.cdc.gov
World Health Organization: www.who.int
International Life Science Institute–India: www.ilsi-india.org

Chapter 25

Food and Agriculture Organization of the United States: http://www.fao.org
Food Safety and Standards Authority of India: http://www.fssai.gov.in
Food Additives: http://www.foodadditivesworld.com
European Food Information Council: http://www.eufic.org
Understanding Food Additives: http://www.understandingfoodadditives.org

Chapter 26

Agency for Toxic Substances and Disease Registry: http://www.atsdr.cdc.gov
Agmarknet, Ministry of Agriculture, Government of India: http://agmarknet.nic.in
Codex Alimentarius: http://www.codexalimentarius.org
World Health Organization: http://www.who.int

Chapter 27

Bureau of Indian Standards: www.bis.org.in
Codex Alimentarius: www.codexalimentarius.org
Export Inspection Council: www.eicindia.gov.in
Food Safety and Standards Authority of India: http://www.fssai.gov.in/
International Organization for Standardization: www.iso.org
Metrology Centre: www.metrologycentre.com
Ministry of Food Processing Industries: mofpi.nic.in
World Trade Organization: www.wto.org

Chapter 28

International Center of Excellence in Food Risk Communication: http://www.foodriskcommunications.com
JIFSAN Foodrisk.org: http://foodrisk.org/
World Health Organization. About Microbiological Risk Assessment (MRA) in Food: http://www.who.int/foodsafety/micro/about_mra/en/

Chapter 29

Food Standards and Safety Authority of India: www.fssai.gov.in/
Hand Hygiene Resource Centre: http://www.handhygiene.org/

Chapter 30

International Standards Organization: iso.org and iso.org/iso/home/standards/management-standards/iso 22000.htm
Centres for Disease Control and Prevention (CDC): www.cdc.gov
world Packaging Organization: www.worldpackaging.org
Genetically Modified Food: http://www.who.int/topics/food_genetically_modified/en/

Annexure 1

Table 1 Recommended Dietary Allowances (RDA) for Energy, Protein, fat and Minerals (Indian Council of Medical Research, 2010)

Group	Category/Age	Body Weight (Kg)	Net Energy (Kcal/day)	Protein (g/day)	Visible Fat (g/day)	Calcium (mg/day)	Iron (mg/day)	Zinc (mg/day)	Magnesium (mg/day)
Men	Sedentary work		2320		25				
	Moderate work	60	2730	60.0	30	600	17	12	340
	Heavy work		3490		40				
Women	Sedentary work		1900		20				
	Moderate work	55	2230	55.0	25	600	21	10	
	Heavy work		2850		30				310
	Pregnant		+350	78.0	30	1200	35	12	
	Lactating 0–6m		+600	74.0	30	1200	21	12	
	6–12m		+520	68.0	30				
Infants	0–6 months	5.4	92kcal/kg/day	1.16g/kg/day	–	500	46μg/kg/day 05	–	30
	6–12 months	8.4	80kcal/kg/day	1.69g/kg/day	19			–	45
Children	1–3 years	12.9	1060	16.7	27		09	5	50
	4–6 years	18.0	1350	20.1	25	600	13	7	70
	7–9 years	25.1	1690	29.5	30		16	8	100
Boys	10–12 years	34.3	2190	39.9	35	800	21	09	120
Girls	10–12 years	35.0	2010	40.4	35	800	27	09	160
Boys	13–15 years	47.6	2750	54.3	45	800	32	11	165
Girls	13–15 years	46.6	2330	51.9	40	800	27	11	210
Boys	16–17 years	55.4	3020	61.5	50	800	28	12	195
Girls	16–17 years	52.1	2440	55.5	35	800	26	12	235

Table 2 Recommended Dietary Allowances (RDA) for Water Soluble and Fat Soluble Vitamins (Indian Council of Medical Research, 2010)

Group	Category/ Age	Body Weight (Kg)	Vitamin A (μg/day)		Thiamin (mg/day)	Riboflavin (mg/day)	Niacin Equivalent (mg/day)	Pyridoxine (mg/day)	Ascorbic Acid (mg/day)	Dietary Folate (μg/day)	Vitamin B_{12} (μg/day)
			Retinol	β-carotene							
Men	Sedentary work				1.2	1.4	16				
	Moderate work	60	600	4800	1.4	1.6	18	2.0	40	200	1.0
	Heavy work				1.7	2.1	21				
Women	Sedentary work				1.0	1.1	12				
	Moderate work	55	600	4800	1.1	1.3	14	2.0	40	200	1.0
	Heavy work				1.4	1.7	16				
	Pregnant		800	6400	+0.2	+0.3	+2	2.5	60	500	1.2
	Lactating 0–6m		950	7600	+0.3	+0.4	+4	2.5			
	6–12m				+0.2	+0.3	+3	2.5	80	300	1.5
Infants	0–6 months	5.4	350	2800	0.2	0.3	710 μg/kg	0.1	25	25	0.2
	6–12 months	8.4			0.3	0.4	650μg//kg	0.4			
Children	1–3 years	12.9			0.5	0.6	08	0.9		80	
	4–6 years	18.0	400	3200	0.7	0.8	11	0.9	40	100	0.2–1.0
	7–9 years	25.1	600	4800	0.8	1.0	13	1.6		120	
Boys	10–12 years	34.3	600	4800	1.1	1.3	15	1.6	40	140	0.2–1.0
Girls	10–12 years	35.0			1.0	1.2	13	1.6			
Boys	13–15 years	47.6	600	4800	1.4	1.6	16	2.0	40	150	0.2–1.0
Girls	13–15 years	46.6			1.2	1.4	14	2.0			
Boys	16–17 years	55.4	600	4800	1.5	1.8	17	2.0	40	200	0.2–1.0
Girls	16–17 years	52.1			1.0	1.2	14	2.0			

Annexure 2

Table 1: Tolerable Upper Limit for Certain Nutrients (National Academy of Sciences, 2011)

Life stage/ Group	Vitamin D (µg/day)	Vitamin E (mg/day)	Sodium (mg/day)	Iodine (mg/day)	Selenium (µg/day)	Copper (µg/day)	Fluoride (mg/day)	Molybednum (µg/day)
Infants								
0–6 months	25	–	–	–	45	–	0.7	–
6–12 months	38	–	–	–	60	–	0.9	–
Children								
1–3 years	63	200	1500	200	90	1000	1.3	300
4–8 years	75	300	1900	300	150	3000	2.2	600
Adolescents								
9–13 years	100	600	2200	600	280	5000	10	1100
14–18 years	100	800	2300	900	400	8000	10	1700
Adults								
19–70 years	100	1000	2300	1100	400	10,000	10	2000
>70 years	100	1000	2300	1100	400	10,000	10	2000
Pregnancy								
14–18 years	100	800	2300	900	400	8000	10	1700
19–50 years	100	1000	2300	1100	400	10,000	10	2000
Lactation								
14–18 years	100	800	2300	900	400	8000	10	1700
19–50 years	100	1000	2300	1100	400	10,000	10	2000

Table 2: Acceptable Intakes for Certain Nutrients (National Academy of Sciences, 2011)

Life stage/ Group	Water (liter/day)	Total Fiber (g/day)	Total Fat (g/day)	Linoleic acid (g/day)	Linolenic acid (g/day)	Biotin (µg/day)	Choline (µg/day)	Potassium (mg/day)	Chromium (µg/day)
Males									
0–6 months	0.7	–	31	4.4	0.5	5	125	400	0.2
7–12 months	0.8	–	30	4.6	0.5	6	150	700	5.5
1–3 years	1.3	19	–	7	0.7	8	200	3000	11
4–8 years	1.7	25	–	10	0.9	12	250	3800	15
9–13 years	2.4	31	–	12	1.2	20	375	4500	25
14–18 years	3.3	38	–	16	1.6	25	550	4700	35
19–30 years	3.7	38	–	17	1.6	30	550	4700	35
31–50 years	3.7	38	–	17	1.6	30	550	4700	35
>51 years	3.7	30	–	14	1.6	30	550	4700	30
Females									
0–6 months	0.7	–	31	4.4	0.5	5	125	400	0.2
7–12 months	0.8	–	30	4.6	0.5	6	150	700	5.5
1–3 years	1.3	19	–	7	0.7	8	200	3000	11
4–8 years	1.7	25	–	10	0.9	12	250	3800	15
9–13 years	2.1	26	–	10	1.0	20	375	4500	21
14–18 years	2.3	26	–	11	1.1	25	400	4700	24
19–30 years	2.7	25	–	12	1.1	30	425	4700	25
31–50 years	2.7	25	–	12	1.1	30	425	4700	25
51–70 years	2.7	21	–	11	1.1	30	425	4700	20
Pregnancy									
14–18 years	3.0	28	–	13	1.4	30	450	4700	29
19–30 years	3.0	28	–	13	1.4	30	450	4700	30
31–50 years	3.0	28	–	13	1.4	30	450	4700	30

Glindex

C

D

E

F

G

I

N

T

U

V

W

X

Y

Z